STRUCTURAL BIOLOGY of VIRUSES

STRUCTURAL
BIOLOGY of
VIRUSES

Edited by

WAH CHIU,

ROGER M. BURNETT,

& ROBERT L. GARCEA

New York   Oxford
Oxford University Press
1997

Oxford University Press

Oxford   New York
Athens   Auckland   Bangkok   Bogota   Bombay   Buenos Aires
Calcutta   Cape Town   Dar es Salaam   Delhi   Florence   Hong Kong
Istanbul   Karachi   Kuala Lumpur   Madras   Madrid   Melbourne
Mexico City   Nairobi   Paris   Singapore   Taipei   Tokyo   Toronto

and associated companies in
Berlin   Ibadan

Published by Oxford University Press, Inc.
198 Madison Avenue, New York, New York 10016

Oxford is a registered trademark of Oxford University Press

Library of Congress Cataloging-in-Publication Data
Structural biology of viruses / edited by Wah Chiu,
Roger M. Burnett, Robert L. Garcea.
p.   cm.
Includes bibliographical references and index.
ISBN 0-19508627-9; ISBN 0-19-511850-2 (pbk.)
1. Viruses—Morphology.   I. Chiu, Wah.   II. Burnett, Roger M.
III. Garcea, Robert L.
QR450.S77   1997
576'.64—dc20   95-26844

Cover illustration:
A cut-away view of the rotavirus showing the
multilayered capsid architecture of the virus.
Provided by Dr. B.V.V Prasad, Baylor College of Medicine.

1 3 5 7 9 8 6 4 2

Printed in the United States of America
on acid-free paper

Contents

Contributors

GILLIAN M. AIR
Department of Biochemistry and
  Molecular Biology
University of Oklahoma Health
  Sciences Center
Oklahoma City, Oklahoma 73190

JAIRO H. AREVALO*
Department of Molecular Biology
The Scripps Research Institute
La Jolla, California 92307

TIMOTHY S. BAKER
Department of Biological Sciences
Purdue University
West Lafayette, Indiana 47907

RAVI BASAVAPPA
Department of Biophysics
University of Rochester Medical
  Center
Rochester, New York 14642

ROGER M. BURNETT
The Wistar Institute
Philadelphia, Pennsylvania 19104

SHERWOOD CASJENS
Department of Oncological Sciences
University of Utah Medical Center
Salt Lake City, Utah 84132

WAH CHIU
Verna and Marrs McLean Department
  of Biochemistry
W.M. Keck Center for Computational
  Biology
Baylor College of Medicine
Houston, Texas 77030

MARIE CHOW
Department of Microbiology and
  Immunology
University of Arkansas for Medical
  Sciences
Little Rock, Arkansas 72205

*Deceased.

GUY D. DIANA
ViroPharma Inc.
Malvern, Pennsylvania 19355

MARY K. ESTES
Division of Molecular Virology
Baylor College of Medicine
Houston, Texas 77030

ROBERT L. GARCEA
Pediatric Hematology/Oncology/Bone
  Marrow Transplantation
University of Colorado School of
  Medicine
Denver, Colorado 80262

JEFFREY M. GREVE
Department of Biotechnology
Bayer Inc.
West Haven, Connecticut 06516

C. NICHOLAS HODGE
Chemical and Physical Sciences, R&D
DuPont Merck Pharmaceutical Co.
Experimental Station
Wilmington, Delaware 19880

LUCAS R. HOFFMAN
Department of Biochemistry
  Biophysics
University of California—San Francisco
San Francisco, California 94143

JAMES M. HOGLE
Department of Biochemistry and
  Molecular Pharmacology
Harvard Medical School
Boston, Massachusetts 02115

ERIC HUNTER
Department of Microbiology
University of Alabama at Birmingham
Birmingham, Alabama 35294

JOHN E. JOHNSON
Department of Molecular Biology
The Scripps Research Institute
La Jolla, California 92037

JONATHAN KING
Biology Department
Massachusetts Institute of Technology
Cambridge, Massachusetts 02139

PRASANNA R. KOLATKAR
Department of Biological Sciences
Purdue University
West Lafayette, Indiana 47907

ROBERT C. LIDDINGTON
Department of Biochemistry
University of Leicester
Leicester, United Kingdom LE1 7RH

MING LUO
Department of Microbiology
Center for Macromolecular
  Crystallography
The University of Alabama at
  Birmingham
Birmingham, Alabama 35294

LEE MAKOWSKI
Institute of Molecular Biophysics
Florida State University
Tallahassee, Florida 32306

J. ANDREW MCCAMMON
Department of Chemistry and
  Biochemistry
Department of Pharmacology
University of California at San Diego
La Jolla, California 92093

MARK A. MCKINLAY
ViroPharma Inc.
Malvern, Pennsylvania 19355

ANNE G. MOSSER
Institute for Molecular Virology
University of Wisconsin
Madison, Wisconsin 53706

NORMAN H. OLSON
Department of Biological Sciences
Purdue University
West Lafayette, Indiana 47907

B.V. VENKATARAM PRASAD
Verna and Marrs McLean Department
of Biochemistry
Baylor College of Medicine
Houston, Texas 77030

MICHAEL G. ROSSMANN
Department of Biological Sciences
Purdue University
West Lafayette, Indiana 47907

ROLAND R. RUECKERT
Institute for Molecular Virology
University of Wisconsin
Madison, Wisconsin 53706

MARJORIE RUSSEL
Laboratory of Genetics
The Rockefeller University
New York, New York 10021

THOMAS J. SMITH
Department of Biological Sciences
Purdue University
West Lafayette, Indiana 47907

PATRICIA G. SPEAR
Department of Microbiology—
Immunology
Northwestern University Medical
School
Chicago, Illinois 60611

ALASDAIR C. STEVEN
National Institute of Arthritis and
Musculoskeletal and Skin Diseases
National Institutes of Health
Bethesda, Maryland 20892

T.P. STRAATSMA
High Performance Computational
Chemistry Environmental Molecular
Science Laboratory
Pacific Northwest Laboratories
Richland, Washington 99352

ADI TREASURYWALA
Allelix Biopharmaceuticals
Mississauga, Ontario L4V1V7

ROBERT A. WELDON, JR.
Department of Microbiology
University of Alabama at Birmingham
Birmingham, Alabama 35294

JUDITH M. WHITE
Department of Anatomy and Cell
Biology
University of Virginia
Health Sciences Center
Charlottesville, Virginia 22908

IAN A. WILSON
Department of Molecular Biology
The Scripps Research Institute
La Jolla, California 92307

ALEXANDER WLODAWER
ABL-Basic Research Program
NCI-Frederick Cancer Research and
Development Center
Frederick, Maryland 21702

Introduction

ROBERT L. GARCEA, ROGER M. BURNETT,
& WAH CHIU

Historically, structural biology and virology have been separate disciplines, and the fields of virology have developed around particular virus families. Rapid advances have occurred recently in the techniques of both structural and molecular biology. These include high performance computing and graphic visualization, X-ray crystallography, electron cryomicroscopy, image reconstruction, molecular cloning, and the expression of recombinant proteins. The convergence of these advances has given detailed information on several viral systems, and "structural virology" now provides some of the most outstanding examples of structure-function relations in biology. Viruses encounter many common problems in their life cycles and the solutions that they have evolved provide instructive contrasts between different biological strategies. These ideas are illustrated by the different chapters, most of which cover a viral system illustrating a particular biological function. The authors have been chosen to integrate both structural and biological viewpoints, and each chapter has been written explicitly to address a broad audience ranging from students to specialists.

One of the central questions in biology is assembly. Viruses confront the architectural paradox of assembling an environmentally stable structure that can be rapidly disassembled upon infection. The assembly process must be efficient, corrected for errors, and have one or more irreversible steps that lead to a specific final product. Using viruses as model systems, investigators have probed how macromolecules are able to recognize each other and interact in an organized and directed fashion to build larger assemblies. Once released, the virus must evade immune surveillance and find another cell with a specific receptor identifying it as permissive for productive replication. The structure must be designed to fulfill all of these biological functions for the particular niche of the virus. Increasingly, structural information on specific steps in viral life cycles is casting light on normal cellular processes. In addition, a new imperative is to develop therapeutic strategies based upon structural information to combat human viral diseases ranging from the common cold to AIDS.

An indication of progress and success in the field of structural virology is the number of new problems and questions that has arisen and is now being addressed experimentally. Most structural information is available for non-enveloped viruses with a protein coat as these are highly symmetric (Chapter 1). This symmetry makes them accessible to current X-ray crystallographic and electron microscopic image reconstruction techniques (Chapter 2). These techniques are becoming increasingly powerful, and a striking feature in recent progress is that the demands imposed by structural virology have driven these two formerly separate fields into mounting a combined attack on some complex virological problems (Chapter 4, 5, 8, 10, and 11). The symmetric viral form arises from the repetitive use of identical building blocks, as necessitated by the relatively small coding capacity of viral genomes (Chapter 1). This genetic economy leads to capsids with helical (Chapter 13) and icosahedral symmetry. Most information is available for the icosahedral viruses (Chapters 4 to 12) and most of the early structural work was aimed at understanding how the coat proteins play their structural role. Remarkably, most have a common core β-barrel motif suggesting its utility as a capsid building block throughout evolution. Knowledge of the constant core domain of these coat proteins draws attention to other features that stabilize the capsid: calcium chelation sites; amphipathic helices (Chapter 7); and accessory glue proteins (Chapter 8). More complex viruses contain several capsid shells (Chapter 9) and questions arise concerning their biological purpose and structural relationship. Despite the beautiful static structures that can be imaged, it should be remembered that macromolecules and their complexes are mobile. Although it is likely that viral capsids are fairly rigid, their surface features could be highly mobile.

The process of capsid assembly itself demonstrates Nature's imagination in attaining the prerequisite goals of efficiency, fidelity, and irreversibility. Here, the genetics of bacteriophage have been the most incisive in dissecting individual steps in assembly (Chapter 11). Parallels to bacteriophage assembly are now being discovered in animal virus systems (Chapter 12). Jigs and templates are often provided by scaffolding proteins which disappear from the final assembled structure. Many different mechanisms can be used to direct unidirectional viral assembly pathways including proteolysis, post-translational modification, calcium-induced structural change, and protein crosslinking. Chaperone proteins are encoded or recruited to prevent premature assembly (Chapters 1 and 14). The role of non-structural proteins in viral morphogenesis remains unclear. What dictates the location where the battery of ingenious viral protein chemistry occurs is another puzzle. Eukaryotic DNA viruses always appear to assemble in the nucleus. Although the nucleus contains the viral genome, the coat proteins are produced in the cytosol. What prevents their premature assembly? How is encapsidation temporally coordinated with replication and transcription? How do the virions escape the nucleus and leave the cell? Similar questions are applicable to the retroviruses.

Although viral capsids are relatively well understood, little is known about the structure of the genome contained within the capsid. Chapter 10 discusses this problem theoretically, and describes plant viruses that have defined structural contacts between their capsids and genomic RNA. Many questions remain to be answered. Is the nucleic acid ordered? Need it be? How is non-genomic nucleic acid excluded? Are there encapsidation initiation sequences? Is the sequence or the structure of the nucleic acid important?

The interaction of viruses with cells has provided numerous insights into normal cell biology. For example, the oncogenes of the DNA viruses are frequently involved in virus production and assembly. Their interaction with cell cycle proteins is certainly related to the need for rapid viral genome replication. The recent identification of a cytosolic-sorting signal for retroviral gag proteins (Chapter 14) undoubtedly will be applicable to normal cell proteins. The important processes of membrane interaction and cell surface recognition have been elucidated by structural studies. The hemagglutinin protein of influenza is a particularly well-studied model for membrane fusion (Chapter 3), which is being further refined with the new structure of the activated molecule.

The interactions of viruses with their receptors illustrate how they exploit different cellular recognition signals to solve their common problem of entering the cell. Surface molecules, such as ICAM-1, CD4, heparan sulfate, and sialic acid, have all been implicated in these events. Much is also known about structural changes in the capsid upon cell binding, particularly with respect to the picornaviruses (Chapters 4 and 6). An important idea that has emerged from structural virology is the "Canyon Hypothesis" (Chapter 4), which explains how the receptor recognition apparatus can evade recognition by the immune system. The discovery of "pocket factors" residing near the binding site shows how these molecules may influence the conformation of the virus and so effect both virus stability and the avidity of attachment to the cellular receptor. How ubiquitous are pocket factors? Do they play an active role in membrane attachment? Further questions abound. Is more than one receptor necessary? How are viruses endocytosed? Do structural changes uniformly accompany the entry process? What directs the virus to the appropriate subcellular compartment, such as the nucleus for DNA viruses (Chapter 12) or the cytoplasm for RNA viruses (Chapter 9)? Are there specific intracellular receptors?

Our knowledge of virus structure has led to a new understanding of viral pathogenesis, particularly with regard to antibody immune surveillance (Chapters 5 and 9). The existence of viral "escape" mutants, so-called because they can escape from antibody neutralization, can now be rationalized. Neutralizing antibody epitopes have been mapped to the virion surface, and cryoelectron micrograph reconstructions of Fab bound to HRV-14 (Chapter 5) and rotavirus (Chapter 9) have now been made. Several mechanisms are possible for how the antibody actually "inactivates" the virus—receptor blockage, aggregation, induction of conformational changes that either stabilize or destabilize the structure. The current view is that all of these mechanisms may be operational, but the structural details in each case remain to be worked out.

Given all of this biological and structural knowledge, can we rationally design anti-viral drugs? Chapters 15 to 17 describe some of the best attempts to date. Most satisfying are the relationships between drug binding and receptor binding sites seen in the rhinovirus example (Chapters 4, 15, and 16). However, as is the case for most anti-microbial agents, resistance can easily be selected and so multi-hit strategies may be required for in vivo therapy. Progress in ''rational'' drug design advances with that for molecular modeling of proteins, but current theoretical and computational techniques are still inadequate to describe the full complexity of the encounter between macromolecule and drug. Pharmacological considerations present further significant obstacles in developing effective new drugs. Certainly this field is rapidly evolving, driven by new algorithms of protein structure prediction, computational power, and the economics of health care.

The editors express their gratitude to the chapter authors, who have devoted enormous efforts to fulfilling the goals of the book. We also acknowledge our debt to many colleagues who have helped in numerous ways, both direct and indirect. An important forum for the development of this new discipline has been the series of FASEB Summer Conferences on Virus Assembly, which has been remarkably successful in bringing together structural biologists with investigators in prokaryotic and eukaryotic virus systems to share the remarkable recent progress in their individual research areas. It is our hope that this book will stimulate and enhance the interdisciplinary interactions and collaborations that will be needed for answering the many major outstanding questions in this fascinating field.

1.

Principles of Virion Structure, Function, and Assembly

SHERWOOD CASJENS

Overview

The fact that all individual virions of a particular virus are identical, or nearly so, makes them ideal subjects for structural studies and for analysis of the mechanisms by which they are assembled. However, because of the great diversity in virus lifestyle and structure, we have found few truly universal principles that apply to the structure and assembly of *every* virus. Perhaps the most important idea to come from this research area is that *macromolecular assembly is not random; rather it is orderly and controlled.* Virion structural proteins typically join growing intermediate structures in an ordered fashion, with the various assembly steps following a particular "assembly pathway." The diversity of viruses has allowed knowledge gained from their study to benefit most areas of biologic structure and their assembly (from protein–protein contact interactions to lipid bilayer structure to nucleic acid condensation to assembly pathway control, to name but a few areas of impact). In the following sections I summarize some of the lessons learned from the study of virus assembly. I attempt to show both the breadth and depth of knowledge by citing examples. References are chosen to allow the reader access to the current literature, rather than to credit those who made discoveries. The reader is urged to glean specific references from the other chapters of this volume.

3

Symmetric Protein Arrays in Virions

Crick and Watson (1956) first realized that there was insufficient genetic information in virus genomes to encode a single asymmetric protein coat molecule that could surround the viral genome, and they suggested that virion coats would have to be made up of symmetric arrays of one or a small number of protein molecules. Indeed, they predicted that these would have to be either helically or cubically symmetric arrays, and it was soon shown by X-ray diffraction and electron microscopy that both helical and icosahedral viruses exist (see Caspar and Klug, 1962). It was reasonably assumed by Crick and Watson (1956) that the proteins in these symmetric arrays would have specific bonding properties and so would make specific intersubunit contacts that would be repeated *exactly* throughout the particle. For helical viruses such as tobacco mosaic virus (TMV), the helical coat protein arrays fulfill this prediction. Each subunit contacts its neighbors essentially identically to all other members of the array with the exception of those at the ends, which have unsatisfied bonding surfaces (Namba et al., 1989). Helical structures can vary in size by virtue of differing lengths, reflected in the number of coat subunits, as well as in having different coat protein helical packing parameters.

The situation with the so-called spherical viruses, those with closed protein shells, is more complex. Although in theory closed shells could be made from subunit arrays with tetrahedral, octahedral, or icosahedral symmetry (the three kinds of cubic symmetry), only ones with icosahedral symmetry have been found. Closed icosahedral shells made up of one type of polypeptide, in which each subunit bonds to its neighbors *identically*, must have 60 subunits—no more and no less (Crick and Watson, 1956). Experimental evidence created a problem for this hypothesis when it was found that apparently icosahedral viruses often have considerably more than 60 identical subunits. Caspar and Klug (1962) devised an ingenious hypothesis to resolve this paradox. Their geometric framework showed that, if intersubunit "bonds [were] deformed," then certain multiples of 60 subunits could be accommodated in a protein shell with icosahedral symmetry. In today's context, the concept of "bond deformations" between proteins is not very meaningful but can be interpreted as changes in protein shape or subtle changes in contact regions between neighboring subunits. Caspar and Klug imagined that all icosahedral viruses have rings of five subunits that interact around each of the 12 fivefold rotationally symmetric icosahedral vertices ($5 \times 12 = 60$ subunits). For viruses in which more subunits are used, these are systematically inserted between the "pentamers" as rings of six subunits or "hexamers." (Space constraints prevent even a summary of this geometric argument here; see Caspar and Klug, 1962, Caspar, 1965, Casjens, 1985a, and Johnson and Fisher, 1994, for detailed developments and discussions of this hypothesis.) These protein subunits were presumed to be subtly different protein conformers and were labeled "quasi-equivalent"; they are not strictly equivalent as they would be in a 60-subunit structure. The allowed multiples of 60 subunits, named for the "triangulation numbers" of their geometric derivation, could be any member of the infinite series $T = 1, 3, 4, 7, \ldots$, corresponding to 60, 180, 240, 420, $\ldots$, subunits (Caspar and Klug, 1962). The number of "conformers" required in each structure by this model is equal to T. Thus, icosahedral viruses are often said to have a shell of a particular T number. In a $T = 3$ shell, for example, the Caspar and Klug theory says there should be 60 subunits each of three conformations of the coat protein, and predicts where the different conformers should lie in the shell. Shell structure is often analyzed by electron microscopy, where individual subunits are not resolved, but "clusters" of subunits, or "capsomeres," are seen when portions of the proteins protrude significantly above the virion surface. Caspar and Klug also derived rules for deducing the apparent clustering, and for deducing T numbers from the arrangement of the clusters.

As numerous atomic structures and high-resolution electron microscopy reconstructions of virions have been elucidated, Caspar and Klug's theory has been scrutinized in detail. A number of variations on their theme have been discovered. An important question has been whether coat proteins actually deform to achieve true quasi-equivalence in the lattice. In reality, protein domains are not continuously "deformable," because the contacts that stabilize folding domains cannot change in a continuous manner or in unison and remain strong. This is also true in virion subunits. However, in $T > 1$ virions, both differences in subunit orientation and changes in the conformation of specific polypeptide arms or loops that extend from the basic structure-forming domains occur among the subunits at the different quasi-equivalent positions. Differences in subunit orientation among quasi-equivalent positions reflect the fact that different or "nonequivalent" bonding surfaces are used in these *analogous* neighbor contacts (see also Conformational Switching and Spatial Control of Icosahedral Capsule Assembly). These nonequivalent interactions can be very different (Liddington et al., 1991; see also Chapter 7) or quite similar (Golmohammadi et al., 1993) in different viruses.

Although the first discussion of the nonequivalence of the various types of intersubunit contacts came through the study of the $T = 3$ single-strand (ss)RNA virions (e.g., Olson et al., 1983; Silva and Rossmann, 1987), a most striking example is in the structure of the $T = 7$ papovavirus simian virus 40 (SV40) (Liddington et al., 1991). In the latter case, each of the hexamer positions predicted by the Caspar and Klug geometry is in fact occupied by a physical pentamer! Thus, each five-member ring interacts with six equally spaced nearest neighbors, and the contacts between each of the five members of the ring and surrounding subunits are different. These observations suggest that during evolution it has been easier or better to devise "nonequivalent" protein interactions that allow $T > 1$ structures to assemble than to devise deformable proteins with unchanging bonding surfaces to perform this function. It may also be true that nonequivalent interactions help to produce a set of interactions that give rise to shells with specific T numbers (see later). The geometric terminology devised by Caspar and Klug (1962) should not be discarded because they emphasized coat protein "deformation," whereas what has been observed is mainly intersubunit bonding differences. It still serves as a useful framework for the discussion of icosahedral virus structure.

Many viruses have been found to be much more complex than was originally envisioned by Crick and Watson (1956) and Caspar and Klug (1962). The more complex virions, such as the poxviruses, herpesviruses, and double-strand (ds)DNA phages, contain tens of different protein types. Whereas some helical viruses, such as TMV, have turned out as predicted, with one coat protein in a simple helical array, other helical viruses, such as the ssDNA filamentous phages, have specialized proteins at both ends. Icosahedral virions that contain only one protein are the exception rather than the rule. Perhaps the most obvious embellishments of the simple one-coat-protein icosahedral shell are the bacteriophage tails, which are themselves made up of many proteins that are built into one vertex of the icosahedral shell, and the enclosure of nucleic acid–containing protein structures (nucleocapsids) within a lipid bilayer envelope, which itself contains other proteins. Many other extensions from simple icosahedral structures have also been found. For example, a number of icosahedral viruses have polypeptides with different amino acid sequences at different ''quasi-equivalent'' positions. These different amino acid sequences can be present either as different protein molecules or as different domains within one molecule. Such structures are said to have ''pseudo triangulation numbers,'' to indicate the fact that the quasi-equivalent positions are not all occupied by the same structural unit. For example, although picornaviruses with 60 molecules each of three different coat proteins could be viewed as $T = 1$ structures composed of heterotrimeric protomers, it is often more useful to view them as pseudo $T = 3$ (Hogle et al., 1985; Rossmann et al., 1985; see also Chapters 6 and 7). Similarly, bean-pod mottle virus, where two β-barrels of one coat protein polypeptide and the single β-barrel of a second coat polypeptide occupy the three ''quasi-equivalent'' positions, is also viewed as pseudo $T = 3$ rather than $T = 1$ (Cheng et al., 1994). The phage T4 capsid has a unique protein at the ''quasi-equivalent'' positions surrounding the fivefold vertices, while the major coat protein occupies the remaining positions (Black and Showe, 1983), so it has a pseudo $T = 13$ structure. Adenoviruses have unique vertex proteins, and the major coat protein contains two β-barrels that occupy adjacent ''quasi-equivalent'' positions in their pseudo $T = 25$ shells (Roberts et al., 1986; Stewart et al., 1993; see also Chapter 8). In many viruses, such as λ, φX174, adenoviruses, and herpesviruses, additional proteins decorate the outside of the basic icosahedral shell, and in SV40, adenoviruses, and possibly T4, internal proteins are present that may not be arranged with icosahedral symmetry (see Chapters 7, 8, and 12).

Arrangement of Nucleic Acid in Icosahedral Virions

The structures of nucleic acids within virions are as varied as the protein arrangements in capsids. In helical viruses such as TMV, each nucleotide is held in a precise position along the helix, whereas, at the other extreme, in the large DNA virions most of the DNA seems not to be held in a precise position. In order to elucidate the pathways by which viruses are assembled, we must understand the structure of the nucleic acids in the virions as well as the protein arrangements. The RNA in TMV follows the helical protein (Namba et al., 1989), and the DNA arrangement in the helical filamentous phages is not well understood, so I limit the discussion here to icosahedral viruses. In no case has the overall nucleic acid arrangement been found to reflect the icosahedral symmetry of the protein capsids, but in some cases portions of the nucleic acids have such an arrangement. Because the imaging methods used to obtain this information involve icosahedral averaging, structures that do not have this symmetry are not visualized.

Icosahedral ssRNA Viruses

The icosahedral ssRNA viruses with $T = 1$, pseudo $T = 3$, $T = 3$, and $T = 4$ protein capsids have small, linear, positive-strand genomes of several thousand nucleotides. They are composed of a protein shell made up of one or a few icosahedrally arranged coat proteins and an internal RNA, which in some cases has a protein linked to the 5′ end. In all atomic structures of these viruses the bulk of the RNA is invisible, and in most cases none of the RNA is seen. However, in three ssRNA viruses, regions of RNA that include 13% to 44% of the genome are observed contacting the inner surface of the capsid at 60 equivalent positions in the icosahedral structures (see Chapter 10). In the pseudo $T = 3$ bean-pod mottle virus, two unpaired strands of seven nucleotides are held near the threefold axis (Chen et al., 1989). Although the strands are not paired, each has bases stacked in a conformation similar to one strand of A-form dsRNA. In the $T = 3$ flockhouse virus (Fisher and Johnson, 1993) and the $T = 1$ satellite TMV (Larson et al., 1993), 10- and 7-bp-long double helices, respectively, are present near the twofold axes. In these cases there are no repeated nucleotide sequences in the genome that could occupy these sites; thus, at different positions in the virion, different nucleotides occupy the otherwise structurally equivalent positions. In addition, when a 19-nucleotide RNA, which forms a stem-loop and is known to bind unassembled coat protein, is soaked into crystals of the $T = 3$ ssRNA phage capsids, one RNA molecule binds to each coat protein dimer in the virion, where it contacts mainly a wall of β-sheet strands (Valegård et al., 1994). In bean-pod mottle virus, at least nine regions of the capsid protein contact the RNA. In flockhouse virus, two α-helices contact the RNA, and in satellite TMV, an extended chain lies along the RNA. There is apparently no universal capsid–RNA contact strategy. In agreement with the earlier conclusion that intravirion RNAs are likely to be tethered to the capsid at a number of locations, they have been observed to be quite immobile within the capsid (Munowitz et al., 1980; McCain et al., 1982). Computer analyses of their nucleotide sequences predict, and spectroscopic measurements confirm, that icosahedral ssRNA virus genome RNAs have a high degree of secondary structure both in and out of the virion (Boedtker and Gesteland, 1975; Fiers, 1979; Yamamoto and Yoshikura, 1986; Bentley et al., 1987; Li et al., 1992). Thus the RNA may not need to be greatly compacted during packaging.

The detailed assembly pathways for icosahedral ssRNA viruses have been difficult to decipher in spite of the apparent simplicity of the components. It should be remembered that this is a rather diverse group of viruses, and they need not all assemble in the same way. Nonetheless, the available information is usually consistent with an assembly strategy in which the coat proteins have high affinity for a particular sequence on the RNA, and the complex of one or a few coat proteins bound to this RNA site nucleates further shell assembly. Additional coat protein molecules then add to the complex. The assembling capsid grows around the already rather compact RNA molecule, perhaps further condensing it by using the RNA contacts seen in the virion (Hogle et al., 1985; Beckett and Uhlenbeck, 1988; Pickett and Peabody, 1993; Fox et al., 1994; Shlesinger et al., 1994).

Icosahedral ssDNA Viruses

In canine parvovirus ($T = 1$) and bacteriophage ϕX174 ($T = 1$), 12% to 13% of their several thousand nucleotide ssDNAs are seen in the atomic structures of the virions. In the parvovirus, a loop of 11 unpaired nucleotides with two associated metal ions is present at 60 equivalent positions at the interior surface of the capsids (Tsao et al., 1991), where it makes mainly van der Waals contacts with the coat protein. In ϕX174, a similar number of nucleotides can be modeled into the electron density beneath the protein shell, where they are in contact with a small basic internal protein through several electrostatic contacts (McKenna et al., 1994). The assembly of ϕX174 and parvoviruses is thought to proceed through the formation of a protein shell followed by insertion of the DNA into that structure (Myers and Carter, 1980; Hayashi et al., 1988).

Icosahedral ds Nucleic Acid Viruses

There are a number of "unrelated" types of dsDNA viruses, whose genomes range in size from 3200 to more than 200,000 bp. Thus, it is perhaps not surprising that they exhibit different intravirion DNA arrangements. The expected extended conformation of free dsDNA in solution means that these encapsidation processes are accompanied by a high degree of DNA compaction (Table 1.1). Little is known about the structure of the nucleic acid in dsRNA virions.

Papovaviruses

Papovavirus intravirion DNAs are nearly covered by nucleosomes built from host-encoded histones (e.g., Dubochet et al., 1986). No well-ordered DNA is seen in the X-ray structure of SV40 (Liddington et al., 1991), but there is no evidence for or against a precise, nonicosahedral arrangement of the DNA in these virions. The in vivo assembly pathways of these viruses have not been rigorously determined, but current evidence suggests that in the papovaviruses

TABLE 1.1 Double-Strand Nucleic Acid Packing Densities in Virions

Virus (Nucleic Acid bp[a])	Triangulation Number[a]	Capsid Inside Radius (Å)[a]	DNA Density (bp/100,000 Å³)[a]
dsDNA			
Papovaviruses[b] (5–8000)	$T = 7^b$	200	16[c]
Adenoviruses (37–39,000)	Pseudo $T = 25$	340	40[d], 22[c]
Herpesviruses (110–230,000)	$T = 16$	450	40[c]
Iridoviruses (150–350,000)	$T > 100?$	620	14[e]
Tailed-bacteriophages			
Short Tail			
P22 (43,500)	$T = 7$	280	45[d], 47[c], 55[f]
T7 (39,400)	$T = 7$	275	45[d], 45[c], 54[f]
φ29 (19,300)	$T = 3^g$	185 × 255	45[d], 46[e]
Contractile tail			
P2 (33,000)	$T = 7$	260	45[c]
P4 (11,600)	$T = 4$	185	44[c]
T4 (170,000)	Pseudo $T = 13^g$	370 × 550	49[d], 48[e]
Noncontractile tail			
λ (48,500)	$T = 7$	290	45[d], 47[c]
Calimoviruses (8000)	$T = 7$	160	47[c]
Hepadnaviruses (3200)	$T = 4$ (and 3?)	125	39[c]
Bacteriophage PRD1 (15,000)	Pseudo $T = 25$	240	26[c]
Bacteriophage PM2 (10,000)	?	300	8[e]
Poxviruses (130–280,000)	Not applicable?	1000 × 1500	2[e]
Baculoviruses (80–170,000)	Not applicable?	300 × 1800	20[e]
dsRNA			
Reoviruses (18–27,000)	$T = 13$	250	38[d], 37[c]
Birnaviruses (6–8000)	$T = 13$	250	11[e]
L-A virus (4600)	$T = 1^h$	180	19[c]
Bacteriophage φ6 (13,500)	$T = 1^h$	200	40[e]
dsDNA crystals	—	—	44–45
dsDNA in solution	—	—	0.008[i]

[a] Individual references are too numerous to list, but many can be found in the other chapters of this volume.
[b] Papovaviruses are unique in that they have proteins bound along the length of the intravirion DNA and have coat protein pentamers occupying the positions expected for hexamers by icosahedral triangulation geometry.
[c] Calculated from interior volume and nucleic acid size using capsid inner radius values from X-ray structures or three-dimensional reconstructions of electron micrographs, assuming the interior is a sphere. Radius values only 10 or 20 Å different from those chosen can change the calculated density by more than 10%; the values chosen are within the experimental error of the measurements.
[d] Calculated from the center-to-center distance between parallel double helices, as measured by low-angle X-ray diffraction. Helices are assumed to be hexagonally packed.
[e] Same as note c, except that inner radii have been less accurately estimated from negatively stained particles in electron micrographs.
[f] Calculated from the measured hydration and size of intravirion DNA, using a partial specific volume of 0.55 cm³/g for NaDNA.
[g] Has an elongated icosahedral capsid.
[h] Coat protein dimers appear to occupy sites predicted to be occupied by monomers by icosahedral triangulation geometry.
[i] The degree of condensation of unpackaged DNA in vivo is not accurately known and certainly varies in different situations. The value given was calculated for the bacteriophage λ chromosome (48.5 kbp) within the interior volume of its *E. coli* host ($\sim 6 \times 10^{11}$ Å³). This represents the theoretical maximally decondensed state for this DNA in the infected cell; note that the phage λ DNA molecule is about 14 μm long, whereas an *E. coli* cell is only 1 to 3 μm in diameter.

the nucleosome-coated DNA is condensed first and the capsid is built around it (Blasquez et al., 1983; Hendrix and Garcea, 1994).

Herpesviruses, Adenoviruses, and the Tailed Bacteriophages

The tailed phages and herpesviruses do not have proteins bound along the length of the intravirion DNA. This first became evident with phages λ and P22, in which no major virion protein components were released with the DNA in vitro, and all major protein components were found to have other structural roles in the virion (e.g., Casjens and Hendrix, 1974). Electron microscopy, spectroscopy, and X-ray scattering indicate that the DNA is essentially all B form and contains few kinks, and that there is "substantial order" in the DNA. In those regions of order, double helices are lying parallel to one another with rather little space between them, so that there probably is not enough space for bound proteins (Richards et al., 1974; Earnshaw and Harrison, 1977; Aubrey et al., 1992). These observations fit well with calculations based on measurements of intravirion DNA hydration ($\sim$1.1 g H_2O/g DNA), in which the sum of the volumes of the DNA and its water of hydration is equal to the interior volume of the virion (Casjens et al., 1992b). Indeed, the density of DNA in the tailed phage and herpes virions is similar to that in crystals of short dsDNAs (Table 1.1; Earnshaw and Casjens, 1980). Herpesvirions, like the phages, are thought to contain DNA in a liquid crystal state (Booy et al., 1991). The arrangement of DNA in adenovirions is unclear; evidence exists for and against virus-encoded "nucleosomes" in the virion (see Devaux et al., 1983; Ruigrok and Schuller, 1993). Other dsDNA viruses, such as hepadna- and calimoviruses, have not been studied, but calculations of intravirion DNA density suggest that they may also have very compact DNA arrangements (Table 1.1).

The detailed arrangement of intravirion dsDNAs is not definitively known. X-ray scattering measurements of interhelix distances in wild-type phage λ and variants with shorter chromosomes found that the interhelix distances increase as the intravirion DNA gets shorter within an invariant capsid, showing that DNA helices are not held in a particular relationship with respect to one another (Earnshaw et al., 1979). DNA observed by electron microscopy of unstained virions in vitreous ice has been interpreted to be in a nematic liquid crystal form (Lepault et al., 1987). Chemical crosslinking experiments have failed to find regions of intravirion DNA that are always close to other DNA segments or to the inside of the capsid (Haas et al., 1982; Windom and Baldwin, 1983; Serwer et al., 1992). Thus, DNA is compacted into the virion interior, where it may not be held in a precise arrangement. The liquid crystal state, the structural model that seems to best fit the available data, is apparently the most energetically favorable in this constricted environment.

In contrast to the previously noted observations on the bulk of the DNA, in the tailed-phage virions there is good evidence that at least one end of these linear DNAs is tethered within the exit channel. In all cases examined, these viruses inject their linear DNA into susceptible hosts beginning with a particular end, and in phage λ that end protrudes from the head portion of the capsid into the tail in the virion (summarized in Earnshaw and Casjens, 1980).

Virion Assembly

Assembly Pathways

Since the reconstitution of TMV four decades ago from separated protein and RNA components (Frankel-Conrat and Williams, 1955), viruses have been found to be excellent models for many aspects of the assembly of macromolecular structures. Historically, the study of assembly of the simple viruses has been driven by structural work, while the study of complex virus assembly has been genetically driven. Currently, each is benefiting from experimental strategies pioneered in the other arena. Virus assembly has been found always to occur in an orderly fashion, rather than by random association of the various components. Even in the simple TMV virion assembly, with only one protein type and one RNA molecule, coat protein assembles onto the RNA genome more readily than it assembles with other coat proteins to form RNA-less particles (Butler et al., 1972), and this assembly initiates at a particular location on the RNA (Zimmern, 1977). The initial coat protein molecules bound to RNA form preferential binding sites for unassembled coat protein molecules. Therefore, binding of the first coat molecules to the RNA nucleates the coat protein assembly process, which then proceeds in an orderly fashion out to both ends of the RNA (reviewed by Stubbs, 1984). In the assembly of more complex viruses, Wood et al. (1968) first found that the many phage T4 structural proteins assemble into the growing particle in a specific, obligate order called an assembly pathway. In this system, over 20 different proteins follow an obligate order of assembly in making just the tail portion of the virion (e.g., Kikuchi and King, 1975). Assembly pathways or portions thereof have now been delineated for many viruses.

Conformational Switching and the Temporal Control of Steps in Virus Assembly Pathways

The early discovery that virus assembly processes follow particular pathways crystallized one of the central mechanistic questions concerning macromolecular assembly. Why do the newly synthesized protein, lipid, and nucleic acid virion components not interact in a haphazard order? How are assembly processes controlled? The full answer to this question, formulated in the 1960s and early 1970s, is not yet known. We know, because control of assembly is maintained in in vitro systems containing purified components, that it is the proteins themselves that exert this control. Typically, newly synthesized virion proteins do not assemble by themselves and often remain well behaved over long periods as soluble monomers or small oligomers. As assembly-naive molecules, they do not associate with the proteins to which they will eventually bond in the completed particle, unless those proteins are already part of a growing virion assembly intermediate.

A generally accepted model for this behavior is that the proteins change conformation upon binding at a site in the growing structure, and in so doing they create a new binding site for another molecule of the same protein in a repeating structure, or a different protein in a more complex structure. Thus, a special nucleation event controls initiation of assembly, and in subsequent assembly steps each protein binds at a newly created site and itself creates a binding site for the next protein. It is this chain of conformational changes that dictates the pathway taken by the assembly process. This *protein-determined cooperative assembly* in pathway control is often referred to as "conformational switching," and has been the subject of several reviews (Wood, 1979; Caspar, 1980; King, 1980; Berget, 1985; Casjens and Hendrix, 1988). The best studied case is the TMV coat protein, wherein structural information is available for coat protein assembled with and without RNA. The nucleation event is the interaction of a self-limited oligomer of coat protein with a particular sequence on the RNA. This triggers a conformational change in the bound oligomer that creates binding sites for the stepwise addition of smaller units of coat protein (monomers and/or small oligomers) in both directions along the RNA from the nucleation complex (e.g., Stubbs, 1984; Potschka et al., 1988; Caspar and Namba, 1990; Butler et al., 1992). Even in this simple case, it is not obvious exactly which of the observed differences between the structure of the coat protein with and without RNA is responsible for the creation of a preferential coat binding site on the growing assembly intermediate (Namba et al., 1989). The detailed molecular mechanisms of such conformational switching events may well be different in different viruses, and their elucidation may ultimately depend upon atomic structure determinations of both the unassembled and assembled conformations of virion proteins as well as study of the assembling systems.

Conformational Switching and Spatial Control of Icosahedral Capsid Assembly

Quasi-equivalent and nonequivalent models for coat protein arrangement in icosahedral shells require that coat proteins occupy T different micro-environments in the virion (or $\sim T$ when the pentamers in the hexamer position of papovaviruses are considered). To accommodate these ideas, one must therefore hypothesize a certain amount of "flexibility" in coat protein bonding properties. I use *flexibility* to mean several structural choices are available, not that each molecule is continuously deformable. It is clear in the atomic structures of $T = 3$, 4, and 7 viruses, and from what we know about protein structure in general, that such flexibility is not due to "deformation" of the core β-barrel domain of those coat proteins. Flexibility can be manifested as conformational differences among arms or loops of the coat protein peptide chain or as differences in contact surfaces utilized between neighboring coat subunits. In the $T > 1$ virion structures that have been determined, major differences in choice of surface used in neighbor contacts (bonding flexibility) and differences in the position of polypeptide arms (conformational differences) are seen among the different ''quasi-equivalent'' positions (e.g., Olson et al., 1983; Silva and Rossmann, 1987; Liddington et al., 1991).

It is challenging to imagine how incorrect structures—for example, ones with vertex pentamers improperly placed in the hexamer lattice—are avoided, given such flexibility in bonding properties. Could coat protein conformational switches be the controlling elements in shape determination? According to a strict quasi-equivalent model, in a $T = 7$ structure, for example, to achieve accurate assembly we would have to imagine that, instead of one conformation for the assembled coat protein, there are seven conformations of the coat protein that are propagated incrementally and precisely through the subunits of the growing particle, so that vertex pentamers are accurately positioned (assembly by ''local rules''; Berger et al., 1994). This concept, although potentially credible for small viruses, becomes harder to imagine for larger viruses.

Such sophisticated, multilevel, cooperative conformational switching has yet to be demonstrated convincingly, and although it remains possible, there appear to be several other mechanisms that achieve accurate assembly of shells larger than $T = 1$. For example, the scaffolding proteins of the dsDNA phages (see later) clearly guide the coat proteins of these viruses into correct assemblies (see Casjens and Hendrix, 1988). Where scaffolding proteins are used, the coat protein assembling units have not been clearly identified, but they may be as small as monomers in phage P22 (Prevelige et al., 1993; see also Chapter 11). There are also strategies that appear to allow coat proteins to assemble into larger shells correctly, without help. For example, those of phage HK97 (Xie and Hendrix, 1995) and papovaviruses (Salunke et al., 1986; Montross et al., 1991) efficiently and accurately assemble into $T = 7$ shells by themselves. We are just beginning to understand the molecular basis of any strategy for attaining accurate shell assembly, but in these two cases it may involve the nonequivalent bonding abilities of these coat proteins. Current information suggests that these coat proteins assemble into self-limited subassemblies, which then aggregate to form icosahedral shells. For example, HK97 coat protein first assembles into self-limited hexamers and pentamers (Xie and Hendrix, 1995). One then can imagine, if free pentamers were to bind free hexamers but pentamers could not bind pentamers and hexamers could not bind hexamers, the spontaneous formation of an icosahedral cap. This cap would be a pentamer surrounded by five hexamers in HK97, or by five additional pentamers for papovaviruses. If, upon binding a pentamer, the hexamer changed conformation such that it could no longer bind other pentamers, this cap would be a ''second-order'' self-limited structure. In addition, if the new hexamer conformation could only bind to other identical hexamers assembled as parts of other caps, this interaction would mediate assembly of 12 caps into a perfect $T = 7$ structure. In the HK97 and papovavirus cases, which are the only known $T = 7$ structures that accurately assemble without a scaffolding protein, it is clear that nonequivalent bonding is used between neighboring subunits. There is severe hexamer skewing in HK97 (Conway et al., 1995), and pentamers replace hexamers in papovaviruses (Liddington et al., 1991). It may be the nonequivalent interactions that allow self-limited intermediates to form. Variations on this theme could in theory allow accurate assembly of shells of different T numbers through other first-order self-limited multimers such as dimers (Rossmann and Erickson, 1985) or trimers (Grimes et al., 1995), and other higher order self-limited structures, such as icosahedral faces. So far, no general way of describing all of these shell assembly pathways has been devised.

It is currently difficult to say exactly why a strategy for shell assembly with self-limited intermediates might arise, but it can be argued that, in addition to providing a kinetic pathway to correct assembly, fewer than T conformations (or fewer bond types) are required in schemes that utilize such intermediates than in a purely quasi-equivalent scheme. For example, in the HK97 case, only three coat protein conformations are needed, those in pentamers, free hexamers, and hexamers assembled into caps. It thus seems possible that this strategy lowers the complexity of the conformational switching and bonding capability required of icosahedral coat proteins. Such strategies have the added advantage of *error checking through use of subassemblies* (see later).

A third way icosahedral viruses have avoided or reduced conformational/ bonding complexity in larger shells is used by the viruses with pseudo T numbers. For example, the picornavirions with pseudo $T = 3$ shells, have three different coat proteins that occupy the three quasi-equivalent positions of a $T = 3$ shell (see Chapter 4). Thus there are three times as many subunits as a $T = 1$ shell with none of the problems associated with quasi- or nonequivalence. Similarly, the multi-β-barrel coat proteins (e.g., adenovirus, bean-pod mottle virus) can in theory contribute to this type of lowering of the required complexity (Burnett, 1984, 1985; see also Chapter 8); indeed the three picornavirus coat proteins are synthesized and initially assemble as a three-barrel polypeptide.

The problem of how icosahedral shell assembly occurs becomes even more complex when one considers the fact that coat proteins of many dsDNA viruses undergo a major conformational change *after* assembly into a shell of the correct T number (Kistler et al., 1978; Prasad et al., 1993; Conway et al., 1995; see also Chapter 11) and many ssRNA plant virus shells can undergo a reversible expansion *after* assembly (Robinson and Harrison, 1982; Speir et al., 1995). The relative instability of the dsDNA virus precursor particles when compared to virions suggests that the coat protein first assembles into structures with weak neighbor contacts, which then shift into a much more tightly bonded structure. The control and mechanism of accurate icosahedral shell assembly may well be as varied as other aspects of virus assembly and remains a challenging unanswered problem.

Catalysis and Enzymes in Virus Assembly

Covalent Alteration of Virion Proteins or Nucleic Acids

Proteolysis   Jacobson and Baltimore (1968), Summers and Maizel (1968), and Laemmli (1970) first discovered that cleavages of structural proteins accompany poliovirus and phage T4 head morphogenesis, and since then such reactions have been found to occur during the formation of infectious virions of many viruses. Cleavages are common in, but not limited to, the more complex virus families. Not all complex viruses have cleaved proteins in the virion; for example, phage T7 has no known maturational cleavages. In general, the reasons for these cleavages are incompletely understood, but several non-mutually-exclusive roles might be to 1) make assembly essentially irreversible, 2) cause spatial or conformational changes required for subsequent assembly or greater stability, 3) alter the function of a protein between virion assembly stages or between virion assembly and nucleic acid delivery, or 4) remove a protein or portion thereof that is no longer required for particle integrity. The proteases that catalyze these cleavages have various origins. The phage T4 maturation protease is encoded by a viral gene (Hintermann and Kuhn, 1992), as are herpesvirus, adenovirus, and picornavirus maturation proteases (Hellen and Wimmer, 1992; Webster et al., 1993; Welch et al., 1993; see also Chapter 12). The flockhouse virus capsid protein cleavage is autocatalytic (Schneeman et al., 1992), and the influenza A virus hemagglutinin cleavage is catalyzed by a host-encoded protease (Walker et al., 1994).

Nucleic Acid Cleavage   In herpesviruses and many-tailed phages, virion-length DNA molecules are cleaved from longer replication intermediates by nucleases whose activities are coupled to the assembly process. Some of these dsDNA phage nucleases have been studied. They are made from two different polypeptides, and ATP binding affects their DNA cleavage properties, but many details of their action are not yet fully understood (e.g., Higgins and Becker, 1994).

Protein Crosslinking   Some viruses have disulfide bonds that covalently connect structural proteins (e.g., Anthony et al., 1992; Zheng et al., 1992). The first nondisulfide crosslink to be found was the covalent joining of proteolytically processed fragments of two proteins in the phage λ head (Hendrix and Casjens, 1974), by a process whose chemistry is not yet understood. Better characterized is the autocatalytic covalent joining of all major coat subunits in phage HK97. In its $T = 7$ capsid, each coat protein molecule is crosslinked to two of its neighbors by asparagine–lysine isopeptide bonds that covalently bind together the subunits within the each of the hexamer and pentamer rings (Duda et al., 1995). The biologic role of the crosslinks is unknown, but they could physically strengthen the virions.

Posttranscriptional Covalent Addition of Chemical Groups to Virion Proteins   Many eukaryotic viruses utilize host-cell protein modification machinery to covalently attach chemical entities such as carbohydrate, lipid, acetyl, sulfate, or phosphate groups to virion assembly proteins. For example, all enveloped viruses have glycoproteins protruding from the exterior surface of their lipid envelopes. The host Golgi complex attaches these carbohydrates during their passage to the cellular membrane (see Chapter 12). Also, major capsid proteins of the papovavirus SV40 and hepadnaviruses are phosphoryl-ated by host kinases (Hendrix and Garcea, 1994; Yu and Summers, 1994), and several viruses have lipids attached to their capsid proteins (e.g., Lee and Chow, 1992; Park and Morrow, 1992; see also Chapter 6). In most cases the biologic roles of these modifications are not yet understood in terms of virus assembly or virion function, but are likely to have important roles in both of these areas.

Assembly Jigs and Templates

Jigs   During attachment of the long, bent-tail fibers to the otherwise com-plete T4 virion, the elbow of these tail fibers binds transiently to the tip of a straight whisker protein that extends from the head−tail junction region. It is thought that this holds the fiber in an orientation that allows its proximal end to join to the tail baseplate, after which it is released from the whisker, which performs the function of a jig (Wood, 1979). Jig function might be more com-mon than this single example suggests, but it is difficult to uncover.

Assembly Templates   A clear example of the use of an assembly template is in the assembly of the phage λ tail. The tail shaft protein normally assembles to form the tubular shaft of the 150-nm-long tail. However, in vitro and in some in vivo situations, it assembles into much longer tail-like structures. Mo-lecular genetic studies found that the length to which the shaft assembles is proportional to the length of a minor tail protein (Katsura and Hendrix, 1984). In mutants missing this minor protein, shaft protein does not assemble in vivo. It is difficult to avoid the conclusion that the shaft protein assembles along a protein template during normal tail assembly.

Scaffolding proteins are a second type of assembly template (see Chapter 11). In herpesviruses and tailed phages, protein procapsids are built first and DNA is subsequently inserted into them. These procapsids are nearly always built with scaffolding proteins in the interior of the particle (phages φX174 and P4 may also have external scaffolds; Hayashi et al., 1988; Marvick et al., 1995). After the coat protein shell is assembled, the scaffold leaves the struc-ture, either as intact disassembled polypeptides or after proteolysis (see Casjens and Hendrix, 1988). Scaffolding proteins, by a poorly understood mechanism, template the assembly of coat proteins into the correct icosahedral shell; indeed, satellite phage P4 scaffolding protein can direct the phage P2 coat protein, which normally forms a $T = 7$ shell with the P2 scaffolding protein, into a $T = 4$ shell (Marvick et al., 1994). In mutants lacking scaffolding proteins, coat protein assembles into aberrant particles in which correct shell closure has not occurred. This is most likely due to incorrectly placed coat protein vertex pen-tamers. The coat proteins themselves appear to carry the information required to assemble into capsidlike structures but must be guided into the precisely correct structure. In most systems, assembly of scaffolds in the absence of coat protein has not been demonstrated, and the results of studies by King, Preve-lige, and coworkers on phage P22 (Prevelige et al., 1993) seem to be best explained by a model in which monomers or small oligomers of scaffolding and coat protein coassemble, even though no stable coat–scaffold mixed oligomers have been found. Thus template assembly and coat assembly are dependent upon one another, like a scaffold grows as a building rises. The structure of these internal scaffolds has remained mysterious; however, in T4 the scaffolds in mutants with elongated procapsids may be helical (Paulson and Laemmli, 1977).

Why are scaffolding proteins required for large viruses? Two ideas have been put forward (Earnshaw and Casjens, 1980): 1) the molecular difficulties in evolving a coat protein that assembles into only a single type of large icosahedral shell are such that a template is required; and 2) because the concentration of biomolecules in cells is very high (>50% macromolecules; Kennel and Reizman, 1977; Zimmerman and Trach, 1991), when attempting to build a hollow shell with only small molecules inside, it is easier to build a structure with a solid interior and excavate it than to devise a mechanism to remove the diverse cytoplasmic macromolecular components that would be trapped inside a hollow structure. It is not possible to know whether either of these is the true evolutionary reason for the existence of scaffolding proteins, but the $T = 7$ polyomavirus capsid and phage HK97 can assemble correctly and efficiently without any apparent internal scaffold (see earlier) showing that coat proteins can exist that assemble $T = 7$ shells by themselves.

Noncovalent Catalysis of Assembly

Assembly "Enzymes"   Do proteins exist that directly catalyze the binding of two other protein components to one another in virus assembly? Few have been found, but the gene 63 protein of phage T4 is the best candidate. Wood and coworkers found that 63 protein is required for addition of the long tail fibers to the particle, that it is not present in the completed virion, and that it can act multiple times. The mechanism by which the 63 protein catalyzes fiber addition is unknown, as is the reason this step might require catalysis, but Wood (1979) hypothesized that it might open a socket to allow entry of a ball at the proximal end of the fiber to create a "ball-and-socket" joint, because in the virion the tail fibers do not extend from the virion at a particular angle.

Scaffolding Proteins   The scaffolding proteins that are not proteolyzed can participate in multiple rounds of procapsid assembly (King and Casjens, 1974; see also Chapter 11). They catalyze virion assembly as they act transiently in a noncovalent manner during assembly, and can perform this function multiple times. In P22 this catalysis is a complex process during which several hundred molecules of scaffolding protein cooperate to form a stable procapsid assembly intermediate and then leave the structure before, or as, DNA is packaged. The route of scaffolding protein exit and the signal that triggers its exit are not known, but there are holes in the coat protein shell of phage P22 procapsids (Prasad et al., 1993). Current data are consistent with a change in coat protein conformation that releases the scaffold.

Molecular Chaperones

One of the original roads to the discovery of molecular chaperones was the isolation of mutations in a host chaperone that block phage λ capsid assembly (Georgopoulos et al., 1973). It was originally thought that the affected host protein might directly aid the assembly of the procapsid, but studies on the mechanism of action of this host molecular chaperone (the *Escherichia coli groEL* and *groES* gene products) make it seem more likely that the role of chaperonins in assembly is the same as with other proteins, namely aiding in the folding of protein monomers or in disassembly of aberrant aggregates (Martin et al., 1993; Todd et al., 1994), and that assembly goes awry when monomers are not properly folded.

In addition to the host chaperonins, phage T4 encodes at least three proteins with the properties of molecular chaperones (Wood, 1979; van der Vies et al., 1994). Each appears to be targeted to one or several specific morphogenetic proteins. The targets may be proteins that are not able to interact with the host chaperonins or need help in folding/assembly in some novel way. For example, the gene 38 protein is not found in finished tail fibers, but is required for the trimerization (Bishop and Wood, 1976; Cerritelli et al., 1994) of the gene 37 protein to form the distal half of the long tail fiber (reviewed by Casjens and Hendrix, 1988). The current model for tail fiber assembly suggests that 37 protein folds as it trimerizes and 38 protein nucleates the process. In addition, the adenoviruses' 100-kDa protein catalyzes the correct folding of the hexon polypeptide (Cepko and Sharp, 1982), and papovaviruses may also encode a targeted chaperone for their coat protein (see Hendrix and Garcea, 1994).

Nucleic Acid Encapsidation

There are three general strategies by which viruses could encapsidate a nucleic acid: (1) condense the nucleic acid first and subsequently build a protein shell around it, (2) build a protein shell and subsequently condense the nucleic acid within it, and (3) progressively condense sections of the nucleic acid within the coat protein shell as the shell grows through co-condensation of nucleic acid and protein. Strategies 1 and 3 might not require external energy, because the energy required to condense the nucleic acid could be provided by the interaction of the various assembly components. Strategy 2, in contrast, might require that external chemical energy be provided to condense the nucleic acid within the procapsid. All three strategies, and variations on them, are used by different viruses. For example, papovaviruses appear to condense the DNA first and build a shell around it (strategy 1). In the ssDNA filamentous phages, the DNA is first condensed into a rod shape by a DNA binding protein and coat protein then replaces the original condensing protein (a variation on strategy 1). Tailed phages and herpesviruses build a protein procapsid and then insert the DNA into it (strategy 2). In the helical ssRNA viruses and nucleocapsids, assembly of the protein on the nucleic acid is the entire assembly pathway, and strategy 3 is used. The icosahedral ssRNA viruses most likely also use a co-

condensation pathway, perhaps with substantial RNA precondensation through intramolecular secondary and tertiary structure interactions.

A universal problem in nucleic acid encapsidation that has been solved the same way by essentially all viruses is how to package only homotypic viral nucleic acid molecules. Where sufficient details are available, specific nucleotide sequences in the viral nucleic acid have been found to mark it for encapsidation. Such sites have been identified and analyzed in ssRNA viruses (e.g., TMV, Sindbis, MS2, human immunodeficiency virus [HIV]), ssDNA viruses (ϕX174, M13), dsRNA viruses (ϕ6), and dsDNA viruses (λ, P22, P1, T3, adenovirus) (see Chapters 11 and 12). These sequences appear to have little in common beyond their use as packaging recognition (*pac*) sites, and the details of how they are utilized range from the first coat protein molecules initiating virion assembly on the TMV RNA site to the complex recognition apparatus of the tailed phages. Interestingly, in the one virus in which substantial effort has so far failed to identify such a sequence (bacteriophage T4), host DNA is completely destroyed during infection so there is no need to discriminate against foreign DNA.

A related problem that all virus assembly must overcome is how to avoid encapsidating nucleic acid–bound proteins, such as the translation, transcription, and replication machinery and regulatory proteins. Solutions to this problem are not clearly understood, but appear to vary from the building of a contiguous array of coat protein molecules that exclude other proteins as the capsid grows (TMV), to the construction of a closed icosahedral shell from which other macromolecules have been excluded and into which DNA is inserted by a mechanism that is thought to peel off DNA-bound proteins (tailed phages).

In the tailed phages, and probably in herpesviruses and adenoviruses, a procapsid is assembled first and then dsDNA is inserted into this preformed container (reviewed by Casjens, 1985b; Black, 1989). Calculations suggest that at least 0.25 kcal/mol of base pairs is required to overcome the bending, charge repulsion, entropy loss, and dehydration in the transition from DNA in solution to the condensed intravirion state (Reimer and Bloomfield, 1978; Rau et al., 1984). ATP hydrolysis is required for successful packaging and is thought to supply the required energy (Guo et al., 1987; Morita et al., 1993). Procapsids contain three essential components, the coat protein shell, an internal scaffold, and a ring of approximately 12 portal protein molecules at one vertex (called the portal vertex; Bazinet and King, 1985). The substrates for DNA packaging are replication-generated concatemers, with a few exceptions such as phage ϕ29 and adenoviruses. Thus DNA must be cleaved into virion-length molecules at the time of packaging. Two proteins form the ''terminase''—the enzyme responsible for recognizing viral DNA and cleaving it near that site to create the terminus that is first inserted into the procapsid. There is a poorly understood device in procapsids that senses when they are filled with DNA and activates a second nucleolytic cleavage that releases the internalized DNA from the concatemer. This device has been genetically linked to the portal protein (Casjens et al., 1992b). After the condensation and final cleavage, the DNA is still unstably packaged. Unless stabilization proteins are added to the portal vertex, possibly to plug the hole through which DNA entered, the DNA can spontaneously come out of the capsid, even in vivo (Lenk et al., 1976).

A number of DNA condensation mechanisms for the dsDNA viruses have been proposed and tested (reviewed by Casjens, 1985b; Black, 1989; Serwer, 1989; Casjens et al., 1992b). The current best working model is that a "DNA translocase" moves the DNA into the procapsid (Hendrix, 1978), but evidence for it is indirect because the function of the putative translocase has not been reproduced in vitro in the absence of a complete DNA packaging system. Details of the putative DNA translocase remain obscure. For example, it is not known if DNA is threaded through the hole in the portal ring during entry (Casjens et al., 1992b; Turnquist et al., 1992). It seems possible that the portal protein in combination with the DNA-bound terminase (procapsid docking complex) forms the DNA translocase molecular machine, which is dismantled after packaging by terminase release. In addition, in φ29 a small RNA is required for DNA packaging (Reid et al., 1994), and in T4 a fragment of coat protein may participate (Xue and Black, 1990).

Do the large animal viruses use dsDNA encapsidation mechanisms similar to those of the tailed phages? The facts that herpesvirions contain liquid crystal DNA (Booy et al., 1991) and herpes- and adenoviruses appear to make procapsids that are subsequently filled with DNA (Hendrix and Garcea, 1994; Thomsen et al., 1994) suggest a mechanism that forces DNA into the capsid. In addition, the partial circular permutation reported for iridovirus DNAs is most easily explained by a packaging pathway that includes overlength DNA substrate, procapsids, headful sensing, and DNA cleavage (Tye et al., 1974; Schnitzler et al., 1987). However, no portal proteins or procapsid docking complexes have yet been characterized for these viruses. Thus it is not yet known whether the tailed phages provide good models for the DNA entry mechanisms used by these eukaryotic viruses, but the possibility remains a reasonable one.

Host Participation in Virus Assembly

Host Proteins That Are Required Transiently for Virion Assembly

Host-encoded chaperonins, proteases, protein kinases, and the like certainly interact with many virion proteins, and these interactions can be essential for successful assembly or infectivity. Several of these have been discussed earlier. In addition, many eukaryotic viruses contain external lipid membrane envelopes that are derived from host membranes. These envelopes include one or more virus-encoded glycoproteins. Details of the creation of virus-specific patches of proteins in a host lipid membrane and the enclosure of nucleic acid–containing nucleocapsids by these patches are not dealt with here (see Doms et al., 1993; see also Chapters 12 and 14). The glycosylation and transport of viral membrane proteins to their target membrane locations are carried out by the host endoplasmic reticulum and Golgi apparatus. Indeed, the study of these and other types of host-modified viral structural proteins has been central in the development of our current understanding of these host processes.

Are There Host Proteins in Virions?

There are few well-characterized instances of host proteins that are *specifically* incorporated into virion particles, and virions are often thought of as containing only virus-encoded proteins. Although the later is clearly true in general, there are examples of specific incorporation of host proteins into virions. The best characterized use of host proteins as essential components in virions is the encapsidation of host-encoded histones bound to the papovavirus DNAs (see earlier). Several studies have reported that small numbers of host membrane proteins are present in some enveloped viruses, but it is difficult to judge whether, as minor components, they are present by evolutionary design or through assembly accidents. Conversely, there are hundreds of molecules of human leukocyte antigen proteins and cyclophilin A in each HIV virion. These seem to be too many to be accidental, and there is evidence that both molecules have roles in creating infectious virions (Arthur et al., 1992; Franke et al., 1994). It therefore cannot be assumed that host proteins are never virion components or, if they are, that they are not biologically relevant.

Virion Function

It must be remembered that virions are not simply inert packages that protect the nucleic acid from physical insult, but are sophisticated molecular mechanisms that deliver the nucleic acid to replication sites inside susceptible cells. Although the mechanisms that attain these goals are as varied as the other aspects of virus assembly, there is at least one universally applicable principle. All viruses have on their exterior surface some protein that binds to receptor molecules on the surfaces of susceptible cells (Haywood, 1994). This protein can be a major structural building block or a specialized protein designed for only this purpose. Atomic structures of receptor–virion complexes (Olson et al., 1993; Stehle et al., 1994) and genetic analyses of virion and receptor proteins (e.g., Colston and Racaniello, 1994) are elucidating the mechanisms by which virions bind cells. The cellular receptors may, or may not, be directly involved in the next stage, penetration of the nucleic acid into the cell. The nucleic acid penetration process is also extremely varied, from the injection mechanism of the dsDNA phages, to lipid membrane fusion, to poorly understood ''direct penetration'' of virions through the plasma membrane. Little is known in any virus about the actual molecular details of nucleic acid transport across the cellular membrane, but in the few cases in which detailed studies have been done, the process is accompanied by extensive changes in the virion structure (Haywood, 1994; Gaudin et al., 1995). Two of the best studied examples are the rearrangement of the phage T4 baseplate during injection (Crowther et al., 1977) and the amazing change in conformation of the influenza virus hemagglutinin protein during membrane fusion (Bullough et al., 1994; see also Chapter 3). Finally, in many cases, the virion delivers, in addition to the nucleic acid, poorly understood pilot proteins, which may guide the nucleic acid to the site of replication in the cell, and enzymes that are required for replication or gene expression of the virus. There are many examples of such enzymes in many virion types, especially in the negative-strand RNA viruses and in the complex dsDNA animal viruses such as herpesviruses and poxviruses. In some cases, such as the reverse transcriptase in retrovirions and the RNA-dependent RNA polymerases of negative-strand ssRNA virions, their requirement and function are obvious, because the cell does not contain such enzymes and their functions are required for viral genes to be expressed (see Chapter 14). In many other cases the role of ''delivered proteins'' is not yet known. Overall, much less is known about virion function in cell entry than about the virus assembly process. The requirement for virions to function in nucleic acid delivery may be the underlying cause of some of the complexities observed in virion assembly, because a very stable particle must be built that is nonetheless able to uncoat spontaneously upon receiving the proper signal.

Genes and Virus Assembly

Regulation of Gene Expression and Virus Assembly

In spite of the extensive work on the nature of virus assembly pathways, it is often still assumed that the genes for the structural proteins are turned on in the order in which their products are needed in the assembly pathway. *This has never been found to be critical to assembly*; the required proteins are all made simultaneously, and it is the properties of the participating proteins themselves that determine their order of assembly (see earlier). Nonetheless, there are connections between gene expression and the assembly process. The most obvious of these is that virus structural proteins are almost invariably made in the ratios required in the formation of completed virions. These synthesis ratios are determined by a combination of transcriptional and translational signal strengths that are evolutionarily set, and only in rare cases are they modulated during infection.

The best studied case in which the synthesis rate of a virus morphogenetic protein is controlled by the assembly process is the phage P22 scaffolding protein. Its rate of synthesis is turned down by scaffolding protein that is not assembled into procapsids (King et al., 1978). Thus just enough scaffolding protein is made to assemble into procapsids with the available coat protein; any excess turns off additional synthesis. This autoregulation is exerted at a posttranscriptional level, but the detailed mechanism remains unknown (Casjens et al., 1985).

There are only a few other examples of gene regulatory activities of capsid proteins. The translational repression of the phage replicase gene by the RNA phage coat protein when it is bound to the packaging site (e.g., Pickett and Peabody, 1993), the action of vesicular stomatitis virus matrix protein in shutoff of host gene expression (Black et al., 1993), and the overexpression of the phage P22 tail protein in the absence of DNA packaging (Adams et al., 1985) are such cases. It is perhaps surprising that so little regulation of gene expression by the virus assembly process is known. Maybe late in infection, when the viral growth cycle is programmed to end, it rarely helps to adjust gene expression rates in response to poor virion assembly. In addition to these involvements in gene expression, virion coat proteins are sometimes involved in nucleic acid replication, but their detailed roles are not completely understood (e.g., ϕX174 [Hayashi et al., 1988]; hepadnaviruses [Hatton et al., 1992]; ssRNA plant ilarviruses [Baer et al., 1994]).

Gene Structure and Virus Assembly

A number of virions contain structural proteins that have identical sequence domains but are otherwise not the same. Thus two different proteins could assemble into a larger structure through homomultimeric assembly of a common domain. The arrangement of the two different proteins in the resulting structure could have dramatic consequences for the assembly process, virion structure, and virion function. There are a number of routes to making such related proteins. A few of the better characterized examples in virus assembly are discussed here.

Proteins with Nested Sequences

Multiple in-frame translation starts or mRNA splicing can give rise to two or more proteins in which one has an extra N-terminal domain relative to the other. For example, the nested phage λ genes *C* and *Nu3* encode two proteins in which the C protein contains the entire Nu3 protein sequence at its C terminus (Shaw and Murialdo, 1980). Both proteins are present in procapsids and are subsequently proteolyzed. In addition, alternate splicing of the $T = 1$ minute virus of mice parvovirus mRNA produces two versions of the capsid protein in which one has an extra N-terminal domain. Both are present in virions, with the longer version accounting for about 16% and the shorter one 84% of the capsid mass (Jogeneel et al., 1986). The reason for such a situation in a $T = 1$ virus, where the coat proteins should occupy equivalent positions, is not known. Likewise, in the internal papovavirion proteins, VP2 has an N-terminal extension relative to VP3 as a result of differential splicing, but their differential roles are unknown (e.g., Griffith et al., 1992). Finally, in cases such as the coronavirus glycoprotein G, structural and nonstructural versions of the same protein are made by initiation at two different in-frame initiation codons (Roberts et al., 1994).

Programmed translational frameshifting and stop codon readthrough can also give rise to related proteins that have different C termini. These mechanisms are widely used in ssRNA and retroviruses, where the frameshift or stop codon readthrough is often located between the regions that encode the major capsid components and the nucleic acid polymerase. It has been proposed that these viruses use this as a method to incorporate polymerase molecules into the virions (Bamford and Wickner, 1994). Other programmed frameshifts, whose biologic roles are less obvious, have been characterized in the phage T7 major capsid protein (Dunn and Studier, 1983) and a phage λ tail assembly protein (Levin et al., 1993).

Incomplete proteolysis can similarly give rise to protein pairs in which one has an extra domain at either end. The cleavage and loss of the N-terminal region from about three fourths of the phage λ gene portal protein molecules is an example of such a situation (Hendrix and Casjens, 1975). Another example may be the herpesvirus scaffolding protein, which is present in procapsids in several C-terminally processed forms (Preston et al., 1992).

Proteins with Alternate Domains

DNA rearrangement, alternate mRNA splicing, and protein rearrangement can be used to produce structural proteins in which alternate domains are attached to a common domain. The clearest example in virus assembly of the first mechanism is found in the phage Mu and P1 families, where a DNA invertase–catalyzed DNA inversion switches two alternate C-terminal domains of the tail fiber protein gene so that two types of virions are made that adsorb to different species of bacteria (see Yarmolinsky and Sternberg, 1988). A clear rationalization for the existence of this switch is the extension of the host range of the virus.

Unfortunately, except in the transcriptase and phage tail fiber cases, we can rarely even speculate on the reasons for the various interesting primary structure relationships mentioned earlier for virion proteins. It seems unlikely that they are "accidents of nature," and the consequences of such arrangements for virion assembly and function should be fertile ground for future study.

Evolution of Virion Structure and Assembly

Evolutionary Relationships Among Virus Coat Proteins

Virus capsids must have arisen at least twice to give rise to the structurally unrelated helical and icosahedral virions. Changes in triangulation number or even nucleic acid type over eons can be imagined, so is it possible that all helical and all icosahedral capsid genes are descendants of one helical and one icosahedral ancestor? Perhaps one of the most surprising findings from the determination of the atomic structure of icosahedral virus coat proteins is the observation that nearly all have coat proteins with virtually identical β-barrel folds (Rossmann and Johnson, 1989; see also Chapter 4). This is true for viruses as different as ssRNA $T = 3$ plant viruses (see Chapter 10), dsDNA $T = 25$ adenoviruses (see Chapter 8), and ssDNA $T = 1$ bacteriophages, even though there is no recognizable amino acid sequence similarity among these coat proteins. It should be noted that the folding of the coat proteins of important groups of icosahedral viruses, such as the herpesviruses and tailed bacteriophages, has not yet been determined. The precise orientation of the β-barrels in the capsid shells varies among the viruses (Rossmann et al., 1983), and in some cases intragene duplications have given rise to viruses with pseudo T numbers (e.g., Roberts et al., 1986; Chen et al., 1989).

The nearly identical protein folds of these different capsid proteins make it very likely that the genes that encode these proteins are descendants from a single, *very ancient* ancestral gene. Even if this were true, it would not mean that the "β-barrel viruses," with their varied lifestyles and assembly pathways, are linear descendants from a single ancestral virus. It is clear that viruses have on many occasions stolen genes from their hosts and exchanged genes with other virus groups (e.g., Casjens et al., 1992a; Sandmeier, 1994), and it is not useful to think about their complete history only in terms of linear evolution. There are rare exceptions to the β-barrel generality; RNA phage and Sindbis virus coat proteins have very different folds that must each have arisen independently (Valegård et al., 1990; Choi et al., 1991). We therefore know there are ways to build a shell that do not involve the β-barrel domain, and that icosahedral capsids arose at least three times during evolution.

Error-Checking Mechanisms in Virus Assembly

An important and poorly understood question in macromolecular assembly is whether there are quality control mechanisms that help avoid assembly errors. Such problems become more serious as the number of subunits required to build the structure becomes larger. One defective subunit can in theory abort the assembly of a whole virion, potentially permanently sequestering many functional subunits into a defective structure. Berget (1985) has reviewed one method by which viruses minimize assembly errors, namely the use of branched pathways in which subassemblies are constructed first and then, if and only if the subassembly is correctly assembled, can it go on to bind other subassemblies to complete the virion. This type of error-checking mechanism avoids making large, inactive structures. However, in spite of the fact that the original pathway studied, phage T4, is extensively branched, some other virion assembly pathways do not appear to be so highly branched.

It is likely that there are additional ways in which viruses minimize assembly errors. For example, assembly templates may have evolved to minimize errors in the assembly of proteins that have the inherent ability to assemble into more than one structure (e.g., scaffolding help in coat protein assembly). In addition, several viruses appear to ensure proper folding of particular capsid proteins by encoding targeted chaperonins (see earlier), or, in the case of membrane proteins, by taking advantage of quality control mechanisms in the host secretory pathways (Doms et al., 1993).

A major source of assembly errors is likely the participation in the assembly process of protein fragments that result from mistakes in synthesis. For example, translation errors can generate at least three types of altered proteins: 1) improper internal starts resulting in C-terminal protein fragments, 2) premature termination resulting in N-terminal fragments, and 3) frameshifts resulting in N-terminal fragments with extra amino acids at their C-termini. Such errors appear to occur at rather significant frequencies (Manley, 1978; Tsung et al., 1989; Jorgenson and Kurland, 1990), and the protein fragments thus produced might bind to growing virions and poison further assembly or make unstable or inactive particles. Such errors would not be so highly detrimental to many cellular gene functions, where the products do not participate in large macromolecular assemblies. One defective monomeric enzyme molecule out of 100 would have little effect on total cellular enzyme activity, whereas a similar error frequency in coat proteins could make most virions imperfect.

Have viruses found mechanisms to avoid these problems? Because initiation of translation is controlled by local mRNA sequence, capsid protein gene sequences could in theory be evolutionarily designed to minimize initiation errors. Generation of N-terminal fragments by premature termination may be more difficult to avoid, because mRNAs could be broken while in the act of translation, and frameshift errors may be difficult to eliminate. However, if virion structural proteins were designed not to fold or assemble without an intact C terminus, the assembly process would be blind to N-terminal fragments of capsid proteins generated by such errors (Casjens et al., 1991). Several of the relatively few cases in which information is available seem to support this idea: removal of a small number of C-terminal amino acids from some virion structural proteins blocks their assembly (e.g., T4 baseplate protein P11 [Plishker and Berget, 1984]; P22 scaffolding protein [L. Sampson and S. Casjens, unpublished results]; hepatitis B core protein [Beames and Lanford, 1993]; polyomavirus VP1 [Garcea et al., 1987]; bluetongue virus VP7 [Le Blois and Roy, 1993]). Conversely, artificially constructed C-terminal fragments are often assembly competent. For example, C-terminal fragments of P22 scaffolding protein can assemble into procapsid-like structures (L. Sampson and S. Casjens, unpublished results), and a peptide near the C terminus of polyomavirus VP3 minor protein is sufficient to bind the major capsid protein (Barouch and Harrison, 1994). Other evidence indicates that the T4 gene 37 protein initiates its simultaneous folding and trimerization near the C terminus (Bishop and Wood, 1976), and the papovavirus X-ray structure and C-terminal mutations in its coat protein are consistent with a crucial role of the C-terminus in papovavirus assembly (Garcea et al., 1987; Liddington et al., 1991). However, perhaps the strongest argument for the use of this evolutionary strategy comes from the many studies of dsDNA phages, in which the N-terminal fragments made by nonsense mutations are almost never incorporated into growing particles, even when the full-length protein is also present (Casjens et al., 1991). In contrast, there are a few nonfunctional N-terminal fragments that are known to assemble into virions (e.g., Adhikari and Berget, 1993), so this cannot be a truly universal principle.

It seems clear that viruses have evolved quality control mechanisms to minimize the construction of imperfect virions, and as more virus assembly systems are studied in this regard, it will be interesting to see if current ideas turn out to be generalizable and if additional mechanisms are found.

Prospects

Not all viruses have similar structures, and those of similar structure are not always assembled by similar pathways. Over the past four decades this great diversity has allowed the study of virus structure and assembly to blossom and be in the vanguard of attaining an understanding of a variety of aspects of macromolecular structure and assembly. But what are the future directions of this endeavor? Much of what has been accomplished to date has led to a general understanding of virion structure and the strategies by which they are assembled, but we must not be trapped by the idea that, when the order of steps are known in an assembly or cell entry pathway, or when an atomic resolution structure of a coat protein shell is known, the "problem has been solved." It would be better to say "new problems have been defined." Very much less is currently known about the detailed molecular mechanisms utilized in individual assembly steps, and about actions of the various virion proteins during cell entry. Understanding in molecular detail the mechanisms by which these steps occur and are controlled is largely in the future, and the study of these questions will certainly lead to new and basic insights. It is clear that study of one system is insufficient; the study of many systems allows the extraction of the underlying principles at different levels as well as the possible variations on these themes. There are many specific problems that are of current interest in this context: for example, understanding the detailed molecular basis for control of assembly steps by conformational switching; spatial control of protein assembly by templates; structure and function of the molecular machine that condenses DNA within dsDNA virus capsids; role of complex gene products in virions; roles of "minor" (in amount only) capsid proteins; functions of covalent modifications of virion structural proteins; and roles of virion structural proteins in delivery of nucleic acid into susceptible cells and the initiation of the infectious cycle once it arrives there.

The study of virus structure and assembly has given rise to a number of the important technical tools that molecular biologists use today, including sodium dodecyl sulfate–polyacrylamide gel electrophoresis (Summers et al., 1965; Laemmli, 1970) and enhancement of electron microscopic images (Crowther et al., 1970), to name only two. In addition, as the field matures, it is becoming possible to use the information gained in many new and sometimes unexpectedly useful ways. The study of the structure and assembly of pathogenic virions is leading to the rational design of drugs that block nucleic acid delivery (Bibler-Muckelbauer et al., 1994; see also Chapter 16) or virion assembly (see Chapter 17); identification of surface peptides that might be useful in immunization, as well as other antiviral strategies (Trono et al., 1989; Sullenger and Cech, 1993; Wilson, 1993); and methods for delivery of genes or genetic devices into cells by packaging them in viruslike particles (Morgenstern and Land, 1990). Several of the best examples of unexpected extensions from this field are the now ubiquitous use of in vitro DNA packaging by bacteriophage λ in the generation of DNA libraries (Hohn and Murray, 1977; Sternberg et al., 1977); the use of virus capsids containing recombinant capsid proteins to present antigens to the mammalian immune system (Brown et al., 1994; Porta et al., 1994); and peptide library display (Marks et al., 1992). None of these later uses was foreseen in the original studies of the mechanisms of assembly and the structures of virions. I expect that such practical extensions from the study of virus structure and assembly can only increase in the future.

Acknowledgments

I thank Roger Hendrix and Jack Johnson for exchange of ideas and for reading portions of this manuscript, and Roger Hendrix and Dennis Bamford for sharing unpublished findings on phage HK97 shell assembly and PRD1 structure, respectively.

References

Adams, M., Brown, H. and Casjens, S. 1985. Bacteriophage P22 tail protein gene expression. J. Virol. 53: 180–184.

Adhikari, P. and Berget, P. 1993. Sequence of a DNA injection gene from Salmonella phage P22. Nucleic Acids Res. 21: 1499.

Anthony, R., Paredes, A. and Brown, D. 1992. Disulfide bonds are essential for the stability of the Sindbis virus envelope. Virology 190: 330–336.

Arthur, L., Bess, J., Sowder, R., Benveniste, R., Mann, D., Chermann, J. and Henderson, L. 1992. Cellular proteins bound to immunodeficiency viruses: implication for pathogenesis and vaccines. Science 258: 1935–1938.

Aubrey, K., Casjens, S. and Thomas, G. 1992. Secondary structure and interactions of the packaged dsDNA genome of bacteriophage P22 investigated by Raman spectroscopy. Biochemistry 31: 11835–11842.

Baer, M., Houser, F., Loesch-Fries, L. and Gehrke, L. 1994. Specific RNA binding by amino-terminal peptides of alfalfa mosaic virus coat protein. EMBO J. 13: 727–735.

Bamford, D. and Wickner, R. 1994. Assembly of double-stranded RNA viruses: bacteriophage ϕ6 and yeast virus L-A. Semin. Virol. 5: 61–70.

Barouch, D. and Harrison, S. 1994. Interactions among the major and minor coat proteins of polyomavirus. J. Virol. 68: 3982–3989.

Bazinet, C. and King, J. 1985. The DNA translocating vertex of dsDNA bacteriophage. Annu. Rev. Microbiol. 39: 109–129.

Beames, B. and Lanford, R. 1993. Carboxy-terminal truncations of the HBV core protein affect capsid formation and the apparent size of encapsidated HBV DNA. Virology 194: 597–607.

Beckett, D. and Uhlenbeck, O. 1988. Ribonucleoprotein complexes of R17 coat protein and a translational operator analog. J. Mol. Biol. 204: 927–938.

Bentley, G., Lweit-Bentley, A., Liljas, L., Skoglund, U., Roth, M. and Unge, T. 1987. Structure of RNA in satellite tobacco necrosis virus: a low resolution neutron diffraction study using $^{1}H_2O/^{2}H_2O$ solvent contrast variation. J. Mol. Biol. 194: 129–141.

Berger, B., Shor, P., Tucker-Kellogg, L. and King, J. 1994. Local rule-based theory of virus shell assembly. Proc. Natl. Acad. Sci. U.S.A. 91: 7732–7736.

Berget, P. 1985. Pathways in viral morphogenesis. In Virus Structure and Assembly (Casjens, S., Ed.), pp. 149–168. Jones and Bartlett, Boston.

Bibler-Muckelbauer, J., Kremer, M., Rossmann, M., Diana, G., Dutko, F., Pevear, D. and McKinlay, M. 1994. Human rhinovirus 14 complexed with fragments of active antiviral compounds. Virology 202: 360–369.

Bishop, R. and Wood, W. 1976. Genetic analysis of T4 tail fiber assembly: I. A gene 37 mutation that allows bypass of gene 38 function. Virology 72: 244–254.

Black, B., Rhodes, R., McKenzie, M. and Lyles, D. 1993. The role of vesicular stomatitis virus matrix protein in inhibition of host-directed gene expression is genetically separable from its function in virus assembly. J. Virol. 67: 4814–4821.

Black, L. 1989. DNA packaging in dsDNA bacteriophages. Annu. Rev. Microbiol. 43: 267–292.

Black, L. and Showe, M. 1983. Morphogenesis of the T4 head. In Bacteriophage T4 (Mathews, C., Kutter, E., Mosig, G. and Berget, P., Eds.), pp. 219–245. ASM Publications, Washington, DC.

Blasquez, V., Beecher, S. and Bina, M. 1983. Simian virus 40 morphogenetic pathway: an analysis of assembly-defective tsB201 DNA-protein complexes. J. Biol. Chem. 258: 8477–8484.

Boedtker, H. and Gesteland, R. 1975. Physical properties of RNA bacteriophages and their RNA. In RNA Phages (Zinder, N., Ed.), pp. 1–28. Cold Spring Harbor Laboratory, Cold Spring Harbor, NY.

Booy, F., Newcomb, W., Trus, B., Brown, J., Baker, T. and Steven, A. 1991. Liquid-crystalline, phage-like packing of encapsidated DNA in herpes simplex virus. Cell 64: 1007–1015.

Brown, C., Welling-Webster, S., Feijlbrief, M., van Lent, J. and Spaan, W. 1994. Chimeric parvovirus B19 capsids for the presentation of foreign epitopes. Virology 198: 477–488.

Bullough, P., Hughson, F., Skehel, J. and Wiley, J. 1994. Structure of influenza haemagglutinin at the pH of membrane fusion. Nature 371: 37–45.

Burnett, R. 1984. In Biological Macromolecules and Assemblies, Vol. 1: Virus Structures and Assemblies (Jurnak, F., and McPhearson, A., Eds.), pp. 337–385. John Wiley & Sons, New York.

Burnett, R. 1985. The structure of the adenovirus capsid. II. The packing symmetry of hexon and its implications for viral architecture. J. Mol. Biol. 185: 125–143.

Butler, P., Bloomer, A. and Finch, J. 1992. Direct visualization of the structure of the "20 S" aggregate of coat protein of tobacco mosaic virus: the "disk" is the major structure at pH 7.0 and the proto-helix at lower pH. J. Mol. Biol. 224: 381–394.

Butler, P., Durham, A. and Klug, A. 1972. Structure and roles of the polymorphic forms of tobacco mosaic virus protein. IV. Control of mode of aggregation of tobacco mosaic virus protein by proton binding. J. Mol. Biol. 72: 1–24.

Casjens, S. 1985a. An introduction to virus structure and assembly. In Virus Structure and Assembly (Casjens, S., Ed.), pp. 1–28. Jones and Bartlett, Boston.

Casjens, S. 1985b. Nucleic acid packaging by viruses. In Virus Structure and Assembly (Casjens, S., Ed.), pp. 75–147. Jones and Bartlett, Boston.

Casjens, S., Adams, M., Hall, C. and King, J. 1985. Assembly-controlled autogenous modulation of bacteriophage P22 scaffolding protein gene expression. J. Virol. 53: 174–181.

Casjens, S., Eppler, K., Sampson, L., Parr, R. and Wyckoff, E. 1991. Fine structure genetic and physical map of the gene 3 to 10 region of the bacteriophage P22 chromosome. Genetics 127: 637–647.

Casjens, S., Hatfull, G., and Hendrix, R. 1992a. Evolution of dsDNA tailed-bacteriophage genomes. Semin. Virol. 3: 383–397.

Casjens, S. and Hendrix, R. 1974. Locations and amounts of the major structural proteins in bacteriophage lambda. J. Mol. Biol. 88: 535–545.

Casjens, S. and Hendrix, R. 1988. Control mechanisms in dsDNA bacteriophage assembly. In The Bacteriophages, Vol. 1 (Calendar, R., Ed.), pp. 1–91. Plenum Press, NY.

Casjens, S., Wyckoff, E., Hayden, M., Sampson, L., Eppler, K., Randall, S., Moreno, E. and Serwer, P. 1992b. Bacteriophage P22 portal protein is part of the gauge that regulates packing density of intravirion DNA. J. Mol. Biol. 224: 1055–1074.

Caspar, D. 1965. Design principles in virus construction. In Viral and Rickettsial Diseases of Man (Horsfall, P. and Tamm, I., Eds.), pp. 51–93. J.B. Lippincott, Philadelphia.

Caspar, D. 1980. Movement and self-control in protein assemblies: quasi-equivalence revisited. Biophys. J. 32: 103–138.

Caspar, D. and Klug, A. 1962. Physical principles in the construction of regular viruses. Cold Spring Harbor Symp. Quant. Biol. 27: 1–24.

Caspar, D. and Namba, K. 1990. Switching in the self-assembly of tobacco mosaic virus. Adv. Biophys. 26: 157–185.

Cepko, C. and Sharp, P. 1982. Assembly of adenovirus major capsid protein is mediated by a non-virion protein. Cell 3: 407–414.

Cerritelli, M., Conway, J. and Steven, A. 1994. Molecular and domainal organization of a hinged viral adhesin. Biophys. J. 68: 20.

Chen, Z., Stauffacher, C., Li, Y., Schmidt, T., Bomu, W., Kamer, G., Shanks, M., Lomonossoff, G. and Johnson, J. 1989. Protein-RNA interactions in an icosahedral virus at 3.0Å resolution. Science 245: 154–159.

Cheng, R., Reddy, V., Olson, N., Fisher, A., Baker, T. and Johnson, J. 1994. Functional implication of quasi-equivalence in a $T = 3$ icosahedral animal virus established by cryo-electron microscopy and X-ray crystallography. Structure 4: 271–282.

Choi, H., Tong, L., Minor, W., Dumas, P., Boege, U., Rossmann, M. and Wengler, G. 1991. Structure of Sinbis virus core protein reveals a chymotrypsin-like serine proteinase and the organization of the virion. Nature 354: 37–43.

Colston, E. and Racaniello, V. 1994. Soluble receptor-resistant poliovirus mutants identify surface and internal capsid residues that control interaction with the cell receptor. EMBO J. 13: 5855–5862.

Conway, J., Duda, R., Cheng, N., Hendrix, R. and Steven, A. 1995. Proteolytic and conformational control of virus capsid maturation: the bacteriophage HK97 case. J. Mol. Biol. 253: 86–99.

Crick, F. and Watson, J. 1956. The structure of small viruses. Nature 177: 473–475.

Crowther, A., Amos, L., Finch, J., DeRosier, D. and Klug, A. 1970. Three dimensional reconstruction of spherical viruses by Fourier synthesis from electron micrographs. Nature 226: 421–425.

Crowther, A., Lenk, E., Kikuchi, Y. and King, J. 1977. Molecular reorganization in the hexagon to star transition of the baseplate of bacteriophage T4. J. Mol. Biol. 116: 489–501.

Devaux, C., Timmins, P. and Berthet-Colominas, C. 1983. Structural studies of adenovirus type 2 by neutron and x-ray scattering. J. Mol. Biol. 167: 119–132.

Doms, R., Lamb, R., Rose, J. and Helenius, A. 1993. Folding and assembly of viral membrane proteins. Virology 193: 545–562.

Dubochet, J., Adrian, M., Schultz, P. and Oudet, P. 1986. Cryo-electron microscopy of vitrified SV40 minichromosomes: the liquid drop method. EMBO J. 5: 519–528.

Duda, R., Hempel, J., Michel, H., Shabanowitz, J., Hunt, D. and Hendrix, R. 1995. Structural transitions during bacteriophage HK97 head assembly. J. Mol. Biol. 247: 618–635.

Dunn, J. and Studier, W. 1983. Compete nucleotide sequence of bacteriophage T7 DNA and the locations of T7 genetic elements. J. Mol. Biol. 166: 477–535.

Earnshaw, W. and Casjens, S. 1980. DNA packaging by the double-stranded DNA bacteriophages. Cell 21: 319–331.

Earnshaw, W. and Harrison, S. 1977. DNA arrangement in isometric phage heads. Nature 268: 598–602.

Earnshaw, W., Hendrix, R. and King, J. 1979. Structural studies of bacteriophage lambda heads and proheads by small angle x-ray diffraction. J. Mol. Biol. 134: 575–586.

Fiers, W. 1979. Structure and function of RNA bacteriophages. Comp. Virol. 13: 69–204.

Fisher, A. and Johnson, J. 1993. Ordered duplex RNA controls capsid architecture in an icosahedral animal virus. Nature 361: 176–179.

Fox, J., Johnson, J. and Young, M. 1994. RNA/protein interactions in icosahedral virus assembly. Semin. Virol. 5: 51–60.

Franke, E., Yuan, H. and Luban, J. 1994. Specific incorporation of cyclophilin A into HIV-1 virions. Nature 372: 359–362.

Frankel-Conrat, H. and Williams, R. 1955. Reconstitution of active tobacco mosaic virus from its inactive protein and nucleic acid components. Proc. Natl. Acad. Sci. U.S.A. 41: 690–695.

Garcea, R., Salunke, D. and Caspar, D. 1987. Site-directed mutation affecting polyomavirus capsid self-assembly in vitro. Nature 329: 86–87.

Gaudin, Y., Ruigrok, R. and Brunner, J. 1995. Low-pH induced conformational changes in viral fusion proteins: implications for fusion mechanisms. J. Gen. Virol. 76: 1541–1556.

Georgopoulos, C., Hendrix, R., Casjens, S. and Kaiser, A. 1973. Host participation in bacteriophage lambda head assembly. J. Mol. Biol. 76: 45–60.

Golmohammadi, R., Valegård, K., Fridborg, K. and Liljas, L. 1993. The refined structure of bacteriophage MS2 at 2.8Å resolution. J. Mol. Biol. 234: 620–639.

Griffith, J., Griffith, D., Rayment, I., Murakami, W. and Caspar, D. 1992. Inside polyomavirus at 25-Å resolution. Nature 355: 652–654.

Grimes, J., Basak, A., Roy, P. and Stuart, D. 1995. The crystal structure of bluetongue virus VP7. Nature 373: 167–170.

Guo, P., Peterson, C. and Anderson, D. 1987. Prohead and gp3-DNA-dependent ATPase activity of the DNA packaging protein gp16 of bacteriophage φ29. J. Mol. Biol. 197: 229–236.

Haas, R., Murphy, R. and Cantor, C. 1982. Testing models of the arrangement of DNA inside bacteriophage λ by crosslinking the packaged DNA. J. Mol. Biol. 159: 71–81.

Hatton, T., Zhou, S. and Standring, D. 1992. RNA- and DNA-binding activities in hepatitis B virus capsid protein: a model for their roles in viral replication. J. Virol. 66: 5232–5241.

Hayashi, M., Aoyama, A., Richardson, D., and Hayashi, M. 1988. Biology of bacteriophage φX174. In The Bacteriophages, Vol. 2 (Calendar, R., Ed.), pp. 1–72. Plenum Press, New York.

Haywood, A. 1994. Virus receptors: binding, adhesion strengthening, and changes in viral structure. J. Virol. 68: 1–5.

Hellen, C. and Wimmer, E. 1992. Maturation of poliovirus capsid proteins. Virology 187: 391–397.

Hendrix, R. 1978. Symmetry mismatch and DNA packaging in large DNA bacteriophages. Proc. Natl. Acad. Sci. U.S.A. 75: 4779–4783.

Hendrix, R. and Casjens, S. 1974. Protein fusion: a novel reaction in bacteriophage λ head assembly. Proc. Natl. Acad. Sci. U.S.A. 71: 1451–1459.

Hendrix, R. and Casjens, S. 1975. Assembly of bacteriophage lambda heads: protein processing and its genetic control. J. Mol. Biol. 91: 187–199.

Hendrix, R. and Garcea, R. 1994. Capsid assembly of dsDNA viruses. Semin. Virol. 5: 15–26.

Higgins, R. and Becker, A. 1994. Chromosome end formation in phage λ, catalyzed by terminase, is controlled by two DNA elements of cos, cosN and cosN and R3, and by ATP. EMBO J. 13: 6152–6161.

Hintermann, E. and Kuhn, A. 1992. Bacteriophage T4 gene 21 encodes two proteins essential for phage maturation. Virology 189: 474–482.

Hogle, J. 1986. Structure and assembly of turnip crinkle virus II. Mechanism of reassembly in vitro. J. Mol. Biol. 191: 639–658.

Hogle, J., Chow, M. and Filman D. 1985. Three-dimensional structure of poliovirus at 2.9Å resolution. Science 229: 1358–1365.

Hohn, B. and Murray, K. 1977. Packaging recombinant DNA molecules into bacteriophage particles in vitro. Proc. Natl. Acad. Sci. U.S.A. 74: 3259–3263.

Jacobson, M. and Baltimore, D. 1968. Polypeptide cleavages in the formation of poliovirus proteins. Proc. Natl. Acad. Sci. U.S.A. 61: 77–84.

Jogeneel, C., Sahli, R., McMaster, G. and Hirt, B. 1986. A precise map of splice junctions in the mRNAs of minute virus of mice, an autonomous parvovirus. J. Virol. 59: 564–573.

Johnson, J. and Fisher, A. 1994. Principles of virus structure. In Encyclopedia of Virology (Webster, R. and Granoff, A., Eds.), pp. 1573–1586. Academic Press Ltd., London.

Jorgenson, F. and Kurland, C. 1990. Processivity errors of gene expression in Escherichia coli. J. Mol. Biol. 215: 511–521.

Katsura, I. and Hendrix, R. 1984. Length determination in bacteriophage λ tails. Cell 39: 73–83.

Kennel, D. and Reizman, H. 1977. Transcription and translation initiation frequencies of the Escherichia coli lac operon. J. Mol. Biol. 114: 1–21.

Kikuchi, Y. and King, J. 1975. Genetic control of bacteriophage T4 baseplate morphogenesis. III. Formation of the central plug and overall assembly pathway. J. Mol. Biol. 99: 695–706.

King, J. 1980. Regulation of structural protein interactions as revealed in phage morphogenesis. In Biological Regulation and Development (Goldberger, E., Ed.), pp. 101–132. Plenum Press, New York.

King, J. and Casjens, S. 1974. Catalytic head assembling protein in virus morphogenesis. Nature 251: 112–117.

King, J., Hall, C. and Casjens, S. 1978. Control of the synthesis of phage P22 scaffolding protein is coupled to capsid assembly. Cell 15: 551–560.

Kistler, J., Aebi, U., Onorato, L., ten Heggler, B. and Showe, M. 1978. Structural changes during transformation of T4 polyheads. I. Characterization of the initial and final states by Fab-fragment labelling of freeze-dried and shadowed preparations. J. Mol. Biol. 126: 571–586.

Laemmli, U. 1970. Cleavage of structural proteins during the assembly of the head of bacteriophage T4. Nature 227: 680–685.

Larson, S., Koszelak, S., Day, J., Greenwood, A., Dodds, J. and McPherson, A. 1993. Double-helical RNA in satellite tobacco mosaic virus. Nature 361: 179–182.

Le Blois, H. and Roy, P. 1993. A single point mutation in the VP7 major core protein of bluetongue virus prevents the formation of core-like particles. J. Virol. 67: 353–359.

Lee, Y. and Chow, M. 1992. Myristate modification does not function as a membrane association signal during poliovirus assembly. Virology 187: 814–821.

Lenk, E., Casjens, S., Weeks, J. and King, J. 1976. Intracellular visualization of precursor capsids in phage P22 mutant infected cells. Virology 68: 182–199.

Lepault, J., Dubochet, J., Baschong, W. and Kellenberger, E. 1987. Organization of double-stranded DNA in bacteriophages: a study by cryo-electron microscopy of vitrified particles. EMBO J. 6: 1507–1512.

Levin, M., Hendrix, R. and Casjens, S. 1993. A programmed translational frameshift is required for the synthesis of a bacteriophage λ tail assembly protein. J. Mol. Biol. 234: 124–139.

Li, T., Chen, Z., Johnson, J. and Thomas, G. 1992. Conformations, interactions and thermostabilities of RNA and proteins in bean pod mottle virus: investigation of solution and crystal structure by laser Raman spectroscopy. Biochemistry 31: 6673–6682.

Liddington, R., Yan, Y., Sahli, R., Benjamin, T. and Harrison, S. 1991. Structure of simian virus 40 at 3.8Å resolution. Nature 354: 278–284.

Manley, K. 1978. Synthesis and degradation of termination and premature termination fragments of β-galactosidase in vitro and in vivo. J. Mol. Biol. 125: 407–432.

Marks, J., Hoogenboom, H., Griffiths, A. and Winter, G. 1992. Molecular evolution of proteins on filamentous phage. J. Biol. Chem. 267: 16007–16010.

Martin, J., Mayhew, M., Langer, T. and Hartl, U. 1993. The reaction cycle of groEL and groES in chaperonin-assisted protein folding. Nature 366: 228–233.

Marvik, O., Dokland, T., Nokling, R., Jacobson, E., Larson, T. and Lindqvist, B. 1995. The capsid size-determining protein sid forms an external scaffold on phage P4 procapsids. J. Mol. Biol. 245: 59–75.

Marvik, O., Sharma, P., Dokland, T. and Lindqvist, B. 1994. Bacteriophage P2 and P4 assembly: alternative scaffolding proteins regulate capsid size. Virology 200: 702–714.

McCain, D., Virudachalam, R., Santini, R., Abdel-Meguid, A. and Markley, J. 1982. Phosphorus-31 nuclear magnetic resonance study of internal motions in ribonucleic acid of southern bean mosaic virus. Biochemistry 21: 5390–5397.

McKenna, R., Ilag, L. and Rossmann, M. 1994. Analysis of the single-stranded DNA bacteriophage φX174 refined at a resolution of 3.0Å. J. Mol. Biol. 237: 517–543.

Montross, L., Watkins, S., Moreland, R., Mamon, H., Caspar, D. and Garcea, R. 1991. Nuclear assembly of polyomavirus capsids in insect cells expressing the major capsid protein VP1. J. Virol. 65: 4991–4998.

Morgenstern, J. and Land, H. 1990. Advanced mammalian gene transfer: high-titer retroviral vectors with multiple drug selection markers and a complementary helper-free packaging cell line. Nucleic Acids Res. 18: 3587–3596.

Morita, M., Tasaka, M. and Fujisawa, H. 1993. DNA packaging ATPase of bacteriophage T3. Virology 193: 748–752.

Munowitz, M., Dobson, C., Griffin, R. and Harrison, S. 1980. On the rigidity of RNA in tomato bushy stunt virus. J. Mol. Biol. 141: 327–333.

Myers, M. and Carter, B. 1980. Assembly of adeno-associated virus. Virology 102: 71–82.

Namba, K., Pattanayek, R. and Stubbs, G. 1989. Visualization of protein-nucleic acid interactions in a virus: refined structure of intact tobacco mosaic virus at 2.9 Å resolution by X-ray fiber diffraction. J. Mol. Biol. 208: 307–325.

Olson, A., Bricogne, G. and Harrision, S. 1983. Structure of tomato bushy stunt viruses. IV. The virus particle at 2.9Å resolution. J. Mol. Biol. 171: 61–93.

Olson, N., Kolatkar, P., Oliveira, M., Cheng, H., Greve, J., McClelland, A., Baker, T. and Rossmann, M. 1993. Structure of a human rhinovirus complexed with its receptor molecule. Proc. Natl. Acad. Sci. U.S.A. 90: 507–511.

Park, J. and Morrow, C. 1992. The nonmyristylated Pr160$^{gag\text{-}pol}$ polypeptide of human immunodeficiency virus type 1 interacts with Pr55^{gag} and is incorporated into virus-like particles. J. Virol. 66: 6304–6314.

Paulson, J. and Laemmli, U. 1977. Morphogenetic core of the bacteriophage T4 head: structure of the core in polyheads. J. Mol. Biol. 111: 459–485.

Pickett, G., and Peabody, D. 1993. Encapsidation of heterologous RNAs by bacteriophage MS2 coat protein. Nucleic Acids Res. 21: 4621–4626.

Plishker, M. and Berget, P. 1984. Isolation and characterization of precursors in bacteriophage T4 baseplate assembly: III. The carboxy termini of protein P11 are required for assembly. J. Mol. Biol. 178: 699–713.

Porta, C., Spall, V., Loveland, J., Johnson, J., Barker, P. and Lomonossoff, G. 1994. Development of cowpea mosaic virus as a high-yielding system for the presentation of foreign epitopes. Virology 202: 949–955.

Potschka, M., Koch, M., Adams, M. and Schuster, T. 1988. Time-resolved solution X-ray scattering of tobacco mosaic virus coat protein: kinetics and structure of intermediates. Biochemistry 27: 8481–8491.

Prasad, B., Prevelige, P., Marietta, E., Chen, R., Thomas, D., King, J. and Chiu, W. 1993. Three-dimensional transformation of capsids associated with genome packaging in a bacterial virus. J. Mol. Biol. 231: 65–74.

Preston, V., Rixon, F., McDougall, I., McGregor, M. and Kobaisi, M. 1992. Processing of the herpes simplex virus assembly protein ICP35 near its carboxy terminal end requires the product of the whole UL26 reading frame. Virology 186: 87–98.

Prevelige, P., Thomas, D. and King, J. 1993. Nucleation and growth phases in the polymerization of coat and scaffolding subunits into icosahedral procapsid shells. Biophys. J. 64: 824–835.

Rau, D., Lee, B. and Parsegian, V. 1984. Measurement of the repulsive force between polyelectrolyte molecules in ionic solution: hydration forces between parallel DNA double helices. Proc. Natl. Acad. Sci. U.S.A. 81: 2621–2625.

Reid, R., Bodley, R. and Anderson, D. 1994. Identification of bacteriophage φ29 prohead RNA domains necessary for in vitro DNA-gp3 packaging. J. Biol. Chem. 269: 9084–9089.

Reimer, S. and Bloomfield, V. 1978. Packaging of DNA in bacteriophage heads: some considerations on energetics. Biopolymers 17: 784–793.

Richards, K., Williams, R. and Calendar, R. 1974. Mode of DNA packing within bacteriophage heads. J. Mol. Biol. 78: 255–263.

Roberts, S., Lichtenstein, D., Ball, A. and Wertz, G. 1994. The membrane-associated and secreted forms of the respiratory syncytial virus attachment glycoprotein G are synthesized from alternative initiation codons. J. Virol. 68: 4538–4546.

Roberts, M., White, J., Grütter, M. and Burnett, R. 1986. Three-dimensional structure of the adenovirus major coat protein. Science 232: 1148–1151.

Robinson, I.K. and Harrison, S.C. 1982. Structure of the expanded state of tomato bushy stunt virus. Nature 297: 563.

Rossmann, M., Abad-Zapatero, C., Murthy, M., Liljas, L., Alwyn Jones, T. and Strandberg, B. 1983. Structural comparisons of some small spherical plant viruses. J. Mol. Biol. 165: 711–736.

Rossmann, M., Arnold, E., Erickson, J., Frankenberger, E., Griffith, J., Hecht, H., Johnson, J., Kamer, G., Lou, M., Mosser, A., Reuckert, R., Sherry, B. and Vriend, G. 1985. Structure of a human common cold virus and functional relationship to other picornaviruses. Nature 317: 147–153.

Rossmann, M. and Erickson, J. 1985. Structure and assembly of icosahedral shells. In Virus Structure and Assembly (Casjens, S., Ed.), pp. 29–73. Jones and Bartlett, Boston.

Rossmann, M. and Johnson, J. 1989. Icosahedral RNA virus structure. Annu. Rev. Biochem. 58: 533–573.

Ruigrok, R. and Schuller, S. 1993. Liquid crystalline DNA in fowl adenovirus. J. Struct. Biol. 110: 177–179.

Salunke, D., Caspar, D. and Garcea, R. 1986. Self-assembly of purified polyomavirus capsid protein. Cell 46: 895–904.

Sandmeier, H. 1994. Acquisition and rearrangement of sequence motifs in the evolution of bacteriophage tail fibers. Mol. Microbiol. 12: 343–350.

Schneemann, A., Zhong, W., Gallagher, T. and Rueckert, R. 1992. Maturation cleavage required for infectivity of a nodavirus. J. Virol. 66: 6728–6734.

Schnitzler, P., Soltau, J., Fischer, M., Reisner, H., Scholtz, J., Delius, H. and Darai, G. 1987. Molecular cloning and physical mapping of the genome of insect iridescent virus type 6: further evidence for circular permutation of the viral genome. Virology 160: 66–74.

Serwer, P. 1989. Double-stranded DNA packaged in bacteriophages: conformation, energetics, and packaging pathway. In Chromosomes: Eukaryotic, Prokaryotic, and Viral, Vol. III (Adolph, K., Ed.), pp. 203–223. CRC Press, Boca Raton, FL.

Serwer, P., Hayes, S. and Watson, R. 1992. Conformation of DNA packaged in bacteriophage T7: analysis by use of ultraviolet light-induced DNA-capsid cross-linking. J. Mol. Biol. 223: 999–1010.

Shaw, J. and Murialdo, H. 1980. Morphogenetic genes C and Nu3 overlap in bacteriophage lambda. Nature 283: 30–33.

Shlesinger, S., Makino, S. and Linial, M. 1994. Cis-acting genomic elements and trans-acting proteins involved in assembly of RNA viruses. Semin. Virol. 5: 39–49.

Silva, A. and Rossmann, M. 1987. Refined structure of southern bean mosaic virus at 2.9Å resolution. J. Mol. Biol. 197: 69–87.

Speir, J., Munshi, S., Wang, G., Baker, T. and Johnson, J. 1995. Structures of the native and swollen forms of cowpea chlorotic mottle virus determined by X-ray crystallography and cryo-electron microscopy. Structure 3: 63–78.

Stehle, T., Yan, Y., Benjamin, T. and Harrison, S. 1994. Structure of murine polyomavirus complexed with an oligosaccharide receptor fragment. Nature 369: 160–163.

Sternberg, N., Tiemeier, D. and Enquist, L. 1977. In vitro packaging of a λ Dam vector containing EcoRI DNA fragments of Escherichia coli and phage P1. Gene 1: 255–280.

Stewart, P., Fuller, S. and Burnett, R. 1993. Difference imaging of adenovirus: bridging the resolution gap between X-ray crystallography and electron microscopy. EMBO J. 12: 2589–2599.

Stubbs, 1984. Macromolecular interactions in tobacco mosaic virus. In Biological Mac-
    romolecules and Assemblies, Vol. 1: Virus Structures and Assemblies (Jurnak, F.
    and McPhearson, A., Eds.), pp. 149–202. John Wiley & Sons, New York.

Sullenger, B. and Cech, T. 1993. Tethering ribozymes to a retroviral packaging signal
    for destruction of viral RNA. Science 262: 1566–1569.

Summers, D. and Maizel, J. 1968. Evidence for large precursor proteins in poliovirus
    synthesis. Proc. Natl. Acad. Sci. U.S.A. 59: 966–971.

Summers, D., Maizel, J. and Darnell, J. 1965. Evidence for virus-specific noncapsid
    proteins in poliovirus-infected HeLa cells. Proc. Natl. Acad. Sci. U.S.A. 54: 505–
    513.

Thomsen, D., Roof, L. and Homa, F. 1994. Assembly of herpes simplex virus (HSV)
    intermediate capsids in insect cells infected with recombinant baculoviruses ex-
    pressing HSV capsid genes. J. Virol. 68: 2442–2457.

Todd, M., Viitanen, P. and Lorimer, G. 1994. Dynamics of the chaperonin ATPase cycle:
    implications for facilitated protein folding. Science 265: 659–666.

Trono, D., Feinberg, D. and Baltimore, D. 1989. HIV-1 gag mutants can dominantly
    interfere with the replication of the wild-type virus. Cell 59: 113–120.

Tsao, J., Chapman, S., Agandje, M., Keller, W., Smith, K., Rossmann, M., Compans,
    R., and Parrish, C. 1991. The three-dimensional structure of canine parvovirus and
    its functional implications. Science 251: 1456–1464.

Tsung, K., Inouye, I. and Inouye, M. 1989. Factors affecting the efficiency of protein
    synthesis in E. coli. J. Biol. Chem. 264: 4428–4433.

Turnquist, S., Simon, M., Egelman, E. and Anderson, D. 1992. Supercoiled DNA wraps
    around the bacteriophage φ29 head-tail connector. Proc. Natl. Acad. Sci. U.S.A.
    89: 10479–10483.

Tye, B., Huberman, J. and Botstein, D. 1974. Non-random circular permutation of phage
    P22 DNA. J. Mol. Biol. 85: 501–532.

Valegård, K., Liljas, L., Fridborg, K. and Unge, T. 1990. The three-dimensional struc-
    ture of the bacterial virus MS2. Nature 345: 36–41.

Valegård, K., Murray, J., Stockley, P., Stonehouse, N. and Liljas, L. 1994. Crystal
    structure of an RNA bacteriophage coat protein-operator complex. Nature 371:
    623–626.

van der Vies, S., Gatenby, A. and Georgopoulos, C. 1994. Bacteriophage T4 encodes
    a co-chaperonin that can substitute for Escherichia coli groES in protein folding.
    Nature 368: 654–656.

Walker, J., Molloy, S., Thomas, G., Sakaguchi, T., Yoshida, T., Chambers, T. and Ka-
    waoka, Y. 1994. Sequence specificity of furin, a proprotein-processing endopro-
    tease, for hemagglutinin of a virulent avian influenza virus. J. Virol. 68: 1213–
    1218.

Webster, A., Hay, R. and Kemp, G. 1993. The adenovirus protease is activated by a
    virus-coded disulphide-linked peptide. Cell 72: 97–104.

Welch, A., McNally, L., Hall, M. and Gibson, W. 1993. Herpesvirus proteinase: site-
    directed mutagenesis used to study maturational, release, and inactivation cleavage
    sites of precursor and to identify a possible catalytic site serine and histidine. J.
    Virol. 67: 7360–7372.

Wilson, T. 1993. Strategies to protect crop plants against viruses: pathogen-derived
    resistance blossoms. Proc. Natl. Acad. Sci. U.S.A. 90: 3134–3141.

Windom, J. and Baldwin, R. 1983. Tests of spool models for DNA packaging in phage
    lambda. J. Mol. Biol. 171: 419–437.

Wood, W. 1979. Bacteriophage T4 assembly and the morphogenesis of subcellular structure. Harvey Lect. 73: 203–213.

Wood, W., Edgar, R., King, J., Henninger, M. and Lielausis, I. 1968. Bacteriophage assembly. Fed. Proc. 27: 1160–1168.

Xie, Z. and Hendrix, R. 1995. Assembly *in vitro* of bacteriophage HK97 proheads. J. Mol. Biol. 253: 74–85.

Xue, M. and Black, L. 1990. Role of the major capsid protein of phage T4 in DNA packaging from structure-function and site-directed mutagenesis studies. J. Struct. Biol. 104: 75–83.

Yamamoto, K. and Yoshikura, H. 1986. Relation between genomic and capsid structures in RNA viruses. Nucleic Acids Res. 14: 389–394.

Yarmolinsky, M. and Sternberg, N. 1988. Bacteriophage P1. In The Bacteriophages, Vol. 1 (Calendar, R., Ed.), pp. 291–438. Plenum Press, New York.

Yu, M. and Summers, J. 1994. Phosphorylation of the duck hepatitis B virus capsid protein associated with conformational changes in the C terminus. J. Virol. 68: 2965–2969.

Zheng, J., Schodel, F. and Peterson, D. 1992. The structure of hepadnaviral core antigens. J. Biol. Chem. 267: 9422–9429.

Zimmerman, S. and Trach, S. 1991. Estimation of macromolecule concentrations and excluded volume effects for the cytoplasm of *Escherichia coli*. J. Mol. Biol. 222: 599–620.

Zimmern, D. 1977. Sequence of the assembly nucleation region of TMA RNA. Cell 11: 455–462.

2.

Principles of Virus Structure Determination

TIMOTHY S. BAKER & JOHN E. JOHNSON

For over 50 years, electron microscopy (EM) and X-ray diffraction have been the most powerful and widely practiced techniques for studying virus structure. EM has always provided a quick way to get a first glimpse of viral morphology. One goal of EM, although not always met in practice, is to record faithful images at 20- to 50-Å resolution of viruses with their ''native'' structures preserved. However, X-ray diffraction is still the only means by which a fully hydrated virus structure can be examined at atomic (2- to 3-Å) resolution.

Pioneering EM work on viruses began with an air-dried sample of the rodlike tobacco mosaic virus (TMV) (Kausche et al., 1939) that was unfixed, unstained, and unshadowed. Although quite crude, the TMV images correlated nicely with other data (e.g., viscosity, diffusion, sedimentation) and assured a major role for EM in structural studies of viruses. Similar studies of other plant viruses, including the helical cucumber strain of tobacco mosaic virus and the spherical tobacco necrosis (TNV) and tomato bushy stunt (TBSV) viruses soon followed (Stanley and Anderson, 1941) as did studies of animal viruses such as vaccinia (Green et al., 1942) and rabbit (Shope) papilloma (Sharp et al., 1942).

Viruses have proved to be excellent model systems for studying numerous aspects of molecular and cellular events. They also have served as invaluable test specimens for development of biological EM methods. TMV has been a paradigm for structural studies of viruses by EM and X-ray diffraction. It was the first virus to be crystallized (Stanley, 1935) and examined by X-ray diffraction (Wyckoff and Corey, 1936), and the first helical virus whose structure was solved at atomic resolution by means of X-ray fiber diffraction technology (Namba and Stubbs, 1986). Single crystals of a spherical virus were first obtained for TBSV (Bawden and Pirie, 1938), and shortly thereafter X-ray photographs from single crystals of a TNV derivative (Crowfoot and Schmidt, 1945) and TBSV (Carlisle and Dornberger, 1948) were recorded. Poliomyelitis virus was the first spherical animal virus crystallized (Schaffer and Schwerdt, 1955) and examined by X-ray crystallography (Finch and Klug, 1959). Virus crystallography ''came of age'' (Harrison, 1980) when the atomic structures of three icosahedral viruses—TBSV (Harrison et al., 1978), southern bean mosaic virus (Abad-Zapatero et al., 1980), and satellite TNV (Unge et al., 1980)—and the 34-subunit disc aggregate of TMV (Bloomer et al., 1978) were determined.

The more than 30-year time lapse between the recording of the first X-ray diffraction photographs and the atomic structure determinations of viruses is a clear testament to the challenges overcome in studying such large macromolecular assemblies with crystallographic techniques. The structure determinations of the plant viruses were soon followed by high-resolution analyses of human rhinovirus (Rossmann et al., 1985; see also Chapter 4) and poliomyelitis virus (Hogle et al., 1985; see also Chapter 6), with the striking result that the tertiary structures of the capsid subunits of the ssRNA picornaviruses closely resemble those of the small, single-strand (ss) RNA plant viruses. Simian virus 40 (SV40) and polyoma virus, which are tumor-forming animal papovaviruses, are currently the largest virions ($\sim$500 Å in diameter) whose structures have been resolved in atomic detail (Liddington et al., 1991; Stehle et al., 1994; see also Chapter 7). A wide spectrum of animal, plant, and bacterial viruses, including examples with double-strand (ds) DNA, ssDNA, dsRNA, and ssRNA genomes, have now been studied with crystallographic techniques. Several of these virus structures are discussed in other chapters in this text.

Structural studies form a basis for understanding viral function. The chemistry and physical properties of a macromolecule and its environment dictate function. Temperature, pH, ionic strength, solvent, ligands and effector molecules, and interactions with other macromolecules are all factors that affect viruses as simple as the 190-Å-diameter satellite TNV or as complex as the over-1000-Å-diameter enveloped human immunodeficiency virus. Decades of virus structure studies have provided key insights into understanding how viruses behave.

Here we outline the principles by which modern methods of EM and X-ray diffraction are used to determine virus structures. The discussion is restricted for convenience to the examination of spherical viruses, with cowpea mosaic virus (CPMV) chosen as an illustrative example. The methods by which structural information is obtained for viruses with helical or lower symmetry, although equally important, are not discussed here mainly because they are not as widely practiced as the techniques used to study spherical viruses. Also, most of the known nonspherical viruses remain accessible to structural analysis only by means of lower resolution EM techniques.

An introduction to the EM and image reconstruction methods used to determine the three-dimensional (3D) structures of spherical viruses precedes a discussion of the well-established X-ray diffraction techniques, with emphasis on recent developments. Enhancements in all these techniques have improved the quality and reliability of structure determinations of numerous biological specimens and have stimulated novel strategies for combining EM and X-ray data in productive ways that yield new insights about viral function that could not be accomplished with any single technique.

Electron Microscopy and Image Reconstruction

Background

Viruses have been a favorite subject for biological microscopy ever since the transmission electron microscope (TEM) was invented and it was possible to image such molecules. The TEM provides a direct way to visualize viruses, and thus has become a valuable tool in virus structural biology (Steven, 1981) and in medical diagnosis (Miller, 1988). The biological and medical literature teems with examples that demonstrate the indispensable role of the microscope in studying viruses. Even the first, quite crude micrographs of virus specimens yielded important and confirmatory data concerning virus sizes and shapes.

Conventional TEM techniques such as metal shadowing, thin sectioning, negative staining, freeze drying, and freeze etching are all widely used to study viruses (Breese, 1981; Nermut et al., 1987a, 1987b). Negative staining is probably the simplest and still most popular way to examine virus morphology (Horne and Wildy, 1979). It was EM images of negatively-stained $T = 3$ plant viruses and $T = 7$ animal viruses, for example, that directly demonstrated that "simple," spherical viruses are composed of icosahedral shells consisting of quasi-equivalently related arrangements of chemically identical protein subunits (Caspar and Klug, 1962).

The first 3D image reconstructions of biological macromolecules were obtained for negatively stained viruses: the tail sheath of the bacteriophage T4 (DeRosier and Klug, 1968) and the icosahedral capsids of human wart virus and TBSV (Crowther et al., 1970a). However, the use of data from viruses that are stained, shadowed, sectioned, or the like is limited mainly by two factors: (1) the distortions and other damaging effects that are induced by dehydration in biological specimens when they are prepared for TEM, and (2) the damage sustained when specimens are imaged with a highly ionizing beam of electrons. These factors normally cause significant loss of the icosahedral symmetry in spherical viruses, and this seriously restricts the success of 3D reconstruction procedures. Even ideal conditions of negative staining or metal shadowing only reveal the distribution of the heavy metal atoms that embed or encase the specimen and do not reveal the density of the specimen itself. In addition, many viruses, especially enveloped ones, are highly susceptible to disruption by many of the most common negative stains because stains can interact strongly with the lipid membrane.

The development of specimen preparation and imaging methods that preserve the hydrated structure of biological molecules in a near-native, vitrified state (Taylor and Glaeser, 1974; Dubochet et al., 1988) and that minimize beam damage to the specimen (Williams and Fisher, 1970; Unwin and Henderson, 1975) has revolutionized EM of viruses (Adrian et al., 1984). Cryo–electron microscopy (cryoEM) vitrifies the specimen, thereby keeping it hydrated, relies on defocus rather than heavy metal atoms to generate image contrast, and produces images under low-dose conditions at near–liquid nitrogen temperatures to give data of unprecedented quality.

The advent of cryoEM sparked renewed interest in the study of icosahedral virus structure with 3D image reconstruction methods because the distortions and artifacts that had previously restricted the use of these methods were eliminated or significantly reduced. However, because cryoEM reveals the projected density of the entire virus with very low contrast, it is usually difficult if not impossible to recognize directly the orientations of particles. In comparison, images obtained with high contrast, negatively stained samples are often easy to interpret directly because only those surfaces accessible to stain are revealed and also because staining is sometimes uneven or one-sided. The low signal-to-noise ratio of cryoEM images thus necessitated the use of computer processing methods. A reformulation of the common lines method, which was developed by Crowther (1971) to study negatively stained virus specimens, was necessary to allow reliable and automatic analysis of cryoEM images (see Image Analysis and 3D Reconstruction) (Fuller, 1987; Baker et al., 1988).

Several animal viruses were among the first spherical viruses to be studied with cryoEM and image reconstruction techniques (Semliki Forest [Vogel et al., 1986], Sindbis [Fuller, 1987], rotavirus [Prasad et al., 1987, 1988; see also Chapter 9], *Nudaurelia capensis* B [Olson et al., 1987], SV40 [Baker et al., 1988]). The 3D structures of over 100 spherical viruses, including those that infect animal, plant, fungal, and bacterial hosts, have now been examined with these techniques (Baker, 1992). All of these studies, performed at 20- to 40-Å resolution, reveal virion morphology and the organization and symmetric arrangement (triangulation number) of the capsid subunits. When individual capsid units are resolved, sometimes even at quite low resolution (≥ 50 Å), the stoichiometry of viral subunits can be determined (Stewart et al., 1991). Systematic changes in subunit conformation that allow the formation of the capsid usually require higher resolution (20- to 30-Å) data (Furcinitti et al., 1991; Stewart et al., 1993). Fundamental relationships between the structures of members of a viral family such as the papovaviruses (Baker et al., 1991) and interactions between the components of the virus that provide a glimpse into the process of virus assembly (Dokland et al., 1992; Dokland and Murialdo, 1993; Dryden et al., 1993; Prasad et al., 1993; Shaw et al., 1993; Yeager et al., 1994; Ilag et al., 1995; see also Chapters 9 and 11) are also revealed by use of cryoEM and reconstruction techniques.

When combined with data obtained by other techniques such as X-ray diffraction, 3D reconstruction is particularly powerful, especially for large viruses or virus complexes that have not been crystallized or are still outside the realm of high-resolution analysis. The structures of a number of isolated viral capsid proteins have been solved to atomic resolution (Wilson et al., 1981; Varghese et al., 1983; Roberts et al., 1986; Choi et al., 1991). Even though the resolution in image reconstructions of noncrystalline virus particles cannot yet compete with that attained by X-ray diffraction, reconstructions can provide a context for diffraction results because the subunit atomic structure can be oriented into the 3D structure of the virion (Choi et al., 1991; Stewart et al., 1991; Olson et al., 1993; Smith et al., 1993a, 1993b; Stewart et al., 1993; Cheng et al., 1994). Antibody and receptor labeling and chemical dissociation studies have been used to locate specific viral components (Prasad et al., 1990; Trus et al., 1992; Newcomb et al., 1993), epitopes (Wang et al., 1992; Smith et al., 1993a, 1993b), and receptor binding sites (Olson et al., 1993). The labeling studies have helped explain key aspects of viral neutralization and attachment. Finally, cryoEM followed by image reconstruction can be performed with heterogeneous populations of particles that allow an exploration of biochemical treatments that have known effects on virus composition or infectivity but that do not cause a transition of the entire population of particles to a defined state (Vénien-Bryan and Fuller, 1994). The ability to accommodate flexibility makes these tools ideal for exploring biologically relevant changes in virus structure and assures that cryoEM techniques will continue to gain popularity.

Preparation of Vitrified Samples

Methods for preparing viruses and other macromolecules for cryoEM are well established (Chiu, 1986; Dubochet et al., 1988). We assume that a highly purified and concentrated suspension ($\sim$1 to 5 mg/ml) of the virus sample can be acquired using standard isolation and purification methods. The key to preparing cryoEM samples is to cool them rapidly enough to produce a solid, vitrified (glasslike, noncrystalline) water layer. Vitrification avoids damage to the biological specimen caused by crystallization of the water, which occurs if the cooling rate is so slow that the water molecules have time to form a highly ordered lattice or if the sample is not kept below about $-140\ °C$ (the devitrification temperature).

Vitrified samples are usually made by first applying a 2- to 5-μl aliquot of the purified virus to an EM grid coated with a holey carbon support film. The grid is blotted nearly dry by pressing a piece of filter paper directly against it and is then rapidly plunged into a liquid cryogen such as ethane or propane. Blotting time, wetting properties of the support, and the humidity near the sample influence the thickness of the vitrified specimen. If dried too much, the specimen gets so thin that the virus particles tend to aggregate at the periphery of the holes in the carbon substrate. Overdrying also leads to increased solute concentrations and dehydration, thereby altering the environment of the specimen and possibly the specimen structure. Conversely, inadequate drying produces a sample so thick that particles may be superimposed or that the electron beam may not be able to penetrate.

After the virus sample is vitrified, the grid is transferred from the cryogen into liquid nitrogen, where it may be stored indefinitely, and then into a cooled cryospecimen holder that is rapidly and carefully inserted into the microscope. All these and subsequent steps, including the recording of images, are carried out with the sample kept at or below -160 °C. The transfer steps must be performed quickly to prevent specimen warming and possible devitrification and to prevent specimen contamination resulting from condensation of water vapor from air in the laboratory. Such condensation can be minimized if the cold sample is continuously bathed in dry nitrogen gas during its transfer to the microscope vacuum.

As is common with most EM techniques, several alternative procedures can be used to achieve desired results. Regardless of which technique is used, the primary goal remains the same: to prepare a uniformly thin, vitrified sample of suitable concentration (Fig. 2.1). The type of support film used may depend on the hydrophilic or hydrophobic properties of the sample. For example, grids may be glow discharged (Dubochet et al., 1982b) or treated with a dilute solution of lipids (S. Fuller, personal communication) before the specimen is applied. Biological samples can be vitrified with a variety of devices and protocols, the choice of which normally depends on the specific requirements of the sample studied. For example, the jet-spray technique, in which the sample is sprayed as droplets onto the coated EM grid as it falls into the cryogen, provides a convenient means to study time-dependent phenomena (Dubochet and McDowell, 1981; Dubochet et al., 1982a).

Relatively high concentrations of buffer salts (>0.1 M) or solutes such as glycerol (5% to 10%) can lead to phase separation upon cooling and structural changes in the specimen. Solutes may be removed without disrupting the virus attached to the grid by floating the grid on a drop of distilled water or an appropriate buffer of low concentration, as is routine for preparing negatively stained samples.

In summary, success in obtaining a vitrified sample for cryoEM depends on many interrelated factors, the most important of which are dictated by the virus (its pI, whether it is enveloped or not, etc.); the solution (pH, ionic strength, etc.); the support film; and the experience of the microscopist. The dimensions and integrity of most viruses are well preserved by cryopreparation techniques (Adrian et al., 1984; Fuller, 1987; Olson and Baker, 1989; Belnap et al., 1993). This is strong evidence that most specimens are fully embedded in the vitreous layer, because those that are not typically distort and shrink (Lepault et al., 1983).

TEM of Vitrified Viruses

The main distinctions between ''conventional'' and cryoEM of biological specimens arise from the requirements to keep the specimen hydrated and cold while minimizing contamination; to minimize the total electron dose; and to

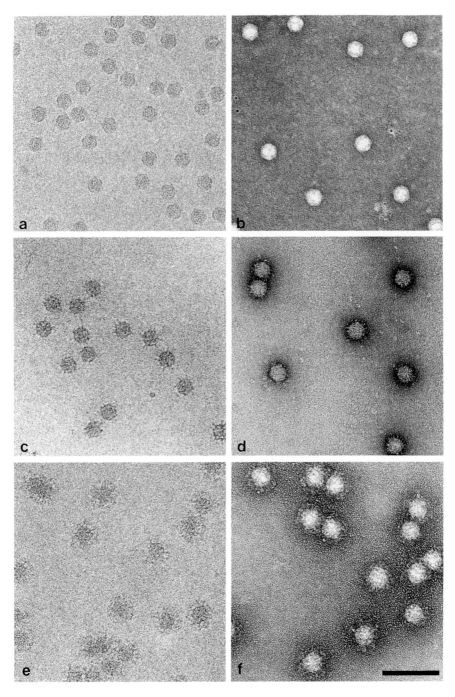

FIGURE 2.1 Micrographs of frozen hydrated (*a*, *c*, *e*) and negatively stained (*b*, *d*, *f*) samples of CPMV. *a*, *b*. Native CPMV. *c*, *d*. CPMV complexed with Fab fragments of monoclonal antibody 5B2. *e*, *f*. CPMV complexed with monoclonal antibody 10B7. The small black dots in *b*, which are part of a negative control experiment, identify goat anti-mouse immunoglobulins conjugated to gold particles, which were used to detect the presence of antibody or Fab bound to CPMV (Porta et al., 1994). Bar, 1000 Å.

enhance the low contrast inherent in unstained specimens. These requirements place additional demands on the microscope, the microscopy, and the microscopist.

Low-Temperature Specimen

The cryoEM specimen must be maintained below the devitrification temperature (~ -140 °C) to assure that water in the sample does not crystallize and damage the specimen. Thus vitrified specimens are kept below the devitrification temperature from the time the specimen is first frozen until all micrographs have been recorded, as described earlier.

A number of commercial and "home-built" holders have been developed to examine cryospecimens. These holders are typically much less stable than normal, room-temperature holders. Cryospecimen holders are subject to greater instabilities caused by the large temperature difference between the cold specimen and the rest of the microscope. Also, vibrations are transmitted to the specimen as a result of boiling of the liquid nitrogen coolant in the Dewar at the end of the holder. Although modern microscopes can achieve an instrumental resolving power of 1 to 2 Å with radiation-resistant, nonbiological specimens, cold holders are currently only stable under ideal conditions to about 3- to 4-Å resolution.

Transfer of the specimen into the high vacuum of the microscope column inevitably leads to release of considerable water vapor from the frost that forms on the holder rod. This water recondenses on the specimen, which, because of its low temperature, acts as an efficient trap for such contaminants. The rapid buildup of condensed water on the specimen thereby reduces the useful time frame for recording images to 30 min or less. Consequently, most cryoEM is now performed with microscopes that have an auxiliary anticontamination device that sandwiches the specimen between a pair of cooled copper blades that have small apertures for the electron beam to pass through. These blades prevent most contamination from reaching the specimen and thereby allow cryoEM to be performed for up to several hours on the same grid. After allowing 15 to 30 min for the specimen stage to stabilize once it is inserted into the microscope, the EM grid is searched to locate a suitable specimen area for photography.

Locating a Suitable Specimen

Cryospecimens are highly sensitive to the damaging effects of the electron beam (see Minimal Irradiation Imaging Reduces Specimen Damage). Thus in the absence of a microscope equipped with an intensified television camera system (e.g., Reynolds, 1988), suitable areas for photography must be located at very low magnification (<2000 to 3000×) to keep the irradiation level very low (<0.05 e$^-$/Å^2/s). At such magnifications it is very difficult and often impossible to see directly unstained viruses smaller than approximately 1000 Å, but it is possible to assess the relative thickness of the vitrified sample. Areas on the grid that are too thick appear opaque, whereas areas that have completely dried are transparent and exhibit high contrast. Regions of the specimen sometimes appear to be vitrified but are actually totally dry, and the buffer salt has crystallized and resembles an "ice" layer. Specimens of optimal thickness tend to have a cloudy appearance.

A convenient way to assess particle concentration is to view (and sacrifice) the sample in one grid square at high magnification (>10,000×) and at a high level of defocus (>10 μm). Alternatively, at low magnification the beam dose can be increased to produce radiation-induced bubbles that first form within the specimen and thereby serve as clear markers of the virus particles (N. Olson, unpublished observations). If the particle concentration is relatively constant in other areas of the EM grid for a vitrified sample of similar thickness, images can then be recorded "blindly" from these areas.

Imaging Conditions

Once a suitable specimen is found, the microscope is switched to high magnification (>100,000×) to facilitate critical focusing and astigmatism corrections. These adjustments are often made at a magnification higher than that used for photography (25,000 to 50,000×) in an area at least 2 μm away from the area to be photographed. Magnification, defocus level, accelerating voltage, beam coherence, and electron dose are among several operating conditions that the cryomicroscopist must control to achieve optimum results. The overall size of the virus studied, the anticipated resolution of the images, and the requirements of the image processing needed to compute a 3D reconstruction to the desired resolution are the most significant factors to be considered. For example, if the expected resolution of the data is 20 to 30 Å, images are typically recorded between 25,000× and 50,000× magnification at 100 kV with a total electron dose of 5 to 20 $e^-/Å^2$ and the objective lens underfocused 0.8 to 1.5 μm. This produces a micrograph with sufficient optical density (0.25 to 1.0) for subsequent image processing (see Image Analysis and 3D Reconstruction). The potential advantages in the use of higher accelerating voltages (e.g., higher resolution, better beam penetration of the sample, and reduced specimen charging), especially with viruses larger than 1000 Å diameter, have been discussed (Zhou and Chiu, 1993).

Phase Contrast Imaging

The inherent "amplitude" (scattering) contrast of unstained biological specimens is quite low because such specimens mainly contain lighter atoms (hydrogen, carbon, nitrogen, oxygen, phosphorus, and sulfur) that scatter electrons to about the same, weak extent. Phase contrast, generated by defocusing the microscope objective lens, thus becomes the dominant form of contrast with unstained vitrified specimens. Compared to the amplitude contrast seen routinely in negatively stained or metal-shadowed specimens, defocus phase contrast is still quite weak. For this reason and because the vitrified sample is so beam sensitive, it is not possible to focus accurately by directly observing specimen details as is possible with stained or shadowed specimens. Instead, focusing must be performed on an area adjacent to the one that will be photographed in order to minimize the electron dose. Critical focusing is achieved either 1) by observing the Fresnel fringes that appear at the edges of holes in the carbon film or that appear around bubbles of volatile molecules that form during specimen irradiation, or 2) by observing the "granularity" of the carbon support film after the layer of vitrified water is burned off.

Unfortunately, the TEM does not produce a faithful image of the projected mass distribution in the specimen. This is caused by imperfections in the electromagnetic lenses (i.e., spherical aberration), the effects of defocusing, the effects of inelastic electron scatter, and the partial coherence of the imaging beam. The relationship between an electron image and the mass distribution in the specimen is described by the contrast transfer function (CTF; Erickson and Klug, 1971), which includes both phase and amplitude contrast effects. The CTF is a characteristic function of the particular microscope used (dictated by the objective lens spherical aberration coefficient), the specimen, and the conditions of imaging (objective lens focal setting, beam coherence, and accelerating voltage).

The microscopist carefully chooses conditions of microscopy to selectively accentuate image features of particular dimensions at the expense of others. Images are typically recorded 0.8 to 2.0 μm underfocus (at 100 kV) to enhance specimen features in the 20- to 40-Å size range. Such defocus levels seem excessive to conventional microscopists, who image shadowed or stained specimens much closer to focus (0.3 to 0.5 μm underfocus). A focal series can provide a useful way to enhance a range of different size features in the cryospecimen. For example, in an enveloped virus such as Sindbis, the overall organization of the external glycoprotein spikes is enhanced at approximately 3 μm or larger defocus, whereas the internal lipid bilayer is best enhanced in images recorded much closer to focus (<1.5 μm) (Fuller, 1987).

Minimal Irradiation Imaging Reduces Specimen Damage

The electron beam rapidly damages vitrified biological specimens. A total dose of 10 to 20 e$^-$/Å^2 typically limits the resolution to 20 to 40 Å in images from such specimens. This dose is comparable to that used during careful analysis of negatively stained specimens (Unwin and Klug, 1974; Baker, 1978) but is 5 to 10 times greater than that used to image unstained specimens at room temperature (Unwin and Henderson, 1975). Thus, cryoEM must be performed in a way that limits specimen exposure to the minimum necessary to record a micrograph with sufficient optical density and signal-to-noise ratio (Williams and Fisher, 1970; Unwin and Klug, 1974; Baker and Amos, 1978). All modern TEMs offer convenient low-dose operation with focusing and astigmatism correction carried out on an area adjacent to the photographed area.

The total electron dose used must be accurately monitored to assure that the specimen is not excessively irradiated. Dose levels may be measured directly with the use of a Faraday cage (Turner et al., 1975; McMath et al., 1982) or they may be estimated with a microscope exposure meter or slow-scan, charge-coupled device (CCD) camera (Krivanek and Mooney, 1993). The exposure meter or CCD must first be calibrated with a Faraday cage or by measuring optical densities on an EM film of known sensitivity exposed to a uniform beam of electrons (without specimen) and developed under controlled conditions (Baker and Amos, 1978).

Image Analysis and 3D Reconstruction

The methods by which a 3D reconstruction of a spherical virus is obtained from a set of projected images are well established (Crowther, 1971; Fuller et al., 1996; Baker and Cheng, 1996). A detailed description of the key aspects of the processing protocol (Fig. 2.2) is beyond the scope of this chapter and is discussed elsewhere (Fuller et al., 1996). The following sections briefly outline salient aspects that affect the quality of the reconstruction.

The icosahedral image reconstruction method produces a 3D structure (density map) by interpreting the separate images of different particles as distinct views of the same structure. If a set of good micrographs are available, the determination of the virus structure in 3D requires three basic steps:

1. Initial estimation of the orientation and center of each particle.
2. Refinement of these parameters by comparison of common data among different views.
3. Combination of the data from a sufficient number of unique views with their relative orientations to produce a 3D density map.

Determination of Particle Orientation and Origin

The main task in computing a 3D reconstruction rests in the determination and refinement of the orientations and origins of the views. Five parameters must be determined for each particle: θ, the inclination of the view vector from a selected twofold axis (typically the z axis in a Cartesian coordinate system); ϕ, the azimuthal angle relative to the line connecting the adjacent fivefolds; ω, the rotation of the projected view relative to the x axis in the scanned image; and x and y, the coordinates of the center of symmetry of the particle where all the symmetry axes cross. Icosahedral symmetry allows one to refer any view to a view in the asymmetric unit (Fig. 2.2). Hence, the unique views comprise a spherical triangle bounded by neighboring fivefold axes ($\theta = 90.0°$, $\phi = \pm31.72°$) and an adjacent threefold axis ($\theta = 69.09°$, $\phi = 0.0°$). Typically, the view parameters must be determined to within about 1° for a low-resolution (>40 Å) reconstruction and about 0.25° for a high-resolution (<25 Å) reconstruction. This is possible to do because dramatic changes occur in the projected image of a large object with changes in its orientation.

A consequence of the icosahedral symmetry of the particle is that 37 pairs of common lines of data are present in any Fourier transform of the projection of an icosahedral particle, and 60 pairs of common lines exist between the

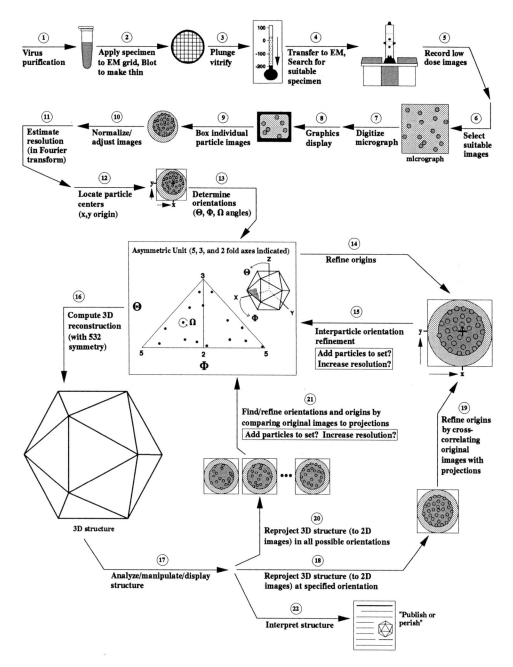

FIGURE 2.2 Schematic diagram of steps involved in cryoEM and image reconstruction of spherical viruses. Steps such as 14 and 15 may be repeated as necessary as the number of images in the data set or the resolution is increased.

transforms of any two projected particle views (Crowther, 1971). The positions of these lines, along which the values of the transform should be identical for objects with perfect icosahedral symmetry, are determined solely by the direction of the view. The orientation angles corresponding to an observed view can be found by stepping θ and ϕ through the asymmetric unit and selecting the values that give the lowest sum of residuals over all 37 pairs of common lines in the transform of the projection. These are the view angles for which the observed view most closely matches that of an ideal icosahedron. The agreement of the phases along the 60 symmetry-related lines between pairs of images, which is called the cross-common lines residual, provides a measure of the agreement between the transforms of different projections, and hence, allows screening for consistency in the data set as well as the refinement of the orientations.

Resolution Limits

The resolution of the reconstruction is always limited by one or more steps in the process, from specimen preparation through imaging, densitometry, view parameter determination, and final calculation of the map (Fig. 2.2). For example, if the specimen was destroyed by the beam prior to the recording of the final image, the information that was lost cannot be retrieved. Thus, the steps of the processing must be designed to preserve the information in the images but not to attempt to recapture information lost by a previous step. Image processing may be powerful, but no magical scheme can recover lost information.

Selection and Digitization of Micrograph

Prior to computer analysis, the available micrographs must be carefully screened to select out the best particle images for including in the 3D analysis. Micrographs are inspected visually to identify those that have the appropriate defocus and minimal astigmatism or movement and also have a suitable particle concentration (i.e., one that is high enough so that a single scanned image will yield an adequate number of views but not so high that images of individual particles overlap). Optical diffraction can be used to help screen images if there is enough carbon substrate in the image to give a strong CTF pattern (Lepault and Leonard, 1985; Jeng et al., 1989; Stewart, 1990). The image of each particle includes the high-contrast, Fresnel fringes at the periphery that arise from the CTF of the defocused microscope. The effects of the CTF are most noticeable in regions of the specimen where there are large differences in density, such as at the interface between the virus surface and the surrounding solvent. The resolutions at which the CTF has its first maximum and minimum provide landmarks for use of the data in the reconstruction.

Careful inspection of the cryoEM images sometimes allows characteristic views to be recognized and also provides some assessment of the heterogeneity of the preparation. A large fraction of the particles may, for example, adopt a preferred orientation, in which case additional images should be recorded to expand the range of particle orientations. A broader range of particle orientations can always be obtained by tilting the specimen holder by a small angle ($\sim 10°$ to $20°$) and this, in combination with in-plane rotational disorder, usually yields a more complete coverage of the asymmetric unit (Dryden et al., 1993).

Micrographs are digitized to convert the photographic data (variations in optical density) into numerical form appropriate for computer analysis. This digitization should be optimized to capture the information contained in the images and is accomplished by scanning the micrograph at a step size *at least* twice as fine as the resolution desired (Shannon, 1949). In practice, it is best to err on the side of caution. Thus, images expected to carry 25-Å information are usually scanned at 5- to 8-Å spacing (i.e., one fifth to one third the desired resolution). A variety of different scanning devices are available for digitizing micrographs. Most operate with a scan step size between 10 and 25 μm, which is useful for digitization of micrographs recorded at magnifications between 20,000× and 50,000×.

Processing the Digitized Micrograph

The common lines algorithm (Crowther, 1971) provides the basis for refining the translational (x, y phase origin) and rotational (θ, ϕ, ω orientation) parameters of the particles and utilizes the Fourier transform of each particle image. Steps involving boxing, masking, background correction, and floating (Fig. 2.2) are designed to minimize artifacts in the calculated transforms. Ideally, each density array to be transformed should contain only a single particle, centered in the array and embedded within a background that is a smooth density of local average value equal to zero. After the scanned, raw micrograph is displayed on an appropriate graphics device, the images of individual particles are centered by eye within a box of the appropriate size and each box is then extracted into a separate file and stored in computer disk memory.

The boxed data often require minor ''massaging'' to correct for artifacts such as nonuniform backgrounds (resulting from differing thickness of the water layer within the field of view) and to mask out the influence of other particles and debris that may also appear within the box (Fig. 2.2). A more accurate estimate of each particle center may be obtained by means of a cross-correlation procedure (Olson and Baker, 1989). Each boxed image is usually then embedded in a larger, square array of pixels (e.g., 128^2, 256^2, 512^2, etc.; power of two dimensions provides computational efficiency). A typical example is that of the top component of CPMV (CPMV-T) with a scanned diameter of 270 Å (or 54 pixels in a 25-μm scan of a 49,000× micrograph) (Baker et al., 1992; Cheng, 1992). CPMV-T was masked within a circular window of 36-pixel radius to exclude other particles and was finally placed in a 128^2 box for Fourier transformation.

Refinement of Particle Parameters and Calculation of 3D Reconstruction

A series of programs that seek to minimize the common lines residuals are used to determine the center and orientation of each view (Fig. 2.2). The best θ, ϕ, and ω parameters are found with respect to the initial x, y center as described earlier, and then the center is refined to give the best common lines residual for that orientation, which is then used to recalculate a new orientation. The process of orientation followed by center refinement is normally repeated for 20 or more particles. Only a small fraction of the particles typically yield reliable orientations because of the low signal-to-noise ratio in the images and systematic variation in common lines degeneracy (Fuller et al., 1996). Once an apparently reliable subset is identified, the transforms are compared by means of the cross–common lines residuals (Crowther, 1971; Fuller, 1987) and the particle orientations and origins are refined against each other. As soon as a small, refined data set is available, a low-resolution, 3D reconstruction is calculated from the available data set. If the reconstruction shows ''reasonable'' features such as continuity and yields projections that correspond with the original images (Baker et al., 1989; Yeager et al., 1990), the set can be used as the basis for addition of further particle images. The additional data are refined against the original set until enough views have been collected to define the reconstruction to the desired resolution (Fuller et al., 1996; Baker and Cheng, 1996).

Reliability of Reconstruction

Aside from subjective criteria, such as exhibiting ''reasonable'' features, the quality of the 3D reconstruction requires quantitative assessment to allow features to be interpreted with confidence. As mentioned earlier, the quality of the final density map is limited by the combined effects of the *worst* steps in the reconstruction process. The number of views needed to define the reconstruction to an isotropic resolution, d, to a first approximation depends on the distribution of views within the asymmetric unit and the diameter, D, of the particle. If the views are spaced evenly, the relation $\pi D/60d$ gives a lower estimate of the number of views required (Crowther et al., 1970b). A more precise answer is found by examining the eigenvalue spectrum for the determination of the coefficients of the expansion functions used to calculate the reconstruction (Crowther et al., 1970b). If all eigenvalues are 1 or greater, the reconstruction is well determined and the data from separate views will be averaged to improve the signal-to-noise ratio in the reconstruction. If eigenvalues are less than 1, the corresponding expansion functions are ill determined and the reconstruction will contain artifacts because of undersampling of data. In practice, enough data should be combined to yield an average eigenvalue significantly greater than 10 and no eigenvalues less than 1.0. For CPMV-T ($D = 270$ Å), more than 15 views were used to achieve a resolution of 23 Å with an acceptable signal-to-noise ratio (Baker et al., 1992), although the formula cited previously would indicate that as few as one sufficiently unique (i.e., not close to an axis of symmetry) view could be used to compute the 3D map to that resolution with no eigenvalue smaller than 10.

The Fourier-Bessel expansion method, which is used to compute the 3D reconstruction, constrains the density map to have D5 point group symmetry (Crowther et al., 1970b). This imposes one fivefold and the five twofolds in the plane perpendicular to the fivefold on the density map. Agreement between the separate views, and hence quality of the reconstruction, is reflected in the additional symmetry (i.e., the nonimposed threefold symmetry found in a truly icosahedral object). The agreement is also reflected by the residual calculated from the cross-common lines between images. The average phase error should be less than 90° for all data out to the resolution limit imposed on the reconstruction.

One method for checking the reliability of the 3D density map is to compare independent reconstructions of the virus from the same or similar samples recorded under comparable conditions of microscopy. Fourier ring correlation (Radermacher et al., 1987) or various types of R-factor comparisons (Baker et al., 1991) or phase residual calculations (Radermacher et al., 1987; Stewart, 1990; Dokland et al., 1992) can be used to assess the reconstruction reliability. If available, information from biochemical data or complementary structural techniques such as X-ray diffraction can help establish or confirm the reliability of the reconstruction. Obviously the interpretation and presentation of the results is very much case dependent. The least squares fitting of the D5 expansion that is used to represent the data averages it so that different regions of the map will show different levels of signal to noise (Crowther et al., 1972). This is particularly important when a preferential orientation of the particles in the water layer leads to an uneven distribution of orientation angles in the data used for reconstruction (Dokland and Fuller, 1994). For this reason, the variation between data sets of the individual features in the map should be examined particularly for those near symmetry axes.

Interpretation of the Electron Density Map

Once the 3D reconstruction is completed and the reliability of the data is assessed, the main goal is to interpret the features in the density map. Initial examination of the density map reveals the external morphology of the virus (Fig. 2.3). Depending on the resolution of the data and the structure of the protein shell, it may be possible to determine the organization of the capsid subunits and also to distinguish approximate subunit boundaries or clustering of subunits into morphologic units (e.g., the capsomeres of the papovaviruses; Baker et al., 1988, 1989, 1991). In most simple capsids, composed solely of identical subunits, the determination of the icosahedral lattice symmetry (T number) generally leads directly to a determination of the subunit stoichiometry (Cheng et al., 1992). However, the papovaviruses are the classic example of a capsid structure whose lattice symmetry ($T = 7$) alone led to an incorrect prediction ($n = 420 = 60*T$) for the stoichiometry of the capsid subunits (Caspar and Klug, 1962). The true stoichiometry ($n = 360$) was later discovered by

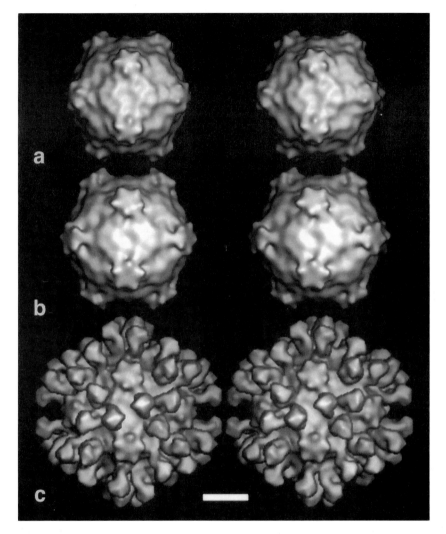

FIGURE 2.3   Gallery of stereo images of 3D structures of native CPMV and CPMV complexed with Fab 10B7. *a*. The 23-Å-resolution X-ray structure of native CPMV computed from atomic model (Stauffacher et al., 1987). *b*. The 23-Å-resolution 3D reconstruction of native CPMV, computed from images shown in Figure 1*a*. *c*. The 23-Å-resolution 3D reconstruction of CPMV/Fab 10B7 complex computed from images shown in Figure 1*b*. Bar, 100 Å.

means of both X-ray crystallographic (Rayment et al., 1982) and EM (Baker et al., 1983, 1988) studies. These results were later confirmed when the atomic structure of SV40 virus was solved (Liddington et al., 1991).

Determination of capsid subunit stoichiometries becomes problematic for the large, complex viruses that have multiple copies each of several different structural proteins. Structural studies of these viruses rely upon the results of a variety of complementary techniques, including biochemical, molecular biology, immunology, virology, and biophysical probes. Examples of large viruses that have been studied with cryoEM and complementary techniques include the rotaviruses, with 6 proteins (Shaw et al., 1993; Yeager et al., 1994; see also Chapter 9); the reoviruses, with 7 proteins (Metcalf et al., 1991; Dryden et al., 1993); the herpesviruses, with 7 proteins (Schrag et al., 1989; Baker et al., 1990; Trus et al., 1992; Newcomb et al., 1993; Zhou et al., 1994; see also Chapter 12); and adenovirus, with least 11 different proteins (Furcinitti et al., 1991; Stewart et al., 1991, 1993; see also Chapter 8).

Because most of the virus mass lies beneath the external surface of the capsid, the entire 3D density map must be examined closely to gain a more complete understanding of the complexities of the structure. Although it is beyond the scope of the discussion here, numerous computational and computer graphics procedures have been devised to aid the visualization and interpretation of subsurface structural features. References cited in the last two paragraphs provide excellent examples that show how reliable information can be obtained about the finer details of subunit morphology and about the interactions between the protein, nucleic acid, and lipid components of viruses. In addition, recent studies have begun to examine aspects of the organization of the nucleic acid genome (Booy et al., 1991; Baker et al., 1992; Dryden et al., 1993; Cheng et al., 1994). The exceptional fidelity of the results obtained by cryoEM as verified by X-ray crystallography of comparable or identical virus structures (see Combination of X-Ray and CryoEM Techniques) provides high confidence in the structural information obtained, and this clearly enhances the ability to ascertain structure–function relationships.

X-Ray Diffraction

X-ray diffraction has provided structural data for viruses ranging from the spherically averaged shape, derived from solution X-ray scattering experiments (e.g., Schmidt et al., 1983), to the determination of positional coordinates for most nonhydrogen atoms in protein capsids, determined from single crystal studies (e.g., Rossmann and Johnson, 1989). Here we discuss classical and modern approaches to structure determination by solution X-ray scattering methods and the methods of 3D structure determination with single crystals. Both techniques are indirect. The initial model is derived from electron density images generated from the experimental amplitudes and derived phases, and then it is refined by objective mathematical procedures that minimize the differences between calculated and observed data. Stereochemical rules for protein structures, established in independent experiments, are enforced during the refinement process. The steps in generating an image of a virus with X-rays correspond to the analogous process with the light microscope (e.g., Glusker and Trueblood, 1985). Optical image formation is described in the next section and serves as a familiar process for comparison.

Image Formation

The scattering of light from an object placed on the stage of a conventional microscope is the first step in magnification and image formation at visible wavelengths. The scattered radiation is gathered by a lens to create an enlarged image at the eyepiece. The optical lens focuses the scattered light by combining the amplitude and phase of the individual waves, and this generates the image. A lens functions by physically altering the path of the scattered radiation, and this is determined by the shape of the lens. In the ideal case, there is no loss of phase information in the process and an image of excellent quality is produced. The magnification depends on the specific properties of the lens, such as its focal length.

The limit of the magnification in a microscope is determined by the wavelength of the radiation; under ideal circumstances, a resolution corresponding to roughly half this wavelength can be approached. For visible light this is about 2000 Å. Higher resolution requires shorter wavelengths, and X-rays of about 1 Å have generated the highest resolution images of hydrated biological specimens. These images are not directly generated, however, because X-rays scattered by an object cannot be focused by a physical lens. Because of their short wavelength, the X-radiation passes through the material of the lens and its path is not altered. An image is created in the X-ray experiment by collecting the amplitude information associated with the first step of the optical experiment described (the diffraction step) and then computationally determining the phase information for the image formation (the focusing step). The two steps are separated in the X-ray experiment, whereas they are directly coupled in the light microscopy experiment. The X-ray image is created as an electron density map because the X-rays are scattered by electrons.

There are two fundamental problems in the X-ray diffraction experiment. First, as in the optical experiment, the object being imaged must be on the same order of size as the wavelength of the X-rays. At optical wavelengths a single object, such as a cell, is an appropriate sample and gives rise to a scattering signal that is readily focused and magnified. It is impossible to image a single virus particle with a size of 180 to 600 Å at 3-Å resolution, because the X-ray scattering signal is not detectable and the molecule is destroyed by the high energy of the X-rays (radiation damage). All X-ray experiments require a large collection of identical molecules in order to obtain a reasonable signal-to-noise ratio for the image formation. These can be randomly oriented in a solution of high concentration or they can be arranged on a crystalline lattice with 3D order. Experiments performed with the first sample type generate a spherically averaged structure, whereas experiments performed with the second type generate a 3D image of the object.

The second problem is the image formation. A method must be used to determine the phase of the individual scattered waves in order to combine the correct amplitude and phase to generate a magnified image. The phase information is lost when the measurement is made because all X-ray detectors are sensitive to the square of the amplitude. This is known as the phase problem

(Stout and Jensen, 1968). Phases for the measured amplitudes in the solution scattering experiment are centrosymmetric and may be assigned as a positive or negative sign to the amplitude value depending on its position along the scattering curve. Three-dimensional data measured from single crystals require a phase accurate to about 45° to generate an image of reasonable fidelity. In each case computational methods must be employed to determine the phase. The final step is the refinement of a high-fidelity model that minimizes the differences between observed and calculated amplitudes to produce coordinates for analysis (e.g., Brunger, 1992).

The theory of X-ray scattering from electrons is one of the most mature areas of physics, and the fidelity of the calculations is remarkably high. Observed and calculated data for small-molecule crystal structures agree within the experimental errors of the measurements (~2% to 3%) and positional accuracy for atoms may be 0.01 Å or better. Although such agreement is not obtained with viruses (typically 14% to 20% with positional accuracy of 0.5 Å or better), the larger discrepancy reflects an inability to accurately model subtle features of the crystal such as bulk solvent and partial disorder, rather than a problem with the theory. Each of these problems is discussed here as they apply to viruses.

Regardless of the X-ray method employed for the analysis, the virus must be in a highly purified and structurally homogeneous state. Viruses considered in this chapter are isometric, with diameters ranging from 180 to 600 Å. They are formed of multiples of 60 copies of one to four protein subunits arranged with the symmetry of an icosahedron. They are spherical to a first approximation, and this property made them attractive samples for early studies by solution X-ray scattering. Plant viruses were of particular interest because they could be isolated in gram quantities with limited effort. Following the development of the Debye theory of molecular scattering of molecules in solution and the early EM work that demonstrated the spherical nature of virions, a number of plant virus structures were analyzed and their average radii were determined with high accuracy (Leonard et al., 1953). The X-ray methods proved to be more accurate than negative-stain EM methods for dimensional analysis because the particles were hydrated and not subject to stain artifacts.

Solution X-Ray Scattering

The solution X-ray scattering experiment is simple in principle. A solution of high virus concentration is placed in a monochromatic X-ray beam and the intensity of scattering from the virus solution is recorded at angular intervals from the direction of the incident X-ray beam. The scattering is circularly symmetric so it can be recorded on a one-dimension detector or, if it is recorded on a two-dimensional detector, the intensity can be circularly averaged to improve the precision of the measurement. The diffraction pattern corresponds to the spherically averaged scattering from all the virus particles in the solution ($\sim 10^{12}$); however, the scattering appears to come from a single spherical par-

TABLE 2.1 Cowpea Mosaic Virus

Icosahedral Particles

Spherically averaged diameter	284 Å
5–5 Dimension	308 Å
3–3 Dimension	272 Å
2–2 Dimension	254 Å
Average thickness of protein shell	39 Å

Capsid Proteins

	# of Amino Acids	M.W.
Large subunit[a] (VP37)	374	41,720
Small subunit[a] (VP23)	213	23,930

Virus Components

Particle	% RNA	M.W.	S	Buoyant Density
Top	0	3.94×10^6	58	1.297
Middle	24	5.16×10^6	98	1.402
Bottom (upper)	34	5.98×10^6	118	1.422
Bottom (lower)	34	5.98×10^6	118	1.470

RNA

Component	M.W.	No. of Ribonucleotides	5′	3′	No. of Proteins
RNA2	1.22×10^6	3482	VPg(4Kd)	polyA(100)	3
RNA1	2.02×10^6	5889	VPg(4Kd)	polyA(87)	5

[a]Sixty copies of each subunit in the capsid.

ticle. The relation between scattering from a single particle and scattering from many particles is elaborated further in the discussion of single crystals.

Diffraction experiments are normally interpreted with an electron density model. If the sample is known to approximate a sphere, as in the case of an icosahedral virus, a model of the approximate dimensions expected can be immediately tested to see if it agrees with the observed pattern. Variations in the spherically averaged electron density can be accommodated in the calculation. The concepts discussed are illustrated by an analysis of CPMV components. CPMV can be separated into four components of different density with a cesium chloride gradient. These components are described in Table 2.1 and correspond to differing quantities of RNA within an identical protein capsid.

The scattering amplitude for a spherical particle of uniform density with radius r is given by

$$F(s) = (\rho - \rho_s)4\pi r^3 \, \frac{\sin m - m \cos m}{m^3} \tag{2.1}$$

where $m = 2\pi rs$, $s = 2 \sin \theta/\lambda$ (θ is half the angle of scattering from the incident beam and λ is the X-ray wavelength), and ($\rho - \rho_s$) is the difference in electron density between the particle and the solvent. The amplitude distribution from spheres containing shells of more than one density is readily calculated by computing the contribution of each shell using equation (2.1) (Harvey et al., 1981). The diffracted intensity is obtained by squaring the computed amplitude. The calculated intensity profiles for CPMV components were compared with the observed profiles and the model parameters were refined to minimize the agreement factor:

$$R = \frac{\sum |I_{obs} - I_{calc}|}{\sum I_{obs}} \tag{2.2}$$

where the sum included evenly spaced points between 250- and 80-Å resolution, about 45 terms. All the data were fitted with a two-shell model using only four adjustable parameters: the radii of the inner (RNA-containing) and outer (protein) shells, the electron density of the inner shell (ρ_i) relative to that of the outer shell (ρ_o) and the solvent (ρ_s), and a multiplication factor to scale the final calculated profile to the measured profile. The relationship between the electron densities is expressed by the parameter Ω:

$$\Omega = (\rho_i - \rho_s)/(\rho_o - \rho_s) \tag{2.3}$$

The fits gave values for the inner and outer protein radii of 101 ± 3 Å and 140 ± 2 Å, with typical R factors of approximately 6%. The spherically averaged capsid protein shell was determined to be approximately 39 Å thick. The fitted electron density of the inner sphere for the top component (empty capsid) is equal, within the expected uncertainty, to the solvent density, as anticipated ($\Omega = 0$). For the middle component, the inner and outer densities were equal (a uniform-density particle, $\Omega = 1$). For the bottom upper and bottom lower components, $\Omega = 1.57 \pm 0.03$ and 1.65 ± 0.03, respectively.

Figure 2.4 shows the X-ray scattering pattern and the one-dimensional electron density computed for each of the components. Each density function was obtained by assigning alternating signs to the observed peaks in the scattering function, assuming the central maximum is positive, and computing the Fourier transform of the amplitude function (Anderegg et al., 1961). The central maximum of the best model was used with the observed data (Earnshaw et al., 1979). Included also are the model density functions providing the best fit to the observed data and the Fourier transforms of the amplitude functions derived from these models. The electron density functions were computed to 50-Å resolution in each case and the data are shown in Figure 2.4. Based on the density required to maximize the fit, the molecular weight of the RNA could be estimated.

The advantage of the solution scattering experiment is the ease and speed at which the size of a hydrated virus can be determined. A modern electronic area detector and a highly collimated intense X-ray source allow the dimensional characterization of a virus particle in a very short time. The disadvantage

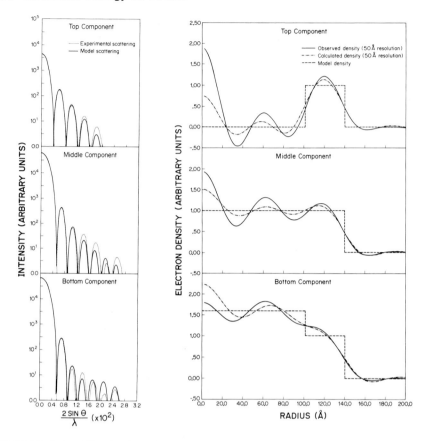

FIGURE 2.4   Diffraction intensity recorded for the three components of CPMV containing different quantities of RNA (see Table 2.1). The data have been corrected for background and plotted on a logarithmic scale. The approximation to a spherical diffraction pattern is emphasized by comparing the theoretical diffraction of spherical shells. *Right.* One-dimensional electron density computed with phases derived from models composed of two spherical shells of different average electron density. The models are step functions shown on the same plot. Note the sensitivity to the differences in the RNA content.

is the one-dimensional nature of the result and, given the techniques previously described for 3D cryoEM, it would be difficult to justify the dedication of an experimental station for solution X-ray scattering studies.

Two features of the experiment maintain interest in solution X-ray scattering. First, it is possible to extract 3D information from high-quality one-dimensional virus data. The shape or envelope of an icosahedral particle can be described by a subset of spherical harmonics that have icosahedral symmetry (Finch and Holmes, 1967; Zheng et al., 1995). The observed X-ray scattering can be used to determine the coefficients of the icosahedral harmonics and thereby generate a 3D envelope with the proper symmetry. The constraints of the icosahedral symmetry allow the extraction of the 3D structure from the one-dimensional data. The first application of the approach was described by Finch and Holmes, and it has had only limited use since then (Jack and Harrison, 1975). The method will probably grow in value with availability of very-high-quality data obtained with area detectors and synchrotron X-ray sources. In principle, a structure comparable to the cryoEM reconstruction could be generated directly from the solution X-ray scattering data (Zheng et al., 1995).

The second advantage of the solution scattering experiment is the possibility of time-resolved studies. At a synchrotron X-ray source it is possible to record a complete virus solution diffraction pattern in 100 ms or less. Fast-readout detectors based on CCDs permit patterns to be recorded at intervals of comparable duration. Experiments of this type permit the structural and kinetic characterization of conformational changes such as particle swelling and assembly. Computational procedures have also improved that allow rapid calculation of scattering curves on the basis of atomic coordinates of particles or intermediates (Lattman, 1989). Stop-flow apparatus for rapidly changing conditions within the sample chamber are readily adaptable for experiments of this type, and it is likely that significant new characterization of the dynamics of virus particles will be made with solution scattering methods.

X-Ray Crystallography

More than 20 different virus structures have been determined at near-atomic resolution in the last 15 years with X-ray crystallography (Stuart, 1993). All of these are nonenveloped viruses, and they range in diameter from 180 to 600 Å. They have been isolated from plants, animals, insects, and bacteria. Purified viruses may be crystallized by methods similar to those used for proteins, and the procedures used for determining their structures are similar as well. In all of the virus structures determined, the symmetry of the particle has been used to improve the phases, which are normally derived initially by isomorphous replacement. In this overview, all the steps in the structure determination are presented.

Crystallization

Well-ordered 3D crystals result when highly purified virions are slowly precipitated. Conditions frequently varied for the production of crystals include virus concentration, pH, buffer, and precipitant. The most successful precipitants have been $(NH_4)_2SO_4$ and various average molecular weights of polyethylene glycol. There is a significant literature on the crystallization of proteins, and most of the methods work well with viruses (Sehnke et al., 1988). As an example, crystallization of CPMV in two different forms is described.

Hexagonal CPMV crystals were grown at 20 °C using both hanging and sitting drop vapor diffusion methods (White and Johnson, 1980). Hexagonal bipyramidal crystals were produced when a virus solution at 20 mg/ml at pH 4.9 in 0.3-M sodium citrate was equilibrated with a 0.6-M sodium citrate solution at pH 4.9. The space group is $P6_122$; $a = 451$ Å, $c = 1038$ Å. The unit cell contains six particles with half of one virus particle in the crystallographic asymmetric unit.

Cubic crystals of CPMV, displaying rhombic dodecahedral morphology, were also grown at 20 °C using vapor diffusion employing the sitting drop method. The virus solution was prepared at 35 mg/ml in 0.05-M potassium phosphate buffer adjusted to pH 7.0. The reservoir solution was 0.4-M ammonium sulfate, 2% polyethylene glycol (PEG) 8000 (w/v), and 0.05-M potassium phosphate buffer at pH 7.0. The solution in the crystallization well was prepared by mixing 25 μl of the reservoir solution with 25 μl of virus solution. Cubic crystals often appeared in 1 to 2 weeks. Crystals could routinely be obtained by seeding. This was performed by crushing a single crystal in 5 μl of 0.05-M potassium phosphate buffer and adding this slurry of crystals to a well containing the crystallization mixture. The contents of the well were thoroughly agitated and 5 μl of the solution was transferred to the next well. The serial dilutions were performed for all nine wells in a tray. Crystals prepared in this way normally grew in 1 to 2 days. The crystal space group was I23 with $a = 317$ Å. The crystal contained two particles per cell with 1/12th of one particle in the crystallographic asymmetric unit.

Data Collection and Processing

Data collection from 3D virus crystals corresponds to the optical analog of gathering the scattered radiation from a sample prior to it passing through the lens of a light microscope. There are two differences. First, because the crystals are 3D, the diffraction pattern is 3D. Because the recorders are two-dimensional, the pattern must be recorded as dozens of two-dimensional planes. Each plane contains a small portion of a 3D pattern projected onto the two-dimensional detector (Fig. 2.5). Therefore, the data collection process requires many exposures with the crystals slightly repositioned (about 0.3° oscillations) for each exposure. A typical complete data set may contain 100 to 200 individual exposures. Because radiation damage is a problem with biological specimens, virus crystals often stop diffracting to high resolution after two or three exposures. Therefore, a typical data set may require 50 to 100 crystals. Recently, flash freezing to liquid nitrogen temperatures has been successful with protein and virus crystals; it dramatically reduces radiation damage and permits an entire data set to be collected from a single crystal.

The second difference is the nature of the pattern. The scattered radiation comes from a single copy of the specimen in a light microscope (e.g., a cell or a portion of a cell). The scattering pattern from such an object is termed *continuous* because the intensity of the scattering can be measured at every position in the recorded pattern. In contrast, a typical crystal will contain 10^{12} virus particles and these are arranged on an ordered 3D lattice. If all virus particles are in the same orientation in the crystal (e.g., the I23 crystal form of CPMV), then the diffraction pattern will appear as if it comes from a single CPMV virus particle, but it will be discontinuous. The sampling observed is a function of the lattice on which the particles are packed. The spacing between

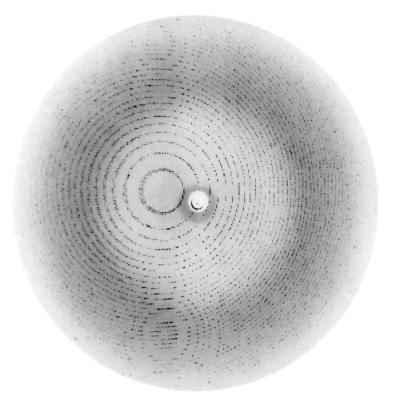

FIGURE 2.5   A high-resolution oscillation photograph from a cubic crystal of CPMV. The outer edge of the pattern is 2.8 Å. Roughly 25,000 diffraction maxima were measured and indexed from this photograph.

the sampling points in the diffraction is inversely related to the spacing between lattice points in the crystal. Because virus particles are large, by the standards of crystalline biological molecules, the spacing between lattice points in the diffraction pattern is exceptionally small. Analysis of the spacing between reflections in the diffraction pattern allows the spacing in the real space lattice to be determined. Analysis of the symmetry in the diffraction pattern permits the symmetry of the crystals to be determined. These crystal characteristics can often be rapidly determined at the beginning of a crystal structure analysis, and from this information the particle packing and orientation may be established, as discussed in the next section.

Frequently virus data collection is performed at a synchrotron X-ray source. The intense, monochromatic, highly collimated, parallel X-radiation has made it possible to resolve diffraction maxima from crystals with lattice constants in excess of 1000 Å. The high intensity of the beam reduces exposure times to seconds compared to many hours with a laboratory X-ray source.

The most common detector for virus data collection is the image plate. It is very sensitive to X-rays (10 to 50 times more sensitive than photographic film for weak exposures) and the optical density on the plate is digitized by a laser scanning process on a raster of about 100 μm in size. Data processing consists of indexing each of the diffraction maxima on an image plate and determining the relative intensity of the reflection (a term used to describe each diffraction maxima because they can be thought of as resulting from X-rays reflecting off specific planes of electron density in the crystal). Integrated intensities of reflections from the 100 to 200 individual patterns are independently measured. They are scaled together by identifying reflections on different plates that are related by crystallographic symmetry and expected to have the same intensity. Scale factors are determined that force these reflections to be equivalent, and the remaining reflections on the same plate are multiplied by the same scale factor. The final data set is a 3D diffraction pattern consisting of a list of indexed diffraction maxima with an estimate of each integrated intensity. Symmetry-equivalent reflections in the data set typically agree within 6% to 10%, on average, after the scaling process. This is a measure of the quality of the data set.

This is half the information required to generate an image with a Fourier synthesis procedure. The other half is the phase of each diffraction maxima. This is required to sum the individual scattered waves to form an image. Procedures for determining these are described in the next sections.

Packing Analysis (Position and Orientation of Particles)

Phase determination requires knowledge of the position and orientation of the virus particle in the unit cell. In some cases the symmetry of the crystal can determine both. CPMV, in the I23 space group, is an exceptionally straightforward case. The unit cell dimension is 317 Å and the lattice points have tetrahedral symmetry. Because the virus particle is 142 Å in diameter (determined by solution X-ray scattering analysis), the cell is only large enough to accommodate two particles and they must be positioned on tetrahedral lattice points (0,0,0 and 1/2,1/2,1/2). These positions are related by the pure translation of body centering, so the two particles are in identical orientations. The virus particles have icosahedral symmetry, and the twofold and threefold symmetry axes of the particle are part of the crystal lattice symmetry. Only the fivefold symmetry of the particle is not included in the lattice symmetry, but its position relative to the other symmetry axes is known. The lattice and particle dimensions, as well as the crystal symmetry, provide the solution to the orientation and translation problem.

Another crystal form of CPMV ($P6_122$, $a = 451$ Å, $c = 1038$ Å) illustrates a different situation. This space group has 12 general positions and it is possible to place 12 particles in one unit cell given the virus size and unit cell size. Preliminary analysis of the diffraction patterns suggested that there were actually six particles per cell. A model of the proposed packing showed channels of a size comparable with the virus particles and an extraordinary honeycomb packing of particles. EM analysis of thin-sectioned, positively stained crystals permitted the confirmation of this unlikely packing (Fig. 2.6). Although the position of the particles was established by diffraction and EM analysis, the particle orientation was determined by rotation function analysis. This procedure systematically examines the 3D diffraction pattern for evidence of non-crystallographic symmetry (NCS; symmetry within a virus particle but not part of the crystal lattice). The hexagonal crystal form of CPMV requires six particles in the unit cell all related by crystallographic sixfold symmetry. The rotation function analysis detects the NCS associated with particles in six different orientations, a total of 186 symmetry elements for the six icosahedral particles. The only means of analyzing such a complex function is the "locked" rotation function wherein all symmetry elements corresponding to an icosahedral set are examined simultaneously. Because the crystallographic symmetry is defined, the symmetry elements corresponding to all six crystallographically related particles can be examined simultaneously. Finally, the packing constrains the particles to lie on a crystallographic twofold axis, so the orientation search is one-dimensional and clearly defined from the analysis.

Solving the Phase Problem

A detailed discussion of the isomorphous replacement method for phase determination is beyond the scope of this chapter; however, an outline of the method is presented and more detail can be found in a number of books on protein crystallography (Blundel and Johnson, 1976; Drenth, 1994).

The formation of heavy atom derivatives is the first step in the phase determination process. Virus crystals are similar to all protein crystals in that they contain large quantities ($\sim$50% by volume) of bulk solvent. Heavy metal derivitizing agents pass through this solvent and can reach the surface of the protein subunits. If there are susceptible, accessible functional groups that will make covalent bonds with the heavy metal compounds, a derivative is formed. Because there are 60 equivalent units in the icosahedral particle, at least 60 positions on the particle are expected to be modified. By determining the position of the heavy atom sites, an estimate of the phase for each of the reflections in the data set can be determined. Heavy atom positions are normally determined with a computed function that is sensitive to vectors between the heavy atoms (difference Patterson function). Because the number of vectors between heavy atoms within one particle will be $60*59$, it is impossible to directly determine the position of the heavy atoms in a virus. The approach taken in virus crystallography is a search procedure that assumes the particle orientation has been determined by the rotation function. A particular position in the viral protein is assumed to be occupied by a heavy atom; the 59 symmetry-equivalent positions are generated and the 3D positions of all the possible vectors between them are calculated. Each of these positions is examined in the difference Patterson function and the values summed. If the sum is relatively small, no heavy atom is located at this position. If it is large, it is a good candidate for a heavy atom site. A grid defining the entire asymmetric

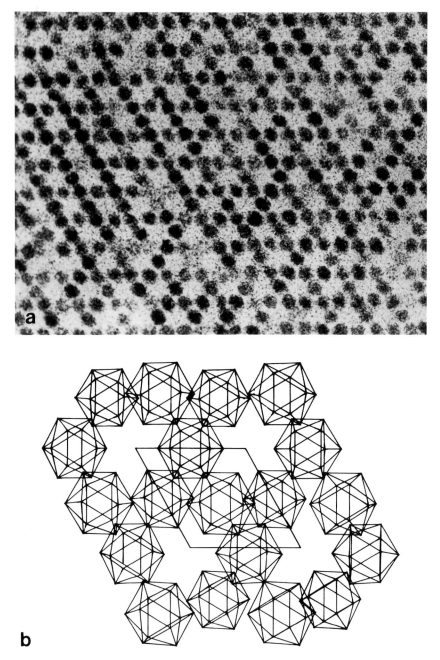

FIGURE 2.6 *a*. An electron micrograph of a CPMV hexagonal crystal sectioned perpendicular to a sixfold axis. The crystals were positively stained with osmium tetroxide. *b*. The packing of icosahedral particles established from the low-resolution diffraction patterns of the hexagonal form of CPMV.

unit of the icosahedron is searched with this procedure. In order to unambiguously determine the 3D position of the heavy atom, the molecular center must be known (described earlier).

When the heavy atom positions are refined, a set of phases can be computed. A single isomorphous replacement (SIR) derivative will provide phases that are ambiguous because there are two mathematically correct solutions to the vector equations that are solved to determine the phase. Only one is a physically correct solution. The NCS in all virus crystals (at least fivefold symmetry, as in the I23 form of CPMV) can be used to improve the phases computed. An image of the structure can be computed with both the physically correct phase and the mathematically correct (but physically nonsense) phase used simultaneously. It is impossible to tell from a single derivative which is correct. This procedure generates a density function that corresponds to the correct image superimposed on a background of noise generated by the set of incorrect phases. The correct image will have the symmetry of the icosahedron; the incorrect set will not generate a physically reasonable image and it will have no symmetry. Averaging the electron density employing the known molecular center and orientation of the symmetry elements will improve the signal-to-noise ratio by $N^{1/2}$, where N is the number of noncrystallographically equivalent positions averaged. The correct signal is enhanced and the background noise is decreased. It is often possible to solve a virus structure (i.e., follow the polypeptide chain of the protein in the electron density) with a SIR derivative because of the NCS. Regardless of the source of the initial phases, the final phases for a virus structure are always improved by averaging with the NCS.

Molecular Replacement Structure Determination

An alternative to the multiple isomorphous replacement method for determining the initial phases of a structure has gained considerable popularity in recent years; it is known as molecular replacement. Assuming that two structures are either known or expected to be similar, and that the high-resolution structure of one has been determined and a set of coordinates are available, it may be possible to use the known structure to determine an initial set of phases for the unknown structure. If the particle position and orientation are known using the methods described earlier, the model structure can be positioned and oriented appropriately in the unit cell of the unknown structure. A set of phases can then be calculated on the basis of the model. These phases can be refined with the NCS and the effect of the phasing model can be minimized. The electron density that results from the phase refinement should have characteristics of the unknown structure. For instance, side chains should correspond to the expected sequence of the unknown structure.

This procedure has recently been used in conjunction with electron density produced from cryoEM and coordinates from previously determined high-resolution structures. The structure of cowpea chlorotic mottle virus was expected to have subunits with the jelly roll β-barrel. The cryoEM density was used as a constraint in building a model of 180 jelly rolls (the southern bean mosaic virus subunit coordinates were used for the initial model) that were trimmed and adjusted to fit the cryoEM density. This phasing model led to high-resolution phases and an electron density map that was modeled in detail (Speir et al., 1995). The use of low-resolution models is feasible because of a process called phase extension. Given phases that are known to 18-Å resolution, as in the case just described, phases computed from the initial model are refined by a cyclic procedure until they converge. The map is then Fourier transformed to 17-Å resolution. Initially data in the map were resolved to 18 Å and, if there was no averaging, the amplitudes beyond 18 Å would all be zero. Because the map was perturbed by the averaging, the amplitudes between 18 and 17 Å have values slightly above zero and phases that are not quite random. These phases are coupled with appropriately weighted observed amplitudes and a map is computed at 17-Å resolution. The phases are refined and the process repeated all the way to high resolution. This method has worked successfully in many virus structure determinations.

Interpretation of the Electron Density

The final step in the structure determination is to construct a chemically reasonable model to fit the electron density image generated in the steps described previously (Fig. 2–7). Generally the structures of viruses are solved at a resolution of 3.5 to 2.5 Å. This is not atomic resolution, but atomic resolution structures have been determined for all the individual amino acids and many small polypeptides. Because a protein is essentially modular with respect to its amino acid components, each residue in the structure can be considered as a unit. There are constraints on distances and angles within the peptide bond and limitations on the dihedral angles that can be maintained in the backbone of the polypeptide. These are described by forbidden areas of the Ramachandran plot, allowing the appropriate protein stereochemistry to be maintained.

The amino acid sequence of the protein is assumed to be known from other methods, and the location of residues in the structure can often be inferred from large, distinctive side chains in the electron density image. There are a variety of procedures for building the detailed model. All of them involve fitting models to the electron density in a graphics system. In recent years the procedures have become considerably automated. A strategy found in the program FRODO and developed further in the program O by A. Jones is the skeletonization of the electron density to define the path of the polypeptide in terms of ''atom'' positions. The atoms used are not related to chemical atoms, but rather define the path of the polypeptide chain in terms of 3D coordinates. The positions are then used in groups to define segments of a polypeptide containing 5 to 10 residues. The general shape of the polypeptide is compared with a library of highly refined 3D structures and the best approximation to the particular segment is found in the database. The builder can then examine this reference set of coordinates for the quality of fit to the density. The coordinates are adjusted to place the C_β atoms in the proper positions and then side chains can be added to fit the density for the

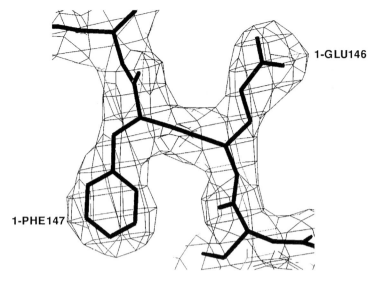

FIGURE 2.7   An example of a portion of the CPMV model placed in the 2.8-Å electron density map. Knowledge of the amino acid sequence and the structures of individual amino acids are used to develop the optimal model. The refined structure is a composite of information derived from the electron density map and amino acid models. These are objectively refined to minimize the difference between observed and calculated (on the basis of the model) diffraction maxima. The portion displayed corresponds to residues in the small subunit.

particular residue. The backbone has reasonable stereochemistry because it was derived from a refined structure. An entire protein can be constructed with this strategy that corresponds to the preliminary model.

Employing one of a number of possible programs, the coordinates are refined to minimize the difference between the calculated and observed structure amplitudes while maintaining reasonable stereochemistry. The so-called constrained refinement incorporates all the known properties of protein stereochemistry as well as maintaining reasonable nonbonded contacts. The program X-PLOR (Brunger, 1992) allows the refinement to proceed with a molecular dynamics component as well in which local maxima in the nonlinear refinement space can be overcome in order to obtain an overall global minimum in the refinement. The final structure will have acceptable stereochemistry and a crystallographic R factor of 15% to 20%. In addition to the atomic coordinates, the thermal factors for the atoms can be refined; this may be related to the mobility of a particular residue. The entire data set of coordinates and thermal factors constitutes the refined structure.

Combination of X-Ray and CryoEM Techniques

The cryoEM and X-ray techniques have each, individually, made important contributions to the study of virus structure and function. Recently, however, sophisticated models have been derived for complex structures that were based on data derived from both methods (Baker and Johnson, 1996). At the highest resolution available with the cryoEM technique (~20 Å), individual domains of proteins can be readily identified, however, these structures cannot establish the location of individual amino acids or even permit density to be assigned to regions of the polypeptide chain. CryoEM determines a shape or envelope with high fidelity, but the details within the envelope cannot be established. In contrast, X-ray crystallography provides the positions of all the atoms within a structure. The method is limited, however, and cannot be used if the structure of interest is too complex or is not sufficiently homogeneous to be crystallized.

An approach that has met with considerable success is exemplified by the structure of CPMV (Fig. 2.8) complexed with the Fab fragment of a monoclonal antibody to the virus. Such a complex is impossible to crystallize because particles with variable numbers of Fab fragments exist in solution. The cryoEM image allows the highly populated particles to be identified, and only

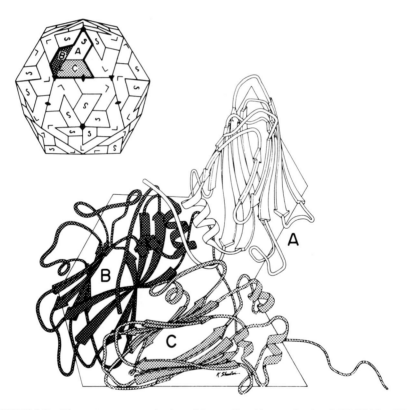

FIGURE 2.8 The quaternary organization of the small and large subunits of CPMV. The ribbon drawing represents the arrangement of the three β-barrels in the virus quaternary structure. The two domains of the large subunit are shown in different shades to emphasize that they form one protein. The small subunit is shown without shading. The heavy outline in the capsid model encloses the suggested precleavage protomer.

these are used in the reconstruction. An excellent image of the complex is derived from the 3D reconstruction when this selection process is used (Fig. 2.3c). This structure clearly shows the location of the Fab binding site relative to the symmetry axes (Fig. 2.9) of the particle, but by itself provides little additional information except for the size and shape of the particle and the Fab fragment.

The structure of CPMV has been previously determined at atomic resolution by X-ray crystallography, and the structures of a number of different Fab fragments have also been determined at atomic resolution. After the cryoEM structure was solved, atomic models of the virus and the Fab fragment were adjusted to fit within the cryoEM density. Combining the cryoEM density with the atomic models permitted the detailed structure of the complex to be modeled (see Color Plate 1). It was possible to identify all the virus residues that were in contact with the Fab fragment and thus to physically define the antigenic site on the viral surface at atomic resolution. It was also possible to position the Fab fragment to visualize how the combining site of the antibody interacted with the viral surface.

A second method of combining X-ray and cryoEM data exploits the sensitivity of these methods to different regions of resolution. It is difficult to measure high-quality, low-resolution data from single crystals. As a result, most high-resolution virus structures contain data between 15 and 2.8 Å. Bulk RNA within virus particles diffracts X-rays to roughly 20-Å resolution; thus most high-resolution virus structures appear to be empty capsids because data sensitive to RNA is not included. CryoEM data are very sensitive to data at this

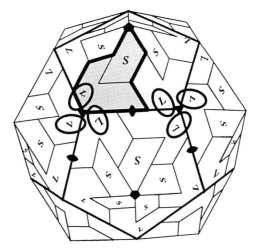

FIGURE 2.9 A schematic representation of the footprint of Fab fragments attached to CPMV. The L subunit is formed of two β-barrels corresponding to picornavirus subunits VP2 at the amino end and VP3 at the carboxyl end, while S corresponds to VP1. Note that the footprint spans a subunit interface with the smaller part interacting with the VP2 portion of one L subunit and the larger part interacting with the VP3 portion of a different L subunit.

resolution, and virus particles analyzed by this method appear full of RNA. It is difficult, however, to discern the boundary between protein and RNA in the EM map. Combining X-ray and EM data clearly defines this boundary and provides insights into understanding protein interactions with bulk RNA. Figure 2.10 shows how difference maps are computed for each component of CPMV where the X-ray electron density, computed at the same resolution as the cryoEM density, is subtracted from the cryoEM structure, leaving density only for the bulk RNA. The distribution of RNA is clearly different in the middle and bottom components as revealed by these methods.

It is likely that the structures of more complex viruses will be determined by studying the structure of the individual proteins of the virus by X-ray crystallography and then assembling the complex virus on the basis of the cryoEM structure. Knowledge of the model for the individual components of the complex affects the final resolution of the structure in a manner that is analogous to the atomic resolution structures derived from 2.0-Å-resolution electron density maps in protein and virus crystallography. If the high-resolutions structures of the individual modules are known (individual amino acids in the X-ray experiment or complete subunits in the cryoEM experiment), they can be fit with much higher accuracy into the envelope defined by the experimental data

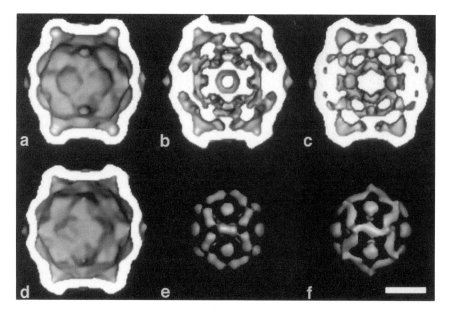

FIGURE 2.10   RNA structure in CPMV. *a.* Cutaway view of far half of empty CPMV-T (top component) 3D reconstruction. *b.* Same as *a* for CPMV-M (middle component: 1.22-MDa RNA). *c.* Same as *a* for CPMV-B (bottom component: 2.02-MDa RNA). *d.* Cutaway view of far half of low-resolution X-ray model. *e.* RNA-M structure computed by subtracting X-ray model (*d*) from CPMV-M (*b*). *f.* RNA-B structure computed by subtracting X-ray model (*d*) from CPMV-B (*b*). Bar, 100 Å. (Reproduced with permission from Baker et al., 1992.)

than the physical resolution of the data would suggest (Liu et al., 1994; Smith et al., 1996).

Acknowledgments

We gratefully thank all the members of our research groups for their diligence and patience, and wish to pay special tribute to N. Olson, W. Wang, and R. Cheng for directing their extraordinary talents in ways that have assured numerous successful cryoEM and image analysis results that have made a major part of this chapter a reality. We thank D. Belnap for preparation of Figure 2.2 and C. Music and R. Cheng for help with photography and computer graphic assistance in the preparation of Figures 2.1, 2.3, and 2.9 through 2.11. We thank Tianwei Lin for preparation of Figure 2.7 and Sharon Fateley for help in preparation of this manuscript.

This work was supported with funds from National Institutes of Health grants GM33050 (T.S.B.) and AI35212 (J.E.J.); National Science Foundation grants MCB-9206305 (T.S.B.) and DMB-8817057A1 (J.E.J.); and a grant from the Lucille P. Markey Trust that supports the Purdue Structural Biology group.

References

Abad-Zapatero, C., Abdel-Meguid, S.S., Johnson, J.E., Leslie, A.G.T., Rayment, I., Rossmann, M.G., Suck, D. and Tsukihara, T. (1980). Structure of southern bean mosaic virus at 2.8 Å resolution. Nature 286: 33–39.

Adrian, M., Dubochet, J., Lepault, J. and McDowall, A.W. 1984. Cryo-electron microscopy of viruses. Nature 308: 32–36.

Anderegg, J.W., Geil, P.H., Beeman, W.W. and Kaesberg, P. 1961. An X-ray scattering investigation of wild cucumber mosaic virus and a related protein. Biophys. J. 1: 657–667.

Baker, T.S. 1978. Preirradiation and minimum beam microscopy of periodic biological specimens. In Proceedings of the Ninth International Congress of Electron Microscopy, The Imperial Press, Ltd, Toronto.

Baker, T.S. 1992. Cryo-electron microscopy and three-dimensional image reconstruction of icosahedral viruses. In Proceedings of the 10th European Congress of Electron Microscopy. Secretariado de Publicaciones de la Universidad de Granada, Granada, Spain.

Baker, T.S. and Amos, L.A. 1978. Structure of the tubulin dimer in zinc-induced sheets. J. Mol. Biol. 123: 89–106.

Baker, T.S. and Cheng, R.H. 1996. A model-based approach for determining orientations of biological macromolecules imaged by cryo-electron microscopy. J. Struct. Biol. 116: 120–130.

Baker, T.S. and Johnson, J.E. 1996. Low resolution meets high: Towards a resolution continuum from cells to atoms. Curr. Opin. Struct. Biol. 6: 585–594.

Baker, T.S., Caspar, D.L.D. and Murakami, W.T. 1983. Polyoma virus "hexamer" tubes consist of paired pentamers. Nature 303: 446–448.

Baker, T.S., Cheng, R.H., Johnson, J.E., Olson, N.H., Wang, G.J. and Schmidt, T.S. 1992. Organized packing of RNA inside viruses as revealed by cryo-electron microscopy and x-ray diffraction analysis. Proc. Elect. Microsc. Soc. Am. 50: 454–455.

Baker, T.S., Drak, J. and Bina, M. 1988. Reconstruction of the three-dimensional structure of simian virus 40 and visualization of the chromatin core. Proc. Natl. Acad. Sci. U.S.A. 85: 422–426.

Baker, T.S., Drak, J. and Bina, M. 1989. The capsid of small papova viruses contains 72 pentameric capsomeres: direct evidence from cryo-electron-microscopy of simian virus 40. Biophys. J. 55: 243–253.

Baker, T.S., Newcomb, W.W., Booy, F.P., Brown, J.C. and Steven, A.C. 1990. Three-dimensional structures of maturable and abortive capsids of equine herpesvirus 1 from cryoelectron microscopy. J. Virol. 64: 563–573.

Baker, T.S., Newcomb, W.W., Olson, N.H., Cowsert, L.M., Olson, C. and Brown, J.C. 1991. Structures of bovine and human papilloma viruses: analysis by cryoelectron microscopy and three-dimensional image reconstruction. Biophys. J. 60: 1445–1456.

Bawden, F.C. and Pirie, N.W. 1938. A plant virus preparation in a fully crystalline state. Nature 141: 513–514.

Belnap, D.M., Grochulski, W.D., Olson, N.H. and Baker, T.S. 1993. Use of radial density plots to calibrate image magnification for frozen-hydrated specimens. Ultramicroscopy 48: 347–358.

Bloomer, A.C., Graham, J., Hovmoller, S., Butler, P.J.G. and Klug, A. 1978. Protein disk of tobacco mosaic virus at 2.8 Å resolution showing the interactions within and between subunits. Nature 276: 362–368.

Blundel, T.L. and Johnson, L.N. 1976. Protein Crystallography. Academic Press, New York.

Booy, F.P., Newcomb, W.W., Trus, B.L., Brown, J.C., Baker, T.S. and Steven, A.C. 1991. Liquid-crystalline, phage-like packing of encapsidated DNA in herpes simplex virus. Cell 64: 1007–1015.

Breese, S.S.J. 1981. Visualization of viruses by electron microscopy. In Electron Microscopy in Biology, (Griffith, J.D., Ed.), pp. 273–303. John Wiley & Sons, New York.

Brunger, A.T. 1992. X-PLOR, Version 3.1: A System for X-Ray Crystallography and NMR. Yale University Press, New Haven, CT.

Carlisle, C.H. and Dornberger, K. 1948. Some X-ray measurements on single crystals of tomato bushy stunt virus. Acta Crystallogr. 1: 194–196.

Caspar, D.L.D. and Klug, A. 1962. Physical principles in the construction of regular viruses. Cold Spring Harbor Symp. Quant. Biol. 27: 1–24.

Cheng, R.H. 1992. Correlation of cryo-electron microscopic and x-ray data and compensation of the contrast transfer function. Proc. Elect. Microsc. Soc. Am. 50: 996–997.

Cheng, R.H., Olson, N.H. and Baker, T.S. 1992. Cauliflower mosaic virus, a 420 subunit (T = 7), multi-layer structure. Virology 186: 655–668.

Cheng, R.H., Reddy, V.S., Olson, N.H., Fisher, A.J., Baker, T.S. and Johnson, J.E. 1994. Functional implications of quasi-equivalence in a T = 3 icosahedral animal virus established by cryo-electron microscopy and x-ray crystallography. Structure 2: 271–282.

Chiu, W. 1986. Electron microscopy of frozen, hydrated biological specimens. Annu. Rev. Biophys. Biophys. Chem. 15: 237–257.

Choi, H.-K., Tong, L., Minor, W., Dumas, P., Boege, U., Rossmann, M.G. and Wengler, G. 1991. Structure of Sindbis virus core protein reveals a chymotrypsin-like serine protease and the organization of the virion. Nature 354: 37–43.

Crowfoot, D. and Schmidt, G.M.J. 1945. X-ray crystallographic measurements on a single crystal of a tobacco necrosis virus derivative. Nature 155: 504–505.

Crowther, R.A. 1971. Procedures for three-dimensional reconstruction of spherical viruses by Fourier synthesis from electron micrographs. Philos. Trans. R. Soc. Lond. B 261: 221–230.

Crowther, R.A., Amos, L.A., Finch, J.T., DeRosier, D.J. and Klug, A. 1970a. Three dimensional reconstructions of spherical viruses by Fourier synthesis from electron micrographs. Nature 226: 421–425.

Crowther, R.A., Amos, R.A. and Klug, A. 1972. Three dimensional image reconstruction using functional expansions. In Proceedings of the Fifth European Reg. Congress on Electron Microscopy, Manchester.

Crowther, R.A., DeRosier, D.J. and Klug, A. 1970b. The reconstruction of a three-dimensional structure from projections and its application to electron microscopy. Proc. R. Soc. Lond. A 317: 319–340.

DeRosier, D.J. and Klug, A. 1968. Reconstruction of three dimensional structures from electron micrographs. Nature 217: 130–134.

Dokland, T. and Murialdo, H. 1993. Structural transitions during maturation of bacteriophage lambda capsids. J. Mol. Biol. 233: 682–694.

Dokland, T.E., Lindqvist, B.H. and Fuller, S.D. 1992. Image reconstruction from cryoelectron micrographs reveals the morphopoietic mechanism in the P2-P4 bacteriophage system. EMBO J. 11: 839–846.

Drenth, J. 1994. Principles of Protein X-Ray Crystallography. Springer-Verlag, New York.

Dryden, K.A., Wang, G., Yeager, M., Nibert, M.L., Coombs, K.M., Furlong, D.B., Fields, B.N. and Baker, T.S. 1993. Early steps in reovirus infection are associated with dramatic changes in supramolecular structure and protein conformation: analysis of virions and subviral particles by cryoelectron microscopy and image reconstruction. J. Cell Biol. 122: 1023–1041.

Dubochet, J., Adrian, M., Chang, J.-J., Homo, J.-C., Lepault, J., McDowall, A.W. and Schultz, P. 1988. Cryo-electron microscopy of vitrified specimens. Q. Rev. Biophys. 21: 129–228.

Dubochet, J., Chang, J.-J., Freeman, R., Lepault, J. and McDowall, A.W. 1982a. Frozen aqueous suspensions. Ultramicroscopy 10: 55–62.

Dubochet, J., Groom, M. and Müller-Neuteboom, S. 1982b. The mounting of macromolecules for electron microscopy with particular reference to surface phenomena and the treatment of support films by glow discharge. Adv. Opt. Elect. Microsc. 8: 107–135.

Dubochet, J. and McDowell, A.W. 1981. Vitrification of pure water for electron microscopy. J. Microsc. 124: RP3–RP4.

Earnshaw, W.C., Hendrix, R.W. and King, J. 1979. Structural studies of bacteriophage lambda heads and proheads by small angle X-ray diffraction. J. Mol. Biol. 134: 575–594.

Erickson, H.P. and Klug, A. 1971. Measurement and compensation of defocusing and aberrations by Fourier processing of micrographs. Philos. Trans. R. Soc. Lond. B 261: 105–118.

Finch, J.T. and Klug, A. 1959. Structure of poliomyelitis virus. Nature 183: 1709–1714.

Finch, J.T. and Holmes, K.C. 1967. Methods in Virology, Vol. III: Structural Studies of Viruses. pp. 351–474. Academic Press, New York.

Fuller, S.D. 1987. The $T = 4$ envelope of Sindbis virus is organized by complementary interactions with a $T = 3$ icosahedral capsid. Cell 48: 923–934.

Fuller, S.D., Butcher, S.J., Baker, T.S. and Cheng, R.H. 1996. Three-dimensional reconstruction of Icosahedral particles—the uncommon line. J. Struct. Biol. 116: 48–55.

Furcinitti, P.S., van Oostrum, J. and Burnett, R.M. 1991. Adenovirus polypeptide IX revealed as capsid cement by difference images from electron microscopy and crystallography. EMBO J. 12: 3563–3570.

Glusker, G.P. and Trueblood, K.N. 1985. Crystal Structure Analysis: A Primer, 2nd ed. Oxford University Press, New York.

Green, R.H., Anderson, T.F. and Smadel, J.E. 1942. Morphological structure of the virus of vaccinia. J. Exp. Med. 75: 651–656.

Harrison, S.C. 1980. Virus crystallography comes of age. Nature 286: 558–559.

Harrison, S.C., Olson, A.J., Schutt, C.E., Winkler, F.K. and Bricogne, G. 1978. Tomato bushy stunt virus at 2.9Å resolution. Nature 276: 368–373.

Harvey, J.D., Belamy, A.R., Earnshaw, W.C. and Schutt, C. 1981. Biophysical studies of reovirus type 3. IV. Low angle X-ray diffraction studies. Virology 112: 240–249.

Hogle, J.M., Chow, M. and Filman, D.J. 1985. Three-dimensional structure of poliovirus at 2.9 Å resolution. Science 229: 1358–1365.

Horne, R.W. and Wildy, P. 1979. An historical account of the development and applications of the negative staining technique to the electron microscopy of viruses. J. Microsc. 117: 103–122.

Ilag, L.L., Olson, N.H., Dokland, T., Music, C.L., Cheng, R.H., Bowen, Z., McKenna, R., Rossmann, M.G., Baker, T.S. and Incardona, N.L. 1995. DNA packaging intermediates of bacteriophage φX174. Structure 3: 353–363.

Jack, A. and Harrison, S.C. 1975. On the interpretation of small-angle X-ray solution scattering from spherical viruses. J. Mol. Biol. 99: 15–25.

Jeng, T.-W., Crowther, R.A., Stubbs, G. and Chiu, W. 1989. Visualization of alpha-helics in tobacco mosaic virus by cryo-electron microscopy. J. Mol. Biol. 205: 251–257.

Kausche, G.A., Pfankuch, E. and Ruska, H. 1939. Die sichtbarnmachung von pflanzlichem virus im übermikroskop. Naturwissenschaften 27: 292–299.

Krivanek, O.L. and Mooney, P.E. 1993. Applications of slow-scan CCD cameras in transmission electron microscopy. Ultramicroscopy 49: 95–108.

Lattman, E.E. 1989. Rapid calculation of the solution scattering profile from a macromolecule of known structure. Proteins Struct. Funct. Genet. 5: 149–155.

Leonard, B.R. Jr., Anderegg, J.W., Schulman, S., Kaesberg, P. and Beeman, W.W. 1953. An X-ray investigation of the sizes and hydrations of three spherical macromolecules in solution. Biochim. Biophys. Acta 12: 499–507.

Lepault, J., Booy, F.P. and Dubochet, J. 1983. Electron microscopy of frozen biological specimens. J. Microsc. 129: 89–102.

Lepault, J. and Leonard, K. 1985. Three-dimensional structure of unstained, frozen-hydrated extended tails of bacteriophage T4. J. Mol. Biol. 182: 431–441.

Liddington, R.C., Yan, Y., Moulai, J., Sahli, R., Benjamin, T.L. and Harrison, S.C. 1991. Structure of simian virus 40 at 3.8-Å resolution. Nature 354: 278–284.

Liu, H., Smith, T.J., Lee, W.M., Mosser, A.G., Rueckert, R.R., Olson, N.H., Cheng, R.H. and Baker, T.S. 1994. The purification, crystallization and structure determination of an F_{ab} fragment that neutralizes human rhinovirus 14 and analysis of the F_{ab}-virus complex. J. Mol. Biol. 240: 127–137.

McMath, T.A., Curzon, A.E. and Frindt, R.F. 1982. A double Faraday cup attachment for relative intensity measurements on an electron microscope. J. Phys. E. Sci. Instrum. 15: 988–990.

Metcalf, P., Cyrklaff, M. and Adrian, M. 1991. The three-dimensional structure of reovirus obtained by cryo-electron microscopy. EMBO J. 10: 3129–3136.

Miller, S.E. 1988. Diagnostic virology by electron microscopy. Am. Soc. Microbiol. News 54: 475–481.

Namba, K. and Stubbs, G. 1986. Structure of tobacco mosaic virus at 3.6 Å resolution: implications for assembly. Science 231: 1401–1406.

Nermut, M.V., Hockley, D.J. and Gelderblom, H. 1987a. Electron microscopy: methods for studies of virus particles and virus-infected cells. Methods for "structural analysis" of the virion. In Animal Virus Structure, (Nemut, M.V. and Seeven, A.C., Eds.), pp. 21–33. Elsevier, Amsterdam.

Nermut, M.V., Hockley, D.J. and Gelderblom, H. 1987b. Electron microscopy: methods for studies of virus particles and virus-infected cells. Methods for study of virus/cell interactions. In Animal Virus Structure, pp. 35–60. Elsevier, Amsterdam.

Newcomb, W.W., Trus, B.L., Booy, F.P., Steven, A.C., Wall, J.S. and Brown, J.C. 1993. Structure of the herpes simplex virus capsid: molecular composition of the pentons and the triplexes. J. Mol. Biol. 232: 499–511.

Olson, N.H. and Baker, T.S. 1989. Magnification calibration and the determination of spherical virus diameters using cryo-microscopy. Ultramicroscopy 30: 281–298.

Olson, N.H., Baker, T.S., Bomu, W., Johnson, J.E. and Hendry, D.A. 1987. The three-dimensional structure of frozen-hydrated Nudaurelia capensis beta virus. Proc. Elect. Microsc. Soc. Am. 45: 650–651.

Olson, N.H., Kolatkar, P.R., Oliveria, M.A., Cheng, R.H., Greve, J.M. McClelland, A., Baker, T.S. and Rossmann, M.G. 1993. Structure of a human rhinovirus complexed with its receptor molecule. Proc. Natl. Acad. Sci. U.S.A. 90: 507–511.

Porta, C., Wang, G., Cheng, R.H., Chen, Z., Baker, T.S. and Johnson, J.E. 1994. Direct imaging of interactions between an icosahedral virus conjugate Fab fragments by cryo-electron microscopy and x-ray crystallography. Virology 204: 777–788.

Prasad, B.V.V., Burns, J.W., Marietta, E., Estes, M.K. and Chiu, W. 1990. Localization of VP4 neutralization sites in rotavirus by three-dimensional cryo-electron microscopy. Nature 343: 476–479.

Prasad, B.V.V., Prevelige, P.E., Marietta, E., Chen, R.O., Thomas, D., King, J. and Chiu, W. 1993. Three-dimensional transformation of capsids associated with genome packaging in a bacterial virus. J. Mol. Biol. 231: 65–74.

Prasad, B.V.V., Wang, G.J., Clerx, J.P.M. and Chiu, W. 1987. Cryo electron microscopy of spherical viruses: an application to rotaviruses. Micron 18: 327–331.

Prasad, B.V.V., Wang, G.J., Clerx, J.P.M. and Chiu, W. 1988. Three-dimensional structure of rotavirus. J. Mol. Biol. 199: 269–275.

Radermacher, M., Wagenknecht, T., Verschoor, A. and Frank, J. 1987. Three-dimensional reconstruction from a single-exposure, random conical tilt series applied to the 50S ribosomal subunit of Eschericia coli. J. Microsc. 146: 113–136.

Rayment, I., Baker, T.S., Caspar, D.L.D. and Murakami, W.T. 1982. Polyoma virus capsid structure at 22.5Å resolution. Nature 295: 110–115.

Reynolds, G.T. 1988. Image intensified microscopy and spectrometry. Comments Mol. Cell Biophys. 5: 297–318.

Roberts, M.M., White, J.L., Grütter, M.G. and Burnett, R.M. 1986. Three-dimensional structure of the adenovirus major coat protein hexon. Science 232: 1148–1151.

Rossmann, M.G., Arnold, E., Erickson, J.W., Frankenberger, E.A., Griffith, J.P., Hecht, H.J., Johnson, J.E., Kamer, G., Luo, M., Mosser, A.G., Rueckert, R.R., Sherry, B. and Vriend, G. 1985. Structure of human cold virus and functional relationship to other picornaviruses. Nature 317: 145–153.

Rossmann, M.G. and Johnson, J.E. 1989. Icosahedral RNA virus structure. Annu. Rev. Biochem. 58: 533–573.

Schaffer, F.L. and Schwerdt, C.E. 1955. Crystallization of purified MEF-1 poliomyelitis virus particles. Proc. Natl. Acad. Sci. U.S.A. 41: 1020–1023.

Schmidt, T., Johnson, J.E. and Phillips, W.E. 1983. The spherically averaged structures of cowpea mosaic virus components by X-ray solution scattering. Virology 127: 65–73.

Schrag, J.D., Prasad, B.V.V., Rixon, F.J. and Chiu, W. 1989. Three-dimensional structure of the HSV1 nucleocapsid. Cell 56: 651–660.

Sehnke, P.C., Harrington, M., Hosur, M.V., Li, Y., Usha, R., Tucker, R.C., Bomu, W., Stauffacher, C.V. and Johnson, J.E. 1988. Crystallization of viruses and virus proteins. J. Crystal Growth 90: 222–230.

Shannon, C.E. 1949. Communication in the presence of noise. Proc. Inst. Radio Engin. 37: 10–20.

Sharp, D.G., Taylor, A.R., Beard, D. and Beard, J.W. 1942. Study of the papilloma virus protein with the electron microscope. Proc. Soc. Exp. Biol. Med. 50: 205–207.

Shaw, A.L., Rothnagel, R., Chen, D., Ramig, R.F., Chiu, W. and Prasad, B.V.V. 1993. Three-dimensional visualization of the rotavirus hemagglutinin structure. Cell 74: 693–701.

Smith, T.J., Olson, N.H., Cheng, R.H., Chase, E.S. and Baker, T.S. 1993a. Structure of a human rhinovirus-bivalently bound antibody complex: implications for viral neutralization and antibody flexibility. Proc. Natl. Acad. Sci. U.S.A. 90: 7015–7018.

Smith, T.J., Olson, N.H., Cheng, R.H., Liu, H., Chase, E.S., Lee, W.M., Leippe, D.M., Mosser, A., Rueckert, R.R. and Baker, T.S. 1993b. Structure of human rhinovirus complexed with F_{ab} fragments from a neutralizing antibody. J. Virol. 67: 1148–1158.

Smith, T.J., Chase, E.S., Schmidt, T.J., Olson, N.H. and Baker, T.S. 1996. Neutralizing antibody to human rhinovirus 14 penetrates the receptor-binding canyon. Nature 383: 350–354.

Speir, J.A., Munshi, S. Wang, G., Baker, T.S. and Johnson, J.E. 1995. Structures of the native and swollen forms of cowpea chlorotic mottle virus determined by x-ray crystallography and cryo-electron microscopy. Structure 3: 63–78.

Stanley, W.M. 1935. Isolation of a crystalline protein possessing the properties of tobacco mosaic virus. Science 81: 644–645.

Stanley, W.M. and Anderson, T.F. 1941. A study of purified viruses with the electron microscope. J. Biol. Chem. 139: 325–338.

Stauffacher, C.V., Usha, R., Harrington, M., Schmidt, T., Hosur, M.V. and Johnson, J.E. 1987. The structure of cowpea mosaic virus at 3.5Å resolution. In Crystallography in Molecular Biology (Moras, D., Drenth, J., Strandberg, B., Suck, D. and Wilson, K., Eds.), pp. 293–308. Plenum Publishing, New York.

Stehle, T., Yan, Y., Benjamin, T.L. and Harrison, S.C. 1994. Structure of murine polyoma virus complexed with an oligosaccharide receptor fragment. Nature 369: 160–163.

Steven, A.C. 1981. Visualization of virus structure in three dimensions. Methods Cell Biol. 22: 297–323.

Stewart, M. 1990. Electron microscopy of biological macromolecules: frozen hydrated methods and computer image processing. In Modern Microscopies: Techniques and Applications, (Duke, P.J. and Michette, A.G., Eds.), pp. 9–39. Plenum Press, New York.

Stewart, P.L., Burnett, R.M., Cyrklaff, M. and Fuller, S.D. 1991. Image reconstruction reveals the complex molecular organization of adenovirus. Cell 67: 145–154.

Stewart, P.L., Fuller, S.D. and Burnett, R.M. 1993. Difference imaging of adenovirus: bridging the resolution gap between X-ray crystallography and electron microscopy. EMBO J. 12: 2589–2599.

Stout, G.H. and Jensen, L.H. 1968. X-Ray Structure Determination: A Practical Guide. Macmillan, New York.

Stuart, D. 1993. Viruses. Curr. Opin. Struct. Biol. 3: 167–174.

Taylor, K.A. and Glaeser, R.M. 1974. Electron diffraction of frozen, hydrated protein crystals. Science 186: 1036–1037.

Trus, B.L., Newcomb, W.W., Booy, F.P., Brown, J.C. and Steven, A.C. 1992. Distinct monoclonal antibodies separately label the hexons or the pentons of herpes simplex virus capsid. Proc. Natl. Acad. Sci. U.S.A. 89: 11508–11512.

Turner, J.N., Hausner, G.G.J. and Parsons, D.F. 1975. An optimized Faraday cage design for electron beam current measurements. J. Phys. E: Sci. Instrum. 8: 954–957.

Unge, T., Liljas, L., Strandberg, B., Vaara, I., Kannan, K.K., Fidborg, K., Nordman, C.E. and Lentz, P.J.J. 1980. Satellite tobacco necrosis virus structure at 4.0 Å resolution. Nature 285: 373–377.

Unwin, P.N.T. and Henderson, R. 1975. Molecular structure determination by electron microscopy of unstained crystalline specimens. J. Mol. Biol. 94: 425–440.

Unwin, P.N.T. and Klug, A. 1974. Electron microscopy of the stacked disc aggregate of tobacco mosaic virus protein I. Three-dimensional image reconstruction. J. Mol. Biol. 87: 641–656.

Varghese, J.N., Laver, W.G. and Colman, P.M. 1983. Structure of the influenza virus glycoprotein antigen neuraminidase at 2.9Å resolution. Nature 303: 35–40.

Vénien-Bryan, C. and Fuller, S.D. 1994. The organization of the spike complex of Semliki Forest virus. J. Mol. Biol. 236: 572–583.

Vogel, R.H., Provencher, S.W., Bonsdorff, C.-H., Adrian, M. and Dubochet, J. 1986. Envelope structure of Semiliki forest virus reconstructed from cryo-electron micrographs. Nature 320: 533–535.

Wang, G.-J., Porta, C., Chen, Z., Baker, T.S. and Johnson, J.E. 1992. Identification of a F_{ab} interaction footprint site on an icosahedral virus by cryoelectron microscopy and x-ray crystallography. Nature 355: 275–278.

White, J.M. and Johnson, J.E. 1980. Crystalline cowpea mosaic virus. Virology 101: 319–324.

Williams, R.C. and Fisher, H.W. 1970. Electron microscopy of tobacco mosaic virus under conditions of minimal beam exposure. J. Mol. Biol. 52: 121–123.

Wilson, I.A., Skehel, J.J. and Wiley, D.C. 1981. Structure of the haemagglutinin membrane glycoprotein of influenza virus at 3Å resolution. Nature 289: 366–373.

Wyckoff, R.W.G. and Corey, R.B. 1936. X-ray diffraction patterns of crystalline tobacco mosaic proteins. J. Biol. Chem. 116: 51–56.

Yeager, M., Berrimann, J.A., Baker, T.S. and Bellamy, A.R. 1994. Three-dimensional structure of the rotavirus hemagglutinin VP4 by cryo-electron microscopy and difference map analysis. EMBO J. 13: 1011–1018.

Yeager, M., Dryden, K.A., Olson, N.H., Greenberg, H.B. and Baker, T.S. 1990. Three-dimensional structure of rhesus rotavirus by cryoelectron microscopy and image reconstruction. J. Cell Biol. 110: 2133–2144.

Zheng, Y., Doerschuk, P. and Johnson, J.E. 1995. Determination of three dimensional low resolution viral structure from solution X-ray scattering data. Biophys. J. 69: 619–639.

Zhou, Z.H. and Chiu, W. 1993. Prospects for using an IVEM with a FEG for imaging macromolecules towards atomic resolution. Ultramicroscopy 49: 407–416.

Zhou, Z.H., Prasad, B.V.V., Jakana, J., Rixon, F.J. and Chiu, W. 1994. Protein subunit structures in the herpes simplex virus A-capsid determined from 400-kV spot-scan electron cryomicroscopy. J. Mol. Biol. 242: 458–469.

3.

Attachment and Entry of Influenza Virus into Host Cells

Pivotal Roles of the Hemagglutinin

JUDITH M. WHITE, LUCAS R. HOFFMAN,

JAIRO H. AREVALO, & IAN A. WILSON

Influenza viruses cause acute and highly contagious respiratory illnesses. It is likely that the viruses have been with us since ancient times, and major epidemics, such as the "Spanish" influenza of 1918, the "Asian" influenza of 1957, and the "Hong Kong" influenza of 1968 led to the deaths of millions of individuals. In addition, nearly annual outbreaks of the virus continue to impart significant morbidity and excess mortality around the world. Hence influenza remains a major unconquered viral disease (Kingsbury, 1990; Murphy and Webster, 1990; Wilson and Cox, 1990).

Influenza is a negative-strand, enveloped RNA virus. As observed with the electron microscope, virus particles are pleomorphic in shape. Whereas some strains produce irregularly shaped filamentous particles, others yield primarily spherical particles, with a diameter of approximately 1200 Å, particularly when propagated in chicken embryos or in tissue cultures (Kingsbury, 1990; Murphy and Webster, 1990). The outer shell of the virus is a glycoprotein-studded lipid bilayer that is acquired as the virus buds from the host cell plasma membrane. The envelopes of influenza A and B viruses contain three integral membrane proteins (Table 3.1): two glycoprotein surface antigens, the hemagglutinin and the neuraminidase proteins, and the matrix 2 (M2) protein. The inner surface of the viral envelope is lined with a peripheral membrane protein called matrix 1 (M1). Inside this shell are eight segments of negative-strand RNA that associate with the nucleoprotein and polymerase proteins (PB1, PB2, and PA) to form ribonucleoprotein (RNP) complexes. (For more detailed discussions of the structure, assembly, and replication of the RNP complexes, see Lamb, 1989;

TABLE 3.1 Proteins Associated with the Influenza Virus Envelope

Protein	Integral/ Peripheral	Type[a]	Oligomer	Approximate No./Virion[b]	Encoded by RNA Segment
Hemagglutinin	Integral	I	Trimer	500	4
Neuraminidase	Integral	II	Tetramer	100	6
M2	Integral	III	Tetramer	10	7
M1	Peripheral	n.a.		3000	7

Data compiled from Kingsbury (1990), Murphy and Webster (1990), and Lamb (1989).
[a]Nomenclature according to von Heijne and Gavel (1988).
[b]Values for HA, NA, and M2 are estimates for the number of oligomers per virion; the value for M1 is based on its monomer molecular weight.

Kingsbury, 1990; Murphy and Webster, 1990.) Influenza C viruses differ from types A and B in having only one glycoprotein surface antigen with hemagglutinin and esterase activity and only seven segments of RNA. Influenza viruses are typed A, B, and C based on relatedness of their matrix and nucleoprotein antigens.

The reason that influenza still remains a perennial problem worldwide is its unparalleled ability to evade the host immune system. The virus accomplishes this evasion by frequently altering the antigenicity of its surface proteins, hemagglutinin and neuraminidase. Influenza A viruses are classified into subtypes based on relatedness of their hemagglutinin and neuraminidase glycoproteins. Currently there are 13 hemagglutinin (H1 through H13) and 9 neuraminidase (N1 through N9) subtypes listed for influenza A viruses. The last worldwide pandemic, a major epidemic caused by an *antigenic shift*, occurred in 1967–1968 when the Hong Kong virus emerged with a new hemagglutinin (H3) but the same neuraminidase (N2) as the previously circulating Japan strains (subtype H2N2). Since 1968, the H3 subtype of hemagglutinin has constantly accumulated mutations that have led to fresh epidemics every 3 to 4 years, by a process known as *antigenic drift*. In 1977, the H1N1 subtype resurfaced and has been cocirculating with the H3N2 strains since then. The current influenza vaccines include a cocktail of most recent strains to protect against both A-type and B-type isolates. (For a detailed discussion of the structural basis for influenza antigenic variation, see Wilson and Cox, 1990.)

Although the major site of infection in host animals is the respiratory tract, influenza viruses can *enter* most cells, including virtually all tissue culture cells. In vivo, the tropism of the virus is limited in large part by whether or not multiple cycles of replication can ensue (Boycott et al., 1994). Virus entry into cells is accomplished by a four-step process (Fig. 3.1):

1. The virus binds to host cell receptors that are sialic acid–containing cell surface glycoproteins and glycolipids.
2. Once tightly bound, the virus is endocytosed into clathrin-coated pits and vesicles and delivered to endosomes. Endosomes are irregularly

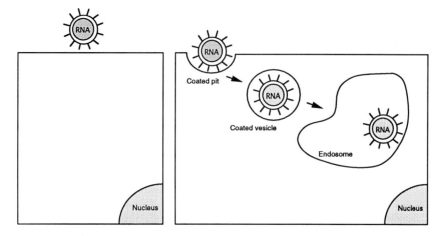

Step 1: Binding   **Step 2: Endocytosis**

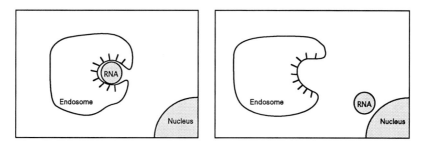

Step 3: Fusion   **Step 4: Genome release**

FIGURE 3.1 Steps in influenza virus entry. *1.* The virus first binds to the cell surface. *2.* The virus is endocytosed in coated pits and coated vesicles and delivered to endosomes. *3.* The virus fuses with the endosome membrane. *4.* The viral RNA, encased in ribonucleoprotein complexes (RNPs), is released into the cytoplasm.

shaped, smooth-surfaced vesicles that maintain a mildly acidic internal pH ($\sim$5 to $\sim$6) because of the presence of a proton pump in their limiting membrane (Marsh and Helenius, 1989).

3. The low endosomal pH triggers fusion between the viral and endosomal membranes. The exact pH at which fusion occurs varies for different strains of the virus, but is generally in the pH 5 to 6 range (White, 1994).

4. Once fusion has occurred, the viral genome (encased as RNP particles) should have access to the host cell cytoplasm. However, evidence has suggested that there are additional requirements, perhaps involving changes in the interior of the virus particle, for release RNPs from the

underlying surface of the recently fused virus membrane so that the viral RNA can move into the cytoplasm and then to the nucleus, where replication occurs (Hay, 1989; Martin and Helenius, 1991; Helenius, 1992; Pinto et al., 1992; Skehel, 1992; Lamb et al., 1994).

The purpose of this chapter is to review the structural basis for the attachment and entry of influenza viruses into host cells. We first briefly review what is known about the structure and function of the three proteins that are integral to the influenza virus envelope, hemagglutinin, neuraminidase, and M2. All of them are involved, either directly or indirectly, in some stage of the entry process. We then discuss the hemagglutinin protein in detail, because this viral surface glycoprotein is pivotal for both virus attachment (Fig. 3.1, step 1) and membrane fusion (Fig. 3.1, step 3) in addition to its role as the major viral antigen.

Proteins of the Influenza Virus Envelope

All three integral membrane proteins of the influenza virus envelope—hemagglutinin, neuraminidase, and M2 (Table 3.1)—are involved in virus attachment and entry into host cells. The structures of the ectodomains of the hemagglutinin protein (Wiley et al., 1981; Wilson et al., 1981) and the globular head domain of the neuraminidase protein (Varghese et al., 1983; see also Chapter 15) are known in atomic detail from high-resolution x-ray crystallographic studies. A model has been proposed for the structure of the M2 protein based on biochemical and electrophysiologic studies (Helenius, 1992; Pinto et al., 1992; Lamb et al., 1994).

Hemagglutinin is a homotrimeric type I integral membrane glycoprotein that projects as a spike, approximately 135 Å in length, from the virus envelope. It is roughly triangular in cross-section, with cross sectional diameters ranging from 35 to 70 Å (Fig. 3.2A). Hemagglutinin is critically involved in binding the virus to the host cell and in inducing the genome-releasing membrane fusion event (Wiley and Skehel, 1987). Hemagglutinin is also a major determinant of the tissue specificity, tropism, and pathogenesis of influenza virus. Hemagglutinin is synthesized as a fusion-inactive precursor, referred to as HA0. Because HA0 must be cleaved, by an appropriate cellular protease, into HA1 and HA2 in order to become fusion competent, virus particles that contain predominantly HA0 are noninfectious. The specific sequence of HA0 around the cleavage site and the presence or absence of an appropriate protease in the host tissue dictate whether or not a particular influenza virus can initiate multiple cycles of infection in a particular cell type of a particular host (Boycott et al., 1994). Hemagglutinin is also a major determinant of both humoral and cell-mediated immune responses to the virus (Wilson and Cox, 1990). The three-dimensional structure as well as the receptor binding and membrane fusion activities of hemagglutinin are described in detail later.

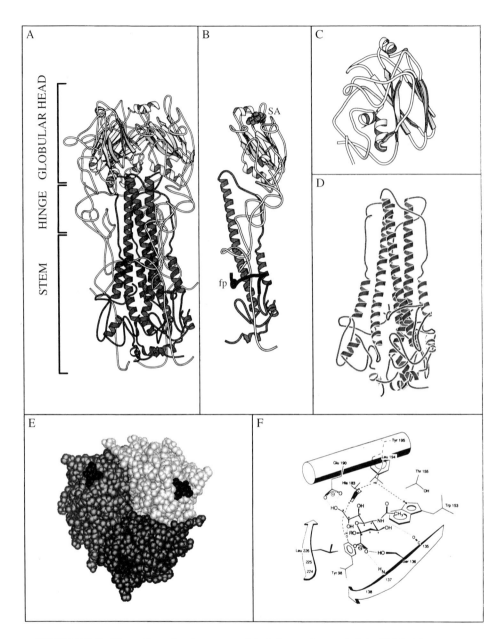

FIGURE 3.2 Three-dimensional structure of the influenza hemagglutinin (X:31). The structure was determined by Wilson et al. (1981). *A*. Ribbon diagram of the hemagglutinin trimer showing the overall domain organization of HA1 (light) and HA2 (dark). *B*. Close-up view of a hemagglutinin monomer highlighting the fusion peptide (fp; black) and the location of the sialic acid binding site (sialic acid, SA; black), which is at the top of the globular domains of the molecule. *C*. Side view of the globular head domain showing the classical jelly-roll fold present in this part of the structure. *D*. HA2 trimer showing the coiled-coil interactions at the trimer interface. *E*. Top view of the trimer highlighting the location of the sialic acid binding sites (darkest region) and the surface boundaries between each of the monomers. *F*. Schematic diagram of the sialic acid binding site showing the possible interactions between hemagglutinin and sialic acid (Sauter et al., 1989). These diagrams were prepared with the program MOLSCRIPT (Kraulis, 1991) using the 4HMG coordinates from the protein data bank. (Panel *F* reproduced with permission from Sauter et al., 1989.)

Neuraminidase is a homotetrameric type II integral membrane glycoprotein that also projects as a spike, approximately 150 Å in length, from the virus envelope. Electron microscopic observation of the neuraminidase protein shows that it is "mushroom shaped." The tetrameric head of neuraminidase is approximately $100 \times 100 \times 60$ Å and is composed of four globular domains. These sit on top of a long, thin (tetrameric), filamentous stem region approximately 100 Å in length (Varghese et al., 1983; Wilson and Cox, 1990). As a potent sialidase, neuraminidase is thought to play two indirect roles in virus entry: 1) to help the virus move through mucus that lines the respiratory tract (so that the virus can reach target cells), and 2) to release progeny virions from the surface of infected cells so that they can infect new cells (Kingsbury, 1990; Murphy and Webster, 1990; see also Chapter 15).

M2 is a homotetrameric type III integral membrane glycoprotein. It is a relatively small protein with a 24-amino-acid domain located outside the virus particle, a 19-amino-acid transmembrane domain, and a 54-amino-acid domain inside the virus particle. M2 is encoded by the same RNA segment (number 7), but from a different open reading frame, as is M1, a peripheral membrane protein (Table 3.1) that is thought to form an inner shell that interacts with the virus envelope as well as the RNPs. Cumulative evidence indicates that M2 forms a low-pH-activated proton channel through the virus membrane (Lamb et al., 1994). Entrance of protons into the virus particle (from the lumen of the endosome) is thought to facilitate the two final stages of the entry process that consist of the membrane fusion reaction (Fig. 3.1, step 3; Bron et al., 1993; Lamb et al., 1994; Wharton et al., 1994) and the release of the RNPs into the cytoplasm (genome penetration; Fig. 3.1, step 4). The influx of protons through the M2 channel may aid these events by weakening interactions between the viral integral membrane proteins and the underlying M1 protein, between the M1 protein and the RNPs, or within the RNPs themselves. For those influenza viruses (e.g., avian viruses) whose hemagglutinins are cleaved intracellularly during transport through the *trans*-Golgi network (TGN) and that induce fusion at a relatively high pH (e.g., 5.5 to 6.0), the M2 protein channel may also play a role late in the viral life cycle. The presence of the M2 proton channel in the membrane of the TGN apparently raises the pH of the lumen of the TGN above pH 5.5 to 6.0, thereby protecting these hemagglutinins from premature low-pH inactivation (Lamb et al., 1994).

Of the three integral components of the influenza virus envelope, the hemagglutinin protein is the only one that is directly involved in the entry process. Moreover, it is critically involved in both *binding* the virus to the host cell (Fig. 3.1, step 1) and inducing membrane *fusion* in endosomes (Fig. 3.1, step 3). The focus of the remainder of this chapter is therefore on the hemagglutinin protein. We first review the structure of the hemagglutinin protein and then demonstrate how this structural information has given insights into how hemagglutinin functions as both an adhesion molecule and a fusogen. We also briefly discuss how information on the structure and function of hemagglutinin is being used in attempts to design novel antiviral agents.

Structure of the Hemagglutinin Protein

Hemagglutinin is synthesized in the rough endoplasmic reticulum (ER) of infected cells as the fusion-inactive HA0 precursor. (For reviews discussing the synthesis, transport, and folding of hemagglutinin along the biosynthetic pathway, see Doms et al., 1993; Helenius, 1994.) The hemagglutinin from a Hong Kong pandemic strain (X:31) of influenza virus, A/Aichi/68 (H3N2), consists of a 16-amino acid signal sequence at its N-terminal end, an ectodomain of 513 amino acids, a transmembrane domain of 27 amino acids, and a cytoplasmic tail of 10 amino acids (Fig. 3.3). The X:31 hemagglutinin is core glycosylated in the ER at seven sites (Asn-X-Ser/Thr) of N-linked glycosylation. The carbohydrate structures are trimmed and modified as the protein traverses the Golgi complex and the TGN such that the final protein contains five complex N-linked carbohydrate structures, at HA1 residues 8, 22, 38, and 285 and at HA2 residue 154, as well as two high-mannose moieties at HA1 residues 81 and 165. The carbohydrates are thought to be involved in the folding (Helenius, 1994) and stability (Wilson et al., 1981) of the hemagglutinin trimer, but vary extensively in number and position in different strains and subtypes of the virus (reviewed in Wilson and Cox, 1990).

The six intramolecular disulfide bonds in HA0 (Fig. 3.3) form in an ATP-dependent fashion while HA0 is still in the ER (Braakman et al., 1992). All evidence suggests that HA0 folds into its correct tertiary and quaternary (trimeric) structure in the ER. Folding and trimerization are prerequisites for movement of hemagglutinin from the ER into the Golgi compartment (Doms et al., 1993; Helenius, 1994). Most hemagglutinins reach the cell surface as the HA0 precursor. If an appropriate protease is present, HA0 is then cleaved into the two disulfide-linked subunits, HA1 and HA2 (Fig. 3.2) (Boycott et al., 1994).

All but nine amino acids of the hemagglutinin ectodomain can be cleaved as an intact molecule from the surface of influenza virions by treatment with the protease bromelain (Fig. 3.3). The resultant, so-called bromelain-released hemagglutinin (BHA), has been crystallized and a high-resolution x-ray structural analysis performed (Fig. 3.2). Based on the crystallographic analysis (Wilson et al., 1981), the hemagglutinin ectodomain is a trimer about 135 Å long (Fig. 3.2A) and 35 to 70 Å in triangular cross section (e.g., Fig. 3.2E). Each monomer has two major domains, a globular head domain and a fibrous stem domain (Fig. 3.2A–D). These two domains are attached to one another via a hinge region. The globular head is composed entirely of residues from HA1 (residues HA1 52 to 275), whereas the fibrous stem contains all of the ectodomain residues of HA2 (HA2 residues 1 to 184) with a few segments from HA1 (HA1 residues 1 to 51 and HA1 residues 276 to 328). Three monomers pack via a central long α-helical coiled-coil structure (Fig. 3.2D) to form the trimeric hemagglutinin spike that projects from the virus envelope.

The dominant feature of the globular head domain is an eight-stranded antiparallel β-structure (Fig. 3.2C). Similar structures, often referred to as ''jelly rolls'' or ''Swiss rolls,'' are found in the capsids of nonenveloped plant

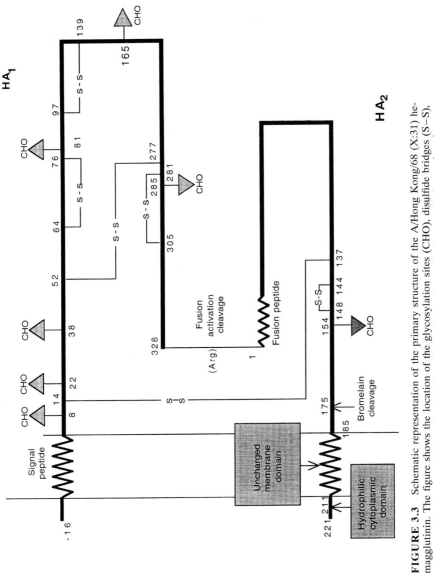

FIGURE 3.3 Schematic representation of the primary structure of the A/Hong Kong/68 (X:31) hemagglutinin. The figure shows the location of the glycosylation sites (CHO), disulfide bridges (S—S), signal peptide (upper tall zigzag line), fusion peptide (short zigzag line), transmembrane domain (lower tall zigzag line), and cytoplasmic domain. Sites of fusion activation cleavage and bromelain cleavage (arrow) are also shown. (Modified from Wilson and Cox, 1990.)

and animal viruses as well as in many other globular proteins (Quiocho, 1986; Branden and Tooze, 1991; Stehle et al., 1994). The receptor binding sites (one per monomer) are located on the outer surface of the globular head domains, distal from the viral membrane (Fig. 3.2B and E). The dominant feature of the proximal fibrous stem region is the triple coiled coil formed by the assembly of the three long (76-Å) α-helices comprising HA2 residues 77 to 126 (Fig. 3.2D). The helices twist 100 Å around one another to form a left-handed superhelix that is close packed (10 Å between helix axes) at its top (N-terminal) end, but more loosely packed (22 Å between helix axes) at its bottom (C-terminal) end. The first few residues of the HA2 N terminus correspond to the start of the hydrophobic fusion peptide (Fig. 3.2B, black), a region critical for the fusion activity of the hemagglutinin trimer (for review, see White, 1992). These, and those in a loop (HA1 residues 14 to 52), are inserted between the long α-helices, about halfway down, and disrupt the close packing of the coiled coil. Three additional short α-helices (from HA2) and seven β-strands (from both HA1 and HA2) surround the coiled coil, forming the outer shell of the fibrous stem (Fig. 3.2D). Energy calculations support the model of the fibrous stem as the major structural framework for the trimer (Eisenberg and Mc-Lachlan, 1986). A more detailed analysis of the structure can be found in Wiley and Skehel (1987), Wiley et al. (1981), Wilson and Cox (1990), and Wilson et al. (1981).

Receptor Binding Activity of the Hemagglutinin Protein

Biologic studies indicated that sialic acid moieties serve as receptors for influenza viruses. Treatment of red blood cells (RBCs) with neuraminidase abolishes their ability to be agglutinated by influenza viruses; resialation or the addition of sialic-acid containing glycolipids restores this capability (Steinhauer et al., 1992; Wiley and Skehel, 1987). The original analysis of the three-dimensional structure of the hemagglutinin trimer (Wilson et al., 1981) led to the hypothesis that the sialic acid binding site consisted of a shallow pocket on the surface of the hemagglutinin trimer (Fig. 3.2E). In contrast to residues on the rest of the surface of the globular head domains, which are highly variable in sequence (Wilson and Cox, 1990), this site contains several residues from the HA1 subunit that are highly conserved among all influenza A and B viruses: Tyr 98, Trp 153, His 183, Glu 190, Leu 194, and Tyr 195 (Fig. 3.2F). Other examples of proteins that contain sialic acid binding sites include the lectin wheat germ agglutinin (Quiocho, 1986; Wilson and Cox, 1990) and the influenza neuraminidase (see Chapter 15), although these are not closely related in three-dimensional structure to hemagglutinin. The polyoma capsid, which has a jelly-roll structural framework, also binds sialic acid, but the protein−carbohydrate interactions are quite different (Stehle et al., 1994).

  Further support for the postulated receptor binding site came from studies demonstrating the pivotal role of a particular residue, HA1 226, of the candidate site in determining receptor specificity (Rogers et al., 1983). Although many cell surface glycoproteins and glycolipids contain terminal sialic acid attached to a penultimate galactose residue, the precise chemistry of the linkage varies. Human isolates of H3 influenza A viruses preferentially agglutinate RBCs with sialic acid attached in an α-2,6 linkage to galactose. The hemagglutinins from these viruses contain leucine at position HA1 226. Selected virus variants that preferentially agglutinate RBCs with sialic acid in an α-2,3 linkage to galactose have Gln at residue HA1 226. Similarly, avian and equine influenza viruses preferentially agglutinate RBCs with sialic acid in an α-2,3 linkage to galactose, and their HAs have Gln at position HA1 226. This strong correlation between the presence of Leu or Gln with α-2,6 versus α-2,3 linkage specificity pinpointed HA1 residue 226 as a key element of the receptor binding pocket (Rogers et al., 1983; Steinhauer et al., 1992).

Final proof that the postulated sialic acid binding pocket is the bona fide receptor binding site came from high resolution crystallographic analysis of a complex between α-2,6 sialyllactose (Neu5Acα(2,6)Galβ(1,4)Glc) and wild-type hemagglutinin from the X:31 strain of influenza. The sialic acid moiety of this trisaccharide fills most of the postulated receptor binding site (Fig. 3.2F). α-2,3-Sialyllactose binds in a similar manner to a variant X:31 hemagglutinin that has Gln at position HA1 226 (Weis et al., 1988). Solution nuclear magnetic resonance studies have corroborated and extended these findings (Sauter et al., 1989).

The receptor binding site is a shallow pocket approximately 12 Å in diameter and approximately 5 Å deep that is located on the surface of the distal tips of each globular head domain (Fig. 3.2E and F). The base of the site is lined by HA1 residues Tyr 98, Trp 153, Thr 155, His 183, Glu 190, Leu 194, and Tyr 195. A β-strand encompassing HA1 residues 224 to 228 lines one side of the site, with another β-strand (HA1 residues 134 to 138) making up the other side. A model has been proposed (Sauter et al., 1989) for how sialic acid interacts in the binding site through a network of hydrogen bonds and hydrophobic interactions (Fig. 3.2F). Because the HA1 side chain atoms at position 226 do not appear to contact the atoms that link sialic acid and galactose, the structural basis for the binding preference of hemagglutinin with Leu or Gln at HA1 226 for α-2,6- or α-2,3-sialyllactose, respectively, is presently not understood.

Despite the striking preferences of hemagglutinins with Leu or Gln at HA1 position 226 to interact with α-2,6- or α-2,3-sialyllactose, respectively, the measured binding affinities for each of these monovalent saccharides are low and not very different for the two isomers. For example, α-2,6-sialyllactose binds to hemagglutinin with Leu at position HA1 226 with a K_d of 2.1 mmol/l; α-2,3-sialyllactose binds to the same hemagglutinin (Leu at HA1 226) with a K_d of 3.2 mmol/l (Sauter et al., 1989). Given these low monomer binding affinities, it is likely that binding to host cell receptors involves multiple interactions. This supposition is supported by two observations. Both naturally occurring and synthetic sialic acid analogs that are polyvalent (i.e., able to bind to more than one receptor site at a time) are superior to monovalent sialic acid analogs in inhibiting the ability of influenza viruses to agglutinate RBCs, a good measure of their receptor binding ability (Steinhauer et al., 1992; Watowich et al., 1994). In addition, the apparent affinity with which sialic acid–bearing liposomes bind to hemagglutinin-expressing cells is 7×10^{10} M^{-1} (Ellens et al., 1990).

A second sialic acid binding site in hemagglutinin was discovered in the process of determining high resolution structures of the BHA fragment of wild-type X:31 hemagglutinin complexed with several sialic acid analogs (Sauter et al., 1992). As with the primary binding site, there is one secondary sialic acid binding site in each monomer of the trimer. The second site is defined by residues 65, 89, 95, 100, 102, 106, and 269 of HA1 and residues 69 to 72 of HA2. This region corresponds to an interface between HA1 monomers where the HA1 globular head domain contacts a loop in HA2 (residues 55 to 76) that is thought to be involved in the fusion-activating conformational change (see later). Certain sialic acid derivatives (e.g., α-2,3-sialyllactose) bind to both the primary and the secondary sialic acid binding site. Others (e.g., α-2,6-sialyllactose) appear to bind only to the primary binding site. The affinity of α-2,3-sialyllactose for the second site is about fourfold lower than for the primary binding site. The role of the second sialic acid binding site in virus entry is not clear. It may participate as an auxiliary receptor during virus binding (Fig. 3.1, step 1) or facilitate the fusion-inducing conformational change, but there is presently no evidence to support either role (Sauter et al., 1992). Alternatively, the second receptor binding site may serve no biologic function.

Membrane Fusion Activity of the Hemagglutinin Protein

The influenza hemagglutinin protein is responsible for the low pH–induced membrane fusion event that occurs between the viral and endosomal membranes (Fig. 3.1, step 3). Fusion is an essential early step in the viral life cycle because it is an obligate prerequisite to the release of the viral RNPs into the cytoplasm (Fig. 3.1, step 4). Because hemagglutinin is both necessary and sufficient to cause membrane fusion, it has been studied extensively as a model membrane fusion protein. (For recent reviews, see Bentz et al., 1993; Siegel, 1993; Stegmann and Helenius, 1993; White, 1995.) In all influenza viruses studied, fusion is activated by low pH, reflecting the intracellular conditions in the endosomal compartment, where fusion occurs. Although the exact pH at which fusion occurs differs somewhat for different virus strains and selected variants, the pH activation profiles for all influenza virus fusion reactions are sharp and generally in the pH 5 to 6 range. Although the presence of receptors in the target membrane can accelerate steps preceding the onset of fusion, influenza viruses fuse efficiently with virtually any target membrane, including pure phospholipid vesicles of varying lipid compositions.

The process of hemagglutinin-mediated membrane fusion can be considered as consisting of four major steps as diagrammed schematically in Figure 3.4. The first step is the low-pH-activated conformational change that converts the hemagglutinin ectodomain to a form capable of binding to target bi-

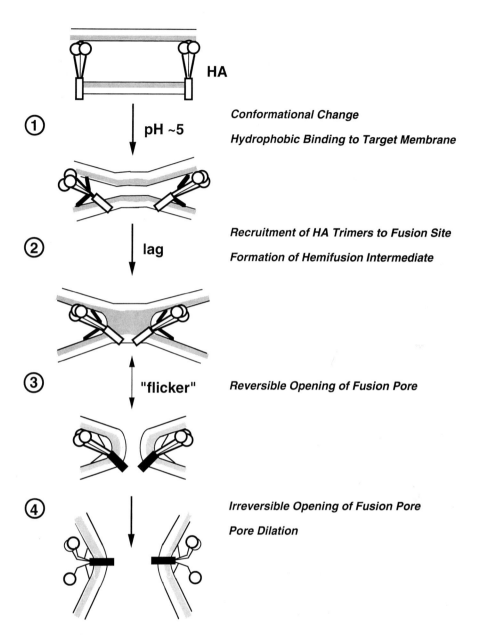

① pH ~5   *Conformational Change*

Hydrophobic Binding to Target Membrane

② lag   *Recruitment of HA Trimers to Fusion Site*

Formation of Hemifusion Intermediate

③ "flicker"   *Reversible Opening of Fusion Pore*

④   *Irreversible Opening of Fusion Pore*

Pore Dilation

FIGURE 3.4 Model for hemagglutinin (HA)-mediated membrane fusion. *Step 1*, A low-pH-induced conformational change exposes the fusion peptides (thickened line) and fosters a hydrophobic association between HA and the target membrane. *Step 2*. Several conformationally altered HA trimers migrate to the fusion site, where they lead to the formation of a hemifusion intermediate. Hemifusion is a state in which the outer, but not the inner, leaflets of the fusing bilayers have mixed. *Step 3*. A narrow fusion pore opens reversibly. The reversible nature of this proposed step could account for the observed flickering seen during electrophysiologic measurements of HA-mediated fusion. *Step 4*. The fusion pore opens irreversibly and dilates. The fusion peptide (thickened line) is thought to be most important for steps 1 and 2. The transmembrane domain (rectangle) is thought to be most important for steps 3 and 4. The model, as shown, invokes the minimal low-pH-induced changes in HA structure that are known to be important for fusion from functional studies. A revised model could incorporate the ''spring-loaded'' conformation of HA (see text and White, 1995 for details). (Modified from White, 1994.)

layers. Target bilayer binding is driven largely by hydrophobic interactions. The second step is formation of the "fusion intermediate," an intermediate that is currently envisioned as a hemifused state in which the outer, but not the inner, leaflets of the fusing bilayers have merged. The third step is the reversible opening of a narrow fusion pore, and the fourth is the irreversible opening and dilation of the fusion pore.

Step 1: The Low-pH-Induced Conformational Change

In terms of biologic activity, the fusion peptide is a very important segment of the hemagglutinin glycoprotein (White, 1992). The first hint of its importance was its observed similarity in position, length, and hydrophobic character to a related sequence in the fusion protein of Sendai virus. Experimental support for the importance of the fusion peptide has come from two lines of evidence: 1) site-specific mutations within the fusion peptide alter or abolish fusion activity (Gething et al., 1986), and 2) preventing fusion peptide exposure through the use of engineered disulfide bonds (Godley et al., 1992; Kemble et al., 1992) or small-molecular-weight inhibitors (Bodian et al., 1993) abolishes fusion activity.

The fusion peptide is found at the N terminus of the HA2 subunit as a result of cleavage of the hemagglutinin precursor, HA0, into its two disulfide-linked polypeptide chains, HA1 and HA2 (Fig. 3.3). This proteolytic processing event is essential for both the fusion activity and infectivity of influenza viruses (Boycott et al., 1994). The fusion peptide sequence is relatively apolar, with a pattern of repeating glycine residues, and is extremely well conserved among all influenza A viruses. The fusion peptide is operationally defined as residues 1 to 24 of the HA2 chain, although the individual residues that are most critical for fusion have yet to be determined. In the native trimer (Fig. 3.2B and D), the N terminus of the fusion peptide lies in the trimer interface. The N terminus is inserted between the triple-stranded coiled coil near HA2 residue 113, with Leu 2 and Phe 3 making contacts around the threefold axis. The fusion peptide then wraps between and around each hemagglutinin subunit and is held tightly in place by a network of hydrogen bonds and other van der Waals contacts (Wilson et al., 1981; Daniels et al., 1985; Weis et al., 1990; Steinhauer et al., 1992). The sequence comprising the first 10 glycine-rich residues makes a series of sharp turns and lies roughly perpendicular to the long axis of the trimer, approximately 100 Å from the distal tip of the HA1 globular head domain and approximately 35 Å from where the hemagglutinin protein inserts into its own viral membrane. Residues 11 to 20, which include three acidic residues, make a tight exposed turn on the surface of the trimer before the peptide reenters the trimer interface, with residues 22 to 29 forming the first strand of a two-stranded antiparallel β-sheet (Fig. 3.2B).

Exposure to low pH causes major irreversible changes in the structure of the hemagglutinin trimer. Most notably, the fusion peptides are released from the trimer interface, as evidenced by increased reactivity with an anti−fusion peptide antibody (White and Wilson, 1987) as well as sensitivity of the hemagglutinin protein to protease digestion (Skehel et al., 1982). Concomitant with fusion peptide exposure (Stegmann et al., 1990), the hemagglutinin ectodomain acquires hydrophobic properties; it will bind to detergent micelles or preformed liposomes. Many other changes occur when hemagglutinin is exposed to low pH, such as alterations in antibody reactivity, protease sensitivity, and tryptophan fluorescence (Doms, 1993). The characterization of mutant hemagglutinins with elevated pH thresholds for fusion (Daniels et al., 1985; Steinhauer et al., 1992) suggests that changes occur throughout the hemagglutinin trimer, with mutations being clustered either in the fusion peptide region or in areas of intersubunit and interdomain contact (e.g., HA1–HA1, HA2–HA2, and HA1–HA2 interfaces; see Fig. 4 in Steinhauer et al., 1992). Nevertheless, the overall secondary structure, as deduced from circular dichroism measurements, does not change significantly. Low-pH-treated hemagglutinin remains trimeric, and the final low-pH form appears to be more thermostable than the native pH 7 structure (Ruigrok et al., 1988). Hence the low-pH-induced conformational change is envisioned to include relative motions of protein domains and local changes in the vicinity of the fusion peptide (Wiley and Skehel, 1987; Weis et al., 1990; Steinhauer et al., 1992; Doms, 1993; White, 1994).

Binding studies with polyclonal and monoclonal anti-peptide antibodies, as well as with monoclonal anti-protein antibodies against known epitopes, have shown that the conformational change in hemagglutinin occurs in two major stages (White and Wilson, 1987; Kemble et al., 1992). During the first stage, changes occur in the stem region of the molecule, most notably release of the fusion peptides. Some effect is also seen in the quaternary structure at the distal tips of the globular head domain interface (Fig. 3.5*A*–*C* and Fig. 3.6). In the second stage, changes occur within in the globular head domain interface as well as in the hinge region (Fig. 3.5*D*–*F* and Fig. 3.6). These latter changes most likely reflect more substantial separation of the globular head domains. In concert with the second stage changes, hemagglutinin loses its sturdy rodlike appearance and becomes wiry and disorganized in appearance when viewed in the electron microscope (reviewed in Stegmann and Helenius, 1993; White, 1994). Whereas the time course of the first stage appears to be relatively temperature independent from 4 °C to 37 °C, the time course for the second stage is highly temperature dependent over this range (White and Wilson, 1987; Stegmann et al., 1990). Studies have supported the two-stage conformational change model (for review, see White, 1994).

A schematic representation of the low- pH-induced conformational changes in hemagglutinin is shown in Figure 3.6. The native hemagglutinin trimer itself is not fusogenic at neutral pHs. Moreover, if X:31 hemagglutinin is pretreated at low pH in the absence of a target membrane, it is rendered fusion inactive. Under these conditions it aggregates (Skehel et al., 1982) and becomes laterally immobile in the plane of the virus membrane (Gutman et al., 1993). Hence, neither the starting nor the end state of X:31 hemagglutinin is fusogenic (Fig. 3.6). Based on the preceding discussion, our current picture is that, once hemagglutinin is exposed to low pH, it passes through at least two conformational intermediates, schematized as I and II in Figure 3.6, before it is once again

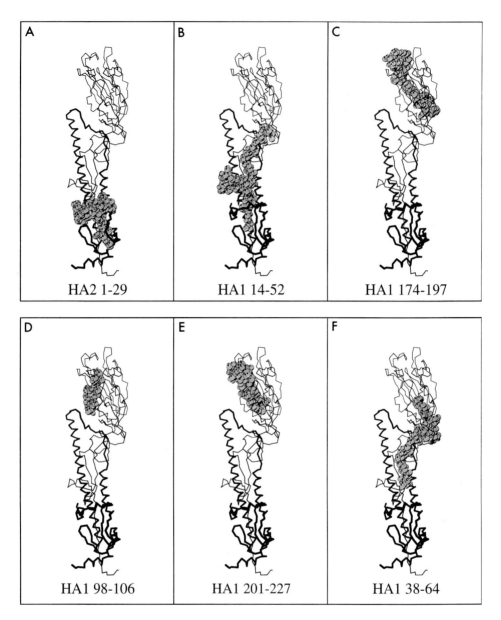

A	B	C
HA2 1-29	HA1 14-52	HA1 174-197

D	E	F
HA1 98-106	HA1 201-227	HA1 38-64

FIGURE 3.5 Low-pH-induced conformational changes in hemagglutinin detected with anti-peptide and monoclonal antibodies. Anti-peptide and monoclonal antibodies against defined epitopes have been used to probe the low-pH-induced conformational change in hemagglutinin. Each panel shown here represents a hemagglutinin monomer with the HA1 subunit drawn with a thin line and the HA2 subunit drawn with a thicker darker line. In panels *A*, *B*, and *D–F*, van der Waals surfaces (gray spheres) highlight peptide sequences that were used to generate anti-peptide antibodies. The van der Waals surfaces (gray spheres) in *C* depict the site B epitope (Wiley et al., 1981). The numbers under each image refer to the peptide sequences recognized by the cognate antipeptide (*A*, *B*, and *D–F*) or anti-protein monoclonal (*C*) antibody employed. Panels *A–C* highlight peptides (*A* and *B*) or epitopes (*C*) whose cognate antibodies showed enhanced reactivity during the first stage of the conformational change. Panels *D–F* highlight peptides whose cognate antibodies showed enhanced reactivity during the second stage of the conformational change. (See White and Wilson, 1987, and Kemble et al., 1992, for details, and Doms, 1993, for a relevant review.)

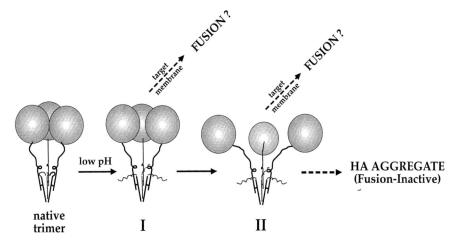

FIGURE 3.6 Schematic representation of the conformational changes as they may relate to hemagglutinin (HA) fusion activity. The conformational change in HA appears to occur in two major stages. During the first (I), there are changes in the stem, including exposure of the fusion peptide (light squiggled line) and some separation of the globular head domains. During the second (II), the globular head domains dissociate substantially from one another. In the absence of a target membrane, low-pH-treated X:31 HA aggregates and becomes inactive for fusion. It is not yet clear which form (I or II), in the presence of a target membrane (dashed arrows), more accurately represents the fusion-active conformation, nor when the fusion peptide is repositioned to the top of the trimer (Carr and Kim, 1993; Bullough et al., 1994). (Reproduced with permission from White, 1994.)

rendered fusion inactive. A major current question is which intermediate form is fusion active.

Because the fusion peptides must be released from the trimer interface in order for hemagglutinin to be fusogenic, and because constraining intersubunit interactions at the top of the hemagglutinin trimer appear to prevent fusion peptide exposure (Godley et al., 1992; Kemble et al., 1992; Bodian et al., 1993), it is generally accepted that hemagglutinin must progress to a stage I–like conformation in order to be fusogenic. However, it is not yet clear whether fusion requires progression to a stage II–like conformation during which further separation of the globular head domains appears to occur. There are three major arguments against a requirement for a stage II–like conformation (Stegmann and Helenius, 1993; White, 1994): 1) the time course of fusion more closely parallels that for the first, rather than the second, set of conformational changes; 2) fusion with X:31 virus proceeds, albeit slowly, at 4 °C, a condition under which substantial separation of the globular head domains is not observed; and 3) for the Japan strain of influenza virus, substantial separation of the globular head domains has not been observed even at 37 °C, a condition under which the virus executes rapid and efficient fusion. There is a caveat to the preceding arguments: Only three to four hemagglutinin trimers appear to be sufficient to initiate the formation of a fusion pore (Danieli et al., 1996). Although we do not yet know how many fusion pores must open in order to

permit the large viral nucleocapsid to enter the host cell cytoplasm, only
a small fraction of the hemagglutinin molecules on the virion surface may
actually participate in fusion. Hence the techniques used to monitor the
conformational state of hemagglutinin may not sample the low number of
hemagglutinin molecules that actively engage in the viral fusion process.

   Two studies have provided further insight into the structure of low-pH
hemagglutinin (Carr and Kim, 1993; Bullough et al., 1994). Both are concerned
with the conformation of HA2 residues 55 to 76 at low pH. In the native pH
7 structure, these residues form a connecting loop between a short α-helix
(residues 38 to 54) and the long central α-helix (residues 77 to 126) of HA2
(Fig. 3.7; left, segment B). An early predictive analysis suggested that this loop
should in fact be part of an extended coiled coil because it was consistent with
a heptad-repeating structure, a signature motif in helical coiled coils (Ward and
Dopheide, 1980). Carr and Kim (1993) showed that synthetic peptides with

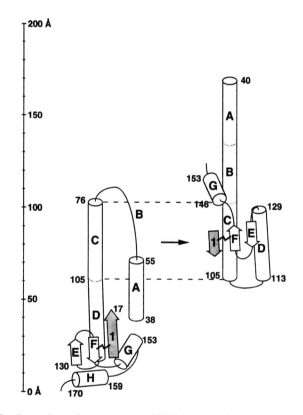

FIGURE 3.7   Comparison of the structure of HA2 in the native and a low-pH form of hemagglutinin. The structure of a proteolyzed derivative (right) of low-pH-treated hemagglutinin,
called TBHA2 (HA1 residues 12 to 16; HA2 residues 40 to 153), is compared with the corresponding region (left) of BHA (HA1 residues 11 to 17; HA2 residues 38 to 175). (From Bullough
et al., 1994; reprinted by permission.)

these loop residues attached to the initial 13 residues of the long α-helix form a coiled coil at low pH. Wiley and coworkers have conclusively shown, by high-resolution X-ray crystallographic analysis of a truncated form of low-pH-treated hemagglutinin (Bullough et al., 1994), that this loop is part of an extended helical coiled coil that includes the small helix, the loop, and the first half of the long α-helix (Fig. 3.7, right). For this analysis, the hemagglutinin protein was treated at pH 4.9 for sufficient time to have undergone the full conformational change (Fig. 3.6) and then exposed to proteases to remove its globular head domains (i.e., most of HA1) and fusion peptides leaving most of HA2 (residues 38 to 175) and a small portion of HA1 (residues 1 to 27). If such a transition occurs in the intact hemagglutinin trimer and represents conformational changes on the fusion pathway, it suggests a mechanism for bringing the fusion peptide nearer to the top of the trimer, and so closer to the target membrane. However, such a "spring-loaded" repositioning of the fusion peptide would not completely solve the fundamental problem of membrane fusion, that is, bringing the two fusing bilayers into close apposition (White, 1992).

This crystallographic analysis of low-pH-treated hemagglutinin revealed another potentially important change in hemagglutinin structure. In the low-pH structure, a central segment of the long α-helix found in native hemagglutinin, HA2 residues 106 to 112, is a loop and is therefore no longer part of the coiled coil. This change near the lower part of the preexisting coiled coil is particularly intriguing in two regards. The first is that residues HA2 106 to 112 define a region where the fusion peptide and a loop encompassing HA1 residues 14 to 52 contact the long α-helix (Fig. 3.2*B*). Both the fusion peptide and the loop (HA1 residues 14 to 52) are exposed very early in the conformational change (White and Wilson, 1987). Their release from the trimer interface may thus destabilize the native coiled coil in this region, allowing it to undergo the helix-to-loop transition. The second relates to the newly appreciated role of the hemagglutinin transmembrane domain (Kemble et al., 1994), and by inference the base of the hemagglutinin trimer, in the overall fusion reaction. In this regard, the helix-to-loop transition of HA1 106 to 112 could play a critical role in overcoming the fusion problem. That is, how are the two membranes brought close together (see next section)?

Despite the provocative implications of the loop-to-helix (Carr and Kim, 1993; Bullough et al., 1994) and helix-to-loop (Bullough et al., 1994) transitions described here, major current challenges are to determine whether either or both of these changes occur in the intact hemagglutinin trimer, and to determine when they occur during the pathway of the proposed conformational changes (Fig. 3.6). Assuming that these changes can occur in the intact hemagglutinin trimer, it will then be of paramount importance to determine whether either change is required for the fusion activity of hemagglutinin, and for which stage of the fusion reaction (Fig. 3.4). Once answers to these questions are known, then this new information can be incorporated into more refined models of hemagglutinin-mediated membrane fusion.

Step 2: Formation of the Hemifusion Intermediate

An early consequence of the low-pH-induced conformational change is that the hemagglutinin ectodomain becomes hydrophobic and binds to target membranes. It is generally accepted that this binding occurs largely, if not exclusively, via the fusion peptides. Indirect evidence suggests that, at some stage of its interaction with target bilayers, the fusion peptide may be in a helical arrangement. Such a motif would position all of its bulky hydrophobic residues on one face of a helix (e.g., Fig. 1c in White, 1992). Many details of the interaction between the fusion peptide and the target membrane remain to be deciphered: What is the secondary structure of the fusion peptide as it engages the target membrane? Does the fusion peptide insert into the target membrane perpendicularly or obliquely with respect to the contacting bilayers, or does it lie in a parallel orientation along the surface of the target bilayer? Which specific residues are critical for fusion? Do fusion peptides interact with the viral as well as the target membrane? Is the fusion peptide involved in the completion, as well as the initiation, of the fusion reaction? (For a more detailed discussion of these questions, see White, 1994.)

Once hemagglutinin-mediated fusion has been triggered by exposure to low pH, there is a discernible lag phase before the onset of lipid mixing or the detection of fusion pore conductances (Fig. 3.4). The length of the lag phase varies with temperature, pH, the presence or absence of a receptor in the target membrane, and hemagglutinin surface density (Stegmann et al., 1990; Stegmann et al., 1991; Stegmann and Helenius, 1993; White, 1994). Interestingly, it appears that hydrophobic, and presumably fusion peptide—mediated, interaction between hemagglutinin and the target membrane (Fig. 3.4, step 1) occurs quite early in the lag, at least for X:31 hemagglutinin (Stegmann et al., 1990, 1991). Several lines of indirect evidence suggest that several hemagglutinin trimers cluster during the lag phase, in the correct orientation with respect to both one another and the target membrane, so as to form a fusion site. Strong support for the implicit cooperative nature of the fusion process has recently been obtained by demonstrating a sigmoidal relationship between the length of the lag phase and hemagglutinin surface density (Danieli et al., 1996).

Our current picture of the initial fusion site is a localized area between the two fusing membranes that is circumscribed by a minimum of three or four conformationally altered hemagglutinin trimers (Danieli et al., 1996) (Fig. 3.4). At this localized site, the two contacting membranes may simply dimple toward one another, or the hemagglutinin trimers may tilt with respect to the plane of the virus membrane. Tilting may be facilitated by the helix-to-loop transition that occurs near the base of the long α-helix (Bullough et al., 1994). Thus, several hemagglutinin trimers would be sandwiched between the two contacting bilayers. In models such as those depicted in Figure 3.4 (see also Siegel, 1993; Stegmann and Helenius, 1993; White, 1994), the fusion peptides engage the target and virus membranes when they are positioned on the sides of the hemagglutinin trimer. This contrasts with the spring-loaded conformational change model, wherein the fusion peptides would be at the top of the hemagglutinin trimer (Carr and Kim, 1993; Bullough et al., 1994; White, 1995). Before we can evaluate these two models, it must be experimentally determined whether the spring-loaded conformational change is important for fusion and, if so, for which stage of the fusion reaction (see Fig. 3.4).

As has been hypothesized, evidence now suggests that hemagglutinin-mediated fusion proceeds via a hemifusion intermediate (Fig. 3.4), a state in which the lipids, but not the aqueous contents, of the fusing compartments mix. We have provided support for the notion of a hemifusion intermediate in hemagglutinin-mediated fusion by showing that a hemagglutinin protein that is anchored solely into the outer leaflet of a bilayer via a glycosylphosphatidyl-inositol tail promotes mixing of lipids but not cytoplasmic contents (Kemble et al., 1994; Melikyan et al., 1995). Based on studies with target membranes of different lipid compositions, it now appears that the phospholipids in the hemi-fusion intermediate are more likely arranged in a ''lipid stalk,'' as opposed to an ''inverted micelle'' (Stegmann, 1993). (For reviews and schematic representations of these possible lipid intermediates, see Siegel, 1993; Zimmerberg et al., 1993.)

Step 3: Reversible Opening of the Fusion Pore

Electrophysiologic studies have indicated that fusion pores approximately 10 to 20 Å in diameter form during hemagglutinin-mediated membrane fusion (Spruce et al., 1989, 1991). Interestingly, these pores share many features with fusion pores that form during regulated exocytosis (White, 1992). One of these similarities is the capacity to flicker, that is, to open and close repeatedly before dilating to a fully open state. We surmise that this flickering state corresponds to the reversible opening of a narrow fusion pore (Fig. 3.4, step 3). What dictates the variable initial conductances of these pores and the number of times that they flicker before opening irreversibly remains to be determined.

*Step 4: Irreversible Opening of the Fusion Pore and
Pore Dilation*

The last step of the fusion reaction is the irreversible opening, or dilation, of the fusion pore. Full dilation of the pore would appear to be necessary for entry of the relatively large viral nucleocapsid into the cytoplasm. It is not yet known what drives the irreversible opening of the fusion pore. A tantalizing possibility is that a different conformational state of hemagglutinin, distinct from that required for the reversible opening of the fusion pore (Fig. 3.4, step 3), drives irreversible pore dilation. Hypothetically, this state may involve the loop-to-helix (Carr and Kim, 1993; Bullough et al., 1994) and helix-to-loop (Bullough et al., 1994) transitions (Fig. 3.6). All we presently know about this final step of the fusion process is that a transmembrane domain is required (Kemble et al., 1994; Melikyan et al., 1995).

Full and irreversible dilation of the fusion pore, although necessary, is not sufficient for virus entry. As discussed previously, the final requirement is dissociation of the RNPs from the viral envelope so that they can be released into the cytoplasm (Fig. 3.1, step 4) and then transit to and into the nucleus in a form suitable for replication (Martin and Helenius, 1991). Release of the viral genome into the cytoplasm appears to require the proton channel activity of the viral M2 protein (Lamb et al., 1994). In this context, it is interesting to note that the M2 channel also appears to modulate the rate of fusion per se. For two avian influenza viruses, A/turkey/Oregon/71 and fowl plague virus, the rate of fusion, measured as mixing of fluorescent lipids, is about twofold higher when the channel is functional than when it is blocked (Bron et al., 1993; Wharton et al., 1994). These latter findings suggest that fusion may proceed more rapidly if the viral glycoproteins are not tethered to the underlying viral matrix.

Perspectives

The three integral proteins of the influenza envelope—hemagglutinin, neuraminidase, and M2—are involved in different aspects of influenza entry into host cells. Investigations of these three proteins are ongoing and will guide studies on the mechanisms of host cell attachment, membrane fusion, and entry of other enveloped viruses, including such devastating human pathogens as the human immunodeficiency virus. Similarly, in light of the tremendous antigenic variability of the influenza virus surface and the consequent difficulty in developing a "universal" influenza vaccine, all three influenza membrane proteins are good targets for the development of novel antiviral agents (see Chapter 15). For example, amantadine and rimantadine, the only anti-influenza agents presently in clinical usage, interfere with the activity of the M2 channel (Lamb et al., 1994). Further analysis of the structure and function of the M2 channel should lead to the development of inhibitors with increased efficacy. Impressively, work using rational, or structure-based, drug design strategies has led to the development of novel sialic acid analogs that are not only potent inhibitors of viral neuraminidase but also potent inhibitors of viral replication (von Itzstein et al., 1993; see also Chapter 15).

A great deal has been learned about the molecular basis for the host cell attachment and membrane fusion functions of hemagglutinin. Consequently, hemagglutinin continues to be an excellent model for both adhesion proteins and membrane fusion proteins. In the arena of drug design, there are several ways to contemplate interfering with the functions of hemagglutinin and hence inhibiting virus entry. One would be to interfere with the receptor binding activity of hemagglutinin. Although monovalent sialic acid analogs do not bind to hemagglutinin with high affinity, it may be possible to develop more potent small molecule inhibitors or polyvalent sialic acid analogs that bridge the primary or secondary sialic acid binding sites (Sauter et al., 1992; Watowich et al., 1994). In terms of fusion, we have identified a family of benzo- and hydroquinones that prevent the fusion-inducing conformational change and hence the fusion activity and infectivity of X:31 influenza virus (Bodian et al., 1993). The first compounds in this category were targeted to bind to a site adjacent to the fusion peptides (see Figs. 1 and 2 in Bodian et al., 1993). Future experimentation will determine whether this is their binding site and whether occupancy of this or other potential small molecule binding sites on the native trimer might have the same effect.

Acknowledgments

Work in the laboratories of Drs. White and Wilson was supported by grants from the National Institutes of Health. We thank Lorraine Hernandez for Figure 3.1.

References

Bentz, J., Ellens, H. and Alford, D. 1993. Architecture of the influenza hemagglutinin fusion site. In Viral Fusion Mechanisms (Bentz, J., Ed.), pp. 163–199. CRC Press, Boca Raton, FL.

Bodian, D.L., Yamasaki, R.B., Buswell, R.L., Stearns, J.F., White, J.M. and Kuntz, I.D. 1993. Inhibition of the fusion-inducing conformational change of influenza hemagglutinin by benzoquinones and hydroquinones. Biochemistry 32: 2967–2978.

Boycott, R., Klenk, H.-D. and Ohuchi, M. 1994. Cell tropism of influenza virus mediated by hemagglutinin activation at the stage of virus entry. Virology 203: 313–319.

Braakman, I., Helenius, J. and Helenius, A. 1992. Role of ATP and disulphide bonds during protein folding in the endoplasmic reticulum. Nature 356: 260–262.

Branden, C. and Tooze, J. 1991. Introduction to Protein Structure. Garland Publishing, New York.

Bron, R., Kendal, A.P., Klenk, H.D. and Wilschut, J. 1993. Role of the M2 protein in influenza virus membrane fusion: effects of amantadine and monensin on fusion kinetics. Virology 195: 808–811.

Bullough, P.A., Hughson, F.M., Skehel, J.J. and Wiley, D.C. 1994. Structure of influenza haemagglutinin at the pH of membrane fusion. Nature 371: 37–43.

Carr, C.M. and Kim, P.S. 1993. A spring-loaded mechanism for the conformational change of influenza hemagglutinin. Cell 73: 823–832.

Danieli, T., Pelletier, S.L., Henis, Y.I. and White, J.M. 1996. Membrane fusion mediated by the influenza virus hemagglutinin requires the concerted action of at least three hemagglutinin trimers. J. Cell Biol. 133: 559–569.

Daniels, R.S., Downie, J.C., Hay, A.J., Knossow, M., Skehel, J.J., Wang, M.L. and Wiley, D.C. 1985. Fusion mutants of the influenza virus hemagglutinin glycoprotein. Cell 40: 431–439.

Doms, R.W. 1993. Protein conformational changes in virus-cell fusion. Meth. Enzymol. 221: 61–82.

Doms, R., Lamb, R., Rose, J. and Helenius, A. 1993. Folding and assembly of viral membrane proteins. Virology 193: 545–562.

Eisenberg, D. and McLachlan, A.D. 1986. Solvation energy in protein folding and binding. Nature 319: 199–203.

Ellens, H., Bentz, J., Mason, D., Zhang, F. and White, J.M. 1990. Fusion of influenza hemagglutinin-expressing fibroblasts with glycophorin-bearing liposomes: role of hemagglutinin surface density. Biochemistry 29: 9697–9707.

Gething, M.-J., Doms, R.W., York, D. and White, J.M. 1986. Studies on the mechanism of membrane fusion: site-specific mutagenesis of the hemagglutinin of influenza virus. J. Cell Biol. 102: 11–23.

Godley, L., Pfeifer, J., Steinhauer, D., Ely, B., Shaw, G., Kaufmann, R., Suchanek, E., Pabo, C., Skehel, J.J. and Wiley, D.C. 1992. Introduction of intersubunit disulfide bonds in the membrane-distal region of the influenza hemagglutinin abolishes membrane fusion activity. Cell 68: 635–645.

Gutman, O., Danieli, T., White, J.M. and Henis, Y.I. 1993. Effects of exposure to low pH on the lateral mobility of influenza hemagglutinin expressed at the cell surface. Biochemistry 32: 101–106.

Hay, A.J. 1989. The mechanism of action of amantadine and rimantadine against influenza viruses. In Concepts in Viral Pathogenesis III (Notkins, A.L. and Oldstone, M.B.A., Eds.), pp. 361–367. Springer-Verlag, New York.

Helenius, A. 1992. Unpacking the incoming influenza virus. Cell 69: 577–578.

Helenius, A. 1994. How N-linked oligosaccharides affect glycoprotein folding in the endoplasmic reticulum. Mol. Biol. Cell 5: 253–265.

Kemble, G.W., Bodian, D.L., Rosé, J., Wilson, I.A. and White, J.M. 1992. Intermonomer disulfide bonds impair the fusion activity of influenza virus hemagglutinin. J. Virol. 66: 4940–4950.

Kemble, G.W., Danieli, T. and White, J.M. 1994. Lipid-anchored influenza hemagglutinin promotes hemifusion, not complete fusion. Cell 76: 383–391.

Kingsbury, D.W. 1990. Orthomyxoviridae and their replication. In Virology, Vol. 1 (Fields, B.N., Knipe, D.M., Chanock, R.M., Hirsch, M.S., Melnick, J.L., Monath, T.P. and Roizman, B., Eds.), pp. 1075–1089. Raven Press, New York.

Kraulis, P.J. 1991. MOLSCRIPT–a program to produce both detailed and schematic plots of protein structures. J. Appl. Cryst. 24: 946–950.

Lamb, R.A. 1989. Genes and proteins of the influenza viruses. In The Influenza Viruses (Krug, R.M., Ed.), pp. 1–87. Plenum Press, New York.

Lamb, R.A., Holsinger, L.J. and Pinto, L.H. 1994. The influenza A virus M2 ion channel protein and its role in the influenza virus life cycle. In Cellular Receptors for Animal Viruses (Wimmer, E., Ed.), pp. 303–321. Cold Spring Harbor Laboratory Press, Cold Spring Harbor, NY.

Marsh, M. and Helenius, A. 1989. Virus entry into animal cells. Adv. Virus Res. 36: 107–151.

Martin, K. and Helenius, A. 1991. Transport of incoming influenza virus nucleocapsids into the nucleus. J. Virol. 65: 232–244.

Melikyan, G.B., White, J.M. and Cohen, F.S. 1995. GPI-anchored influenza hemagglutinin induces hemifusion to both red blood cell and planar bilayer membranes. J. Cell Biol. 131: 679–691.

Murphy, B.R. and Webster, R.G. 1990. Orthomyxoviruses. In Virology, Vol. 1 (Fields, B.N., Knipe, D.M., Chanock, R.M., Hirsch, M.S., Melnick, J.L., Monath, T.P. and Roizman, B., Eds.), pp. 1091–1152. Raven Press, New York.

Pinto, L.H., Holsinger, L.J. and Lamb, R.A. 1992. Influenza virus M2 protein has ion channel activity. Cell 69: 517–528.

Quiocho, F.A. 1986. Carbohydrate-binding proteins: tertiary structures and protein-sugar interactions. Annu. Rev. Biochem. 55: 287–315.

Rogers, G.N., Paulson, J.C., Daniels, R.S., Skehel, J.J., Wilson, I.A. and Wiley, D.C. 1983. Single amino acid substitutions in influenza haemagglutinin change receptor binding specificity. Nature 304: 76–78.

Ruigrok, R.W.H., Aitken, A., Calder, L.J., Martin, S.R., Skehel, J.J., Wharton, S.A., Weis, W. and Wiley, D.C. 1988. Studies on the structure of the influenza virus haemagglutinin at the pH of membrane fusion. J. Gen. Virol. 69: 2785–2795.

Sauter, N.K., Bednarski, M.D., Wurzburg, B.A., Hanson, J.E., Whitesides, G.M., Skehel, J.J. and Wiley, D.C. 1989. Hemagglutinin from two influenza virus variants bind to sialic acid derivatives with millimolar dissociation constants: a 500-MHz proton nuclear magnetic resonance study. Biochemistry 28: 8388–8396.

Sauter, N.K., Glick, G.D., Crowther, R.L., Park, S.-J., Eisen, M.B., Skehel, J.J., Knowles, J.R. and Wiley, D.C. 1992. Crystallographic detection of a second ligand binding site in influenza virus hemagglutinin. Proc. Natl. Acad. Sci. U.S.A. 89: 324–328.

Siegel, D.P. 1993. Modeling protein-induced fusion mechanisms: insights from the relative stability of lipidic structures. In Viral Fusion Mechanisms (Bentz, J., Ed.), pp. 475–512. CRC Press, Boca Raton, FL.

Skehel, J.J. 1992. Influenza virus: amantadine blocks the channel [News]. Nature 358: 110–111.

Skehel, J.J., Bayley, P.M., Brown, E.B., Martin, S.R., Waterfield, M.D., White, J.M., Wilson, I.A. and Wiley, D.C.J. 1982. Changes in the conformation of influenza virus hemagglutinin at the pH optimum of virus-mediated membrane fusion. Proc. Natl. Acad. Sci. U.S.A. 79: 968–972.

Spruce, A.E., Iwata, A. and Almers, W. 1991. The first milliseconds of the pore formed by a fusogenic viral envelope protein during membrane fusion. Proc. Natl. Acad. Sci. U.S.A. 88: 3623–3627.

Spruce, A.E., Iwata, A., White, J.M. and Almers, W. 1989. Patch clamp studies of single cell-fusion events mediated by a viral fusion protein. Nature 342: 555–558.

Stegmann, T. 1993. Influenza hemagglutinin-mediated membrane fusion does not involve inverted phase lipid intermediates. J. Biol. Chem. 268: 1716–1722.

Stegmann, T., Delfino, J.M., Richards, F.M. and Helenius, A. 1991. The HA2 subunit of influenza hemagglutinin inserts into the target membrane prior to fusion. J. Biol. Chem. 266: 18404–18410.

Stegmann, T. and Helenius, A. 1993. Influenza virus fusion: from models toward a mechanism. In Viral Fusion Mechanisms (Bentz, J., Ed.), pp. 89–111. CRC Press, Boca Raton, FL.

Stegmann, T., White, J.M. and Helenius, A. 1990. Intermediates in influenza-induced membrane fusion. EMBO J. 9: 4231–4241.

Stehle, T., Yan, Y., Benjamin, T.L. and Harrison, S.C. 1994. Structure of murine polyomavirus complexed with an oligosaccharide receptor fragment. Nature 369: 160–163.

Steinhauer, D.A., Sauter, N.K., Skehel, J.J. and Wiley, D.C. 1992. Receptor binding and cell entry by influenza viruses. Semin. Virol. 3: 91–100.

Varghese, J.N., Laver, W.G. and Colman, P.M. 1983. Structure of the influenza virus glycoprotein antigen neuraminidase at 2.9 A resolution. Nature 303: 35–40.

von Heijne, G. and Gavel, Y. 1988. Topogenic signals in integral membrane proteins. Eur. J. Biochem. 174: 671–678.

von Itzstein, M., Wu, W.-Y., Kok, G.B., Pegg, M.S., Dyason, J.C., Jin, B., Phan, T.V., Smythe, M.L., White, H.F. and Oliver, S.W. 1993. Rational design of potent sialidase-based inhibitors of influenza virus replication. Nature 363: 418–423.

Ward, C.W. and Dopheide, T.A. 1980. Influenza virus haemagglutinin: structural predictions suggest that the fibrillar appearance is due to the presence of a coiled-coil. Aust. J. Biol. Sci. 33: 449–455.

Watowich, S.J., Skehel, J.J. and Wiley, D.C. 1994. Crystal structures of influenza virus hemagglutinin in complex with high-affinity receptor analogs. Structure 2: 719–731.

Weis, W., Brown, J.H., Cusack, S., Paulson, J.C., Skehel, J.J. and Wiley, D.C. 1988. Structure of the influenza virus hemagglutinin complexed with its receptor, sialic acid. Nature 333: 426–431.

Weis, W.I., Cusack, S.C., Brown, J.H., Daniels, R.W., Skehel, J.J. and Wiley, D.C. 1990. The structure of a membrane fusion mutant of the influenza virus haemagglutinin. EMBO J. 9: 17–24.

Wharton, S.A., Belshe, R.B., Skehel, J.J. and Hay, A.J. 1994. Role of virion M2 protein in influenza virus uncoating: specific reduction in the rate of membrane fusion between virus and liposomes by amantadine. J. Gen. Virol. 75: 945–948.

White, J.M. 1992. Membrane fusion. Science 258: 917–924.

White, J.M. 1994. Fusion of influenza virus in endosomes: role of the hemagglutinin. In Cellular Receptors for Animal Viruses (Wimmer, E., Ed.), pp. 281–301. Cold Spring Harbor Laboratory Press, Cold Spring Harbor, NY.

White, J.M. 1995. Membrane fusion: the influenza paradigm. CSHSQB 60: 581–588.

White, J.M. and Wilson, I.A. 1987. Anti-peptide antibodies detect steps in a protein conformational change: low-pH activation of the influenza virus hemagglutinin. J. Cell Biol. 105: 2887–2896.

Wiley, D.C. and Skehel, J.J. 1987. The structure and function of the hemagglutinin membrane glycoprotein of influenza virus. Annu. Rev. Biochem. 56: 365–394.

Wiley, D.C., Wilson, I.A. and Skehel, J.J. 1981. Structural identification of the antibody-binding sites of Hong Kong influenza haemagglutinin and their involvement in antigenic variation. Nature 289: 373–378.

Wilson, I.A. and Cox, N.J. 1990. Structural basis of immune recognition of influenza virus hemagglutinin. Annu. Rev. Immunol. 8: 737–771.

Wilson, I.A., Skehel, J.J. and Wiley, D.C. 1981. Structure of the haemagglutinin membrane glycoprotein of influenza virus at 3 A resolution. Nature 289: 366–372.

Zimmerberg, J., Vogel, S.S. and Chernomordik, L.V. 1993. Mechanisms of membrane fusion. Annu. Rev. Biophys. Biomol. Struct. 22: 433–466.

4.

Rhinovirus Attachment and Cell Entry

MICHAEL G. ROSSMANN, JEFFREY M. GREVE,

PRASANNA R. KOLATKAR, NORMAN H. OLSON,

THOMAS J. SMITH, MARK A. MCKINLAY,

& ROLAND R. RUECKERT

Viral Receptors

Unlike plant viruses, most animal, insect, and bacterial viruses attach to specific cellular receptors that, in part, determine host range and tissue tropism. Viruses have adapted themselves to utilize a wide variety of cell-surface molecules as their receptors, including proteins, carbohydrates, and glycolipids. Some viruses recognize very specific molecules (e.g., a large group of rhinoviruses recognize intercellular adhesion molecule-1 [ICAM-1]), whereas other viruses recognize widely distributed chemical groups (e.g., influenza viruses recognize sialic acid moieties; see Chapter 3). The tissue distribution of the receptor will in part determine the tropism of the virus and, hence, the symptoms of the infection. Similarly, species differences between receptor molecules can limit host range. For instance, only humans and apes have been shown to be susceptible to rhinovirus infections, a property correlated to the inability of human rhinoviruses to bind to the receptor ICAM-1 molecule in other species.

Although sequence alignments and similarities of picornavirus genomes show that these viruses have evolved from a common ancestor (Palmenberg, 1989), nevertheless they recognize a variety of receptors (Table 4.1). Possibly the primordial virus had the ability to weakly bind to a large number of different molecules. With time, different viruses evolved that became progressively more efficient and specialized toward recognizing one particular molecule as a way of infecting specific cells. Indeed, the grouping of viruses might suggest such a scenario. Thus, all polioviruses appear to recognize the same

TABLE 4.1 Receptors Families for Picornaviruses based on Virus Competition for Cell Receptors

Virus	Receptor Molecule	Receptor Family	References
Human rhinovirus major group 78 serotypes, including 3, 5, 9, 12, 14 15, 22, 32, 36, 39, 41, 51, 58, 59, 60 66, 67, 89	ICAM-1	Immunoglobulin (Ig) (5 Ig domains)	Abraham and Colonno (1984), Greve et al. (1989), Staunton et al. (1989)
Human rhinovirus minor group 11 serotypes, including 1A, 2, 44, 49	Low-density lipoprotein receptor (LDLR)	LDLR	Abraham and Colonno (1984), Hofer et al. (1994)
Polioviruses	Poliovirus receptor	Ig (3 Ig domains)	Mendelsohn et al. (1989)
Coxsackievirus A13, 18, 21	ICAM-1	Ig (5 Ig domains)	Colonno et al. (1986), Roivainen et al. (1991)
Coxsackievirus A2, 5, 13, 15, 18	?	?	Colonno et al. (1986), Roivainen et al. (1991), Schultz and Crowell (1983)
Echovirus 1	VLA-2	Integrin	Bergelson et al. (1992)
Echovirus 6	?	?	Crowell (1966)
Foot-and-mouth disease viruses, types $A_{12}119$, O_{1B}, C_{3Res}, SAT_{1-3}	RGD integrin	Integrin	Sekiguchi et al. (1982), Mason et al. (1993)
Mengo virus	?	Glycophorin (?)	Burness (1981), Burness and Pardoe (1983)

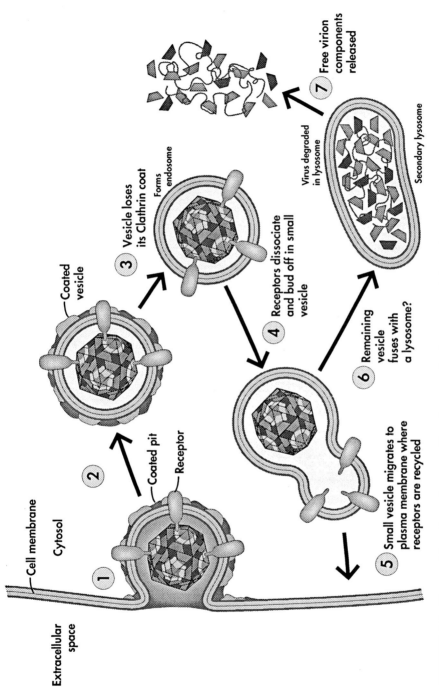

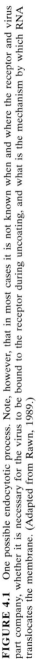

FIGURE 4.1 One possible endocytotic process. Note, however, that in most cases it is not known when and where the receptor and virus part company, whether it is necessary for the virus to be bound to the receptor during uncoating, and what is the mechanism by which RNA translocates the membrane. (Adapted from Rawn, 1989.)

receptor, and most coxsackie A viruses recognize their own receptor, whereas coxsackie B viruses recognize another receptor. Yet, surprisingly, rhinovirus serotypes can be divided into three groups that recognize different receptors (Abraham and Colonno, 1984; Uncapher et al., 1991). Furthermore, the receptor for the major group of rhinoviruses, ICAM-1, belongs to the immunoglobulin superfamily (Greve et al., 1989; Staunton et al., 1989), whereas the receptor for the minor group has been reported to be the low density lipoprotein receptor (Hofer et al., 1994).

Receptor binding is only the first, albeit essential, step in the infection process. The virus, or the virus genome alone, then has to enter the cell, a process that requires translocation of the viral genome or a subviral particle across the membrane into the cytoplasm, and, in some cases, into the nucleus. Because delivery of the viral genome into the cell involves major rearrangements of the capsid structure, entry must be a tightly regulated process that is triggered at the cell. The limiting membrane may be at the cell surface (as in the case of paramyxoviruses and herpesviruses) or in the endocytotic pathway (Fig. 4.1) (as in the case of Semliki Forest virus and influenza virus). The mechanism by which protein-encapsidated viruses such as picornaviruses (Rueckert, 1990) enter the cytoplasm has not been elucidated, but must differ significantly in detail from the membrane-fusion strategy demonstrated by enveloped viruses in that RNA must be translocated through the membrane, possibly through a protein pore formed by capsid proteins. Several reports indicate that rhinovirus entry requires a low-pH step, suggesting entry through the endosomal pathway. Similar data exist for foot-and-mouth disease virus (FMDV) and poliovirus, although it is not universally accepted that poliovirus enters through the endosomal pathway (Gromeier and Wetz, 1990).

Rhinovirus Structure and the Canyon Hypothesis

The genus *Rhinovirus* is composed of a group of over 100 serologically distinct viruses, which are a major cause of the common cold in humans (Rueckert, 1990). These viruses belong to the picornavirus family, which also contains the genera *Enterovirus*, *Aphthovirus*, *Cardiovirus*, and hepatitis A virus. The picornaviruses are small, icosahedral, nonenveloped, single-strand RNA viruses. X-ray crystal structures have been determined for at least one member in each picornavirus genus except for hepatitis A viruses (Hogle et al., 1985; Rossmann et al., 1985; Luo et al., 1987; Acharya et al., 1989; Filman et al., 1989; Kim et al., 1989). Polioviruses (genus *Enterovirus*) (see Chapter 6) are structurally the most similar to rhinoviruses. Unlike the enteroviruses, rhinoviruses are unstable below pH 6. The infectious virion has a molecular mass of about 8.5×10^6 Da and an external diameter of around 300 Å.

Each of the 60 icosahedral protomers in picornaviruses contains four viral polypeptides, VP1 to VP4. VP1, VP2, and VP3 reside on the exterior of the virus and make up its protein shell (Fig. 4.2). These three peptides, each having a molecular mass of roughly 35 kDa, contain a common eight-stranded, anti-

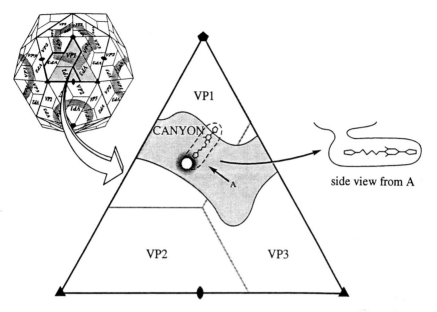

FIGURE 4.2 Diagrammatic view of picornavirus with enlargement of one icosahedral asymmetric unit showing the outline of the canyon and the entrance to the WIN pocket. The terms "north" (top) and "south" rims of the canyon refer to this standard orientation. The 6S protomeric assembly unit (which differs from the geometric definition of the asymmetric unit) is shown in heavy outline on the icosahedron. (Reprinted with permission from Oliveira et al. (1993). Copyright by Current Biology Ltd.)

parallel, β-barrel motif (Rossmann et al., 1985) (Fig. 4.3). Their amino termini intertwine to form a network on the interior of the protein shell. Five VP3 amino termini form a five-stranded helical β-cylinder on the virion's interior about each icosahedral fivefold axis (Fig. 4.4). This β-cylinder stabilizes the pentamer and is thought to be important for its assembly (Hogle et al., 1985; Arnold et al., 1987).

VP4 is smaller than the other viral polypeptides and resides inside the virion's protein shell. VP4 is lost from the capsid as a result of virus uncoating, although the specific role of VP4 in uncoating or entry has not been elucidated. A mutant of human rhinovirus serotype 14 (HRV14) defective in VP4–VP2 cleavage (Lee et al., 1993) is able to bind to receptor and undergo cell-induced conformational transitions but is unable to initiate a new round of replication, suggesting that mature VP4 is required for targeting of RNA within the cell. The amino terminus of VP4 is myristylated, which may promote its association with lipid membranes during viral assembly or uncoating (Chow et al., 1987). In poliovirus (see Chapter 6), the myristylate moiety lies inside the virion coat close to the β-cylinder. The first 25 to 28 amino-terminal residues of VP4 are disordered in the rhinovirus structures, but a density consistent with myristylate is seen internally near the center of the pentamer in rhinoviruses 14, 1A, and 16 (Kim et al., 1989; Arnold and Rossmann, 1990; Oliveira et al., 1993) (Fig. 4.4).

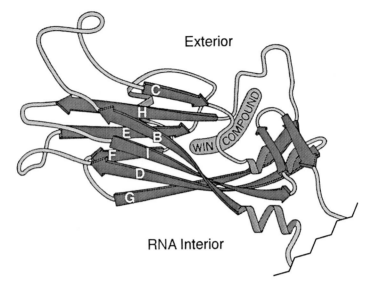

FIGURE 4.3 Schematic representation of the VP1 fold of HRV14. The folding topology of the two sheets ''BIDG'' and ''CHEF'' is the same in VP2 and VP3 as well as in most other viral capsid proteins. The binding site of antiviral WIN compounds within the hydrophobic interior of VP1 is also shown.

Each of the three larger capsid proteins has various insertions between the β-strands of the basic folding motif. These insertions mostly decorate the viral exterior and form ''puffs'' and loops that are hypervariable and have been shown to be the binding site of neutralizing antibodies (Rossmann et al., 1985; Sherry and Rueckert, 1985; Sherry et al., 1986; see also Chapter 5). The surfaces of rhinoviruses (and polioviruses) contain a series of remarkably deep crevices or ''canyons,'' unlike anything observed in plant virus structures. The canyon is formed roughly at the junction of VP1 (forming the ''north'' rim) with VP2 and VP3 (forming the ''south'' rim). The GH loop in VP1 (often referred to as the ''FMDV loop'' because of its immunodominance in the homologous FMDV structure) forms much of the floor of the canyon. Together with the carboxyl termini of VP1 and VP3, the GH loop of VP1 also forms much of the ''south'' rim of the canyon.

It was hypothesized (Rossmann et al., 1985) that the canyon (one around each fivefold vertex; Figs. 4.2 and 4.4) in HRV was the site of receptor attachment, largely inaccessible to the broad antigen binding region seen in antibodies. Thus, residues in the lining of the canyon, which should be resistant to accepting mutations that might inhibit receptor attachment, would avoid presenting an unchanging target to neutralizing antibodies. Indeed, the neutralizing immunogenic sites that had been mapped by escape mutations were not in the canyon, but on the most exposed and variable parts of the virion both in HRV (Rossmann et al., 1985; Sherry and Rueckert, 1985; Sherry et al.,

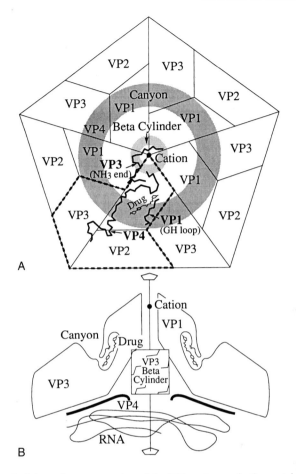

FIGURE 4.4 *A*. Schematic representation of 1 of 12 pentamers in the protein shell of rhino-viruses. A single cation, probably Ca^{2+}, is located at the center on the outer surface. The large shaded region is the canyon, a 12-Å-deep depression on the surface of the virus, encircling the fivefold axis. The bold lines indicate structural features (GH loop in VP1 residues 217 to 225, VP3 residues 1 to 20, and VP4 residues 29 to 65) belonging to one protomer (dashed outline), which may be important in the uncoating process. VP4, located on the interior surface of the protomer, has its amino end near the β-cylinder. The β-cylinder (see also *B*) spans the interior third of the capsid and is composed of five pentamerically related amino ends of VP3 encircling the fivefold axis. The GH loop of VP1 forms part of the floor of the canyon and also the ceiling of the drug binding pocket. *B*. Cross section containing the fivefold axis showing some of the features described in *A*. (Reprinted with permission from Giranda et al. (1992). Copyright by the National Academy of Sciences.)

1986) and in poliovirus (Hogle et al., 1985; Page et al., 1988). The ''canyon hypothesis'' suggests that one strategy for viruses to escape the host's immune surveillance is to protect the receptor attachment site in a surface depression (Fig. 4.5). Similar depressions related to host cell attachment have also been found on the surface of the hemagglutinin spike of influenza virus (Wilson et al., 1981; Weis et al., 1988; see also Chapter 3).

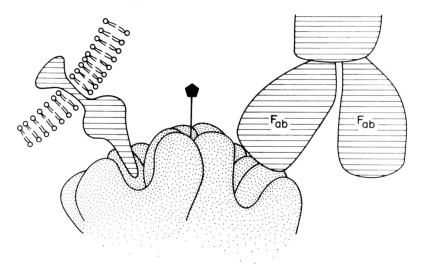

THE CANYON HYPOTHESIS

FIGURE 4.5   The presence of depressions on the picornavirus surface suggests a strategy for the evasion of immune surveillance. The dimensions of the putative receptor binding sites, the "canyon," sterically hinder an antibody's (top right) recognition of residues at the base of the site, while still allowing recognition and binding by a smaller cellular receptor (top left). This would allow conservation of receptor specificity while permitting evolution of new serotypes by mutating residues on the viral surface, outside the canyon. (Reprinted with permission from Luo et al. (1987). Copyright by the American Association for the Advancement of Science.)

A number of lines of evidence emerged to support the canyon hypothesis. First, a comparison of the variability of surface-exposed residues between a number of picornaviruses indicated that amino acid residues lining the canyon are significantly more conserved than other surface-exposed residues (Rossmann and Palmenberg, 1988; Chapman and Rossmann, 1993). Second, the hypothesis rationalized the contrast between many vertebrate virus structures and plant viruses. That is, animal viruses tend to have surface depressions, with a notable exception in FMDV (Acharya et al., 1989), but viruses whose hosts do not have immune systems tend to have smooth surfaces or protrusions on their surfaces. Third, site-directed mutagenesis of HRV14 indicated that modification of several amino acid residues located in the base of the canyon has an impact upon virus–receptor affinity (Colonno et al., 1988). Specifically, mutants with substitutions at residues 1273,[1] 1223, 1103, and 1220 exhibited an alteration in virus–receptor affinity. Fourth, certain capsid-binding "WIN" antiviral compounds (see Chapter 16) block the binding of some of the major

[1]Residues are numbered sequentially for each of VP1, VP2, VP3, and VP4, but start at 1001, 2001, 3001, and 4001, respectively.

receptor rhinoviruses, including HRV14 (Pevear et al., 1989). These compounds bind to many picornaviruses in a hydrophobic pocket located under the canyon floor (Fig. 4.2) and, in most cases, block virus from uncoating once inside the cell (Smith et al., 1986; Badger et al., 1988; Kim et al., 1993). Upon binding to HRV14, a conformational change occurs in the roof of the pocket, which is also the floor of the canyon (Fig. 4.6). Several amino acid residues are displaced by as much as 4 Å. These findings suggested that the conformational changes at the base of the canyon prevent viral attachment to cells.

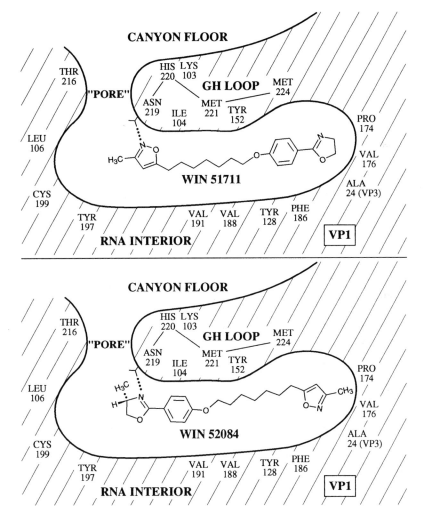

FIGURE 4.6 Schematic representation of the binding of the antiviral agents WIN 51711 and 52084 into a pocket underneath the canyon in HRV14. This causes enlargement of the pocket and conformational changes in the floor of the canyon, inhibiting attachment of the virus to HeLa cells in some cases, and also increasing the stability of the virus in all cases. (Reprinted with permission from Dutko et al. (1989). Copyright by Springer-Verlag New York Inc.)

Although the observations for rhinovirus were consistent with the canyon being the receptor binding site, they did not provide conclusive proof nor did they identify a complete footprint of the receptor on the virus surface. A structural and sequence comparison between HRV14 (major receptor) and HRV1A (minor receptor) indicated significant differences between these two viruses in the shape of the base of the canyon and, most importantly, the charge distribution (Kim et al., 1989; Chapman and Rossmann, 1993). Clearly, the basis of recognition between virus and receptor is complex, and elucidation of their contacts required a more direct approach.

Binding of ICAM-1, the Major Group Rhinovirus Receptor, to Virus Surface

There are at least 78 serotypes (Tomassini et al., 1989) that bind to ICAM-1, the major group rhinovirus receptor (Greve et al., 1989; Staunton et al., 1989). The ICAM-1 molecule has five immunoglobulin-like domains (D1 to D5, numbered sequentially from the amino end), a transmembrane portion, and a small cytoplasmic domain (Simmons et al., 1988; Staunton et al., 1988). Domains D2, D3, and D4 are glycosylated (Fig. 4.7). Unlike immunoglobulins, ICAM-1 appears to be monomeric (Staunton et al., 1989). Mutational analysis of ICAM-1 has shown that domain D1 contains the primary binding site for rhinoviruses as well as the binding site for its natural ligand, lymphocyte function–associated antigen-1 (LFA-1) (Lineberger et al., 1990; Staunton et al., 1988, 1990; McClelland et al., 1991). Other surface antigens within the immunoglobulin superfamily that are used by viruses as receptors include CD4 for human immunodeficiency virus type 1 (Dalgleish et al., 1984; Klatzmann et al., 1984; Maddon et al., 1986; Robey and Axel, 1990), the poliovirus receptor (Mendelsohn et al., 1989), and the mouse coronavirus receptor (Williams et al., 1991). In ICAM-1, in the poliovirus receptor (Freistadt and Racaniello, 1991; Koike et al., 1991), and in CD4 (Arthos et al., 1989), the primary receptor–virus binding site is domain D1. The structures of the two amino-terminal domains of CD4 have been determined to atomic resolution (Ryu et al., 1990; Wang et al., 1990; Brady et al., 1993). Truncated proteins corresponding to the two amino-terminal domains of ICAM-1 (D1D2, consisting of 185 amino acids) as well as the intact extracellular portion of ICAM-1 (D1 to D5, consisting of 453 amino acids) have been expressed in CHO cells (Greve et al., 1991). The desialated form of D1D2 has been crystallized (Kolatkar et al., 1992).

The structure of the complexes of D1D2 with HRV16 (Olson et al., 1993) and with HRV14, and of D1D5 with HRV16 (P. R. Kolatkar, N. H. Olson, C. Music, J. M. Greve, T. S. Baker and M. G. Rossmann, unpublished results), has been determined using cryoelectron microscopy and image reconstruction procedures (see Color Plate 2). The position of the ICAM-1 molecule relative to the icosahedral symmetry axes of the virus is unambiguous and shows the receptor binding into the canyon (see Color Plate 3). Each D1D2 molecule has

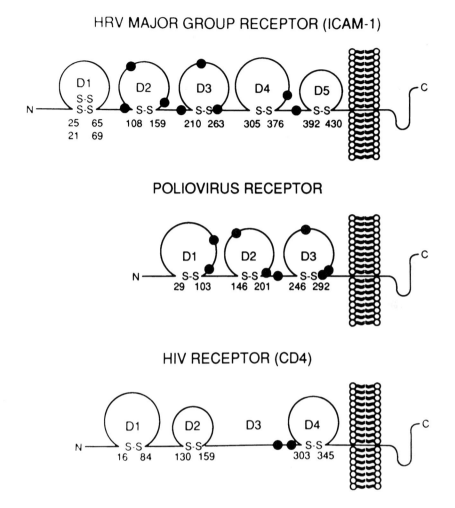

FIGURE 4.7 Schematic diagram of viral receptors. The relative size and distribution of immunoglobulinlike domains are shown. The black circles show the position of potential glycosylation sites. Numbers indicate the amino acid positions of Cys residues involved in predicted disulfide (S–S) bridges. (Reprinted with permission from Colonno (1992). Copyright by Academic Press Limited.)

an approximate dumbbell shape, consistent with the presence of a two-domain structure. A difference map between the electron microscopic density and the 20-Å resolution HRV16 or HRV14 densities confirmed that the D1D2 molecule binds to the central portion of the canyon roughly as predicted by Giranda et al. (1990). There are some small differences in orientation of D1D2 when complexed to HRV16 or HRV14 that may relate to the change in length of the VP1 BC loop forming the north rim of the canyon. The D1D2 ICAM fragment is oriented roughly perpendicular to the viral surface and extends to a radius of about 205 Å. Its total length is about 75 Å.

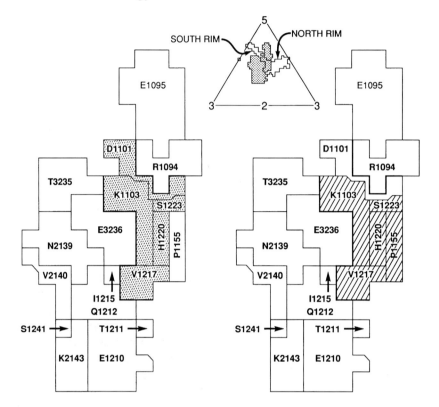

FIGURE 4.8 *Top.* View of the icosahedral asymmetric unit bounded by adjacent five- and threefold axes, outlining residues on the HRV14 surface. The limits of the canyon are shown, arbitrarily demarcated by a 138-Å radial distance from the viral center (Rossmann and Palmenberg, 1988). Residues under the ICAM-1 footprint are stippled. Improved resolution of the electron density could only marginally alter the HRV residues at the virus–receptor interface. *Left* and *Right.* Enlarged view of the residues in the ICAM-1 footprint showing the residues (hatched areas) that, when mutated, affect viral attachment (*right*) (Colonno et al., 1988), and the residues (stippled areas) altered in structure by the binding of antiviral compounds that inhibit attachment and uncoating (*left*) (Smith et al., 1986). (Reprinted with permission from Olson et al. (1993). Copyright by the National Academy of Sciences.)

An inability to produce domain D1 in isolation and the sequence alignment between ICAM-1 and CD4 suggested that domains D1 and D2 of ICAM-1 are intimately associated through a common, extended β-strand, as is seen in the structure of CD4 (Ryu et al., 1990; Wang et al., 1990). Confirmation of structural similarity between D1D2 of ICAM-1 and CD4 was also shown by means of a cross-rotation function between the known structure of D1D2 for CD4 and the crystal diffraction data for ICAM-1 D1D2 (P. R. Kolatkar and M. G. Rossmann, unpublished results). Thus, it seemed reasonable to use the known structures of CD4 for fitting the reconstructed density map (see Color Plate 2), although there was slightly too little density for domain D1 and too much density for D2. A better assessment of the fit of domain D1 to the density was obtained by taking the predicted D1 structure of ICAM-1, including all side chains, and superimposing it onto the fitted Cα backbone of CD4. One major difference is that, although domain D1 of CD4 resembles a variable, immunoglobulin-like domain with two extra β-strands, the ICAM-1 prediction is based on a more likely analogy to an immunoglobulin constant domain. This gives domain D1 of ICAM-1 a sleeker appearance, consistent with the observed difference density. The extra density in D2 (in the region farthest away from the virus) compared with domain D2 of CD4 is probably due to the four associated carbohydrate groups located in this region.

The footprint of ICAM-1 onto the HRV14 structure (Fig. 4.8) correlates very well with Colonno's mutational studies of residues in the canyon that alter affinity of the virus to HeLa cell membranes (Colonno et al., 1988). All the residues are part of the canyon floor and lie centrally within the footprint of the D1D2 molecule binding site. Similarly, there is excellent agreement between the ICAM-1 footprint and residues on the virus surface whose conformation is changed by antiviral agents (Smith et al., 1986; Heinz et al., 1989; Pevear et al. 1989; see also Chapter 16).

Properties of ICAM-1

Immunoglobulinlike domains consist of seven β-strands (βA to βG) arranged into two β-sheets that form a β-sandwich (see Color Plate 3). The sequence of the first domain of ICAM-1 (D1) has two unusual features for an immunoglobulinlike domain: 1) it is relatively short, being 88 residues instead of the more typical size of approximately 100 residues; and 2) instead of the typical two cysteine residues, located in the βB strand and the βF strand, there are four cysteines (Fig. 4.7). The βB and the βF cysteines usually participate in an intrachain disulfide bond across the β-sandwich in most members of the immunoglobulin supergene family. However, the additional two cysteine residues in ICAM-1 D1 have an $i + 4$ spacing relative to Cys 21 and Cys 65, which in a β-strand would place them in proper register for forming a second disulfide bond between the βB and βF strands. These sequence features are conserved in domains present in a small family of cell surface receptors that are related by sequence homology and by their ability to serve as ligands for various members of the heterodimeric integrin receptor family. ICAM-1, ICAM-2, and ICAM-3 all serve as counter-receptors for the integrin receptor LFA-1 (Springer, 1990). The amino acid sequences of ICAM-2 D1, ICAM-3 D1, the vascular cell adhesion molecule-1 (VCAM-1) D1, and VCAM-1 DAS are all closely related to ICAM-1 D1, and all possess the novel four-cysteine sequence motif. Because these domains all have been either implicated or directly shown to contain the binding sites for their integrin counter-receptors, it is possible that ICAM-1 D1 represents a prototype for an integrin binding domain. The three-dimensional structure of another integrin binding domain, the fibronectin type III domain (Leahy et al., 1992), has indicated that it also possesses an immunoglobulin-type fold. The tripeptide sequence Arg-Gly-Asp (RGD), which contains the critical site for recognition by many integrin recep-

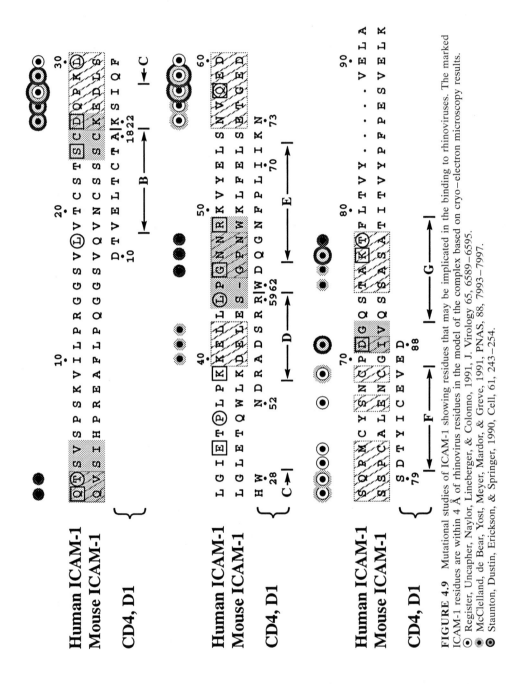

FIGURE 4.9 Mutational studies of ICAM-1 showing residues that may be implicated in the binding to rhinoviruses. The marked ICAM-1 residues are within 4 Å of rhinovirus residues in the model of the complex based on cryo–electron microscopy results.
⊙ Register, Uncapher, Naylor, Lineberger, & Colonno, 1991, J. Virology 65, 6589–6595.
⦿ McClelland, de Bear, Yost, Meyer, Mardor, & Greve, 1991, PNAS, 88, 7993–7997.
◉ Staunton, Dustin, Erickson, & Springer, 1990, Cell, 61, 243–254.

118

tors, is exposed on the loop between the βF and the βG strands (Leahy et al., 1992; Main et al., 1992). The RGD sequence is also found in adenovirus (see Chapter 8). A number of the mutations that reduce rhinovirus binding to ICAM-1 also cluster in the βF-βG loop (Fig. 4.9).

The parts of the predicted ICAM-1 structure that contact HRV14 or HRV16 are the amino-terminal four residues and loops BC (residues 24 to 26), DE (residues 45 to 49), and FG (residues 71 and 72; see Fig. 4.3 for nomenclature). Staunton et al. (1990), McClelland et al. (1991), and Register et al. (1991) have examined the effects of a number of site-directed mutations and mouse–human substitutions in domain D1 of ICAM-1 on rhinovirus binding. On the basis of these reports, seven regions in D1, corresponding roughly to the amino terminus (residues 1 and 2), loop BC (residues 26 to 29), strand D (residues 40 and 43), loop DE (residues 46 to 48), strand F (residue 67), loop FG (residues 70 to 72), and the G strand (residues 75 to 77), have been implicated in virus binding (Fig. 4.9). There is correspondence to, or significant overlap between, the four regions of ICAM-1 seen here to be in contact with rhinovirus and four of the seven regions identified by site-directed mutagenesis. Thus, although there appears to be reasonable agreement between the mutational studies of ICAM-1 and the observed virus–receptor contacts of the complex, what constitutes the binding site remains cryptic. Mutational analyses, the structure determination of ICAM-1, and the high-resolution structure of ICAM-1 and a major group rhinovirus are required for clarification.

Virus Entry and Uncoating

Productive viral uncoating requires that the RNA move from inside the viral protein shell, through a cellular membrane, into the cytosol. Such displacement probably requires large conformational changes in the rhinovirus coat. For poliovirus (see Chapter 6) or rhinovirus, acidification of endosomes may be required for an infection to proceed normally as measured by either progeny virus production or cytopathic effects (Madshus et al., 1984a, 1984b; Zeichhardt et al., 1985; Neubauer et al., 1987; Gromeier and Wetz, 1990), although Pérez and Carrasco (1993) concluded that acidification is not essential.

Rhinovirus and poliovirus 149S infectious virions undergo several progressive transformations (Lonberg-Holm and Korant, 1972; Everaert et al., 1989) when bound to cells (Fig. 4.10), which can be followed by sedimentation through sucrose gradients. The 149S virions are initially converted to 135S to 125S particles, which have lost VP4 but retain RNA (altered or ''A''-particles). Subsequently, the RNA is released with the formation of 80S empty capsids as well as small capsid fragments.

A-particles have a number of properties that suggest a role in virus entry. They have been shown to be hydrophobic and bind to liposomes (Korant et al., 1975; Hoover-Litty and Greve, 1993). It has also been shown that the formation of poliovirus A-particles is associated with externalization of the amino terminus of VP1 and that removal of approximately 30 residues from

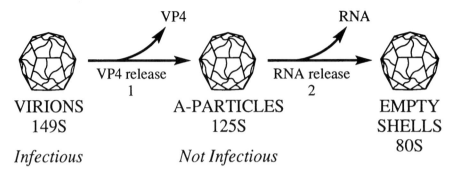

VP4   RNA

VP4 release   RNA release
1   2

VIRIONS   A-PARTICLES   EMPTY
149S   125S   SHELLS
                                               80S
Infectious   *Not Infectious*

FIGURE 4.10   Two steps in the uncoating of picornaviruses. The first step, release of VP4, is mediated by interaction with viral receptor, by decreasing pH, or by heating to 52 °C. The second step, release of RNA, may be caused by acidification of the membrane-bound particle. (Reprinted with permission from Giranda et al. (1992). Copyright by the National Academy of Sciences.)

the amino terminus of VP1 by proteolysis abolishes the ability of poliovirus to bind to liposomes (Fricks and Hogle, 1990). The sequence of the amino-terminal 23 residues of VP1 suggests that it could form an amphipathic α-helix and, thus, could promote interactions with lipid bilayers. Nevertheless, A-particles are not infectious, do not bind to receptor, and are capable of translocating RNA across a membrane. Thus, the A-particle may be either a fusion-competent uncoating intermediate or a nonproductive side product in uncoating (Madshus et al., 1984b; Gromeier and Wetz, 1990).

A-like particles can be generated under certain conditions in vitro for rhino- and polioviruses (Koike et al., 1992; Hoover-Litty and Greve, 1993; Yafal et al., 1993). HRV14 incubated at pH 5 to 6, the pH likely to be found in endosomes, is converted to 135S A-particles. HRV14 incubated with soluble ICAM-1 is converted, through a virus−receptor complex intermediate, to 80S empty capsids, suggesting that receptor binding can destabilize the virion (Hoover-Litty and Greve, 1993). However, the efficiency of this process is low and varies considerably from serotype to serotype; for instance, HRV16 is quite stable when complexed with soluble receptor. It is clear that uncoating per se is not sufficient for successful entry into cells; Lee et al. (1993) have shown that an HRV14 mutant deficient in VP4−VP2 cleavage can be uncoated in cells but is not infectious, suggesting that the released RNA is not targeted to the correct intracellular compartment. Presumably, low-pH-catalyzed uncoating of virus in endosomes while the virus is anchored close to the membrane is required for successful uncoating.

Because the conformational changes required for uncoating that occur on acidification are probably similar to those that occur on viral interaction with a receptor, a structural determination of these changes could be useful. It has been possible to study the initial changes that occur in wild-type HRV14 crystals upon lowering the pH by using a very-high-intensity synchrotron X-ray source (Giranda et al., 1992), permitting the rapid recording of the diffraction pattern before the crystals completely disintegrated. It was found that an ion binding site on the icosahedral fivefold axes (Fig. 4.4), the interior of the virus shell near the fivefold axes including the amino-terminal residue of VP3, much of the ordered part of VP4, and the GH loop of VP1 all became disordered. Furthermore, the magnitude of the disorder increased as the time of acid exposure increased. An expansion of the β-cylinder (even beyond the first residue) and cation release, therefore, may be among the first events permitting eventual escape of VP4s, possibly along the fivefold axial channels. There are parallels to this process in the externalization of VP1 through the fivefold axial channels of canine parvovirus (Tsao et al., 1991) and the ejection of single-strand DNA through the fivefold ion channel of φX174 (McKenna et al., 1992). An alternative proposal made by Fricks and Hogle (1990) based on mutational analyses suggests that the first step in uncoating and the externalization of VP1 is a weakening of the contacts between the assembly protomeric unit.

Inhibition of Uncoating and the Pocket Factor

The capsid-binding, antiviral agents such as the WIN compounds bind into a hydrophobic pocket in VP1 below the canyon floor (see Chapter 16). Not only do they inhibit attachment in HRV14 and other major group rhinoviruses, but they also stabilize major and minor group rhinoviruses in vitro to acidification (Gruenberger et al., 1991) and heat (Fox et al., 1986). HRV14 differs from other picornaviruses in that its pocket is empty in the native structure. For example, there is electron density in the homologous pockets of poliovirus Mahoney 1, poliovirus Sabin 3, and a chimera of poliovirus 2 (Hogle et al., 1985; Filman et al., 1989; Yeates et al., 1991; see also Chapter 6). This density has been interpreted as a sphingosine- or palmitate-like molecule because of the hydrophobic nature of the pocket and the polar environment at one end of the pocket. Similarly, the somewhat smaller electron density in the pocket of HRV1A (Kim et al., 1989, 1993) and HRV16 (Oliveira et al., 1993) has been tentatively interpreted as a fatty acid, eight or more carbon atoms long (Fig. 4.11). Smith et al. (1986) imply, and Filman et al. (1989) explicitly propose, that the pocket might be the site for binding of a cellular pocket factor to regulate viral assembly and uncoating.

Binding of WIN compounds to HRV14 causes major conformational changes in the pocket and, hence, also in the canyon floor (the receptor attachment site). These changes were correlated to inhibition of attachment in the presence of the antiviral compounds (Heinz et al., 1989; Pevear et al., 1989). In contrast, in HRV1A (a minor receptor group virus) and polioviruses, where the WIN compounds merely displace the pocket factor without a correspondingly large conformational change, there is inhibition of uncoating but not of attachment. Preliminary results suggested that rhinoviruses of the minor receptor group exhibited no inhibition of attachment, whereas those of the major receptor group behaved like HRV14, for which attachment is inhibited.

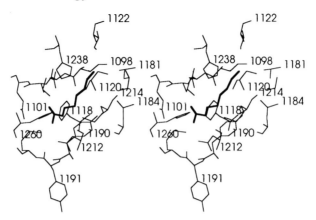

FIGURE 4.11   Stereo view of the putative fatty acid in the hydrophobic interior of VP1 in its atomic environment. The view is from outside the HRV16 virus and is similar to that shown in Figure 4.2. (Reprinted with permission from Oliveira et al. (1993). Copyright by Current Biology Ltd.)

Thus, it was a surprise to find ''cofactor'' electron density in the HRV16 pocket that altered the shape of the pocket to resemble that of the ''WIN-filled'' form of HRV14 (Kim et al., 1989, 1993). In HRV16, the height for the density of the fatty acid is comparable to that of amino acid side chains, indicating that most pockets are fully occupied. However, the height decreases beyond the sixth carbon atom, suggesting that the density might represent a mixture of fatty acids 6, 8, or 10 carbon atoms long.

In HRV1A and HRV16, the more active antiviral compounds tend to have an aliphatic chain five or fewer carbon atoms long (Mallamo et al., 1992), correlating with the available space within the binding pocket (Diana et al., 1990, 1992; Kim et al., 1993). In HRV14, the most active antiviral agents tend to be longer, with seven-carbon aliphatic chains. For example, WIN 56291 contains an aliphatic chain of only three carbons and is equally active against HRV16 and HRV1A, but less active against HRV14. Thus, for each serotype there is an optimal drug size that displays the greatest activity and binding affinity (Diana et al., 1990, 1992) and best fills the volume of the pocket. It follows that the smaller pocket factors, which can be easily displaced by WIN compounds in HRV16 and HRV1A (Kim et al., 1993; Oliveira et al., 1993), bind with less affinity than the antiviral compounds. Nevertheless, the pocket factors seen in the electron densities remain in the pocket even after extensive dialysis of the virus sample. The WIN compounds have a binding constant comparable to their minimal inhibitory concentrations of approximately 10^{-8} mol/l (Fox et al., 1986, 1991).

The Role of the Pocket Factor

The natural pocket factors found in HRV16, HRV1A, and polioviruses, like WIN compounds, increase the thermal stability of the virus (Heinz et al., 1990) by filling an internal hydrophobic cavity (Eriksson et al., 1992a,b). Similarly, drug-dependent mutants of poliovirus depend on WIN compounds to maintain their stability (Mosser and Rueckert, 1993). The pocket factor may therefore be required to stabilize the virus in transit from one cell to another. However, the delivery of the infectious RNA into the cytoplasm must require a destabilizing step, which might be effected by expulsion of the pocket factor during receptor-mediated uncoating.

Because ICAM-1 binds to HRV14 and to HRV16, the shape of the canyon for HRV16 should be similar to that in HRV14 when ICAM-1 binding occurs. Because soluble ICAM-1 binds to purified HRV14, which does not contain any pocket factor, presumably the pocket is empty when ICAM-1 binds to HRV16 (Fig. 4.12). However, the structure of HRV16 shows the presence of a pocket factor in the purified virus (Oliveira et al., 1993). Hence, it must be assumed that the pocket factor is displaced before the receptor can seat itself into the canyon. In essence, there are two competing equilibria: the binding of ICAM-1 and the binding of the pocket factor to the virus. Although the sites of binding of ICAM-1 and of the pocket factor are not the same, they are in close proximity and interfere with each other. The floor of the canyon is also the roof of the pocket for the cofactor or WIN compounds. When ICAM-1 binds, the floor is depressed downward, which is possible only when there is nothing in the pocket. Conversely, when there is a compound in the pocket, its roof is raised upward. The displacement of the pocket factor per se does not cause the virus to fall apart. For instance, when HRV14 is crystallized, it does not contain a pocket factor, and the complex of HRV16 with ICAM-1 is reasonably stable. Nevertheless, the absence of pocket factor increases the potential for disruption by lowered pH or by formation of the receptor–virus complex.

Presumably, the destabilization of the virus on cell attachment is made possible by the displacement of a sufficient number of pocket factors when the receptor competes for the overlapping binding site. Progressive recruitment of receptors is then sufficient to trigger release of the VP4s. The terminal myristate moieties of VP4 and the exposure of the amino terminus of VP1 will permit entry through the cell membrane, possibly by creating a channel along the fivefold axes of the virus (Giranda et al., 1992).

The natural pocket factor, found in HRV16, is small and has fewer interactions in the drug binding pocket than do the larger WIN compounds. This is consistent with the observation that WIN compounds can displace the pocket factor. The relative and competing affinities of WIN compounds for the pocket and of ICAM-1 for the receptor binding site will determine the inhibitory properties of the drug for a particular rhinovirus serotype (Fig. 4.12).

A class of HRV14 drug-resistant (compensation) mutants can be selected by growing the virus in the presence of antiviral WIN compounds. Such mutants occur at a frequency of about $1:10^4$ virions. They have been shown to be mostly single mutations (Heinz et al., 1990; Shepard et al., 1993), and six of the seven characterized to date are situated near the walls and floor of the canyon. WIN compounds bind into the pocket of these mutant viruses and deform the canyon floor in a manner similar to their effect on wild-type viruses

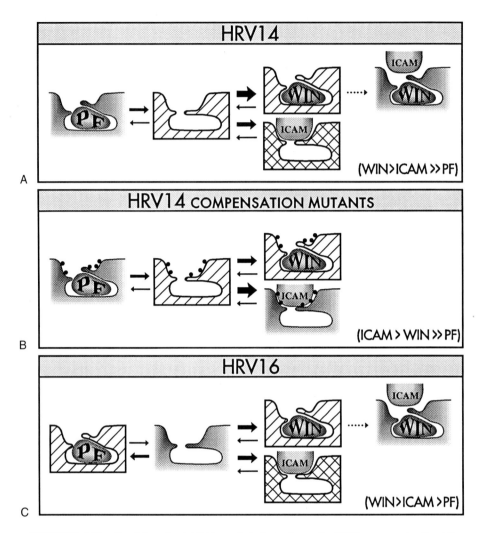

FIGURE 4.12 Conditions for inhibition of viral attachment by WIN compounds. Crystallographically and electron microscopically determined structures are shown with diagonal lines and cross-hatching, respectively, while hypothetical structures are stippled. *A.* In wild-type HRV14, the pocket factor (PF) binds weakly and is not observed in crystallographic studies. When WIN compounds bind into the pocket, they deform the roof of the pocket, which is also the floor of the canyon. This inhibits the attachment of the virus to the ICAM-1 receptor and, hence, presumably the binding affinity of WIN is greater than that of ICAM-1. When ICAM-1 recognizes the canyon floor, the putative pocket factor must be displaced by ICAM-1 and, hence, the binding affinity of ICAM-1 is greater than that of pocket factor. *B.* Drug-resistant compensation mutants of HRV14 cluster around the canyon walls and floor (●) and increase the affinity of ICAM-1 for the virus. Although WIN compounds can bind to the virus, they do not inhibit infectivity. Thus, the binding affinity of the mutant virus to ICAM-1 is greater than that of WIN. *C.* Wild-type HRV16 contains a pocket factor. This can be replaced by WIN compounds, which inhibit attachment. Hence, in this case the affinity of HRV16 for WIN is greater than that of ICAM-1, which is greater than that of pocket factor.

(Hadfield et al., 1995). In one of these mutants (Ser 1223 → Gly), the affinity of ICAM-1 for the virus is enhanced (M. P. Fox, D. C. Pevear and F. J. Dutko, unpublished data). Thus it is reasonable to conclude that ICAM-1 binds better to these mutant viruses than do the WIN compounds (Fig. 4.12*B*).

In the case of poliovirus or HRV1A (a minor group rhinovirus), only uncoating is inhibited by WIN compounds, and not attachment. If the pocket factor must be absent for the virus to uncoat, binding of receptor to these viruses should lead to displacement of the pocket factor, just as is the case for the major group rhinoviruses. Similarly, the WIN compounds must also be displaced by the receptor because there is no inhibition of attachment, thus requiring the remaining WIN compounds to stabilize the virus sufficiently to inhibit uncoating.

Conclusions

Receptor Attachment Site

The canyon hypothesis, which suggested that the receptor binding site can be hidden from immune surveillance in a ''canyon'' on the surface of the capsid, has been verified for the major group of rhinoviruses. Low-resolution images of soluble ICAM-1–virus complexes have shown that receptor binds directly into the canyon, and capsid residues have been identified under the footprint of ICAM. Capsid-binding compounds that bind to a pocket under the canyon prevent virus binding to receptor, presumably by inducing a conformational change in the canyon floor that precludes receptor binding. Although the amino acid residues lining the canyon are more conserved between rhinovirus serotypes than other surface-exposed residues, no clear relationship has been established between sequence and receptor specificity.

Mutational analyses indicate that the canyon is also the receptor attachment site for poliovirus (Racaniello, 1992). The receptor for the minor group of rhinovirus serotypes has been reported to be the low-density lipoprotein receptor (Hofer et al., 1994), a vastly different type of molecule than ICAM-1. It is yet to be determined whether the minor group of rhinoviruses also uses the canyon for receptor attachment. Sequence comparisons (Chapman and Rossmann, 1993) of the minor group HRV2 suggest that the site of attachment of receptor to HRV2 is different than the homologous site in the major group HRV14 (Duechler et al., 1993). That does not exclude the possibility that other parts of the canyon are involved in minor group rhinovirus recognition.

There may be alternative ways of hiding the receptor attachment site. For instance, there are two regions of FMDV that have been implicated in receptor binding. These are an RGD sequence in the GH loop, which is conserved in most strains of FMDV, and the carboxyl terminus of VP1 (Fox et al., 1989). Because the surface of FMDV is smoother than that of HRV14, Acharya et al. (1990) suggested that the RGD sequence might be protected from immune recognition by surrounding it with hypervariable residues. The RGD region is disordered in the native structure, but can be partially ordered by reduction of a disulfide in the GH loop (Parry et al., 1990), a conformational change that might also be required for receptor recognition. A similar order–disorder phenomenon, associated with receptor recognition, has been observed in Mengo virus (Luo et al., 1987; Kim et al., 1990). Conditions in Mengo virus crystals that order the GH loop in VP3 also correspond to conditions that permit the virus to bind and infect cells. In Mengo virus and also in FMDV, the regions implicated in receptor binding are both of variable sequence and disordered structure. This disorder might hinder the immune system from learning a recognizable epitope. Perhaps direct interaction with a host cell or the indirect effect of the chemical environment immediately surrounding the host cell might be required for the adoption of a less disordered structure ready for receptor recognition.

Virus Entry

A virus must be stable in the extracellular environment during transit between hosts, but also must be destabilized once it has bound to or entered the host cell, shedding its protein coat to allow infection to proceed. A number of conformational changes in the virion have been observed in infected cells, although the precise role of these changes and the molecular mechanism of translocation of the RNA into the cytoplasm is unknown. HRV can be destabilized in vitro by both low pH and binding to ICAM-1, both of which are required for infection in vivo. The amino terminus of VP1 in poliovirus and probably the myristate moiety at the amino end of VP4 become exposed after viral binding to receptor. The amino end of VP1 has some properties in common with a "fusion peptide." Some low-pH-induced conformational changes in HRV14 have been observed around the fivefold symmetry axes and in VP4. It is possible that one of the fivefold axes will be selected as the site at which the viral RNA exits the virion and enters the cell.

In rhinoviruses and polioviruses, the need for reversible stabilization appears to be fulfilled by the binding of a small cellular aliphatic molecule, called the "pocket factor," into a hydrophobic pocket in VP1. In the major group of rhinovirus serotypes, the binding site for ICAM-1, the virus receptor, overlaps with the binding site of the stabilizing pocket factor. Virus attachment is therefore a competition between two equilibria: 1) binding of the pocket factor into the pocket, and 2) binding of the receptor into the canyon. Provided that the receptor competes successfully with the pocket factor, many pocket factors will be lost as receptor molecules are recruited, destabilizing the virus as a prelude for uncoating (Oliveira et al., 1993).

Certain antiviral compounds also bind in the hydrophobic pocket, displacing the pocket factor. If the affinity of an antiviral compound for the pocket is higher than that of ICAM-1, the antiviral compound will prevent receptor attachment and uncoating. Drug escape mutations in VP1 that improve binding affinity for ICAM-1 can shift this balance, overcoming the antiviral effect (Fig. 4.12) (Oliveira et al., 1993).

Acknowledgments

We would like to thank Sharon Wilder for her careful work in the preparation of this manuscript, and Mark O'Neil and Robert Werberig for help with the production of some of the figures. We also thank our various foundations for financial support: P.R.K. was supported by a Jane Coffin Childs Foundation postdoctoral fellowship; N.H.O. by a National Institutes of Health research grant to Timothy S. Baker; M.G.R., R.R.R. and T.J.S. were all supported by their respective National Institutes of Health research grants; M.G.R. also received support from the Sterling-Winthrop Pharmaceuticals Research Division; and M.G.R. and R.R.R. obtained support from the Lucille P. Markey Foundation for the development of structural studies at Purdue University and the University of Wisconsin, respectively.

References

Abraham, G. and Colonno, R.J. 1984. Many rhinovirus serotypes share the same cellular receptor. J. Virol. 51: 340–345.

Acharya, R., Fry, E., Logan, D., Stuart, D., Brown, F., Fox, G. and Rowlands, D. 1990. The three-dimensional structure of foot-and-mouth disease virus. In New Aspects of Positive-Strand RNA Viruses (Brinton, M. A. and Heinz, F.X., Eds.), pp. 211–217. American Society for Microbiology, Washington, DC.

Acharya, R., Fry, E., Stuart, D., Fox, G., Rowlands, D. and Brown, F. 1989. The three-dimensional structure of foot-and-mouth disease virus at 2.9 Å resolution. Nature 337: 709–716.

Arnold, E. and Rossmann, M.G. 1990. Analysis of the structure of a common cold virus, human rhinovirus 14, refined at a resolution of 3.0 Å. J. Mol. Biol. 211: 763–801.

Arnold, E., Luo, M., Vriend, G., Rossmann, M.G., Palmenberg, A.C., Parks, G.D., Nicklin, M.J.H. and Wimmer, E. 1987. Implications of the picornavirus capsid structure for polyprotein processing. Proc. Natl. Acad. Sci. U.S.A. 84: 21–25.

Arthos, J., Dean, K.C., Chaikin, M.A., Fornwald, J.A., Sathe, G., Sattentau, Q.J., Clapham, P.R., Weiss, R.A., McDougal, J.S., Pietropaolo, C., Axel, R., Truneh, A., Maddon, P.J. and Sweet, R.W. 1989. Identification of the residues in human CD4 critical for the binding of HIV. Cell 57: 469–481.

Badger, J., Minor, I., Kremer, M.J., Oliveira, M.A., Smith, T.J., Griffith, J.P., Guerin, D.M.A., Krishnaswamy, S., Luo, M., Rossmann, M.G., McKinlay, M.A., Diana, G.D., Dutko, F.J., Fancher, M., Rueckert, R.R. and Heinz, B.A. 1988. Structural analysis of a series of antiviral agents complexed with human rhinovirus 14. Proc. Natl. Acad. Sci. U.S.A. 85: 3304–3308.

Bergelson, J.M., Shepley, M.P., Chan, B.M.C., Hemler, M.E. and Finberg, R.W. 1992. Identification of the integrin VLA-2 as a receptor for echovirus 1. Science 255: 1718–1720.

Brady, R.L., Dodson, E.J., Dodson, G.G., Lange, G., Davis, S.J., Williams, A.F. and Barclay, A.N. 1993. Crystal structure of domains 3 and 4 of rat CD4: relation to the NH_2-terminal domains. Science 260: 979–983.

Burness, A.T.H. 1981. Glycophorin and sialylated components as receptors for viruses. In Virus Receptors, Part 2 (Lonberg-Holm, K. and Philipson, L., Eds.), pp. 64–84. Chapman and Hall, London.

Burness, A.T.H. and Pardoe, I.U. 1983. A sialoglycopeptide from human erythrocytes with receptor-like properties for encephalomyocarditis and influenza viruses. J. Gen. Virol. 64: 1137–1148.

Chapman, M.S. and Rossmann, M.G. 1993. Comparison of surface properties of picornaviruses: strategies for hiding the receptor site from immune surveillance. Virology 195: 745–756.

Chow, M., Newman, J.F.E., Filman, D., Hogle, J.M., Rowlands, D.J. and Brown, F. 1987. Myristylation of picornavirus capsid protein VP4 and its structural significance. Nature 327: 482–486.

Colonno, R.J. 1992. Molecular interactions between human rhinoviruses and their cellular receptors. Semin. Virol. 3: 101–107.

Colonno, R.J., Callahan, P.L. and Long, W.J. 1986. Isolation of a monoclonal antibody that blocks attachment of the major group of human rhinoviruses. J. Virol. 57: 7–12.

Colonno, R.J., Condra, J.H., Mizutani, S., Callahan, P.L., Davies, M.E. and Murcko, M.A. 1988. Evidence for the direct involvement of the rhinovirus canyon in receptor binding. Proc. Natl. Acad. Sci. U.S.A. 85: 5449–5453.

Crowell, R.L. 1966. Specific cell-surface alteration by enteroviruses as reflected by viral-attachment interference. J. Bacteriol. 91: 198–204.

Dalgleish, A.G., Beverley, P.C.L., Clapham, P.R., Crawford, D.H., Greaves, M.F. and Weiss, R.A. 1984. The CD4 (T4) antigen is an essential component of the receptor for the AIDS retrovirus. Nature 312: 763–767.

Diana, G.D., Kowalczyk, P., Treasurywala, A.M., Oglesby, R.C., Pevear, D.C. and Dutko, F.J. 1992. CoMFA analysis of the interactions of antipicornavirus compounds in the binding pocket of human rhinovirus-14. J. Med. Chem. 35: 1002–1006.

Diana, G.D., Treasurywala, A.M., Bailey, T.R., Oglesby, R.C., Pevear, D.C. and Dutko, F.J. 1990. A model for compounds active against human rhinovirus-14 based on X-ray crystallography data. J. Med. Chem. 33: 1306–1311.

Duechler, M., Ketter, S., Skern, T., Kuechler, E. and Blaas, D. 1993. Rhinoviral receptor discrimination: mutational changes in the canyon regions of human rhinovirus types 2 and 14 indicate a different site of interaction. J. Gen. Virol. 74: 2287–2291.

Dutko, F.J., McKinlay, M.A. and Rossmann, M.G. 1989. Antiviral compounds bind to a specific site within human rhinovirus. In Concepts in Viral Pathogenesis III (Notkins, A.L. and Oldstone, M.B.A., Eds.), pp. 330–336. Springer-Verlag, New York.

Eriksson, A.E., Baase, W.A., Wozniak, J.A. and Matthews, B.A. 1992a. A cavity-containing mutant of T4 lysozyme is stabilized by buried benzene. Nature 355: 371–373.

Eriksson, A.E., Baase, W.A., Zhang, X.-J., Heinz, D.W., Blaber, M., Baldwin, E.P. and Matthews, B.W. 1992b. Response of a protein structure to cavity-creating mutations and its relation to the hydrophobic effect. Science 255: 178–183.

Everaert, L., Vrijsen, R. and Boeyé, A. 1989. Eclipse products of poliovirus after cold-synchronized infection of HeLa cells. Virology 171: 76–82.

Filman, D.J., Syed, R., Chow, M., Macadam, A.J., Minor, P.D. and Hogle, J.M. 1989. Structural factors that control conformational transitions and serotype specificity in type 3 poliovirus. EMBO J. 8: 1567–1579.

Fox, G., Parry, N.R., Barnett, P.V., McGinn, B., Rowlands, D.J. and Brown, F. 1989. The cell attachment site on foot-and-mouth disease virus includes the amino acid sequence RGD (arginine-glycine-aspartic acid). J. Gen. Virol. 70: 625–637.

Fox, M.P., McKinlay, M.A., Diana, G.D. and Dutko, F.J. 1991. Binding affinities of structurally related human rhinovirus capsid-binding compounds are correlated to their activities against human rhinovirus type 14. Antimicrob. Agents Chemother. 35: 1040–1047.

Fox, M.P., Otto, M.J. and McKinlay, M.A. 1986. The prevention of rhinovirus and poliovirus uncoating by WIN 51711: a new antiviral drug. Antimicrob. Agents Chemother. 30: 110–116.

Freistadt, M.S. and Racaniello, V.R. 1991. Mutational analysis of the cellular receptor for poliovirus. J. Virol. 65: 3873–3876.

Fricks, C.E. and Hogle, J.M. 1990. Cell-induced conformational change in poliovirus: externalization of the amino terminus of VP1 is responsible for liposome binding. J. Virol. 64: 1934–1945.

Giranda, V.L., Chapman, M.S. and Rossmann, M.G. 1990. Modeling of the human intercellular adhesion molecule-1, the human rhinovirus major group receptor. Proteins 7: 227–233.

Giranda, V.L., Heinz, B.A., Oliveira, M.A., Minor, I., Kim, K.H., Kolatkar, P.R., Rossmann, M.G. and Rueckert, R.R. 1992. Acid-induced structural changes in human rhinovirus 14: possible role in uncoating. Proc. Natl. Acad. Sci. U.S.A. 89: 10213–10217.

Greve, J.M., Davis, G., Meyer, A.M., Forte, C.P., Yost, S.C., Marlor, C.W., Kamarck, M.E. and McClelland, A. 1989. The major human rhinovirus receptor is ICAM-1. Cell 56: 839–847.

Greve, J.M., Forte, C.P., Marlor, C.W., Meyer, A.M., Hoover-Litty, H., Wunderlich, D. and McClelland, A. 1991. Mechanisms of receptor-mediated rhinovirus neutralization defined by two soluble forms of ICAM-1. J. Virol. 65: 6015–6023.

Gromeier, M. and Wetz, K. 1990. Kinetics of poliovirus uncoating in HeLa cells in a nonacidic environment. J. Virol. 64: 3590–3597.

Gruenberger, M., Pevear, D., Diana, G.D., Kuechler, E. and Blaas, D. 1991. Stabilization of human rhinovirus serotype 2 against pH-induced conformational change by antiviral compounds. J. Gen. Virol. 72: 431–433.

Hadfield, A., Oliveira, M.A., Kimk, H., Minor, J., Kremer, M.J., Heinz, B.A., Shepard, D., Pevear, D.L., Ruechert, R.R. and Rossmann, M.G. 1995. Structural studies on human rhinovirus 14 drug-resistant compensation mutants. J. Mol. Biol. 253: 61–73.

Heinz, B.A., Rueckert, R.R., Shepard, D.A., Dutko, F.J., McKinlay, M.A., Fancher, M., Rossmann, M.G., Badger, J. and Smith, T.J. 1989. Genetic and molecular analyses of spontaneous mutants of human rhinovirus 14 that are resistant to an antiviral compound. J. Virol. 63: 2476–2485.

Heinz, B.A., Shepard, D.A. and Rueckert, R.R. 1990. Escape mutant analysis of a drug-binding site can be used to map functions in the rhinovirus capsid. In Use of X-ray Crystallography in the Design of Antiviral Agents (Laver, W.G. and Air, G.M., Eds.), pp. 173–186. Academic Press, San Diego.

Hofer, F., Gruenberger, M., Kowalski, H., Machat, H., Huettinger, M., Kuechler, E. and Blaas, D. 1994. Members of the low density lipoprotein receptor family mediate cell entry of a minor-group common cold virus. Proc. Natl. Acad. Sci. U.S.A. 91: 1839–1842.

Hogle, J.M., Chow, M. and Filman, D.J. 1985. Three-dimensional structure of poliovirus at 2.9 Å resolution. Science 229: 1358–1365.

Hoover-Litty, H. and Greve, J.M. 1993. Formation of rhinovirus-soluble ICAM-1 complexes and conformational changes in the virion. J. Virol. 67: 390–397.

Kim, K.H., Willingmann, P., Gong, Z.X., Kremer, M.J., Chapman, M.S., Minor, I., Oliveira, M.A., Rossmann, M.G., Andries, K., Diana, G.D., Dutko, F.J., McKinlay, M.A. and Pevear, D.C. 1993. A comparison of the anti-rhinoviral drug binding pocket in HRV14 and HRV1A. J. Mol. Biol. 230: 206–225.

Kim, S., Boege, U., Krishnaswamy, S., Minor, I., Smith, T.J., Luo, M., Scraba, D.G. and Rossmann, M.G. 1990. Conformational variability of a picornavirus capsid: pH-dependent structural changes of Mengo virus related to its host receptor attachment site and disassembly. Virology 175: 176–190.

Kim, S., Smith, T.J., Chapman, M.S., Rossmann, M.G., Pevear, D.C., Dutko, F.J., Felock, P.J., Diana, G.D. and McKinlay, M.A. 1989. Crystal structure of human rhinovirus serotype 1A (HRV1A). J. Mol. Biol. 210: 91–111.

Klatzmann, D., Champagne, E., Chamaret, S., Gruest, J., Guetard, D., Hercend, T., Gluckman, J.C. and Montagnier, L. 1984. T-lymphocyte T4 molecule behaves as the receptor for human retrovirus LAV. Nature 312: 767–768.

Koike, S., Ise, I. and Nomoto, A. 1991. Functional domains of the poliovirus receptor. Proc. Natl. Acad. Sci. U.S.A. 88: 4104–4108.

Koike, S., Ise, I., Sato, Y., Mitsui, K., Horie, H., Umeyama, H. and Nomoto, A. 1992. Early events of poliovirus infection. Semin. Virol. 3: 109–115.

Kolatkar, P.R., Oliveira, M.A., Rossmann, M.G., Robbins, A.H., Katti, S.K., Hoover-Litty, H., Forte, C., Greve, J.M., McClelland, A. and Olson, N.H. 1992. Preliminary X-ray crystallographic analysis of intercellular adhesion molecule-1. J. Mol. Biol. 225: 1127–1130.

Korant, B.D., Lonberg-Holm, K., Yin, F.H. and Noble-Harvey, J. 1975. Fractionation of biologically active and inactive populations of human rhinovirus type 2. Virology 63: 384–394.

Leahy, D.J., Hendrickson, W.A., Aukhil, I. and Erickson, H.P. 1992. Structure of a fibronectin type III domain from tenascin phased by MAD analysis of the selenomethionyl protein. Science 258: 987–991.

Lee, W.M., Monroe, S.S. and Rueckert, R.R. 1993. Role of maturation cleavage in infectivity of picornaviruses: activation of an infectosome. J. Virol. 67: 2110–2122.

Lineberger, D.W., Graham, D.J., Tomassini, J.E. and Colonno, R.J. 1990. Antibodies that block rhinovirus attachment map to domain 1 of the major group receptor. J. Virol. 64: 2582–2587.

Lonberg-Holm, K. and Korant, B.D. 1972. Early interaction of rhinoviruses with host cells. J. Virol. 9: 29–40.

Luo, M., Vriend, G., Kamer, G., Minor, I., Arnold, E., Rossmann, M.G., Boege, U., Scraba, D.G., Duke, G.M. and Palmenberg, A.C. 1987. The atomic structure of Mengo virus at 3.0 Å resolution. Science 235: 182–191.

Maddon, P.J., Dalgleish, A.G., McDougal, J.S., Clapham, P.R., Weiss, R.A. and Axel, R. 1986. The T4 gene encodes the AIDS virus receptor and is expressed in the immune system and the brain. Cell 47: 333–348.

Madshus, I.H., Olsnes, S. and Sandvig, K. 1984a. Different pH requirements for entry of the two picornaviruses, human rhinovirus 2 and murine encephalomyocarditis virus. Virology 139: 346–357.

Madshus, I.H., Olsnes, S. and Sandvig, K. 1984b. Mechanism of entry into the cytosol of poliovirus type 1: requirement for low pH. J. Cell Biol. 98: 1194–1200.

Main, A.L., Harvey, T.S., Baron, M., Boyd, J. and Campbell, I.D. 1992. The three-dimensional structure of the tenth type III module of fibronectin: an insight into RDG-mediated interactions. Cell 71: 671–678.

Mallamo, J.P., Diana, G.D., Pevear, D.C., Dutko, F.J., Chapman, M.S., Kim, K.H., Minor, I., Oliveira, M. and Rossmann, M.G. 1992. Conformationally restricted analogues of disoxaril: a comparison of the activity against human rhinovirus types 14 and 1A. J. Med. Chem. 35: 4690–4695.

Mason, P.W., Baxt, B., Brown, F., Harber, J., Murdin, A. and Wimmer, E. 1993. Antibody-complexed foot-and-mouth disease virus, but not poliovirus, can infect normally insusceptible cells via the Fc receptor. Virology 192: 568–577.

McClelland, A., deBear, J., Yost, S.C., Meyer, A.M., Marlor, C.W. and Greve, J.M. 1991. Identification of monoclonal antibody epitopes and critical residues for rhinovirus binding in domain 1 of ICAM-1. Proc. Natl. Acad. Sci. U.S.A. 88: 7993–7997.

McKenna, R., Xia, D., Willingmann, P., Ilag, L.L., Krishnaswamy, S., Rossmann, M.G., Olson, N.H., Baker, T.S. and Incardona, N.L. 1992. Atomic structure of single-stranded DNA bacteriophage φX174 and its functional implications. Nature 355: 137–143.

Mendelsohn, C.L., Wimmer, E. and Racaniello, V.R. 1989. Cellular receptors for poliovirus: molecular cloning, nucleotide sequence, and expression of a new member of the immunoglobulin superfamily. Cell 56: 855–865.

Mosser, A.G. and Rueckert, R.R. 1993. WIN 51711-dependent mutants of poliovirus type 3: evidence that virions decay after release from cells unless drug is present. J. Virol. 67: 1246–1254.

Neubauer, C., Frasel, L., Kuechler, E. and Blaas, D. 1987. Mechanism of entry of human rhinovirus 2 into HeLa cells. Virology 158: 255–258.

Oliveira, M.A., Zhao, R., Lee, W.M., Kremer, M.J., Minor, I., Rueckert, R.R., Diana, G.D., Pevear, D.C., Dutko, F.J., McKinlay, M.A. and Rossmann, M.G. 1993. The structure of human rhinovirus 16. Structure 1: 51–68.

Olson, N.H., Kolatkar, P.R., Oliveira, M.A., Cheng, R.H., Greve, J.M., McClelland, A., Baker, T.S. and Rossmann, M.G. 1993. Structure of a human rhinovirus complexed with its receptor molecule. Proc. Natl. Acad. Sci. U.S.A. 90: 507–511.

Page, G.S., Mosser, A.G., Hogle, J.M., Filman, D.J., Rueckert, R.R. and Chow, M. 1988. Three-dimensional structure of poliovirus serotype 1 neutralizing determinants. J. Virol. 62: 1781–1794.

Palmenberg, A.C. 1989. Sequence alignments of picornaviral capsid proteins. In Molecular Aspects of Picornavirus Infection and Detection (Semler, B.L. and Ehrenfeld, E., Eds.), pp. 211–241. American Society for Microbiology, Washington, DC.

Parry, N., Fox, G., Rowlands, D., Brown, F., Fry, E., Acharya, R., Logan, D. and Stuart, D. 1990. Structural and serological evidence for a novel mechanism of antigenic variation in foot-and-mouth disease virus. Nature 347: 569–572.

Pérez, L. and Carrasco, L. 1993. Entry of poliovirus into cells does not require a low-pH step. J. Virol. 67: 4543–4548.

Pevear, D.C., Fancher, M.J., Felock, P.J., Rossmann, M.G., Miller, M.S., Diana, G., Treasurywala, A.M., McKinlay, M.A. and Dutko, F.J. 1989. Conformational change in the floor of the human rhinovirus canyon blocks adsorption to HeLa cell receptors. J. Virol. 63: 2002–2007.

Racaniello, V.R. 1992. Interaction of poliovirus with its cell receptor. Semin. Virol. 3: 473–482.

Rawn, J.D. 1989. Biochemistry. Neil Patterson Publishers, Burlington, NC.

Register, R.B., Uncapher, C.R., Naylor, A.M., Lineberger, D.W. and Colonno, R.J. 1991. Human-murine chimeras of ICAM-1 identify amino acid residues critical for rhinovirus and antibody binding. J. Virol. 65: 6589–6596.

Robey, E. and Axel, R. 1990. CD4: collaborator in immune recognition and HIV infection. Cell 60: 697–700.

Roivainen, M., Hyypiä, T., Piirainen, L., Kalkkinen, N., Stanway, G. and Hovi, T. 1991. RGD-dependent entry of coxsackievirus A9 into host cells and its bypass after cleavage of VP1 protein by intestinal proteases. J. Virol. 65: 4735–4740.

Rossmann, M.G. and Palmenberg, A.C. 1988. Conservation of the putative receptor attachment site in picornaviruses. Virology 164: 373–382.

Rossmann, M.G., Arnold, E., Erickson, J.W., Frankenberger, E.A., Griffith, J.P., Hecht, H.J., Johnson, J.E., Kamer, G., Luo, M., Mosser, A.G., Rueckert, R.R., Sherry, B. and Vriend, G. 1985. Structure of a human common cold virus and functional relationship to other picornaviruses. Nature 317: 145–153.

Rueckert, R.R. 1990. Picornaviridae and their replication. In Virology, 2nd ed., Vol. 1 (Fields, B.N. and Knipe, D.M., Eds.), pp. 507–548. Raven Press, New York.

Ryu, S.E., Kwong, P.D., Truneh, A., Porter, T.G., Arthos, J., Rosenberg, M., Dai, X., Xuong, N., Axel, R., Sweet, R.W. and Hendrickson, W.A. 1990. Crystal structure of an HIV-binding recombinant fragment of human CD4. Nature 348: 419–425.

Schultz, M. and Crowell, R.L. 1983. Eclipse of coxsackievirus infectivity: the restrictive event for a non-fusing myogenic cell line. J. Gen. Virol. 64: 1725–1734.

Sekiguchi, K., Franke, A.J. and Baxt, B. 1982. Competition for cellular receptor sites among selected aphthoviruses. Arch. Virol. 74: 53–64.

Shepard, D.A., Heinz, B.A. and Rueckert, R.R. 1993. WIN compounds inhibit both attachment and eclipse of human rhinovirus 14. J. Virol. 67: 2245–2254.

Sherry, B. and Rueckert, R. 1985. Evidence for at least two dominant neutralization antigens on human rhinovirus 14. J. Virol. 53: 137–143.

Sherry, B., Mosser, A.G., Colonno, R.J. and Rueckert, R.R. 1986. Use of monoclonal antibodies to identify four neutralization immunogens on a common cold picornavirus, human rhinovirus 14. J. Virol. 57: 246–257.

Simmons, D., Makgoba, M.W. and Seed, B. 1988. ICAM, an adhesion ligand of LFA-1, is homologous to the neural cell adhesion molecule NCAM. Nature 331: 624–627.

Smith, T.J., Kremer, M.J., Luo, M., Vriend, G., Arnold, E., Kamer, G., Rossmann, M.G., McKinlay, M.A., Diana, G.D. and Otto, M.J. 1986. The site of attachment in human rhinovirus 14 for antiviral agents that inhibit uncoating. Science 233: 1286–1293.

Springer, T.A. 1990. Adhesion receptors of the immune system. Nature 346: 425–434.

Staunton, D.E., Dustin, M.L., Erickson, H.P. and Springer, T.A. 1990. The arrangement of the immunoglobulin-like domains of the ICAM-1 and the binding site of LFA-1 and rhinovirus. Cell 61: 243–254.

Staunton, D.E., Marlin, S.D., Stratowa, C., Dustin, M.L. and Springer, T.A. 1988. Primary structure of ICAM-1 demonstrates interaction between members of the immunoglobulin and integrin supergene families. Cell 52: 925–933.

Staunton, D.E., Merluzzi, V.J., Rothlein, R., Barton, R., Marling, S.D. and Springer, T.A. 1989. A cell adhesion molecule, ICAM-1, is the major surface receptor for rhinoviruses. Cell 56: 849–853.

Tomassini, J.E., Maxson, T.R. and Colonno, R.J. 1989. Biochemical characterization of a glycoprotein required for rhinovirus attachment. J. Biol. Chem. 264: 1656–1662.

Tsao, J., Chapman, M.S., Agbandje, M., Keller, W., Smith, K., Wu, H., Luo, M., Smith, T.J., Rossmann, M.G., Compans, R.W. and Parrish, C.R. 1991. The three-dimensional structure of canine parvovirus and its functional implications. Science 251: 1456–1464.

Uncapher, C.R., DeWitt, C.M. and Colonno, R.J. 1991. The major and minor group receptor families contain all but one human rhinovirus serotype. Virology 180: 814–817.

Wang, J., Yan, Y., Garrett, T.P.J., Liu, J., Rodgers, D.W., Garlick, R.L., Tarr, G.E., Hussain, Y., Reinherz, E.L. and Harrison, S.C. 1990. Atomic structure of a fragment of human CD4 containing two immunoglobulin-like domains. Nature 348: 411–418.

Weis, W., Brown, J.H., Cusack, S., Paulson, J.C., Skehel, J.J. and Wiley, D.C. 1988. Structure of the influenza virus haemagglutinin complexed with its receptor, sialic acid. Nature 333: 426–431.

Williams, R.K., Jiang, G.S. and Holmes, K.V. 1991. Receptor for mouse hepatitis virus is a member of the carcinoembryonic antigen family of glycoproteins. Proc. Natl. Acad. Sci. U.S.A. 88: 5533–5536.

Wilson, I.A., Skehel, J.J. and Wiley, D.C. 1981. Structure of the haemagglutinin membrane glycoprotein of influenza virus at 3 Å resolution. Nature 289: 366–373.

Yafal, A.G., Kaplan, G., Racaniello, V.R. and Hogle, J.M. 1993. Characterization of poliovirus conformational alteration mediated by soluble cell receptors. Virology 197: 501–505.

Yeates, T.O., Jacobson, D.H., Martin, A., Wychowski, C., Girard, M., Filman, D.J. and Hogle, J.M. 1991. Three-dimensional structure of a mouse-adapted type 2/type 1 poliovirus chimera. EMBO J. 10: 2331–2341.

Zeichhardt, H., Wetz, K., Willingmann, P. and Habermehl, K.O. 1985. Entry of poliovirus type 1 and mouse Elberfeld (ME) virus into HEp-2 cells: receptor-mediated endocytosis and endosomal or lysosomal uncoating. J. Gen. Virol. 66: 483–492.

5.

Antibody-Mediated Neutralization of Picornaviruses

THOMAS J. SMITH & ANNE G. MOSSER

Introduction

Structure of Human Rhinovirus 14

Human rhinovirus is a major cause of the common cold, and is a member of the picornavirus family (Rueckert, 1990). These small (~280-Å diameter), nonenveloped viruses have a single-stranded RNA genome. Other members of this family include viruses such as poliovirus, coxsackievirus, hepatitis A virus, and foot-and-mouth disease virus, some of which cause serious human and animal diseases. The structures of several picornaviruses have been determined to atomic resolution (Rossmann and Johnson, 1989; Stuart, 1993; see also chapters 4 and 6). The icosahedral shell of these viruses is composed of four viral proteins; VP1, VP2, VP3, and VP4. The structures of the first three capsid proteins share a similar eight-stranded, antiparallel, β-barrel motif. This β-barrel structure is highly conserved among the picornavirus capsid proteins, and is also shared with DNA (Tsao et al., 1991), plant (Harrison et al., 1978; Abad-Zapatero et al., 1980; Stauffacher et al., 1987), bacterial (McKenna et al., 1992), and insect (Hosur et al., 1987) viruses. The major differences between the viral protein structures appear on the loops that connect the β-strands and at the amino- and carboxyl-terminal extensions.

Antigenic Sites

The antigenicity of the viral surface has been studied by producing a panel of neutralizing monoclonal antibodies and then selecting virus variants ("antibody escape mutants") capable of forming plaques in the presence of neutralizing quantities of monoclonal antibodies. When the amino acid substitutions in these mutants were mapped onto the known structure of the virus capsid, it was found that viruses typically had three or four immunogenic patches on their surfaces (Minor et al., 1986; Sherry et al., 1986; Page et al., 1988). These were called NIm sites, for *n*eutralizing *im*munogenic sites. As additional studies have been performed, the areas of the NIm sites have grown (Uhlig et al., 1990; Wiegers and Dernick, 1991) to include much of the accessible parts of the external virus surface, with protruding sites being the most immunogenic.

Using the procedure described previously, the antigenic sites of human rhinovirus 14 (HRV14) have been mapped onto the three-dimensional structure (Fig. 5.1). A total of 35 neutralizing monoclonal antibodies against HRV14 were produced and characterized by the isolation of 62 antibody-escape viral mutants that were cross-tested against the entire panel of antibodies (Sherry et al., 1986). The isolates, and the antibodies that failed to neutralize them, formed four groups. Comparisons of the sequences of their genomic RNAs and correlation with the atomic structure showed that each group had amino acid substitutions in a tight cluster (NIm site) on exposed and protruding portions of the virus surface (Fig. 5.1, Table 5.1) (Rossmann et al., 1985).

Canyon Hypothesis

Once the three-dimensional structure of HRV14 had been determined, the NIm sites were found to be positioned about large depressions on the capsid that encircle each of the fivefold axes (Rossmann et al., 1985; see also Chapter 4).

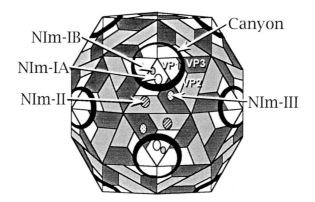

FIGURE 5.1 Icosahedral model of HRV14. Each of the three major viral proteins is shown in different shades of gray. Also depicted are two of the 60 sets of neutralizing immunogenic sites (NIm sites). About each fivefold axis is a dark circle representing the canyon.

TABLE 5.1 Amino Acid Substitutions in Natural Escape Mutants of Antibodies from the Four Immunogenic Sites on HRV14

	VP1	VP2	VP3
NIm-IA	95, 95		
NIm-IB	83, 85, 138, 139		
NIm-II	210	136, 158, 159, 161, 162	
NIm-III	287		72, 75, 78, 203

The residues lining the canyon were more conserved among the rhinoviruses than the residues on the rim. In addition, the canyon dimensions made it seem unlikely that antibodies bind to the deeper recesses. Therefore, it was postulated that the floor of this canyon may be involved in viral recognition of the cell receptor protein. In this way, the virus can tolerate mutations about the rim of the canyon to thwart antibody binding that do not affect the manner in which it recognizes the host cell (Rossmann et al., 1985).

The hypothesis that the canyon is the site of receptor binding has been substantiated in several ways. Site-directed mutagenesis of the canyon floor identified several residues critical for cell receptor binding (Colonno et al., 1988). Subsequently, it was demonstrated that the deformations in the canyon floor caused by antiviral drug binding also abrogated cell receptor attachment (Pevear et al., 1989). Finally, visual proof of this hypothesis was demonstrated by the cryo-electron microscopy study of HRV16 complexed with the two amino-terminal immunoglobulinlike domains of its receptor, intercellular adhesion molecule (ICAM)-1(Olson et al., 1992, 1993; see also Chapter 4).

Structure of Immunoglobulin

Antibodies, produced by B lymphocytes, represent a humoral component of the adaptive immune response. There are several structurally and functionally distinct types of antibodies produced by mammals: immunoglobulins (Igs)A, D, E, G, and M. IgA is mostly found in nasal and intestinal secretions and therefore plays a major role in preventing pathogen entry at mucosal surfaces. IgE binds to mast cells and basophils, causing allergic and inflammatory responses, but is thought to be important in parasite immunity. The function of IgD is less clear, it is probably involved in B cell stimulation and development. IgM and IgG are the major serum immunoglobulins that are usually secreted by immature and mature B cells, respectively.

IgG molecules have two antigen binding arms (Fabs) and one region that does not bind antigen (Fc) that are all connected at a hinge region (Fig. 5.2). The Fab and Fc arms both have dimensions of approximately $40 \times 40 \times 70$ Å. There are two light ($\sim$25 kDa) and two heavy ($\sim$50 kDa) chains in an IgG molecule. The Fab arms are composed of a light chain and part of the heavy chain, whereas the Fc region is made from portions of the two heavy chains.

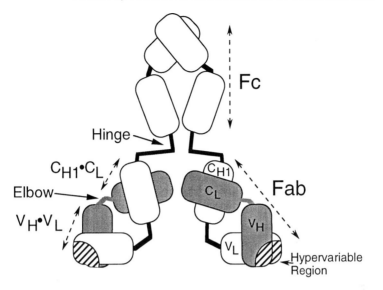

FIGURE 5.2   Model of a typical IgG protein. The light chain is shown in gray and the heavy in white. Each of the immunoglobulin domains is depicted as a rounded rectangle. The V_H and V_L immunoglobulin domains combine to form the variable module and the C_H1 and C_L domains form the constant module of the Fab arm.

Both the heavy and light chains are composed of β-barrels, whose structures are highly conserved among many proteins of immunologic function. Each Fab arm has four such β-barrels (domains), two from the heavy chain and two from the light chain. One β-barrel at the amino-terminus of the light chain combines with the amino-terminal β-barrel from the heavy chain to form the variable (V) module ($V_H \cdot V_L$) The subsequent β-barrels from the heavy and light chains combine to form the next region called the Fab constant (C) module ($C_{H1} \cdot C_L$). The connection between these two modules is called the elbow region. The modules are delineated as constant and variable on the basis of the degree of conservation of amino acid sequences between different antibodies. Three loops from each of the heavy and light chains comprise the six-loop antigen binding region within the variable module. This region is very heterologous among different antibodies, hence the name ''hypervariable region.''

The structures of many Fab and IgG molecules have been determined using electron microscopy and X-ray crystallography (for reviews, see Kabat et al., 1987; Harris et al., 1993). From these studies, it has been shown that the hinge region is a highly flexible structure that ''tethers'' the Fab and Fc regions together. The hinge is rich in cysteine and proline residues and has an extended structure that pushes the Fab arms away from the Fc region. This region, composed of both heavy chains, has a different number of disulfide bonds between the two chains depending upon the antibody isotype. The hinge region has been shown to play an important role in antibody binding to Fc receptors and complement $C1_{qrs}$ protein (the first serum protein involved in the classical complement activation pathway; Klein et al., 1981). The elbow region is also thought to be somewhat flexible because it has been found to vary among the atomic structures of Fab fragments (e.g., Ely and Herron, 1989).

Mechanisms of Antibody-Mediated Neutralization

Neutralization is the reduction in infectivity of a virus preparation either in solution or when in contact with susceptible cells. Possible mechanisms of neutralization include (1) aggregating virions, which simply reduces the number of infectious units in solution; (2) induction of conformational changes in the virion that reduce its infectivity; and (3) abrogation of some step in the virus infection cycle. Theoretically, each monoclonal antibody may be capable of neutralization by one or more mechanism. An antibody's ability to neutralize may depend on its binding affinity, on the location of its binding site on the viral surface, and on its binding orientation.

Aggregation

The binding of almost all monoclonal antibodies leads to some degree of virus aggregation. Some antibodies precipitate virions over a wide range of virus: antibody ratios, whereas others cause very little aggregation. However, even these poorly precipitating antibodies can form immune complexes containing two or three virions, as observed using sucrose gradients. An antibody's tendency to aggregate virions probably depends on the site to which it binds on the virus and the direction in which its other Fab arm points once the first arm has bound (Mosser et al., 1989). If a second binding site is located at the proper distance and direction, this IgG arm will tend to bind bivalently to a single virion and will be a poor aggregator. If the second arm points away from any appropriately spaced, icosahedrally related sites, the antibody will be a good precipitator of virus. Aggregation is conceptually the simplest method of neutralization by reducing the number of independent infectious units. This phenomenon has been observed in several studies that have shown that the residual infectivity of antibody-bound virion dimers and trimers isolated from sucrose gradients was approximately one half and one third, respectively, of the native virions (Icenogle et al., 1983; Thomas et al., 1985).

Induction of Conformational Changes

Another possible effect of antibody binding is the alteration of the conformation of the virus coat. Antibody-induced conformational change was initially thought to be the prime mechanism of neutralization, because early observations suggested that (1) antiserum neutralized virus with one-hit kinetics (Dulbecco et al., 1956), and (2) the pI of picornaviruses dropped from 7 to 4 upon treatment with antiserum (Mandel, 1976). These observations led to the hypothesis that the binding of a single antibody molecule caused a concerted conformational change in the virus capsid leading to loss of infectivity and a change in the surface charge. For picornaviruses, the shift in pI after antibody binding has been duplicated in several laboratories (Emini et al., 1983; Colonno et al., 1989), although the causal relationship between the pI shift and neutralization has been questioned (Icenogle, et al., 1983; Brioen et al., 1985b). In another approach to testing this theory, if it could be shown that a single antibody is sufficient to neutralize infectivity, then the proposed process of antibody-induced conformational changes would be strongly supported. However, when the number of antibodies required for neutralization of unaggregated virions was carefully quantitated, no monoclonal antibodies were found to be capable of neutralization at a ratio of one antibody per virion (Icenogle et al., 1983; Leippe, 1991). These measurements are not without debate because there has been one report of single-antibody-mediated neutralization using polyclonal antibodies (Wetz et al., 1986).

More recently, several laboratories reported that antibodies specific for protein sequences buried inside the virus capsid are capable of precipitation (Roivainen et al., 1993) or neutralization (Li et al., 1994). These authors hypothesize that the virus "breathes," exposing residues normally buried. It was suggested that antibodies reacting with these previously buried residues might trap the virus in a noninfectious conformation.

Finally, some poliovirus-specific monoclonal antibodies have been found to mediate RNA release from virions when binding at low ionic strength or elevated temperatures (Brioen et al., 1985a; Delaet and Boeye, 1993). These authors speculated that virus neutralization by these antibodies may be activated by fever.

Interference with Virus Replication Cycle

Once virions have been exposed to cells, they must attach, penetrate into the cell and uncoat, thus delivering their intact RNA to the cytoplasm. Each of these steps has been shown to be susceptible to antibody interference.

Attachment   Fab fragments from four monoclonal antibodies against HRV14 blocked attachment (Colonno et al., 1989). Because these antibodies also caused a pI shift in the virus, it was not clear whether attachment was blocked by steric interference of Fab binding or by a conformational change in the virion. In one study with poliovirus type 1, attachment was inhibited by all monoclonal antibodies tested but, with one exception, this inhibition was insufficient to explain the degree of neutralization (Emini et al., 1983).

Penetration   Viral "penetration" can be experimentally quantitated by measuring the amount of cell-attached virus released from cells by treatment with 8-M urea. Emini et al. (1983) found that, although six anti-poliovirus type 1 monoclonal antibodies inhibited penetration somewhat, this inhibition could not account for the extent of neutralization. However, these authors repeated the previously published results that hyperimmune antiserum could completely block penetration (Holland and Hoyer, 1962). Because mixtures of monoclonal antibodies were not any more effective at blocking penetration than single antibodies (Emini et al., 1983), it is possible that other components of serum, such as complement, act synergistically with antibodies.

Uncoating   The uncoating of picornaviruses is a multistep process, beginning with conversion of the approximately 150S mature virion into a 125S to135S particle lacking VP4 (Everaet et al., 1989; Fricks and Hogle, 1990). A minor subset of antibodies to poliovirus type 1 were found to be capable of neutralizing virus already attached to cells and blocking this first step in uncoating (Vrijsen et al., 1993). Because these antibodies were generally weak aggregators, it was speculated that bivalent attachment of both Fab arms of the antibody to the same virion was necessary for prevention of uncoating.

Antibodies have even been shown to protect cells against viral cytopathic effects when added after production of new virions has begun (Tolskaya et al., 1992). Because viral replication was terminated without cell lysis, it was speculated that the antibodies may have entered the cell to exert their effects. An alternative interpretation is that virus may be released from cells without lysis and that antibodies in the extracellular media may prevent virally induced cytopathic effects.

Although the availability of monoclonal antibodies has allowed researchers to study the effects of individual antibodies on viral functions and perform a variety of studies that depend on chemical homogeneity (e.g., structural and stoichiometric studies), antibodies in an infected host are only one part of the total immune response. The relationship between antibodies and other aspects of the adaptive and innate immune response are only now being explored. For example, some monoclonal antibodies have been shown to act synergistically with interferon to inhibit poliovirus (Langford et al., 1991).

Neutralization Studies of HRV14

Classes of Neutralizing Antibodies

Thirty-two HRV14-specific neutralizing antibodies were classified by measuring the ability of increasing amounts of antibody to neutralize a constant amount of virus (Leippe, 1991) (Fig. 5.3). Ten of the antibodies, all strongly neutralizing antibodies that recognized site 1A, exhibited neutralization curves like that shown in Figure 5.3A. As the antibody:virus ratio increased, the surviving infectivity fell by 4 to 5 logs. At high antibody concentrations, the residual infectivity leveled out, revealing the presence of neutralization-resistant mutants. Because subsequent testing revealed that most of these antibodies were poor viral aggregators (see later), it was proposed that these antibodies are capable of binding both Fab arms to a single virion (inset in Fig. 5.3A).

Twelve other antibodies, associated with all four antigenic sites, gave curves typical of that shown in Figure 5.3B. Here, viral infectivity was never reduced by more than 3 logs, and there was *increased* infectivity at extremely

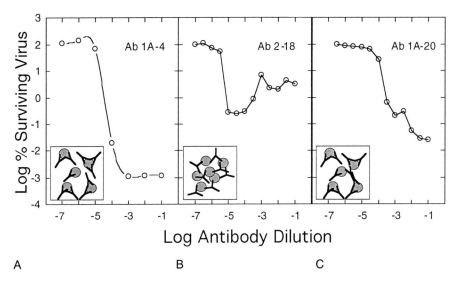

FIGURE 5.3 Neutralization profiles of the three major groups of neutralizing monoclonal antibodies to HRV14. HRV14 (10^8 plaque-forming units/ml) and dilutions of ascites fluids containing monoclonal antibodies were mixed and incubated for 1 h at room temperature and then overnight at 4 °C. Surviving virus infectivity was determined by plaques assay. *A*. Strongly neutralizing, weakly aggregating antibody 1A-4. Cartoon inset shows neutralization by a subsaturating amount of bivalently bound antibody. *B*. Weakly neutralizing, strongly aggregating antibody 2–18. Cartoon inset shows antibody-induced viral aggregation that fails at extremely high antibody: virus ratios because virus coated with antibodies no longer precipitates. *C*. Intermediate neutralizing, weakly aggregating antibody 1A-20. Cartoon inset shows postulated hybrid mechanism of neutralization.

high antibody:virus ratios. All of these antibodies precipitated virus from solutions at most antibody:virus ratios, and therefore aggregation appeared to be the main mechanism of neutralization.

Ten antibodies, associated with sites 1A, 2 and 3, gave curves that had shapes intermediate between these two groups (e.g., Fig. 5.3*C*). As a group, the neutralization curves of these 10 antibodies formed a continuum with the other two groups, so assignment to an "intermediate" group was somewhat arbitrary. We assume that their interactions with virions had some of the characteristics of both of the other two groups.

Aggregation Studies

A key to understanding these neutralization curves was the construction of precipitation curves for each of the 32 antibodies (Leippe, 1991). Increasing concentrations of antibody were added to a standard amount of radiolabeled virus and the percentage of radioactivity in the pellet after a brief microcentrifugation was quantified. Eight of the 10 strongly neutralizing antibodies were found to be weakly precipitating, and all 12 of the weakly neutralizing antibodies were strong precipitators. This result agrees with the hypothesis that the strongly neutralizing antibodies are primarily those capable of binding bivalently to a single virion and therefore less likely to form large aggregates. The U-shaped curves (Fig. 5.3B) of the weakly neutralizing antibodies show reduction in titer as a result of aggregation at moderate antibody:virion ratios (equivalence zone of antibody–antigen precipitation) and restoration of infectivity at high antibody:virion ratios (antibody excess zone) when antibody-coated individual virions are present. It should be noted that, although these antibodies are characterized as "weak" neutralizers in solution, they are fully capable of inhibiting viral plaque formation and may be quite important in antiviral protection in vivo, where antibodies are working in concert with other components of the immune system.

Stoichiometry of Antibody Binding and Neutralization

In order to measure the number of antibodies needed to neutralize a virion and to determine the maximum number of antibodies that can bind to a virion, virus aggregates had to be separated from monomeric antibody-virus complexes. Therefore, radiolabeled antibodies and virus were mixed at various ratios and centrifuged on a sucrose gradient. The gradient profile showed a series of peaks representing virus monomers, dimers, trimers, and higher aggregates (Icenogle et al., 1983; Leippe, 1991). The strongly precipitating antibodies proved difficult to study because 1) precipitation occurred at optimal neutralization, 2) most of the strongly precipitating antibodies were never found associated with the monomeric virus peak except at extremely high antibody:virus input ratios, and 3) these monomeric species were very unstable. For these antibodies, the number of antibodies bound to virus at saturation could be determined, but not the neutralization stoichiometry.

Weakly precipitating antibodies formed stable complexes with unaggregated virus. Therefore, the average number of antibodies bound per virion could be calculated and the infectivities of these monomeric complexes were determined (Leippe, 1991). The Poisson formula predicts that when there is an average of one antibody per virion, then $1/e$, or about 37%, of the virions would *not* have an antibody bound. If one bound antibody were capable of neutralizing a virion, then the infectivity of a mixture with a ratio of one antibody per virion would be $1/e$ of a nonneutralized control. Therefore, the neutralization number was defined as the average number of antibodies attached per monomeric virion when the infectivity of this population has been reduced to $1/e$ of control infectivity. The neutralization numbers of four neutralizing anti-HRV14 antibodies ranged from 6 to 20. None was capable of neutralizing virus with a single antibody.

The saturation number is the number of antibodies bound to monomeric virus at extremely high-input antibody:virus ratios. For the weakly precipitating antibodies, the saturation numbers ranged from 32 to 49, supporting the hypothesis that these antibodies bind both Fab arms to the same virion. These numbers are slightly higher than the predicted 30 antibodies per virion, probably because, at very high antibody concentrations, some antibodies might be binding with only one Fab arm. Strongly precipitating antibodies, in contrast, demonstrated saturation numbers of 58 to 72, consistent with the model proposing that these antibodies are unlikely to bind both Fab arms to a single virion. For the strongly precipitating antibodies, resedimentation of the antibody-saturated monomeric virus peaks showed that the antibody-virus complex was unstable. This instability was made apparent by extensive aggregation and the presence of unbound antibody sedimenting at the top of the sucrose gradients.

Cell Attachment Studies

A representative series of antibodies that bound to each of the four NIm sites of HRV14 were tested for their ability to block cell attachment (Colonno et al., 1989). These authors found that a monoclonal antibody (MAb34-IA) to the NImIA site reduced attachment to HeLa cell membranes in a dose-dependent fashion. Antibody-induced aggregation of virus by antibodies binding to other sites made the precipitation-based assay uninformative. However, high concentrations of papain-derived Fab fragments of these antibodies were shown to block attachment to HeLa membranes. At high concentrations, the Fab fragments also decreased the pI of HRV14 from 7.0 to values between 2.0 and 3.6.

In the case of MAb17-IA, which binds to the NImIA site and is a weakly precipitating antibody, 16 antibodies per virion were required to inhibit attachment by 90%. In addition, 14 antibodies per virion were required to neutralize 90% of the input infectivity (Lee, 1992). This result suggests that, for at least this antibody, inhibition of attachment may be the primary mechanism of neutralization.

Studies on Stabilization at Low pH

HRV14, like other rhinoviruses, is unstable at a pH below 6.0 (Rueckert, 1990). Twelve different monoclonal antibodies that bound to the NIm-IA site were shown to be capable of stabilizing HRV14 infectivity at pH 5.0, and three of these stabilized at pH 4.5 (Lee, 1992). Four antibodies, binding to other NIm sites, could not protect HRV14 from inactivation after incubation at pH 5.0. Fab fragments of one of the NImIA antibodies, MAb17-IA, did not protect viral infectivity at pH 5.0 even if present at a concentration of 200 Fab molecules per virion, suggesting that bivalent binding of the antibody was necessary for pH stabilization. However, MAb1-IA, a strong aggregator and therefore unlikely to bind bivalently, was also able to protect the virions against low pH effects. Therefore, bivalent binding may not be necessary for stabilization. Perhaps antibody affinity or the antibody binding site are more important determinates for stabilization than bivalent binding.

Fab17-IA–HRV14 Complex Structure

Cryo–Electron Microscopy Structure Determination

Monoclonal antibody 17-IA strongly neutralizes HRV14 (Leippe, 1991), does not precipitate the virions, and is likely to bind bivalently to the capsid surface (Mosser et al., 1989). Because MAb17-IA is a strongly neutralizing antibody that is known to induce a pI shift in the capsid (Colonno et al., 1989), it was chosen for structural studies. The goal was to look for possible antibody-induced conformational changes in the capsid and to examine the binding contacts between the antibody and the virus.

The cryo–electron microscopy image reconstruction of the Fab17-IA–HRV14 complex has several distinctive features (Fig. 5.4). A dimple at the Fab elbow region allows a clear demarcation between the constant and variable regions. The Fab molecules did not bind radially to the virion surface, but rather bound tangentially from fivefold axes toward twofold axes. The Fabs bind to the NIm-IA site on the north rim of the canyon, lay across the canyon, make contact with the south rim, and meet an icosahedrally related Fab at the nearest twofold axis. Contrary to some of the proposed models for antibody-mediated neutralization, the HRV14 capsid features are nearly identical to image reconstructions of the virus alone. The lack of observable conformational changes suggests that, if Fab binding does cause conformational changes in the capsid, the magnitude of these changes is smaller than can be detected at the resolution of these studies. Therefore, the change in pI of the capsid upon antibody binding (Mandel, 1976; Emini et al., 1983) is probably not due to large, concerted conformational changes in the virion.

Fit of Atomic Structures into Electron Density

The variable domain makes extensive contact with the surface of HRV14 about the NIm-IA site. By examining the intersection of the Fab density with the HRV14 capsid density, the "footprint" of the bound Fab can be visualized (Fig. 5.5). The interaction between the Fab and the surface of HRV14 encompasses not only the NIm-IA site but part of the NIm-IB site as well. Therefore, the position of Fab binding is as expected, but the area of contact is much larger than was implied by the natural mutations alone.

The structures of both Fab17-IA (Liu et al., 1994) and HRV14 (Rossmann et al., 1985) have been determined to atomic resolution and were used to interpret the Fab–virus image reconstruction (Fig. 5.6). Although cryo–electron microscopy image reconstruction studies do not yield atomic-resolution electron density maps, they have proven to be remarkably accurate. A previously reported Fab–virus complex described the positioning of the Fab molecule to be accurate to within approximately 4 Å (Wang et al., 1992). Recent electron microscopy studies of the bacteriophage φχ174 accurately described small protrusions that were approximately 9 Å across at the base and approximately 23 Å long found at the threefold axes of the atomic structure (McKenna et al.,

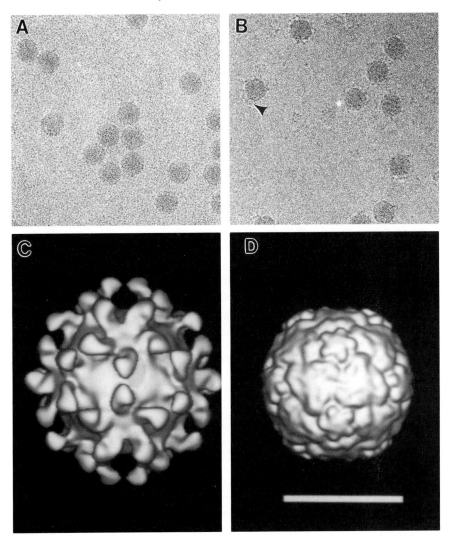

FIGURE 5.4   Electron microscopy studies of the Fab17-IA–HRV14 complex. Shown here are the raw cryo–electron microscopy images of HRV14 alone (*A*) and the Fab17-IA–HRV14 complex (*B*). The arrow in *B* notes the clear presence of bound Fab molecules on the surface of the normally smooth HRV14 particle. *C*. A surface rendering of the Fab17-IA–HRV14 complex image reconstruction. *D*. The HRV14 particle alone. This latter panel was generated from the atomic coordinates of HRV14. The bar represents 200 Å. (Reproduced with permission from Smith et al., 1993b.)

1992). These results suggest that we should be able to discern features with dimensions on the order of at least 10 Å. Therefore, although it is not reasonable to assign exact conformations to side chains and peptide strands, we can be certain of their general location.

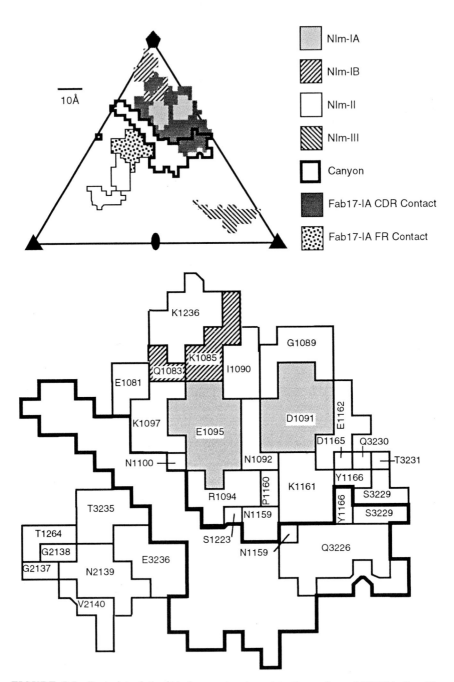

FIGURE 5.5 Footprint of the Fab fragment as bound to the surface of HRV14. *Top.* The triangular icosahedral asymmetric unit, with each of the major recognition sites depicted in various gray patterns. *Bottom.* Exploded view of the Fab17-IA contact area. Note that bound Fab17-IA actually overlaps the NIm-IB as well as the NIm-IA site. Residues are designated by the one-letter code; the first number represents the viral protein and the last three numbers designate the amino acid number. (Reproduced with permission from Smith et al., 1993b.)

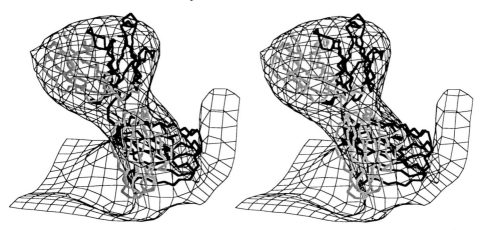

FIGURE 5.6 Stereo diagrams of the atomic structure of Fab17-IA fit into the cryo–electron microscopy density. The thicker lines represent the α-carbon backbone of the Fab fragment. These thick gray and black lines depict the light chain and heavy chain, respectively. The virion surface is toward the bottom and the nearest icosahedral five-fold axis is toward the right side of the figure. (Reproduced with permission from Smith et al., 1993b.)

Accuracy of Method

Because the incorporation of high-resolution structures into electron micros-copy density is a rather new technique, the accuracy of such a method had to be ascertained. As a test, the Fab model was systematically displaced from its optimal position in the Fab envelope and the resulting correlation between the observed and model density was calculated (Liu et al., 1994). Significant changes in agreement between the calculated and observed electron densities were observed when the Fab molecule was translated as little as 3 to 5 Å about the surface of the virion or when the Fab molecule was rotated 3° to 6°.

Potential Electrostatic Interactions

These results suggest that this method can be used to determine the general location of contact residues to within a few Ångstroms. With this limitation in mind, the antibody–antigen interface was examined (Smith et al., 1993b; Liu et al., 1994). The first observation was that the β-structure loop of VP1 that contains the two residues that previously defined the NIm-IA site (Asp 1091 and Glu 1095) fits well into the cleft between the heavy and light chain hy-pervariable regions. This virus–Fab interaction was predicted in earlier mod-eling studies because acidic Asp 1091 and Glu 1095 on the NIm-IA site are complemented by a basic cleft between the hypervariable regions of the Fab heavy and light chains. In addition, these modeling studies suggested that the CDR2 loop of the Fab heavy chain might interact with a cluster of lysines on HRV14 (Lys 1236, Lys 1097, and Lys 1085). To test this hypothesis, these residues on HRV14 were individually mutated to glutamine and glutamate (Lee, 1992; Smith et al., 1993a, 1993c). As predicted, changes of all three residues affected antibody binding. Mutations of Lys 1236 and Lys 1085 showed marginal effects on antibody binding (2.4- and 3.9-fold decreases, respectively), whereas the Lys 1097 mutant was nearly equivalent to a natural escape mutant (36-fold decrease in affinity). When the atomic models of HRV14 and Fab17-IA were placed into the electron microscopy density, residues Asp 555 and Asp 557 of the heavy chain CDR2 loop were observed to lie directly over Lys 1097. Lys 1085 and Lys 1236, which seemed to be less important for antibody binding, were somewhat distal to this CDR2 loop. Their role in antibody binding may be more to maintain the positively charged nature of this region or to set up a water–lysine network to maintain the Lys 1097 side chain position. From the extensive charge complementarity of the two surfaces, it is clear that charge is an important criterion for antigen recognition (see Color Plate 5).

Modeling of Bivalently Attached Antibodies

Bound Fab molecules nearly touched at icosahedral twofold axes, suggesting that MAb17-IA might bind bivalently to HRV14. It was found that a bivalently bound antibody could be easily modeled by simply rotating the constant domains of the bound Fab molecules about the elbow axes toward the virion surface (Smith et al., 1993c). The resulting interactions between icosahedrally twofold-related, Fab constant domains were nearly identical to those observed in the structure of the intact antibody Kol (Matsushima et al., 1980; Marquart et al., 1980). This similarity in the constant domain interactions between the modeled mAb17-IA and the known Kol structures suggested that the elbow region might "flex" as the antibody binds bivalently to the surface of the virion.

Fab17-IA did not cause large conformational changes in the capsid, but its position on the capsid occludes the cell receptor binding site. As described in Chapter 4, the structures of the cell receptor (ICAM)–HRV16 complex (Olson et al., 1993) and ICAM–HRV14 (N.H. Olson et al., unpublished results) have been determined. The ICAM site is immediately to the "south" of the Fab17-IA binding site. The binding positions of the Fabs and ICAM are, in fact, so close that it is impossible to have both bound concomitantly. This conclusion was substantiated by binding studies showing abrogation of HRV14 binding to host cells by MAb17-IA (Smith et al., 1993c).

MAb17-IA–HRV14 Complex Structure

Fit of Atomic Structure into Complex Density

To prove the bivalent model proposed in the Fab–virus study, image reconstruction studies were performed on the IgG–virus complex. Using a combination of high sodium chloride concentrations (~0.3 to 0.5 M), excess MAb17-IA, and ultrasonic homogenization to keep the antibody in solution, high concentrations of monomeric virions saturated with antibody were obtained. Cryo–electron microscopy and image reconstruction techniques yielded the structure of HRV14 with MAb17-IA clearly bound bivalently to the surface (Fig. 5.7) (Smith et al., 1993a).

The antibody–virus structure was strikingly similar to the Fab–virus structure with some important differences. The Fab arms of the IgG bound in a orientation similar to that of the Fab fragments except there was now a strong connection at the icosahedral twofold axes joining the two arms together. Whereas sodium dodecyl sulfate–polyacrylamide gel electrophoresis analysis demonstrated that the antibody contained the Fc portion, it was not visible in the image reconstruction. However, weak density was observed directly above the icosahedral twofold axes, suggesting that the Fc region is so flexible that it "smears" out in the reconstruction. The appearance of weak density for the Fc region concurs with previous crystallographic and electron microscopy studies that demonstrated the extremely dynamic nature of the hinge region. Conformational changes were still not observed in the capsid upon bivalent binding of the antibody. Therefore, it is also unlikely that bivalent attachment of the antibody induces large conformational changes in the capsid. This implies that the previously observed pI shift is not correlated with a large conformational change in the virus or that the pI shift requires, in addition to antibody binding, the presence of nonphysiologic conditions such as those needed for isoelectric focusing (e.g., low ionic strength and unusual buffers). Alternatively, the pI changes may be due to alterations in surface accessibility and not to any structural changes.

Evidence of Antibody Flexibility

The Fab17-IA structure, as bound to the virion surface in the Fab complex, is not well contained by the MAb17-IA envelope (Fig. 5.8A). Although the variable domain fits well into the density, the constant domain protrudes at the top. However, if the constant domain is rotated toward the virion surface by approximately 16° to 18° about the elbow axis, then the entire Fab arm fits within the IgG envelope (Fig. 5.8B). In this position, the light chain of the constant domain fills most of the connective density between the two Fab arms. Elbow angle variation among the crystallographically determined Fab structures has been used to suggest that this region is flexible (Davies and Metzger, 1983; Amit et al., 1986; Colman et al., 1987; Ely and Herron, 1989). However, this is the first time that the elbow region has been shown to actively participate in immunoglobulin binding.

In both the Fab and IgG complex structures, portions of the variable domain other than the hypervariable region are observed to bind in close proximity to the "south wall" of the canyon (Smith et al., 1993b). These contacts probably do not contribute to the specificity of the antibody binding, but rather

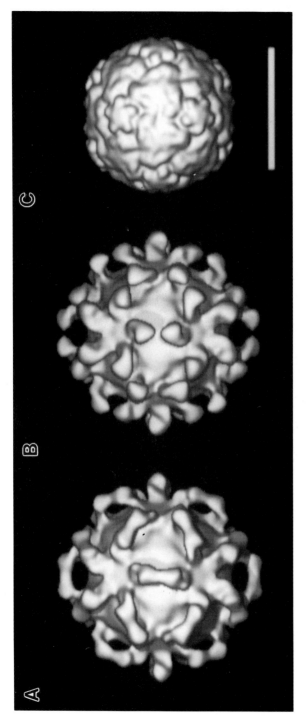

FIGURE 5.7 Image reconstruction of the MAb17-IA–HRV14 complex (A), the Fab17-IA–HRV14 complex (B), and HRV14 alone (C). The bar represents 200 Å. Note that there is a strong connection between the Fab arms in the IgG reconstruction (A) and that the exposed viral surface is very similar in all three panels. This latter observation strongly suggests that gross conformational changes do not occur in the capsid upon antibody binding. (Reproduced with permission from Smith et al., 1993a.)

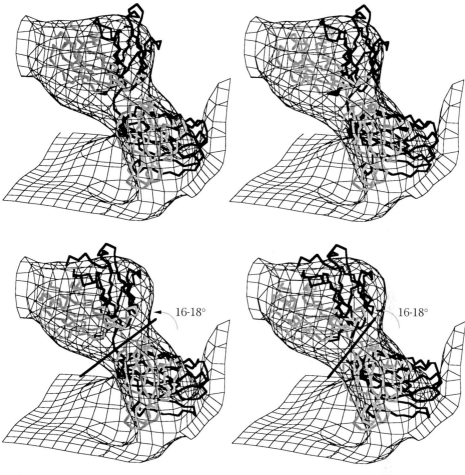

B

FIGURE 5.8 Stereo diagrams of original Fab model placed into the IgG density before (*A*) and after (*B*) rotating the constant domain about the elbow axis by approximately 16°. The view here is the same as that in Color Plate 5, with the light chain depicted with gray lines and the heavy chain in black. (Reproduced with permission from Smith et al., 1993a.)

arise from the angle at which the Fab fragment binds to the virion surface. Future crystallographic (Smith and Chase, 1992) and mutational studies should be able to discern the importance of these contacts. Currently, crystals of the Fab17-IA–HRV14 complex diffract to approximately 4- to 5-Å resolution. At this resolution, potential main chain deviations in the two structures should be clearly visible. In addition, electron microscopy studies are underway to examine other antibodies that bind to the same and different antigenic sites to ascertain whether this antibody is unusual in its inability to produce large conformational changes in the capsid.

Conclusions

This study demonstrates that, as has been observed with other Fab−protein complexes, the interface between antigen and antibody is extensive, and much of the hypervariable region plays an important role in binding. Furthermore, the contact area between the antibody and antigenic site is much larger than expected from mutational analysis. Although only two residues were identified as crucial for antibody neutralization by selection of naturally occurring mutants, many other residues were found to be in contact with Fab17-IA. The probable reason for this discrepancy is that only those mutants with viability similar to the native virion were originally selected.

Although there has been speculation that the elbow region of antibodies is flexible, most attention in the literature has been paid to the hinge region. In our studies it was clear that the elbow affords the antibody an additional degree of rotational freedom so that it can optimize antibody binding. Thus one can visualize the Fab regions of the immunoglobulin as ''arms,'' the hinge region representing shoulders, and the elbow region, elbows. Using both of these regions, the arms of the antibody can flex and twist until the best hypervariable contact is made with the antigen. Such motion is likely to occur not only when the antibodies bind to symmetric viruses, but also when they bind to asymmetric cell or bacterial surfaces. Also, such flexibility might allow some antibodies to bind bivalently to the virion surface without necessarily binding across icosahedral twofold axes. By allowing the antibody to bind bivalently, the elbow and hinge regions can theoretically help increase the affinity (avidity) of the antibody for antigen by as much as a factor of 10^3.

The electron microscopy results suggest that, although this antibody might cause a change in the pI of the capsid, such a change does not necessarily signal gross conformational changes in the capsid structure. Instead, the correlation between the stoichiometries of antibody binding and neutralization and the position of antibody binding strongly suggest that this antibody neutralizes by sterically blocking cell receptor attachment. Structural studies on other antibodies that recognize the same and different antigenic sites are necessary to determine whether this behavior is typical of other neutralizing antibodies.

The experiments reviewed here suggest a few of the most likely in vitro mechanisms of neutralization but do not address the most relevant in vivo mechanisms. If a crucial element of antibody-mediated neutralization is antibody-mediated stabilization or destabilization of the capsid, then distal, compensatory mutations might be expected to arise when virus is grown in the presence of antibodies. Such mutations have been isolated only in the presence of capsid-stabilizing WIN compounds (Heinz et al., 1989; see also Chapters 6 and 16), but not with antibodies. The WIN mutations were isolated in the presence of intermediate concentrations of WIN compounds, whereas antibody escape mutants have been mostly isolated at only high antibody concentrations. If distal, compensatory mutations can be isolated, perhaps in the presence of intermediate concentrations of antibody, then the importance of putative capsid effects would be established. Also, if antibody stabilization or destabilization of the capsid is a major component of neutralization, then we might expect that more viruses would evolve large, flexible, immunodominant protuberances such as those found on foot-and-mouth disease virus (Logan et al., 1993) that would enable them to evade neutralization by isolating and limiting the effects of antibody binding. Perhaps, because of the remarkable interplay between immune system components, the most biologically relevant antibody-mediated processes in vivo may be those pertaining to only antibody binding, such as inhibition of cell attachment, aggregation, and opsonization.

Acknowledgments

We would like to thank the many people who played crucial roles in the work reviewed here: E. Chase, R.H. Chen, W. Lee, D. Leippe, H. Liu, and N. Olson. We wish to especially thank M.G. Rossmann, R.R. Rueckert, and T.S. Baker for all of their advice, support, and contributions.

This work was supported by grants from the National Institutes of Health (GM10704 to T.J.S., AI31960 to R.R. Rueckert, and GM33050 to T.S.B.), from the National Science Foundation (MCB 9206305 to T.S.B.), and from the Lucille P. Markey Charitable Trust (Purdue Structural Biology Center).

References

Abad-Zapatero, C., Abdel-Meguid, S.S., Johnson, J.E., Leslie, A.G.W., Rayment, I., Rossmann, M.G., Suck, D. and Tsukihara, T. 1980. Structure of southern bean mosaic virus at 2.8Å resolution. Nature 286: 33–39.

Amit, A.G., Mariuzza, R.A., Phillips, S.E.V. and Poljak, R.J. 1986. Three-dimensional structure of an antigen-antibody complex at 2.8 Å resolution. Science 233: 747–753.

Brioen, P., Rombaut, B. and Boeyé, A. 1985a. Hit-and-run neutralization of poliovirus. J. Gen. Virol. 66: 2495–2499.

Brioen, P., Thomas, A.A.M. and Boeyé, A. 1985b. Lack of quantitative correlation between the neutralization of poliovirus and the antibody-mediated pI shift of the virions. J. Gen. Virol. 66: 609–613.

Colman, P.M., Laver, W.G., Varghese, J.N., Baker, A.T., Tulloch, P.A., Air, G.M. and Webster, R.G. 1987. Three dimensional structure of a complex of antibody with influenza virus neuraminidase. Nature 326: 358–363.

Colonno, R.J., Callahan, P.L., Leippe, D.M. and Rueckert, R.R. 1989. Inhibition of rhinovirus attachment by neutralizing monoclonal antibodies and their Fab fragments. J. Virol. 63: 36–42.

Colonno, R.J., Condra, J.H., Mizutani, S., Callahan, P.L., Davies, M.-E. and Murcko, M.A. 1988. Evidence for the direct involvement of the rhinovirus canyon in receptor binding. Proc. Natl. Acad. Sci. U.S.A. 85: 5449–5453.

Davies, D.R. and Metzger, H. 1983. Structural basis of antibody function. Annu. Rev. Immun. 1: 87–117.

Delaet, I. and Boeyé, A. 1993. Monoclonal antibodies that disrupt poliovirus only at fever temperatures. J. Virol. 67: 5299–5302.

Dulbecco, R., Vogt, M. and Strickland, A.G.R. 1956. A study of the basic aspects of neutralization of two animal viruses, western equine encephalitis and poliomyelitis virus. Virology 2: 162–205.

Ely, K.R. and Herron, J.N. 1989. Three-dimensional structure of a light chain dimer crystallized in water: conformational flexibility of a molecule in two crystal forms. J. Mol. Biol. 210: 601–616.

Emini, E.A., Ostapchuk, P. and Wimmer, E. 1983. Bivalent attachment of antibody onto poliovirus leads to conformational alteration and neutralization. J. Virol. 48: 547–550.

Everaet, L., Vrijsen, R. and Boeyé, A. 1989. Eclipse products of poliovirus after cold-synchronized infection of HeLa cells. Virology 171: 76–82.

Fox, M.P., Otto, M.J. and McKinlay, M.A. 1986. Prevention of rhinovirus and poliovirus uncoating by WIN 51,711, a new antiviral drug. Antimicrob. Agents Chemother. 30: 110.

Fricks, C.E. and Hogle, J.M. 1990. Cell-induced conformational change in poliovirus; externalization of the amino terminus of VP1 is responsible for liposome binding. J. Virol. 64: 1934–1945.

Harris, L.J., Larson, S., Hasel, K.W., Day, J., Greenwood, A. and McPherson, A. 1993. The three-dimensional structure of an intact monoclonal antibody for canine lymphoma. Nature 360: 369–372.

Harrison, S.C., Olson, A.J., Schutt, C.E., Winkler, F.K. and Bricogne, G. 1978. Tomato bushy stunt virus at 2.9Å resolution. Nature 276: 368–373.

Heinz, B.A., Rueckert, R.R., Shepard, D.A., Dutko, F.J., McKinlay, M.A., Francher, M., Rossmann, M.G., Badger, J. and Smith, T.J. 1989. Genetic and molecular analysis of spontaneous mutants of human rhinovirus 14 resistant to an antiviral compound. J. Virol. 63: 2476–2485.

Holland, J.J. and Hoyer, B.H. 1962. Early stages of enterovirus infection. Cold Spring Harbor Symp. Quant. Biol. 27: 101–112.

Hosur, M.V., Schmidt, T., Tucker, R.C., Johnson, J.E., Gallagher, T.M., Selling, B.H. and Rueckert, R.R. 1987. Structure of an insect virus at 3.0 Å resolution. Proteins 2: 167–176.

Icenogle, J., Shiwen, H., Duke, G., Gilbert, S., Rueckert, R. and Anderegg, J. 1983. Neutralization of poliovirus by a monoclonal antibody: kinetics and stoichiometry. Virology 127: 412–425.

Kabat, E.A., Wu, T.T., Reid-Miller, M., Perry, H.M., and Gottesman, K.S. 1987. Sequences of Proteins of Immunological Interests. Public Health Service, Bethesda, MD.

Klein, M., Haeffner-Cavaillon, N., Isenman, D.E., Rivat, C., Navia, M.A., Davies, D.R. and Dorrington, K.J. 1981. Expression of biological effector functions by immunoglobulin G molecules lacking the hinge region. Proc. Natl. Acad. Sci. U.S.A. 78: 524–528.

Langford, M.P., Crainic, R. and Wimmer, E. 1991. Antibodies may act synergistically or additively with interferon to inhibit poliovirus. Microb. Pathog. 10: 419–427.

Lee, W.M. 1992. Human rhinovirus 14: synthesis and characterization of a molecular cDNA clone which makes highly infectious transcripts, PhD. Thesis, University of Wisconsin.

Leippe, D.M. 1991. Stoichiometry of picornavirus neutralization by murine monoclonal antibodies, PhD. Thesis, University of Wisconsin.

Li, Q., Yafal, A.G., Lee, Y.M.H., Hogle, J. and Chow, M. 1994. Poliovirus neutralization by antibodies to internal epitopes of VP4 and VP1 results from reversible exposure of the sequences at physiological temperatures. J. Virol. 68: 3965–3970.

Liu, H., Smith, T.J., Lee, W.M., Leippe, D., Mosser, A. and Rueckert, R.R. 1994. The purification, crystallization, and structure determination of an Fab fragment that neutralizes human rhinovirus 14. J. Mol. Biol. 240: 127–137.

Logan, D., Abu-Ghazaleh, R., Blakemore, W., Curry, S., Jackson, T., King, A., Lea, S., Lewis, R., Newman, J., Perry, N., Rowlands, D., Stuart, D. and Fry, E. 1993. Structure of a major immunogenic site on foot-and-mouth disease virus. Nature 362: 566–568.

Mandel, B. 1976. Neutralization of poliovirus: a hypothesis to explain the mechanism and the one-hit character of the neutralization reaction. Virology 69: 500–510.

Marquart, M., Deisenhofer, J., Huber, R. and Palm, W. 1980. Crystallographic refinement and atomic models of the intact immunoglobulin molecule Kol and its antigen-binding fragment at 3.0 Å and 1.9 Å resolution. J. Mol. Biol. 141: 369–391.

Matsushima, M., Marquart, M., Jones, T.A., Colman, P.M., Bartle, K., Hubert, R. and Palm, W. 1978. Crystal structure of the human Fab fragment Kol and its comparison with the intact Kol molecule. J. Mol. Biol. 121: 441–459.

McKenna, R., Xia, D., Willingman, P., Ilag, L.L., Krishaswamy, S., Rossmann, M.G., Olson, N.H., Baker, T.S. and Incardona, N.L. 1992. Atomic structure of single-stranded DNA bacteriophage fX174 and its functional implications. Nature 355: 137–143.

Minor, P.B., Ferguson, M., Evans, D.M.A., Almond, J.W. and Icenogle, J.P. 1986. Antigenic structures of polioviruses of serotype 1, 2, 3. J. Gen. Virol. 67: 1283–1291.

Mosser, A.G., Leippe, D.M. and Rueckert, R.R. 1989. Neutralization of picornaviruses: support for the pentamer bridging hypothesis. In Molecular Aspects of Picornavirus Infection and Detection (Semler, B.L. and Ehrenfeld, E., Eds.), pp. 155–167. American Society for Microbiology, Washington, DC.

Olson, N.H., Kolatkar, P.R., Oliveira, M.A., Cheng, R.H., Greve, J.M., McClelland, A., Baker, T.S. and Rossmann, M.G. 1993. Structure of a human rhinovirus complexed with its receptor molecule. Proc. Natl. Acad. Sci. U.S.A. 90: 507–511.

Olson, N.H., Smith, T.J., Kolatkar, P.R., Oliveira, M.A., Rueckert, R.R., Greve, J.M., Rossmann, M.G. and Baker, T.S. 1992. Cryoelectron microscopy of complexes of human rhinovirus with a monoclonal Fab and the viral cellular receptor. Proc. Elect. Microsc. Soc. Am. 50: 524–525.

Page, G.S., Mosser, A.J., Hogle, J.M., Filman, D.J., Rueckert, R.R. and Chow, M. 1988. Three dimensional structure of poliovirus serotype 1 neutralization determinants. J. Virol. 62: 1781–1794.

Pevear, D.C., Fancher, F.J., Feloc, P.J., Rossmann, M.G., Miller, M.S., Diana, G., Treasurywala, A.M., McKinlay, M.A. and Dutko, F.J. 1989. Conformational change in the floor of the human rhinovirus canyon blocks adsorption to HeLa cell receptors. J. Virol. 63: 2002–2007.

Roivainen, M., Piirainen, L., Rysä, T., Närvänen, A. and Hovi, T. 1993. An immunodominant N-terminal region of VP1 protein of poliovirion that is buried in crystal structure can be exposed in solution. Virology 195: 762–765.

Rossmann, M.G., Arnold, E., Erickson, J.W., Frankenberger, E.A., Griffith, J.P., Hecht, H.J., Johnson, J.E., Kamer, G., Luo, M., Mosser, A.G., Rueckert, R.R., Sherry, B. and Vriend, G. 1985. Structure of a human common cold virus and functional relationship to other picornaviruses. Nature 317: 145–153.

Rossmann, M.G. and Johnson, J.E. 1989. Icosahedral RNA virus structure. Annu. Rev. Biochem. 58: 533–573.

Rueckert, R.R. 1990. Picornaviridae and Their Replications. Raven Press, New York.

Sherry, B., Mosser, A.G., Colonno, R.J. and Rueckert, R.R. 1986. Use of monoclonal antibodies to identify four neutralization immunogens on a common cold picornavirus, human rhinovirus 14. J. Virol. 57: 246–257.

Smith, T.J. and Chase, E.S. 1992. Purification and crystallization of intact human rhinovirus complexed with a neutralizing Fab. Virology 191: 600–606.

Smith, T.J., Olson, N.H., Cheng, R.H., Chase, E.S. and Baker, T.S. 1993a. Structure of a human rhinovirus-bivalent antibody complex: implications for virus neutralization and antibody flexiblity. Proc. Natl. Acad. Sci. U.S.A. 90: 7015–7018.

Smith, T.J., Olson, N.H., Cheng, R.H., Liu, H., Chase, E., Lee, W.M., Leippe, D.M., Mosser, A.G., Ruekert, R.R. and Baker, T.S. 1993b. Structure of human rhinovirus complexed with Fab fragments from a neutralizing antibody. J. Virol. 67: 1148–1158.

Smith, T.J., Olson, N.H., Cheng, R.H., Liu, H., Chase, E.S., Lee, W.M., Leippe, D.M., Mosser, A.G., Rueckert, R.R. and Baker, T.S. 1993c. Structure of human rhinovirus complexed with Fab fragments from a neutralizing antibody. J. Virol. 67: 1148–1158.

Stauffacher, C.V., Usha, R., Harrington, M., Schmidt, T., Hosur, M.V. and Johnson, J.E. 1987, The structure of cowpea mosaic virus at 3.5 Å resolution. In Crystallography in Molecular Biology (Moras, D., Drenth, J., Strandberg, B., Suck, D. and Wilson, K., Eds.), pp. 293–308. Plenum, New York.

Stuart, D. 1993. Virus structure. Curr. Opin. Struct. Biol. 3: 167–174.

Thomas, A.A.M., Brioen, P. and Boeyé, A. 1985. A monoclonal antibody that neutralizes poliovirus by cross-linking virions. J. Virol. 54: 7–13.

Tolskaya, E.A., Ivannikova, T.A., Kolesnikova, M.S., Drozdov, S.G. and Agol, V.I. 1992. Post-infection treatment with antiviral serum results in survival of neural cells productively infected with virulent poliovirus. J. Virol. 66: 5152–5156.

Tsao, J., Chapman, M.S., Agbandje, M., Keller, W., Smith, K., Wu, H., Luo, M., Smith, T.J., Rossmann, M.G., Compans, R.W. and Parrish, C.R. 1991. The three-dimensional structure of canine parvovirus and its functional implications. Science 251: 1456–1464.

Uhlig, J., Wiegers, K. and Dernick, R. 1990. A new antigenic site of poliovirus recognized by an intertypic cross-neutralizing monoclonal antibody. Virology. 178: 606–610.

Vrijsen, R., Mosser, A. and Boeyé, A. 1993. Post-absorption neutralization of poliovirus. J. Virol. 67: 3126–3133.

Wang, P., Porta, C., Chen, Z., Baker, T.S. and Johnson, J.E. 1992. Identification of a Fab interaction site (footprint) on an icosahedral virus by cryo-electron microscopy and X-ray crystallography. Nature 355: 275–278.

Wetz, K., Willingmann, P., Zeichhardt, H. and Habermehl, K.O. 1986. Neutralization of poliovirus by polyclonal antibodies requires binding of a single IgG molecule per virion. Arch. Virol. 91: 207–220.

Wiegers, K. and Dernick, R. 1991. Molecular basis of antigenic structures of poliovirus: Implications for their evolution during morphogenesis. J. Virol. 66: 4597–4600.

6.

The Role of Conformational Transitions in Poliovirus Pathogenesis

MARIE CHOW, RAVI BASAVAPPA,

& JAMES M. HOGLE

In the 10 years since a high-resolution structure for poliovirus was obtained, many studies have added significantly to our understanding of the molecular biology of poliovirus and picornaviruses in general. Thus, a more detailed perspective is now available in regard to the molecular mechanisms underlying many of the biologic properties of poliovirus, including assembly, immune recognition, cell entry, and virus tropism and host range. In particular, the use of infectious clones, the characterization of site-specifically constructed or randomly mutagenized viral mutants, and the development of in vitro assay systems have provided insights into the mechanisms of viral RNA replication, protein synthesis and processing, and the regulation of these processes (reviewed in Racaniello, 1990; Wimmer et al., 1993; Rueckert, 1996). Genetic analyses of capsid protein mutations have contributed greatly to the characterization of capsid protein folding and proteolytic processing, the capsid assembly pathway, the antigenic structure of the virus, domains influencing capsid stability, and the host range of the virus. The identification, cloning, and characterization of the human poliovirus receptor has provided additional insights into the host range and tissue specificities of the virus and initiated the molecular dissection of the virus−receptor interaction (Racaniello, 1990). Finally, the more recent determinations of the structures of the Sabin 3 vaccine strain and a poliovirus chimera, the poliovirus empty capsid, and poliovirus−antiviral drug complexes have provided an additional structural framework to complement these genetic and molecular biologic studies (Yeates et al., 1988; Filman et al., 1989; Basavappa et al., 1994; Grant et al., 1994; Hiremath et al., 1995).

These studies in aggregate have begun to provide insights into the structural contributions that underlie the multifaceted phenotypes that characterize different aspects of poliovirus pathogenesis and that are discussed here.

Roles of the Capsid Proteins during the Virus Life Cycle

Poliovirus initiates infection by binding to a receptor that is expressed on susceptible cells (Fig. 6.1). The virus particle is taken up and uncoated and the plus-stranded, approximately 7400-nucleotide viral RNA genome is released into the cytoplasm. The viral genome is translated and the protein covalently modified with myristic acid to generate the poliovirus polyprotein (Chow et al., 1987). This polyprotein is the precursor of all known virus-specific proteins and is proteolytically processed by virally encoded proteases to yield the mature viral proteins. The nonstructural viral proteins forming the replication complexes necessary for synthesis of minus-stranded RNAs, mRNAs, and progeny RNA genomes are located within the carboxyl-terminal two thirds of the polyprotein. The capsid proteins are located in the amino-terminal third of the polyprotein in the order VP4—VP2—VP3—VP1. The polyprotein is initially cleaved to yield the P1 protein, which is the capsid precursor. Virion assembly is initiated by cleavage of P1 to yield a heterotrimeric complex of VP0, VP3, and VP1 (called the protomer), which is then assembled into pentamers. The cleavage of P1 by the 3CD protease requires the complete and correct folding of the P1 precursor (Ypma-Wong et al., 1988). Thus, this processing step plays a dual role, controlling the timing of the initiation of assembly and serving as an editing mechanism to ensure that only correctly folded protomers enter the assembly pathway. Pentamers can further associate in the absence of RNA to form empty capsids. It is not known whether the viral RNA is encapsidated directly into empty capsids or if pentamers nucleate around the RNA. Either during or subsequent to packaging of the RNA, the immature capsid protein precursor, VP0, is cleaved to form VP4 and VP2 and the mature virion is formed. This cleavage of VP0 is associated with a significant increase in the stability of the particle. Thus this maturational cleavage may be viewed as a step that drives assembly to completion and makes assembly irreversible. Evidence demonstrates that, in the closely related rhinovirus 14, this cleavage is also required for the infectivity of progeny virus (Lee et al., 1993). Viral infection leads to cell lysis and release of the progeny virus into the surrounding environment.

During the infectious life cycle, the virus particle and capsid proteins display multiple functional roles. Upon translation of the viral genome, the proteins must fold and efficiently self-assemble into capsid structures. Concomitant with capsid assembly, the viral capsid proteins must specifically recognize and encapsidate only VPg-linked, plus-stranded RNA molecules and thus must contain RNA sequence recognition and binding capabilities. In the assembled virus, the capsid proteins must provide sites for specific recognition and binding of the cellular receptor on the cell surface. In addition, the assembled capsid

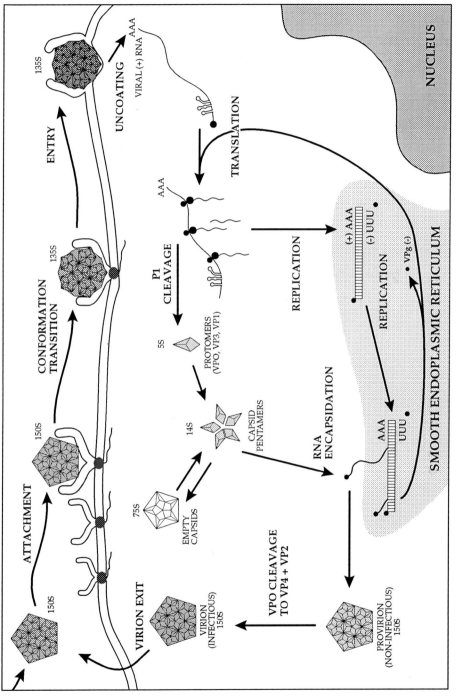

FIGURE 6.1 Replication cycle of poliovirus.

proteins must provide, for the RNA genome, a protective coat that is sufficiently stable to survive the rigors of passage from cell to cell and host to host. Because poliovirus is an enteric virus, the coat must be sufficiently stable to resist the low pH of the stomach and the proteases present in the small intestine. However, to fulfill its delivery role, upon binding the cellular receptor, the capsid must undergo a set of conformational rearrangements that result in uptake into the cell and release of the RNA genome into the cell cytoplasm. Thus a balance must be maintained between stability of the virion and its ability to initiate infection. These multiple roles have been amply demonstrated by studies of mutants obtained by site-directed mutagenesis of the infectious cDNA clone. In contrast to studies with small proteins, in which scanning mutagenesis or even deletion/insertion mutagenesis is possible, many substitutions, presumed to be benign, are lethal to the virus for poorly understood reasons. In addition, upon characterization of the site-directed mutants, many of the viable capsid mutations have multiple and often unexpected phenotypes (Compton et al., 1990; Kirkegaard, 1990; Moscufo and Chow, 1992; Reynolds et al., 1992; Moscufo et al., 1993). Although the structure has not always allowed prediction of the phenotypes for a given mutation, it has provided a useful tool for rationalizing the observed properties. There is growing evidence that the different functional roles are often intimately linked with conformational changes of the capsid proteins (see later). This is particularly clear during cell entry and capsid assembly stages. The growing body of mutation data is now beginning to provide exciting clues as to the conformational mechanics of the virion.

Structure of Mature Virion

The general principles of virus structures have been reviewed extensively in this volume and elsewhere (Harrison et al., 1996; see also Chapter 1). The brief description provided here is intended to highlight several important structural features that are essential in the interpretation of the genetic studies discussed in detail later in this chapter. The three larger capsid proteins (VP1, VP2, and VP3) share a common ''core'' structure that can be described as an eight-stranded, wedge-shaped β-barrel or jelly roll (Fig. 6.2). This core motif is shared with a number of other spherical viruses of plant and animal origin. Each of the capsid proteins has a unique set of loops that connect the regular secondary structural elements of this core, and each of the proteins has a unique, rather long extension at its amino terminus. In the virion, the capsid proteins pack on a $T = 1$ icosahedral surface. The cores of the proteins form the closed shell of the virus, with the narrow end of VP1 pointing toward the particle fivefold axes and the narrow ends of VP2 and VP3 alternating around the particle threefold axes. This shell is relatively thin (averaging approximately 30 Å) and encloses a large central cavity in which the viral RNA is located. Unfortunately, the RNA lacks the symmetry of the virion and is therefore transparent in the X-ray structure.

The tilts of the cores around the symmetry axes result in protrusions at the fivefold axes and somewhat smaller protrusions at the threefold axes. These protrusions are separated by a broad circular depression surrounding the fivefold axes and a saddle-shaped depression across the twofold axes. This overall shape is characteristic of most of the viruses whose capsid protein share the β-barrel core. The loops, which connect the eight strands of the β-barrel, and the carboxyl termini of the capsid proteins decorate the surface of the virion, giving each virus a characteristic shape. In poliovirus, the connecting loops at the narrow ends of the capsid proteins accentuate the protrusions at the fivefold and threefold axes, and the loop connecting the G and H strand of VP1, together with a large double loop connecting the E and F strands of VP2, form a large surface protrusion near the particle twofold axes (Fig. 6.2). These surface features accentuate the depression surrounding the fivefold axes. The prominent surface loops have been shown to contain the antigenic sites of the virus (Page et al., 1988), and in the closely related rhinoviruses the depression surrounding the fivefold axis has been shown to contain the binding site for the virus receptor (Olson et al., 1993; see also Chapter 4). The N-terminal extensions, together with the VP4 (which in many ways can be considered as the detached portion of the amino-terminal extension of VP2) decorate the inner surface of the protein shell of the virus. The amino termini form an elaborate network that directs the assembly of the virus and plays a crucial role in stabilizing the mature virion (see Color Plate 18) (Hogle et al., 1985). One particularly striking feature of this network is a five-stranded tube of parallel β-structure that is formed by five copies of the amino terminus of VP3 as they intertwine around fivefold axes (Fig. 6.3). This β-tube is flanked by five copies of a small three-stranded β-sheet formed from the extreme amino termini of VP4 and VP1. The amino terminus of VP4 is myristylated, and the fatty acid moiety mediates the interaction between VP4 and VP3 sequences (Chow et al., 1987). The β-tube structure can only form during the formation of pentameric assembly intermediates. Additional network interactions involving the C, H, E, and F strands of VP3 and the amino-terminal extensions of VP1 and VP2 form a seven-stranded β-structure that links pentamers in the virion (Fig. 6.3) (Filman et al., 1989). This feature can only form as pentamers assemble into larger structures.

Conformational Rearrangements

The mature virus structure has provided a particularly useful framework for describing those properties that are functions of the intact virus itself, specifically recognition of the virus by antibodies and early events in receptor recognition. However, the description does have the limitation of providing a model for only one of several capsid protein−containing structures that have been described. Thus the virion structure provides a static view of one stage within a dynamic process.

a

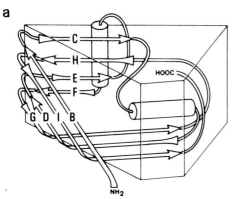

b

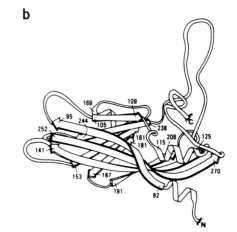

c

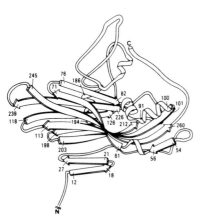

d

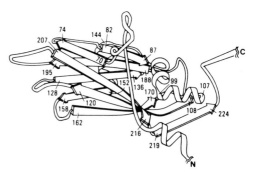

e

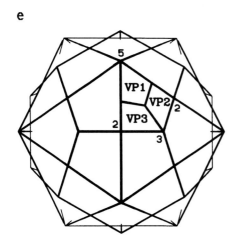

f

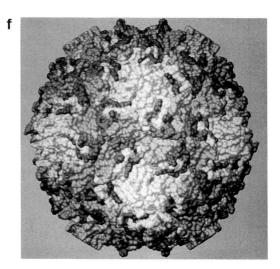

Significant conformational rearrangements of the capsid proteins appear to play a role at several stages of the virus life cycle. During the assembly process, cleavage of the precursor protein P1 results in significant rearrangements of the newly generated carboxyl termini of VP0 and VP3 and amino termini of VP3 and VP1. These rearrangements generate the interfaces on the protomer that interact in the pentameric assembly intermediate (Flore et al., 1990). The rearrangements also generate those portions of the internal network that stabilize this intermediate (most notably the intertwined β-tube formed by five copies of the amino terminus of VP3 at fivefold axes). Moreover, the recently described structure of poliovirus empty capsids demonstrates that the maturation cleavage of VP0 to VP2 and VP4 potentiates further conformational changes that complete the formation of the internal network (Basavappa et al., 1994). Except for the positions of VP4 and of the amino-terminal extensions of VP1 and VP2 on the inner surface, the empty capsid structure (i.e., the packing of the β-cores and the outer surface of the particle) is identical to that of the mature virus. Comparison of the empty capsid structure with that of the mature virus reveals that the final cleavage of VP0 (perhaps in conjunction with the encapsidation of the viral RNA) results in the formation of a number of intraprotomer and intrapentamer interactions involving the newly released carboxyl terminus of VP4 and the amino-terminal extensions of VP1 and VP2, and completes the formation of the interpentamer interactions, including the seven-strand β-structure. This completion of the internal network results in a net gain of a large number of stabilizing interactions in the mature virion (located at intraprotomer, intrapentamer, and interpentamer boundaries) and explains the role of VP0 cleavage in finalizing the assembly process.

During cell entry, interaction of the virion with the cellular receptor at physiologic temperatures produces a particle with an altered sedimentation coefficient of 135S (versus 160S for the mature virion). This alteration in sedimentation behavior is accompanied by disruption of the internal network and the relocation and exposure of VP4 and most of the amino-terminal regions of

---

FIGURE 6.2   Structure and organization of the major capsid proteins of poliovirus. *a.* Schematic representation of the common wedge-shaped eight-stranded antiparallel β-barrel core motif shared by VP1, VP2, and VP3. Individual β-strands are shown as arrows and are labeled alphabetically. Flanking helices are indicated by cylinders. *b−d.* Ribbon diagrams of VP1 (*b*), VP2 (*c*), and VP3 (*d*). Residue numbers of P1/Mahoney strain have been included as landmarks. VP4 is not shown, and the amino-terminal and carboxyl-terminal extension of VP1 and VP3 have been truncated for clarity. *e.* A geometric representation of the outer surface of the poliovirus virion is generated by superimposing an icosahedron and a dodecahedron. The symmetry axes of the particle and the positions of VP1, VP2, and VP3 in a reference protomer are indicated. Like the virion, the geometric figure has large radial projections at the fivefold axes and somewhat smaller projections at the threefold axes. *f.* This space-filling representation of the outer surface of the intact virus is in the same orientation as the geometric figure in *e.* In the center of the picture, VP1 is light gray, VP2 is medium gray, and VP3 dark gray. Depth cueing causes all three proteins to appear darker at the periphery of the view. (Reproduced with permission from Filman et al., 1989.)

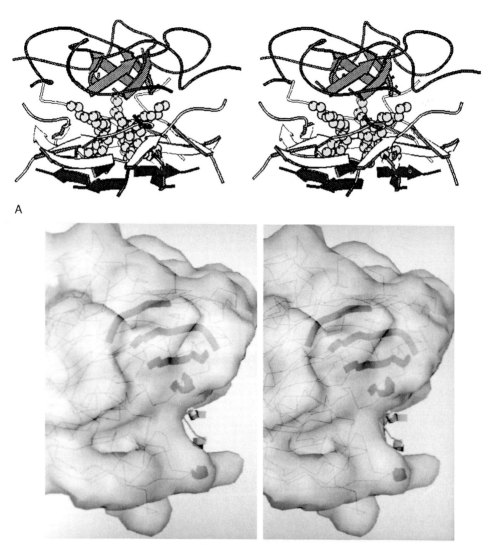

A

B

FIGURE 6.3 Secondary structural elements formed by the amino-terminal extensions of the capsid proteins and VP4. *A.* The β-tube structure at the fivefold axis. In this stereo representation, the outside of the virus particle is at the top and the inside of the virus is at the bottom. The five copies of the amino terminus of VP3 (dark gray at the top of the picture) intertwine to form a parallel β-tube. The tube is flanked by five copies of a three-stranded β-sheet, with two of the strands contributed by VP4 (white) and the third strand formed by residues near the amino terminus of VP1 (black). The interactions between VP4 and VP3 are mediated by the myristate moieties (gray spheres). This tube structure stabilizes interactions between protomers within a pentamer.

B and *C.* Stereo representation of a seven-stranded cross-pentamer β-interaction with the outside of the virus at the top and inside of the virus at the bottom. In these representations, the structure of a reference protomer is represented as an α-carbon backbone (thin lines) enclosed with a semitransparent surface with the outside of the virus up and the inside down. Figure continues.

VP1 from the inner to the outer surface of the virion shell. Exposure of the amino terminus of VP1 has been shown to confer membrane binding properties to the 135S particle (Fricks and Hogle, 1990). A very similar, if not identical, conformation rearrangement is observed when the mature 160S virion is exposed briefly to slightly elevated temperatures under conditions of low ionic strength (2 min at 45 °C) (Wetz and Kucinski, 1991). Like the receptor-induced 135S particle, the heat-induced 135S particles appear to relocate VP4 and amino-terminal VP1 sequences to the particle surface. The similarity between the receptor-induced and heat-induced 135S particles suggests that the 160S-to-135S transition requires energy, may not be uniquely dependent on receptor

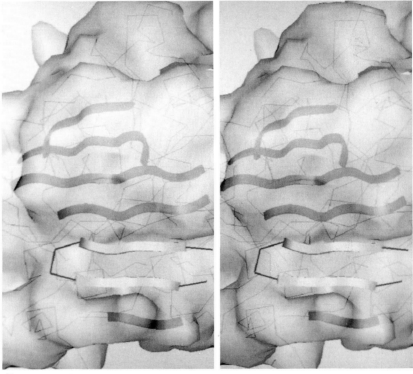

C

FIGURE 6.3 *Continued*   The outer four strands are contributed by the C, H, E, and F strands of the β-barrel of VP3 of the reference protomer (gray ribbons at top); the fifth and sixth strands are contributed by a two-stranded β-structure near the amino terminus of VP2 from a threefold related protomer (white ribbons at center); and the (innermost) seventh strand is contributed by residues form the amino-terminal extension of VP1 of the reference protomer (gray ribbon at bottom). In *B*, the structure is a view of a pentamer interface (front surface) looking toward the fivefold axis. The twofold axis is at the left and the threefold axis is at the right. In *C*, the interface is shown from the side as viewed from the twofold axis toward the threefold axis. Note how the fifth and sixth strands (contributed by the amino terminus of VP2 from the neighboring pentamer) penetrate into a deep groove in the reference protomer. This structure would be expected to stabilize the interactions between two pentamers.

interaction, and might be induced by a variety of other environmental stimuli. A more drastic transition is induced upon incubation for longer periods or at more elevated temperatures (e.g., 55 °C) (Rueckert, 1976). This particle sediments at 80S, has released the RNA, and has externalized VP4 and portions of the amino terminus of VP1. Comparison of the sedimentation coefficients of the mature virus, the 135S particle, and the 80S particle suggests that the 135S particle must be expanded by approximately 10% in radius compared to the mature virion.

Recent evidence demonstrates that, in addition to the concerted and irreversible conformational changes described above, the virion has considerable conformational flexibility at physiologic temperatures (Li et al., 1994). Thus, incubation of virions at 37 °C results in the transient exposure of VP4 and the amino terminus of VP1, which can be detected immunologically. The immunologic data demonstrate that the exposure is temperature dependent (it is not detected at 25 °C) and is reversible. The reversible exposure of sequences (which are irreversibly exposed upon specific environmental cues) is consistent with the view that the virion is poised to undergo the irreversible conformational rearrangements at physiologic temperatures but is prevented from doing so by an energy barrier. This barrier may be overcome either by increasing the temperature or by interaction with the receptor. In this context, the receptor induces the conformational rearrangement by a mechanism very similar to that by which enzymes catalyze reactions. The data also suggest that the virion is metastable. The formation of a metastable particle can be explained only if the pathway to the minimum energy configuration is blocked by a kinetic barrier. Covalent modification such as the cleavage of VP0 to VP4 and VP2 is a classic mechanism for trapping an assembly intermediate in a local minimum. Thus the mature virus can be thought of as an intermediate that links the assembly and cell entry processes. This picture of the role of the virion is entirely consistent with several groups of mutants containing single substitutions in the capsid proteins that exhibit defects in both assembly and cell entry (Compton et al., 1990; Moscufo et al., 1993).

Structures of conformationally altered particles such as the 135S and 80S particles may eventually become available. However, these structures at best would provide a series of snapshots along a dynamic pathway. Further information about the nature of the dynamic pathways, that link these structural snapshops must be provided by a combination of genetic analysis and computational simulations. Fortunately, a series of genetic studies, when evaluated within the context of the mature virion structure alone, has begun to provide intriguing insights into the nature of the conformational rearrangements associated with virus maturation and cell entry. These genetic studies were originally designed to address quite disparate phenotypes associated with pathogenesis, and include analyses of 1) mutations associated with temperature sensitivity of the Sabin (vaccine) strain of serotype 3 poliovirus (P3/Sabin); 2) mutations that modify the host range or the ability of the virus to persist in cultured neuronal cells; 3) mutations that confer resistance to neutralization by solubilized receptor; and 4) mutations that confer resistance to (and in some cases dependence on) antiviral drugs that bind the capsid. These studies point to a surprisingly limited number of residue clusters that regulate the ability of the virus to undergo conformational rearrangements.

Thermosensitivity and Attenuation of Sabin 3 and Non-Temperature-Sensitive Mutants

The Sabin attenuated vaccine strain of type 3 poliovirus (P3/Sabin) was derived from a neurovirulent parental strain, P3/Leon, by multiple passages under suboptimal growth conditions. The vaccine strain is attenuated for its ability to cause paralysis in both monkeys and humans. In addition, it is temperature sensitive for growth in cell culture (titers are generally 5 to 6 logs lower when the virus is grown at 39.5 °C vs. 35 °C). Recombinant viruses generated by manipulation of the cDNA clones of the P3/Sabin and P3/Leon genomes identified two mutations that were dominant determinants of attenuation of P3/Sabin (Minor et al., 1989; Westrop et al., 1989). One of these mutations occurred in the 5′ noncoding region of the viral genome in a region that has since been identified as a site for the internal initiation of cap-independent translation of the viral genome (Pelletier and Sonenberg, 1989). The second mutation coded for the substitution of a phenylalanine (P3/Sabin) in place of a serine (P3/Leon) at residue 91 of VP3. This mutation was also shown to be the primary determinant of the temperature-sensitive (ts) phenotype of P3/Sabin (Minor et al., 1989). In the virion, VP3-91 is located at the surface of the particle in the depression surrounding the fivefold axis and in the vicinity of the interface between fivefold-related protomers (Fig. 6.4). In P3/Sabin, the phenylalanine side chain of VP3-91 is exposed on the surface of the virion. The exposure of this large hydrophobic side chain is energetically very unfavorable, and we have proposed that the destabilization caused by exposure of the side chain would promote thermally induced conformational changes that would allow the phenylalanine side chain to be buried (Filman et al., 1989). It has not been possible to convincingly demonstrate a difference in the thermal stability of the mature virion of P3/Sabin versus P3/ Leon. The destabilization must therefore be expressed at some other stage of the virus life cycle. In a series of temperature jump experiments, Macadam et al. (1991) have shown that the temperature sensitivity is expressed during assembly at the level of pentamer formation. Moreover, Macadam et al. demonstrated that a temperature shift to the nonpermissive temperature early in assembly resulted in increased turnover of P1, suggesting that P1 was being sequestered into classical degradation pathways within the cell, which normally serve to eliminate misfolded proteins. Thus at nonpermissive temperatures, these mutations appear to affect the correct folding and stability of the capsid proteins and early assembly intermediates.

In a separate study, Minor, Macadam, and colleagues characterized a series of non-ts revertants of P3/Sabin either isolated from healthy vaccinees or selected by forced growth in cell culture at 39 °C of a P3/Leon-derived virus that was engineered to contain a phenylalanine at VP3-91 (Macadam et al., 1989;

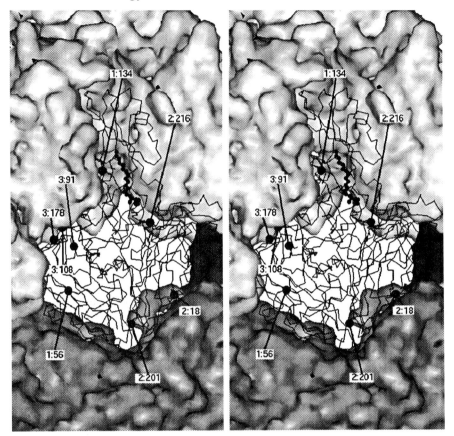

FIGURE 6.4 Location of mutations that regulate temperature sensitivity in P3/Sabin (see Table 6.1). In this stereo representation, the locations of mutations are indicated by solid black spheres on an α-carbon backbone of a reference protomer. The reference protomer is shaped like a downward-pointing arrow, with the shaft of the arrow originating from a fivefold axis (top center), the point of the arrow at a threefold axis (bottom center), and the barbs of the arrow at particle twofold axes (left and right center). The lipid binding pocket is occupied by a putative sphingosinelike lipid ligand (thick black line). Fivefold-related subunits are shown as light gray surface representations. Subunits from neighboring pentamers are shown as dark gray surface representations. Labels indicate the capsid protein number:residue number of the identified mutation.

Minor et al., 1989). The in vitro−selected revertants invariably contained either the genotypic reversion to the parental (P3/Leon) serine at VP3-91 or a suppressor mutation at a second site that introduced a leucine in place of the phenylalanine at VP1-132. The side chain of VP1-132 is in a pocket that binds an endogeneous lipid ligand (which has become known as ''pocket factor'') in the hydrophobic center in the core of VP1 (see Chapters 4 and 16). The location of a non-ts suppressor in this pocket suggests that binding of a natural lipid ligand at this site plays a role in stabilizing the mature virion (Filman et al., 1989).

The non-ts revertants isolated from vaccinees generally contain multiple mutations in the capsid protein region (Macadam et al., 1989; Minor et al., 1989). Macadam et al. have constructed each of these mutations singly and in combination back onto the P3/Sabin background and investigated their ability to grow at the nonpermissive temperature. They identified a number of mutations that were capable of either partially or completely suppressing the ts phenotype (Table 6.1). The second-site suppressor mutations from the vaccine isolates that have been characterized thus far mapped to two general locations: 1) in the interface between fivefold-related protomers, close to the position of Phe 3091, and 2) in the internal network present on the inner surface of the particle (Fig. 6.4). The fivefold (intrapentamer) interface is analogous to an interface in structurally related plant viruses that is disrupted when the viruses expand upon depletion of divalent cations at alkaline pH (Incardona and Kaesberg, 1964). By analogy, the mutations in this interface in P3/Sabin may counteract the destabilizing effect of the primary mutation at VP3-91, by stabilizing an interface that is unstable at nonpermissive temperatures. The mutations in the amino-terminal network include mutations in residues that make critical intraprotomer and interpentamer interactions on the inner surface of the particle. Again the location of these suppressor mutations suggests that they may counter destabilization caused by the phenylalanine at VP3-91 by increasing the stability of assembly products.

Further support for this model in which destabilizing effects of one mutation are counteracted by stabilizing effects of mutations at spatially distant sites comes from the analysis of the mutations identified in a non-ts variant that contained three point mutations at VP2-200, VP2-215 and VP1-54 (Macadam et al., 1989, 1991). Individually the mutation at either VP2-215 or VP2-54 was able to partially suppress the ts phenotype of P3/Sabin, whereas the mutation at VP2-200 exacerbated the ts phenotype. Together the mutations at VP2-215 and VP1-54 produced an extreme non-ts virus (despite the presence of the phenylalanine at VP3-91). This virus grew well at 39 °C but was incapable of growing at 32 °C. According to our model discussed previously, the combination of two stabilizing mutations (VP2-215 and VP1-54) together with two destabilizing mutations (VP2-200 and VP3-91) in the original isolate results in a wild-type (non-ts) phenotype.

Mouse Adaptation

Most poliovirus strains, including the well-characterized P1/Mahoney, are highly specific for humans and old world primates. However, all strains will undergo a single round of replication within nonsusceptible cells upon transfection of the RNA genome and will infect murine cells that have been transfected with the gene for the human poliovirus receptor. Several poliovirus strains have been identified that are either naturally able to replicate in the mouse central nervous system or can be adapted to grow there by multiple passage in rodent brains. These mouse-adapted strains have been used to study

TABLE 6.1 TS Phenotype of P3/Sabin

Strain	Capsid Protein	Residue Number[a]	Amino Acid Parental	Amino Acid Mutant	Structural Location[b]	Phenotype
P3/Leon	VP3	91 (91)	Ser	Phe	Surface accessible near fivefold interface	ts at 39
P3/Leon	VP1	132 (134)	Phe	Leu	Lipid binding pocket	ts suppressor in combination with Phe at VP3-91
P3/Sabin	VP3	108 (108)	Thr	Ala	Inaccessible at fivefold interface	ts suppressor in combination with Phe at VP3-91
	VP3	178 (178)	Gln	Leu	Inaccessible at fivefold interface	
	VP2	18 (18)	Leu	Ile	Internal network, cross-pentamer β-structure	
P3/Sabin	VP1	54 (56)	Ala	Val	Internal network, amino-terminal VP1 loop in contact with VP4	Partial ts suppressor individually when paired with Phe at VP3-91; extreme ts resistance when together in combination with Phe at VP3-91
	VP2	215 (216)	Leu	Met	Inaccessible at fivefold interface	
P3/Sabin	VP2	200 (201)	Arg	Lys	Inaccessible, near VP1 loop and cross-pentamer β-structure	Increased ts phenotype

[a]The residue numbers in parentheses refer to the equivalent residue in P1/Mahoney.

[b]Surface accessibility determined by evaluating the maximum radius of a sphere that contacts any atom within the residue without contacting another atom in the structure. Residues were scored as accessible if they could be contacted by a sphere greater than or equal to 3 Å in radius.

poliovirus pathogenesis and to probe the nature of the receptor–virus interaction.

The Lansing strain of type 2 poliovirus (P2/Lansing) is a particularly well-characterized mouse-adapted strain. This virus is capable of causing paralysis in mice upon intracerebral injection but does not grow in murine cell lines. Using chimeric viruses constructed by recombinant methods from the cDNA clones of P2/Lansing and the primate-specific P1/Mahoney strains, La Monica and Racaniello (La Monica et al., 1986, 1987) demonstrated that the mouse adaptation phenotype resided within the capsid proteins of P2/Lansing. Subsequent studies showed that chimeras expressing residues 94 to 104 of VP1 from P2/Lansing in a P1/Mahoney background were nearly as neurovirulent as the parental Lansing strain in mice (Martin et al., 1988; Murray et al., 1988). These residues correspond to the BC loop of the capsid protein VP1 and are highly exposed at the fivefold axes of the virion (Fig. 6.5). The location of these residues on the surface of the virion suggests that they contribute significantly to binding of a receptor in the mouse central nervous system. However, recombinant viruses either lacking the BC loop or containing a variety of exogenous sequences can still replicate in human cell lines (Evans and Almond, 1991). This suggests that the receptor interaction in the mouse by these mouse-adapted strains is fundamentally different from that in human cells.

The sequence of the BC loop of VP1 for P2/Lansing and P1/Mahoney differs in 6 of the 10 residues (Table 6.2). Structural studies of the mouse-adapted chimera demonstrated that the conformation of the BC loop in the chimera differs significantly from that of the parental P1/Mahoney and indicated that several of the sequence differences contributed to the structural differences (Yeates et al., 1988). Based on the assumption that the conformation of the BC loop was the major determinant of mouse virulence, Moss and Racaniello (1991) probed the role of these sequence differences by constructing P2/Lansing chimeras containing P1/Mahoney sequences at these positions. The chimeras expressing Mahoney sequences at these structurally critical residues were attenuated. These investigators were able to isolate neurovirulent revertants of these mouse attenuated strains by intracerebral inoculation at very high titer. Sequence analyses of these neurovirulent strains revealed that they maintained the attenuating sequence differences in the BC loop in VP1 but contained substitutions in the amino-terminal extension of VP1 (Fig. 6.5). Reconstruction of these mutations individually into the P1/Mahoney background resulted in significant increase in mouse neurovirulence (Moss and Racaniello, 1991). In other studies, Couderc et al. (1993) have selected for mouse virulent variants of P1/Mahoney by passage at high titer in mouse brains. They identified two mutations that displayed the mouse neurovirulent phenotype when reconstructed individually into the P1/Mahoney background (Table 6.2). Once again the two mutations were located in the amino-terminal network on the inner surface of the virion (Fig. 6.5). In both sets of studies, the location of all of these residues in the internal network is inconsistent with their direct involvement in receptor binding and strongly suggests that they confer mouse virulence

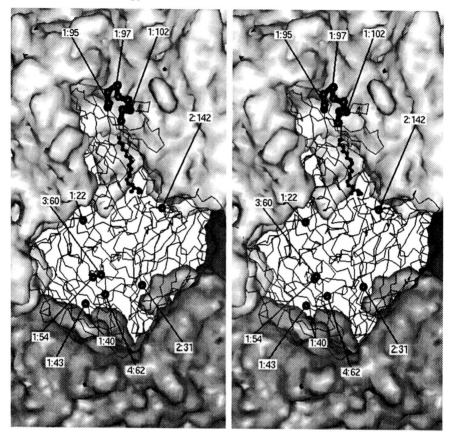

FIGURE 6.5 Location of mutations that mediate the ability of P2/Lansing and P1/Mahoney to infect mice upon intracerebral inoculation (see Table 6.2). Note that several of the mutations, including 1:43, 2:142, 3:60, and 4:62, were originally selected by screening for viral mutants that persistently infect human neuroblastoma cell lines. The organization of the figure is the same as that described for Figure 6.4.

by facilitating the receptor-induced conformational transitions necessary for infection.

In a third set of studies, Couderc et al. (1994) selected for poliovirus variants of P1/Mahoney that were capable of establishing persistent infections in human neuroblastoma cell lines. In addition to their ability to persist in human neuronal cells, all of the variants studied had become neurovirulent in mice. Sequence analysis of one of the variants revealed five missense mutations within the capsid protein region of the genome (Table 6.2). When reconstructed individually into the P1/Mahoney background, four of the five substitutions were found to confer mouse neurovirulence. Two of these mutations (VP2-142 and VP3-60) are located in highly exposed loops on the surface of the virion. These two mutations along with the BC loops of VP1 surround the deep de-

TABLE 6.2 Mouse-Adapted Neurovirulence

Strain	Capsid Protein	Residue Number[a]	Amino Acid Parental	Amino Acid Mutant	Structural Location[b]	Phenotype
P1/Mahoney	VP1	94–104 (94–104)	Mahoney	Lansing	Surface accessible near fivefold axes	Acquisition of mouse neurovirulence
P2/Lansing	VP1	95 (95)	Asp	Pro	Surface accessible near fivefold axes	Attenuation of mouse neurovirulence
	VP1	97 (97)	Pro	Ser		
	VP1	102 (102)	Ser	Asp		
P2/Lansing	VP1	40 (40)	Glu	Gly	Internal network, amino-terminal VP1 loop contacting VP4	Neurovirulent revertant in combination with attenuating substitutions at VP-95, 97, 102
	VP1	54 (54)	Pro	Ser		
P1/Mahoney	VP1	40	Gly	Glu	Internal network, VP1 loop contacting VP4	Increased mouse neurovirulence
	VP1	54	Pro	Ser		
P1/Mahoney	VP1	22	Thr	Ile	Internal network, near fivefold interface, contacting VP4	Increased mouse neurovirulence
	VP2	31	Ser	Thr	Internal network, near VP1 loop, contacting VP4	
P1/Mahoney	VP2	142	His	Tyr	Surface accessible, within receptor binding footprint for rhinovirus 16	Persistent infection of a human neuroblastoma cell line, increased mouse neurovirulence
	VP3	60	Thr	Ile	Surface accessible	
	VP1	43	Ala	Val	Internal network, VP1 loop contacting VP4	
	VP4	62	Ile	Val	Internal network, contacting VP1 loop	

[a]The residue numbers in parentheses refer to the equivalent residue in P1/Mahoney.
[b]Surface accessibility determined by evaluating the maximum radius of a sphere that contacts any atom within the residue without contacting another atom in the structure. Residues were scored as accessible if they could be contacted by a sphere greater than or equal to 3 Å in radius.

pressions or canyons that ring the fivefold axes of the virion (Fig. 6.5). An image reconstruction analysis of cryo–electron micrographs of complexes between rhinovirus 16 and its receptor, intercellular adhesion molecule-1 (ICAM-1), demonstrate that the receptor binding site for the rhinovirus major group is located within this canyon (Olson et al., 1993; see also Chapter 4). The BC loop and VP2-142 in poliovirus are located within a region analogous to the binding "footprint" of ICAM-1 on rhinovirus 16 as defined by these studies. However, residue VP3-60 of poliovirus is located well outside of this footprint, suggesting either that the murine poliovirus receptor binds to a significantly different region of the canyon or that VP3-60 plays an as yet undefined role in events subsequent to receptor binding. The other two mutations (residues VP1-43 and VP4-62) are located within the internal network of the virion, very close to the residues defined in the previous two studies (Moss and Racaniello, 1991; Couderc et al., 1993). In each of these three studies, the clustering of these independently isolated, "mouse-adapted" substitutions in VP4 and in the amino-terminal regions of VP1 that contact VP4 suggest that these residues may mediate mouse virulence by affecting the abilities of VP4 and the amino-terminal extension of VP1 to be externalized during cell entry.

These observations suggest a general model for mouse adaptation and for the role of receptors in poliovirus cell entry. In this model, the receptor not only serves as a targeting signal that allows the virus to attach to cells, but also provides the energy necessary to induce structural changes (presumably the 160S-to-135S conformational transition) that are required for subsequent steps in cell entry and uncoating. Specific proteins on the surface of mouse neuronal cells (perhaps the mouse homolog [Mph] to the human poliovirus receptor [Morrison and Racaniello, 1992]) must serve as receptors for the mouse-adapted virus strains. However, interaction of the mouse receptor with the "nonadapted" polio strains may not provide sufficient energy to induce the necessary conformational changes. The strength of the virus–receptor interaction can be increased by altering residues on the virus surface within the receptor "footprint" that are directly involved with receptor binding. Alternatively, virus–receptor interactions with weak binding energies can be compensated by lowering the energy barrier for the receptor-induced transition.

Soluble Receptor–Resistant Mutants

The poliovirus receptor from human cells (PVR) has been cloned and sequenced (Mendelsohn et al., 1989). This receptor (PVR) is a member of the immunoglobulin (Ig) superfamily whose cellular function is unknown. As a result of alternate splicing patterns, four forms of the receptor protein are expressed from the PVR receptor gene—two membrane-bound and two secreted forms (Koike et al., 1990). Sequence analysis of the PVR gene predicts that the extracellular portion of the protein contains one V-like and two C2-like Ig domains. Deletion analyses of PVR have demonstrated that the V-like domain (domain 1) is required for virus binding with the second Ig-like domain, enhancing efficiency of productive binding (Freistadt and Racaniello, 1991; Koike et al., 1991; Zibert et al., 1992). The requirement of PVR domain 1 for virus binding is confirmed by studies of site-directed mutations and chimeric receptors made between PVR and its murine homolog (Mph) (Aoki et al., 1994; Bernhardt et al., 1994; Morrison et al., 1994). Curiously, the pattern of receptor expression in human tissues is much broader than virus tropism (Mendelsohn et al., 1989; Freistadt et al., 1990; Koike et al., 1990). This suggests that additional proteins (perhaps coreceptors) may also be required for efficient infection in vivo.

PVR protein has been expressed in recombinant systems and produced in solubilized forms. These solubilized forms of the receptor bind virus and are capable of blocking infectivity and inducing the 160S-to-135S transition (Kaplan et al., 1990a; Zibert et al., 1992). This provides an in vitro system in which to study in detail the receptor–virus interaction and the conversion of the infectious 160S particle to the noninfectious 135S particle. Racaniello and colleagues have used this system to isolate viral mutants that resist neutralization upon binding to the solubilized receptor but are still capable of initiating productive infections in susceptible cells (Kaplan et al., 1990b). Characterization of these soluble receptor–resistant (SRR) poliovirus mutants further extends the correlation of the ability of the receptor to induce the 160S- to-135S transition with mutations that modulate either the stability of the particle, or the affinity of the particle for the receptor. Furthermore, these mutants reemphasize the importance of the lipid binding pocket and the fivefold interfaces in mediating the conformational changes related to cell entry.

The vast majority of these soluble receptor mutants contain a single mutation and display one of two phenotypes: 1) greatly reduced affinity for the receptor on the surface of cells with or without reduced ability to undergo the 160S-to-135S transition after binding to cells; or 2) moderately reduced affinity for the receptor coupled with a reduced ability to undergo the 160S-to-135S transition after binding to cells (Table 6.3) (Colston and Racaniello, 1994). Resistance to neutralization by the soluble receptor appears to require either significant reduction in receptor affinity or a combination of reduced receptor affinity and increased particle stability. An exception is a double mutant that displays a very unusual phenotype. This double mutant binds receptor almost as well as wild-type virus and is *more* efficiently converted to the 135S form upon interaction with cells. The relationship between the observed phenotype in the double mutant and its resistance to neutralization by soluble receptor is not clear. With the exception of one of the two mutations in the double mutant, all of the identified mutations map to either the lipid binding pocket or the nearby interface between fivefold-related protomers (Fig. 6.6). As might be expected, several of the mutants that display greatly reduced affinity for the receptor involve residues that are accessible on the virus surface for direct interactions with the receptor. However, many of the mutants, including several with greatly impaired receptor binding ability, involve mutations in residues that are not accessible for receptor contact.

How is it that residues that are inaccessible for direct contact with the receptor can modulate receptor binding? In many receptor–ligand interactions, tight binding includes a slow step that requires conformational rearrangements.

TABLE 6.3 Resistance to Neutralization by Solubilized PVR

Strain	Capsid Protein	Residue Number	Amino Acid Parental	Amino Acid Mutant	Structural Location[a]	Phenotype
P1/Mahoney	VP1	132	Met	Ile	Inaccessible in lipid binding pocket	Greatly reduced binding to PVR; normal 160S-to-135S alteration
	VP1	226	Asp	Gly	Accessible in G-H loop	
	VP1	234	Leu	Pro	Accessible in G-H loop, near opening of lipid binding pocket	
	VP1	236	Asp	Gly	Partially accessible in G-H loop, near opening of lipid binding pocket	
	VP1	241	Ala	Thr	Inaccessible in H strand, in fivefold interface	
	VP3	183	Ser	Gly	Partially accessible in fivefold interface	
P1/Mahoney	VP1	228	Leu	Phe	Inaccessible, near surface in G-H loop	Greatly reduced binding to PVR, 160S-to-135S transition is reduced upon binding to receptor on cell surface
	VP2	215	Ser	Cys	Inaccessible, in fivefold interface	
P1/Mahoney	VP1	225	Gly	Asp	Inaccessible, near surface in G-H loop	Partial reduction in binding to PVR; 160S-to-135S transition appears normal
P1/Mahoney	VP1	226	Asp	Asn	Accessible, in G-H loop	Partial reduction in binding to PVR; 160S-to-135S transition is reduced upon binding to receptor
	VP1	231	Ala	Val	Partially accessible, in G-H loop	
	VP1	241	Ala	Val	Inaccessible, in H strand, in fivefold interface	
	VP1	265	His	Arg	Internal surface, near fivefold interface, contacting VP4	
	VP3	178	Gln	Leu	Inaccessible, in fivefold interface	
P1/Mahoney	VP2	231	Ile	Met	Inaccessible, intraprotomer interface	In combination, appears to bind PVR with normal affinities, but 160S-to-135S transition is increased upon binding to receptor
	VP3	178	Gln	Arg	Inaccessible, in fivefold interface	

[a]Surface accessibility determined by evaluating the maximum radius of a sphere that contacts any atom within the residue without contacting another atom in the structure. Residues were scored as accessible if they could be contacted by a sphere greater than or equal to 3 Å in radius.

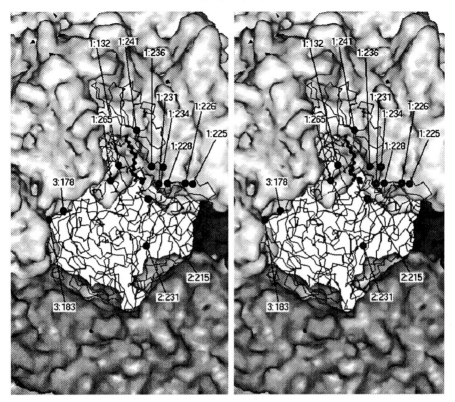

FIGURE 6.6   Location of mutations that confer resistance to neutralization by solubilized PVR (see Table 6.3). The organization of the figure is the same as that described for Figure 6.4.

Thus, tight binding can be modulated by steric interference for the primary interaction or by interfering with the ability of the receptor or ligand (in this instance poliovirus) to undergo the necessary conformational rearrangements necessary to reach the final bound state. All of the buried mutations in the SRR mutants are located in regions of the structure that have been shown to regulate thermal stability of type 3 poliovirus (see earlier). The dual role of these specific regions of the capsid in modulating thermal stability and receptor affinity suggests that tight receptor binding requires conformational adjustments on the virus in the vicinity of the receptor binding site. Within the context of the general receptor model presented earlier, the tight binding complex takes on the role of the transition state in the conversion of 160S to 135S. By inducing the formation of this transition state, receptor binding plays a role akin to formation of an enzyme–substrate complex in catalyzing the 135S transition. By extension of this analogy, one would expect that it would be possible to inhibit infectivity with an inhibitor that either blocks the formation of the transition state or prevents the decay into product.

Drug Resistance Mutants

In the late 1970s investigators at Sterling-Winthrop Pharmaceuticals described a novel antiviral agent, called arildone, that bound directly to poliovirus. With the drug bound, the virus attached to cells and was internalized but was unable to uncoat (Diana et al., 1977; McSharry et al., 1979; Caliguiri et al., 1980). Virus–drug complexes were also shown to be very resistant to thermal inactivation, suggesting that the drugs acted by stabilizing the viral capsid. Subsequent studies showed that arildone was effective against a broad range of picornaviruses, including enteroviruses (polioviruses and coxsackieviruses) and rhinoviruses. Based on this lead, a number of pharmaceutical companies, including Sterling-Winthrop and Janssen Pharmaceutica, have pursued development of families of capsid-binding compounds as potential antiviral agents. Structural studies of rhinoviruses and polioviruses complexed with these antiviral agents demonstrate that the drugs bind in a pocket in the hydrophobic core of VP1 (reviewed in Grant et al., 1994; see also Chapters 4 and 16). The drug binding site is identical to the site occupied by the lipidlike ligand in poliovirus. The observed stabilizing effect of a suppressor mutation located within this site in a non-ts revertant of P3/Sabin would then suggest that these drugs usurp a site that is normally used to modulate viral stability.

Mosser and Rueckert (1993) have selected a number of variants of P3/Sabin that are resistant to inactivation by disoxaril (also called WIN 51711), which is the prototype capsid binding drug from Sterling-Winthrop (Table 6.4). The drug-resistant variants could be segregated into two different groups, which differ in the extent to which they are resistant to the drug (Mosser and Rueckert, 1993; Mosser et al., 1994). The first group were resistant to low levels of disoxaril but could still be inactivated at high concentrations of the drug. The second group were resistant to all drug concentrations tested and, in fact, were incapable of growing in the absence of drug. Further characterization of the drug-dependent variants showed that they accumulated virions to normal titers in the absence of drug but were quantitatively converted to the 135S form of the virus at physiologic temperatures (37 °C) upon cell lysis.

Sequence analyses demonstrated that several of the drug-resistant variants contained mutations in the drug/lipid binding pocket (Table 6.4). These mutations are all in VP1. One of these mutations results in the replacement of a Met (parental) with a Leu at VP1-260. Wild-type P1/Mahoney, which also contains a Leu side chain at the corresponding position, also is relatively resistant to inactivation by disoxaril. Models based on the structural analyses of drug–virus complexes suggested that the presence of the Leu side chain at VP1-260 would interfere with the conformational adjustment of a neighboring Met side chain (at residue VP1-130) that is required for efficient binding of disoxaril (Fig. 6.7). The remaining two mutations are in the G strand of VP1, which lines the lipid binding pocket and is located on the inner surface of the interface between fivefold-related protomers. Even relatively conservative substitutions at these residues would be expected to have profound steric effects on the ability of the drug to bind and could in addition potentially alter the

TABLE 6.4 Resistance to Capsid Binding Drugs

Strain	Capsid Protein	Residue Number[a]	Amino Acid Parental	Amino Acid Mutant	Structural Location[b]	Phenotype
P3/Sabin	VP1	105 (107)	Met	Thr	Accessible in C strand, contacts E-F loop	Disoxaril (WIN 51711) resistance
	VP1	129 (131)	Asp	Val	Inner surface, contacts VP4	
	VP1	192 (194)	Ile	Phe	Inaccessible, lining lipid binding pocket near fivefold interface	
	VP1	194 (196)	Tyr	Leu	Inaccessible, lining lipid binding pocket near fivefold interface	
	VP1	260 (261)	Met	Leu	Inaccessible, near opening of lipid binding pocket	
P3/Sabin	VP4	46 (47)	Ser	Leu	Internal network, contacts VP1	Disoxaril dependence
	VP1	49 (51)	Ala	Val	Internal network, VP1 loop contacts VP4	
	VP1	51 (53)	Asn	Ser	Internal network, VP1 loop contacts VP4	
	VP1	52 (54)	Pro	Ser	Internal network, VP1 loop contacts VP4	
	VP2	205 (206)	Ala	Val	Internal surface near VP1 loop, contacting VP4	
	VP2	208 (209)	Val	Ile	Internal surface near VP1 loop, contacting VP4	
	VP3	49 (49)	Ile	Met	Internal surface, contacts VP1 loop and VP4	
	VP4	53 (54)	Thr	Ala	Internal network, contacts VP1 loop	

[a]The numbers in parentheses refer to the equivalent residues in P1/Mahoney. Note that VP4 residue 1 of P1/Mahoney is the myristate moiety, which is covalently attached to the amino-terminal glycine residue. Although the VP4 of P3/Sabin is also myristoylated, the numbering scheme of Mosser and Rueckert (1993) begins with the amino-terminal glycine.

[b]Surface accessibility determined by evaluating the maximum radius of a sphere that contacts any atom within the residue without contacting another atom in the structure. Residues were scored as accessible if they could be contacted by a sphere greater than or equal to 3 Å in radius.

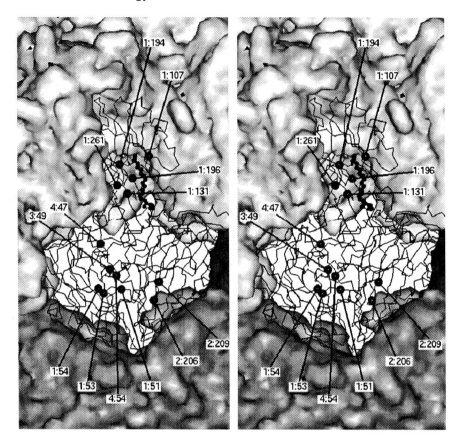

FIGURE 6.7   Location of mutations that confer resistance to capsid binding drugs (see Table 6.4). The organization of the figure is the same as that described for Figure 6.4.

protomer interface. Sequence analyses of the drug-resistant variants also revealed several substitutions outside the drug binding site (Table 6.4). These include mutations at residues VP1-105, VP1-129, and VP4-47 (P3/Sabin). Residue VP1-105 is located at the carboxyl-terminal end of the C strand and interacts with the EF loop of VP1. Although the role of this mutation is not entirely clear, the EF loop of VP1 is part of the interface between fivefold-related protomers. Mutations at residue VP1-105 may indirectly alter interactions at this interface that involve the EF loop. The remaining substitutions are located on the inner surface of the virus (Fig. 6.7). Residue VP1-129 interacts with several VP4 residues within the same protomer and fivefold-related protomers. Similarly, residue VP4-47 interacts with fivefold-related copies of VP4. Like the mouse adaptation mutations, mutations at these two positions (VP1-129 and VP4-47) may be playing a role by promoting the structural transitions that result in the externalization of VP4 during cell entry. Thus these observations continue a theme wherein a stabilizing interaction (e.g., drug binding) at one site of the virion can be counteracted by destabilizing interactions in the internal network.

This theme is further reinforced by the location of the mutations observed in the drug-dependent variants (Fig. 6.7). Mutations in these variants are located either within the amino-terminal network or in residues within VP2 that interact with portions of the network. These sites are closely clustered in space. The mutations in the amino-terminal region of VP1 reside in a loop that interacts with VP4 and mediates the interactions between the cores of VP2 and VP3. This loop also has been implicated in mouse adaptation. Mutations in the amino terminus of VP3 and VP4 (VP3-49 and VP4-54) are in van der Waals contact with each other and are located very close to this VP1 loop. Residues in the core of VP2 (VP2-205 and VP2-207) are in a portion of the G strand whose residues interact with the amino-terminal extension of VP1. This segment of VP1 forms the seventh strand of the interpentamer β-sheet described earlier. All of these residues participate in critical interactions that are formed after VP0 cleavage and contribute to the net increase in stabilizing interactions within the mature virion. The location of the drug-dependent mutations in obviously critical areas of the network is consistent with a model in which the drug-dependent phenotype is an extreme case of drug resistance resulting from mutations outside of the binding pocket. Both classes of mutants destabilize the virus to compensate for the stabilization provided by the binding of the drug. However, although the drug-resistant mutants remain sufficiently stable to maintain viability in the absence of drug, the destabilization caused by the drug-dependent mutations is sufficiently severe to render the virus nonviable in the absence of the stabilizing effects of the drug.

Concluding Discussion

In the four sets of mutants discussed here, the studies were designed to probe four very different phenotypes of the virus. Analyses of these mutants have introduced several common themes. The four studies point to a restricted set of localized regions within the structure that regulate the stability and conformational mechanics of the capsid protein structures. These include 1) a hydrophobic pocket in the β-barrel of VP1, which naturally binds an endogeneous ligand and experimentally binds antiviral drugs; 2) the interface between fivefold-related protomers, which is critical during pentamer formation; and 3) the amino-terminal network, which is located on the internal surface of the virus capsid. Mutations in these key structures that either stabilize or destabilize their interactions can compensate for, respectively, destabilizing or stabilizing interactions at remote sites. Thus the virus appears to have evolved a window in which the balance between stability (required for survival in the extracellular environment) and flexibility (required for replication) is optimized under physiologic conditions. The four sets of mutants described here demonstrate that mutations that alter this window of stability have profound effects on the biologic phenotypes of the resultant viruses.

In addition, all four sets of mutations point to two critical elements in the network: 1) the loop comprising VP1 residues 44 to 56 and portions of VP4 in the immediate vicinity of the VP1 loop (see Color Plate 18) and 2) the seven-stranded cross-pentamer β-sheet (Fig. 6.3*B* and *C*). These two elements of the structure as they exist in the mature virion are dependent on the cleavage of VP0 and may account in large part for the great increase in stability that is observed upon VP0 cleavage in the final stages of assembly. Thus, VP0 cleavage can be viewed as the step that cocks the trigger mechanism for the conformational alterations induced by receptor interaction. This leads to the view proposed in the introduction that the virion is a metastable structure that links assembly and cell entry.

Although the structure of the 135S cell entry intermediate is not known, the observations discussed here lead to a testable working model of the nature of the conformational rearrangements that lead to the formation of the 135S particle. This model is perhaps best presented by analogy with the well-characterized expansion of plant viruses (Incardona and Kaesberg, 1964). The expansion takes place upon depletion of divalent cations at alkaline pH. These conditions are typical of the intracellular environment of plant cells, and the expanded form has been proposed to represent an intermediate in either the assembly or entry of the virus. Like the 135S transition in poliovirus, the expansion of the plant viruses produces a marked reduction in sedimentation coefficient and results in the externalization of portions of the amino termini of some but not all of the capsid proteins (Golden and Harrison, 1982). Furthermore, the expansion of plant viruses results in the disruption of several interfaces (including, in poliovirus, the interface that is analogous to the five-fold interface between protomers and the interface between pentamers that is stabilized by the cross-pentamer seven-stranded β-sheet) and produces large fenestrations on the virus surface (Robinson and Harrison, 1982). These openings may be sufficiently large to permit the externalization of the viral RNA. Extension of this structural analogy suggests that, in poliovirus, externalization of VP4 and the amino termini of VP1 functionally leads to a fenestrated particle that ultimately releases the viral RNA and collapses to form the 80S species.

The combination of genetic, biochemical, and structural studies of a diverse group of mutant polioviruses has begun to provide a unified dynamic perspective of the biologic functions of the virus at the molecular level. The insights gained from studies of this very simple virus will almost certainly provide general principles that are applicable to more complex viral systems and to the host themselves.

Acknowledgments

We would like to thank many colleagues who provided information prior to publication, in particular B. Blondel, T. Couderic, A.J. Macadam, A.G. Mosser, and V.R. Racaniello. We thank M. Wien for critical reading of the manuscript.

Many of the concepts presented here evolved from research supported by Public Health Service grants AI20566, AI22627, and AI32480 from the National Institute of

Allergy and Infectious Diseases, and program project grant P50 NS16998 from the National Institute of Neurological and Communicative Disorders and Stroke.

References

Aoki, J., Koike, S., Ise, I., Sato-Yoshida, Y. and Nomoto, A. 1994. Amino acid residues on human poliovirus receptor involved in interaction with poliovirus. J. Biol. Chem. 269: 8431–8438.

Basavappa, R., Syed, R., Flore, O., Icenogle, J.P., Filman, D.J. and Hogle, J.M. 1994. Role and mechanism of the maturation cleavage of VP0 in poliovirus assembly: structure of the empty capsid assembly intermediate at 2.9Å resolution. Protein Sci. 3: 1651–1669.

Bernhardt, G., Harber, J., Zibert, A., DeCrombrugghe, M. and Wimmer, E. 1994. The poliovirus receptor: identification of domains and amino acid residues critical for virus binding. Virology 203: 344–356.

Caliguiri, L.A., McSharry, J.J. and Lawrence, G.W. 1980. Effect of arildone on modification of poliovirus in vitro. Virology 105: 86–93.

Chow, M., Newman, J.F.E., Filman, D., Hogle, J.M., Rowlands, D.J. and Brown, F. 1987. Myristylation of picornavirus capsid protein VP4 and its structural significance. Nature 327: 482–486.

Colston, E. and Racaniello, V.R. 1994. Soluble receptor-resistant poliovirus mutants identify surface and internal capsid residues that control interaction with the cell receptor. EMBO J. 13: 5855–5862.

Compton, S.R., Nelson, B. and Kirkegaard, K. 1990. Temperature-sensitive poliovirus mutant fails to cleave VP0 and accumulates provirions. J. Virol. 64: 4067–4075.

Couderc, T., Guedo, N., Calvez, V., Pelletier, I., Hogle, J., Colbere-Garapin, F. and Blondel, B. 1994. Substitutions in the capsids of poliovirus mutants selected in human neuroblastoma cells confer on the Mahoney type 1 strain a phenotype neurovirulent in mice. J. Virol. 68: 8386–8391.

Couderc, T., Hogle, J., Le Blay, H., Horaud, F. and Blondel, B. 1993. Molecular characterization of mouse-virulent poliovirus type 1 Mahoney mutants: involvement of residues of polypeptides VP1 and VP2 located on the inner surface of the capsid protein shell. J. Virol. 67: 3808–3817.

Diana, G.D., Salvador, U.J., Zalay, E.S., Johnson, R.E., Collins, J.E., Johnson, D., Hinshaw, W.B., Lorenz, R.R., Thielding, W.H. and Pancic, F. 1977. Antiviral activity of some beta-diketones. 1. Aryl alkyl diketones. in vitro activity against both RNA and DNA viruses. J. Med. Chem. 20: 750–756.

Evans, D.J. and Almond, J.W. 1991. Design, construction and characterization of poliovirus antigen chimeras. Methods Enzymol. 203: 386–400.

Filman, D.J., Syed, R., Chow, M., Macadam, A.J., Minor, P.D. and Hogle, J.M. 1989. Structural factors that control conformational transitions and serotype specificity in type 3 poliovirus. EMBO J. 8: 1567–1579.

Flore, O., Fricks, C.E., Filman, D.J. and Hogle, J.M. 1990. Conformational changes in poliovirus assembly and cell entry. Semin. Virol., 1: 429–438.

Freistadt, M.S., Kaplan, G. and Racaniello, V.R. 1990. Heterogeneous expression of poliovirus receptor-related proteins in human cells and tissues. Mol. Cell. Biol. 10: 5700–5706.

Freistadt, M.S. and Racaniello, V.R. 1991. Mutational analysis of the cellular receptor for poliovirus. J. Virol. 65: 3873–3876.

Fricks, C.E. and Hogle, J.M. 1990. Cell-induced conformational change in poliovirus: externalization of the amino terminus of VP1 is responsible for liposome binding. J. Virol. 64: 1934–1945.

Golden, J. and Harrison, S.C. 1982. Proteolytic dissection of turnip crinkle virus subunit in solution. Biochemistry 21: 3862–3866.

Grant, R.A., Hiremath, C.N., Filman, D.J., Syed, R., Andries, K. and Hogle, J.M. 1994. Structures of poliovirus complexes with antiviral drugs: implications for viral stability and drug design. Curr. Biol. 4: 784–797.

Harrison, S.C., Wiley, D.C. and Skehel, J.J. 1996. Virus structure. In Virology, 3rd ed. (Fields, B.N. and Knipe, D.M., Eds.), pp. 59–100. Lippincott-Raven, New York.

Hiremath, C.N., Grant, R.A., Filman, D.J. and Hogle, J.M. 1995. The binding of the antiviral drug Win51711 to the Sabin strain of type 3 poliovirus: structural comparison with drug-binding in rhinovirus 14. Acta Crystallogr. D. 51: 473–489.

Hogle, J.M., Chow, M. and Filman, D.J. 1985. The three dimensional structure of poliovirus at 2.9Å resolution. Science 229: 1358–1365.

Incardona, N.L. and Kaesberg, P. 1964. A pH-induced structural change in bromegrass mosaic virus. Biophys. J. 4: 11–21.

Kaplan, G., Freistadt, M.S. and Racaniello, V.R. 1990a. Neutralization of poliovirus by cell receptors expressed in insect cells. J. Virol. 64: 4697–4702.

Kaplan, G., Peters, D. and Racaniello, V.R. 1990b. Poliovirus mutants resistant to neutralization with soluble cell receptors. Science 250: 1596–1599.

Kirkegaard, K. 1990. Mutations in VP1 of poliovirus specifically affect both encapsidation and release of viral RNA. J. Virol. 64: 195–206.

Koike, S., Horie, H., Ise, I., Okitsu, A., Yoshida, M., Iizuka, N., Takeuchi, K., Takegami, T. and Nomoto, A. 1990. The poliovirus receptor protein is produced both as membrane-bound and secreted forms. EMBO J. 9: 3217–3224.

Koike, S., Ise, I. and Nomoto, A. 1991. Functional domains of the poliovirus receptor. Proc. Natl. Acad. Sci. U.S.A. 88: 4104–4108.

La Monica, N., Kupsky, W.J. and Racaniello, V.R. 1987. Reduced mouse neurovirulence of poliovirus type 2 Lansing antigenic variants selected with monoclonal antibodies. Virology 161: 429–437.

La Monica, N., Meriam, C. and Racaniello, V.R. 1986. Mapping of sequences required for mouse neurovirulence of poliovirus type 2 Lansing. J. Virol. 57: 515–525.

Lee, W.-M., Monroe, S.S. and Rueckert, R.R. 1993. Role of maturation cleavage in infectivity of picornaviruses: activation of an infectosome. J. Virol. 67: 2110–2122.

Li, Q., Gomez Yafal, A., Lee, Y.M.-H., Hogle, J. and Chow, M. 1994. Poliovirus neutralization by antibodies to internal epitopes of VP4 and VP1 results from reversible exposure of these sequences at physiological temperature. J. Virol. 68: 3965–3970.

Macadam, A.J., Ferguson, G., Arnold, C. and Minor, P.D. 1991. An assembly defect as a result of an attenuating mutation in the capsid proteins of the poliovirus type 3 vaccine strain. J. Virol. 65: 5225–5231.

Macadam, A.J., Pollard, S.R., Ferguson, G., Dunn, G., Skuce, R., Almond, J.W. and Minor, P.D. 1989. Reversion of attenuated and temperature-sensitive phenotypes of the Sabin type 3 strain of poliovirus in vaccines. Virology 174: 408–414.

Martin, A., Wychowske, C., Conderc, T., Crainic, R., Hogle, J. and Girard, M. 1988. Engineering a poliovirus type 2 antigenic site on a type 1 capsid results in a chimaeric virus which is neurovirulent for mice. EMBO J. 7: 2839–2847.

McSharry, J.J., Caliguiri, L.A. and Eggers, H.J. 1979. Inhibition of uncoating of poliovirus by arildone, a new antiviral drug. Virology 97: 307–315.

Mendelsohn, C., Wimmer, E. and Racaniello, V.R. 1989. Cellular receptor for polio-virus: molecular cloning, nucleotide sequence and expression of a new member of the immunoglobulin superfamily. Cell 56: 855–865.

Minor, P.D., Dunn, G., Evans, D.M.A., Magrath, D.I., John, A., Howlett, J., Phil-lips, A., Westrop, G., Wareham, K., Almond, J.W. and Hogle, J.M. 1989. The temperature-sensitivity of the Sabin type 3 vaccine strain of poliovirus: molecular and structural effects of a mutation in the capsid protein VP3. J. Gen. Virol. 70: 1117–1123.

Morrison, M.E., He, Y.-J., Wien, M.W., Hogle, J.M. and Racaniello, V.R. 1994. Homolog-scanning mutagenesis reveals poliovirus receptor residues important for virus binding and replication. J. Virol. 68: 2578–2588.

Morrison, M.E. and Racaniello, V.R. 1992. Molecular cloning and expression of a murine homolog of the human poliovirus receptor gene. J. Virol. 66: 2807–2813.

Moscufo, N. and Chow, M. 1992. Myristate-protein interactions in poliovirus: interac-tions of VP4 threonine-28 contribute to the structural conformation of assembly intermediates and the stability of assembled virions. J. Virol. 66: 6849–6857.

Moscufo, N., Gomez Yafal, A., Rogove, A., Hogle, J.M. and Chow, M. 1993. A mu-tation in VP4 defines a new step in the late stages of cell entry by poliovirus. J. Virol. 67: 5075–5078.

Moss, E.G. and Racaniello, V.R. 1991. Host range determinants located on the interior of the poliovirus capsid. EMBO J. 5: 1067–1074.

Mosser, A.G. and Rueckert, R.R. 1993. WIN 51711-dependent mutants of poliovirus type 3: evidence that virions decay after release from cells unless drug is present. J. Virol. 67: 1246–1254.

Mosser, A.G., Sgro, J.-Y. and Rueckert, R.R. 1994. Distribution of drug resistance mutations in type 3 poliovirus identifies three regions involved in uncoating func-tions. J. Virol. 68: 8193–8201.

Murray, M.G., Bradley, J., Yang, X.-F., Wimmer, E., Moss, E.G. and Racaniello, V.R. 1988. Poliovirus host range is determined by a short amino acid sequence in neu-tralization antigenic site 1. Science 241: 213–215.

Olson, N.H., Kolatkar, P.R., Oliveira, M.A., Cheng, R.H., Greve, J.M., McClelland, A., Baker, T.S. and Rossman, M.G. 1993. Structure of a human rhinovirus complexed with its receptor molecule. Proc. Natl. Acad. Sci. U.S.A. 90: 507–511.

Page, G.S., Mosser, A.G., Hogle, J.M., Filman, D.J., Rueckert, R.R. and Chow, M. 1988. Three dimensional structure of poliovirus serotype 1 neutralizing determi-nants. J. Virol. 62: 1781–1794.

Pelletier, J. and Sonenberg, N. 1989. Internal binding of eucaryotic ribosomes on poliovirus RNA: translation in HeLa cell extracts. J. Virol. 63: 441–444.

Racaniello, V. (Ed.). 1990. Curr. Top. Microbiol. Immunol. 161.

Reynolds, C., Birnby, D. and Chow, M. 1992. Folding and processing of the capsid protein precursor P1 is kinetically retarded in neutralization site 3B mutants of poliovirus. J. Virol. 66: 1641–1648.

Robinson, I.K. and Harrison, S.C. 1982. Structure of the expanded state of tomato bushy stunt virus. Nature 297: 563–568.

Rueckert, R.R. (Ed.). 1976. On the structure and morphogenesis of picornaviruses. Compr. Virol. 6.

Rueckert, R.R. 1996. Picornaviridae: The viruses and their replication. In Virology, 3rd ed. (Fields B.N. and Knipe, D.M., Eds.), pp. 609–654. Lippincott-Raven, New York.

Westrop, G.D., Wareham, K.A., Evans, D.M.A., Dunn, G., Minor, P.D., Magrath, D.I., Taffs, F., Marsden, S., Skinner, M.A., Schild, G.C. and Almond, J.W. 1989. Genetic basis of attenuation of the Sabin type 3 oral poliovirus vaccine. J. Virol. 63: 1338–1344.

Wetz, K. and Kucinski, T. 1991. Influence of different ionic and pH environments on structural alterations of poliovirus and their possible relation to virus uncoating. J. Gen. Virol. 72: 2541–2544.

Wimmer, E., Hellen, C.U.T. and Cao, X. 1993. Genetics of poliovirus. Annu. Rev. Genet. 27: 353–436.

Yeates, T.O., Jacobson, D.H., Martin, A., Wychowski, C., Girard, M., Filman, D.J. and Hogle, J.M. 1988. Three-dimensional structure of a mouse-adapted type 2/type 1 poliovirus chimera. EMBO J. 10: 2331–2341.

Ypma-Wong, M.F., Filman, D.J., Hogle, J.M. and Semler, B.L. 1988. Structural domains of the poliovirus polyprotein are major determinants for proteolytic cleavage in gln-gly pairs. J. Biol. Chem. 263: 17846–17856.

Zibert, A., Selinka, H.-C., Elroy-Stein, O. and Wimmer, E. 1992. The soluble form of two N-terminal domains of the poliovirus receptor is sufficient for blocking viral infection. Virus Res. 25: 51–61.

7.

Structural Biology of Polyomaviruses

ROBERT L. GARCEA & ROBERT C. LIDDINGTON

Murine polyomavirus (polyoma) and simian virus 40 (SV40), genera of the *Papovaviridae* family, have been studied for almost 40 years, mainly because of their ability to transform cells in culture and cause tumors in animals. Not only has the knowledge of the biology of these viruses illuminated the fields of cell cycle control and tumorigenesis, but their structures have provided several surprises. Now, with atomic-resolution structural data available for both polyoma and SV40 capsids, along with the ability to assemble virus capsids in vitro from recombinant proteins, these viruses are at the forefront of structural virology experimentation. In this chapter, we describe the structure of polyomaviruses in the context of their biology, and we propose a brief model of the steps in polyomavirus infection. We hope that these viruses will be appreciated as model systems for intracellular protein trafficking, the interaction of structural proteins with eukaryotic chromatin, and the coordinated, accurate assembly of protein subunits in large macromolecular structures.

Genomic Structure

Polyomaviruses have similar genomic structures (Fig. 7.1). The early transcribed proteins are termed the T (tumor) antigens (identified by their relative size); these are the viral regulatory proteins and also the oncogenes. The late proteins are termed VP (viral protein) 1, 2, and 3, and are the structural proteins of the viral capsid. The murine and hamster polyomaviruses have three T an-

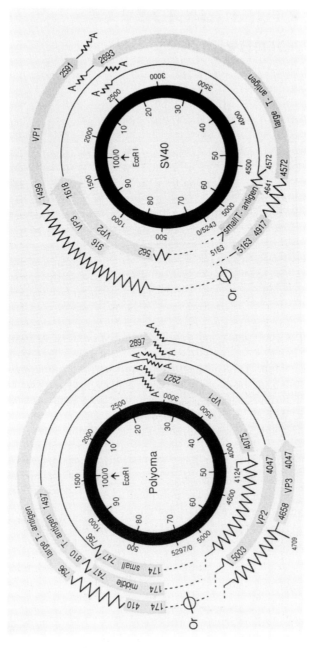

FIGURE 7.1 Genomes of SV40 (*right*) and polyomavirus (*left*). Polyomavirus has a middle T antigen in addition to the large and small T antigens. Polyomavirus host-range nontransforming (*hr-t*) mutant viruses map to the intervening sequence of large T antigen. VP3 is a subset of the VP2 reading frame and is initiated internally from the same mRNA. (Reproduced with permission from Fields et al., 1990.)

188

tigens (small, middle, and large) while all other members, including SV40, lack a middle T antigen. VP3 is an internally initiated form of VP2, and shares the same C terminus. The origin of replication, transcriptional start sites, and enhancer sequences are located similarly.

Virion Structure

Mature polyomaviruses are icosahedrally symmetric particles of 500-Å diameter. A protein capsid surrounds a minichromosome in which circular double-stranded viral DNA is folded into nucleosomes by cellular core histones. The capsid contains 360 copies of the major structural protein, VP1, arranged as 72 pentamers on a $T = 7d$ icosahedral surface lattice (Rayment et al., 1982), and smaller quantities of the minor structural proteins, VP2 and VP3 (Tooze, 1980), in the approximate ratio of one VP2 or VP3 molecule per VP1 pentamer. The crystal structures of SV40 and polyoma have been determined at atomic resolution, and complete models of VP1 have been built. Because the minichromosome and the minor capsid proteins VP2 and VP3 do not share the icosahedral organization of the capsid, structural information is much less detailed for them. In the 3.8-Å SV40 structure, additional density within the hollow of each VP1 pentamer was interpreted as a segment of one of the minor proteins, VP2 or VP3 (Liddington et al., 1991). Low-resolution (25-Å) difference maps between polyoma virions and empty particles revealed density that filled the conical hollow of each pentamer and extended to the virion core (Fig. 7.2) (Griffith et al., 1992). This location supports a role for VP2 and VP3 in directing assembly of VP1 around the minichromosome. Image analysis of electron micrographs of frozen hydrated SV40 reveals continuous density from the center of the particle to a radius of 165 to 170 Å, indicating that the

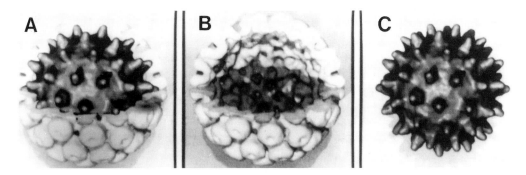

FIGURE 7.2   Computer reconstruction of full (A) and empty (B) particles and non-VP1 density (C) in the polyoma virion. Low-resolution (25-Å) electron density maps of full and empty polyoma particles were "subtracted" to yield the non-VP1 density. As shown in panel C, 72 pronglike projections extend from the virion core through the central fivefold axis of each capsomere. This density should correspond to VP2, VP3, and/or the viral minichromosome. (Courtesy of D.L.D. Caspar; adapted from Griffith et al., 1992.)

nucleohistone core is highly condensed (Baker et al., 1988). There is an apparent break (~10 Å) in the density between the capsid and the chromatin, suggesting that the interactions between VP1, VP2, and VP3 and the minichromosome may be flexible. Indeed, the SV40 structure shows that the N-terminal arm of VP1 is disordered as it enters the virion core. Both SV40 and polyoma VP3 have been shown biochemically to have a domain near their carboxyl terminus that interacts with VP1 pentamers (Gharakhanian et al., 1988; Barouch and Harrison, 1994).

Structure and Assembly of the Viral Minichromosome

By cryo–electron microscopy, unencapsidated SV40 minichromosomes are globular, with a mean diameter of 300 to 400 Å (Dubochet et al., 1986). These globular structures collapse when spread on grids in negatively stained preparations, and then appear as beads-on-a-string (Müller et al., 1978). Nucleosomes are nonrandomly distributed over the genomic DNA, although their location is not unique to specific DNA sequences (Ambrose et al., 1989). Possible stacking geometries have been proposed, but no unique configuration has been identified experimentally.

In contrast to the phage model of injecting genomic DNA into a preformed capsid, polyoma appears to assemble by incremental addition of capsid proteins onto the viral minichromosome (Garber et al., 1978; Baumgartner et al., 1979; Fernandez-Munoz et al., 1979; Jakobovits and Aloni, 1980; Blasquez et al., 1983; Garcea and Benjamin, 1983; Ng and Bina, 1984). By extracting "subvirion" assembly intermediates from virus-infected cells, a pattern of intermediates was detected: 95S "replicating" complexes, 75S "minichromosomes," 200S "pre"-virions, and 240S virion forms. The fast-sedimenting complexes were further separated into salt-sensitive and salt-resistant forms (Fanning and Baumgartner, 1980). These data suggested that capsid proteins are added to the 75S minichromosome to form a labile structure that matures into a stable virion. Histone H1 is found in the 75S minichromosome fraction but is lost as capsid proteins are added (Coca-Prados and Hsu, 1979; LaBella and Vesco, 1980), suggesting that viral structural proteins displace histone H1. In addition, the core histones bound to the viral genome are more acetylated than those bound to cellular DNA, and their acetylation state increases, because of decreased deacetylation activity, as encapsidation is completed (Chestier and Yaniv, 1979; LaBella and Vesco, 1980). Histone hyperacetylation is observed for nucleosomes located on rapidly transcribing or replicating DNA, and such a modification that neutralizes the basic amino termini of the histones may also facilitate coat protein binding to DNA. Nuclease digestion has revealed that a 400-bp region near the viral origin of replication, extending into the late coding region, appears to be free of nucleosomes (Varshavsky et al., 1979; Jakobovits et al., 1980; Saragosti et al., 1980). This region has been implicated as a signal for SV40 virion encapsidation, although it is uncertain whether the specific nucleotide sequence or the accessibility of the DNA in this region is the packaging determinant (Oppenheim et al., 1992).

Capsid Structure

The capsid organization initially presented a structural puzzle (Rayment et al., 1982; Salunke et al., 1986), because the all-pentamer arrangement did not conform with the Caspar-Klug theory of quasi-equivalence that had been successful in rationalizing the structures of other viruses (see Chapter 1). As illustrated in Figure 7.3, one can distinguish two classes of VP1 pentamers, according to their location in the shell. Twelve pentamers lie on the 12 fivefold rotation axes of the icosahedron, each surrounded by 5 other pentamers ("pentavalent"); the remaining 60 pentamers do not lie on symmetry axes, and they are surrounded by 6 other pentamers ("hexavalent"). Studies of recombinant polyoma VP1 demonstrated that capsid assembly is an intrinsic property of the VP1 protein, and is not dictated by either the minor proteins or posttranslational modifications. The high-resolution crystal structure of SV40 virions (Liddington et al., 1991) resolved the puzzle of how the different types of contact are made, and what elements of flexibility and alternative bonding accommodate the mismatch of symmetries. The core of a VP1 pentamer has an invariant structure, defining a standard building block for the capsid; long C-terminal arms tie the pentamers together, with additional contributions from divalent cations and disulfide bonds. The arms extend from one pentamer and fit into binding sites on ad-

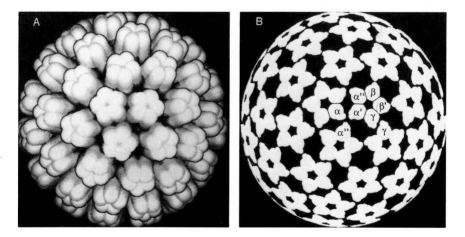

FIGURE 7.3 Overall view of the polyomavirus virion. The shell is composed of 360 copies of VP1 organized into 72 pentamers. The pentamers are situated at the points of a $T = 7d$ icosahedral surface lattice. There are 12 five-coordinated pentamers and 60 six-coordinated pentamers. The three kinds of interpentamer clustering are indicated. Three subunits (α, α', α'') form a threefold interaction, the subunits labeled β and β' form one sort of twofold interaction, and the γ subunits form another kind of twofold interaction across an icosahedral twofold axis. (Adapted from Salunke et al., 1986.)

jacent pentamers. Flexibility is limited to the direction an arm takes as it emerges from the core of its subunit and heads toward a neighbor.

Subunit and Pentamer Structure

A VP1 monomer consists of an N-terminal extended arm, an antiparallel β-sandwich, and a long C-terminal arm (Figs. 7.4 and 7.5). The central β-sandwich has the same topology as the capsid proteins of picornaviruses and plant viruses (Rossmann and Johnson, 1989; Harrison, 1991; see also Chapter 4), and the same strand labeling (B to I) has been used. However, the lengths of the strands and the relative positions of the two sheets impart a quite different overall shape. The structure of the VP1 monomer is the same in all six unique environments in the viral capsid, with the only major differences occurring in the C-terminal arms.

Five VP1 subunits form an intimately associated pentamer: the N-terminal half of the G strand contributes a fifth strand to the CHEF sheet of the clockwise neighbor, and the D-E loop nestles into the small jelly roll (EF loop) and the BC and HI loops of the clockwise neighbor. The pentamer is roughly cylindrical, about 80 Å in diameter and 70 Å tall, extending from a radius of 180 to 250 Å in the virus particle. It has a hollow conical interior of approximately 50-Å diameter at its base, tapering to about 12 Å at a neck formed by the short FG corners (see Fig. 7.2). This neck opens up on the top surface into a dimple about 20 Å wide and 12 Å deep. The large BC, DE, and HI loops from each subunit interlock to form the top surface of the pentamer.

The N-terminal arm of each VP1 monomer extends across the inside of the pentamer beneath the clockwise neighboring subunit (Fig. 7.5). The arm contains the VP1 nuclear localization signal (Wychowski et al., 1986; Moreland and Garcea, 1991), and has also been shown to mediate DNA binding in vitro (Moreland et al., 1991). It is completely internal in a virus particle, and it therefore cannot contribute to nuclear transport of an intact virion during the initial stage of infection (Clever et al., 1991). However, the arm would be exposed on a free pentamer, and could contribute to its nuclear transport during assembly. The N-terminal arm in polyoma is similar to that in SV40 but has a somewhat different conformation. The first ordered residue is Lys 17, and Cys 19 makes a disulfide bridge to the CD loop (see below).

Pentamer-Pentamer Contacts

The C-terminal arms (Fig. 7.6; see Color Plate 8) form the principal interpentamer contacts, by extending away from their pentamer of origin and invading a subunit of an adjacent pentamer. They permit the conformational "bond switching" that allows pentameric capsomeres to reside in both five- and six-fold symmetric positions on the capsid surface. The arms can be subdivided into three parts: the "C-helix," the "C-insert" (including the J strand locked between the A and B strands; see Fig. 7.5), and the terminal "C-loop." In all cases, the C-insert interacts identically with the invaded pentamer. The flexi-

```
            10        20         30          40            50            60           70        80
sv  APTKRKGS  CPGAAPKKPKEPVQVPKLVIKGGIEVLGVKTGVDSFTEVECPLNPQMGNPD
                                                                        EHQKGLSKSLAAEKQFTDDSPD
    < disordered >< N-arm >×<β-A>< α-A>< AB >< β-B >< BC loop

py  APKRKSGVSKCETKCTKACPRPAPVPKLLIKGGMEVLDLVTGPDSVTEIEAFLNPRMGQPTPESLTEGGQYGWSRGINLATSDTDDSPE
            10        20         30          40            50            60           70        80        90

            90          100          110          120            130          140          150          160
sv  KEQLPCYSVARIPLPNINEDLTCGNILMWEAVTVKTEVIGVTAMLNLH               SGTQKTHENGAGKPIQSNFHFPAVGGEPLELQGVLANY
    >< β-C     >< CD   >< β-D  >< α-B><                DE loop    >< β-E   >< EF loop

py  NNTLPTWSMAKLQLPMLNEDLTCDTLQMWEAVSVKTEVGSGSLLDVHGFNKPTDTVNTKGISTPVEGSQYHVFAVGGEPLDLQGLVTDA
            100          110          120            130          140          150          160          170          180

            170                   180          190          200          210          220          230          240          250
sv  RTKYPAQ       TVTPKNATVDSQQMNTDHKAVLDKDNAYPVECWVPDSKNENTRYFGTYTGGENVPPVLHITNTATTVLLDEQGVG
                 EF loop                   >< β-F   >  < β-G1 > <β-G2><  GH

py  RTKYKEEGVVTIKTITKKDMVNKDQVLNPISKAKLDKDGMYPVEIWHPDPAKNENTRYFGNYTGGTTTPPVLQFTNTLTTVLLDENGVG
            190          200          210          220          230          240          250          260          270

            260          270          280          290          300          310          320          330
sv  PLCKADSLYVSAVDICGLFTNTSGT  QQWKGLPRYFKITLRKRRSVKNPYPISFLLSDLINRRTQRVDGQPMIGMSSQVEEVRVYEDTEE
    loop >< β-H   >< HI loop><               β-1              >< ... < α-C > ... C arm ...  < β-J > ..

py  PLCKGEGLYLSCVDIMGWRITRNYDVHHWRGLPRYFKITLRKRWVKNPYPMASLISSLFNNMLPQVGGQPMEGENTQVEEVRVYDGTEP
            280          290          300          310          320          330          340          350

            340          350          360
sv  LPGDPDMIRYIDEFGQTTRMQ
    C arm .. < C-loop >>

py  VPGDPDMTRYVDRFGKTKTVFPGN
            360          370          380
```

FIGURE 7.4 Correspondence between the SV40 and polyoma VP1 protein sequences, with structural features noted. The major differences are insertions in the BC and EF loops, which are located on the outer surface of the capsid. (Reproduced with permission from Liddington et al., 1991.)

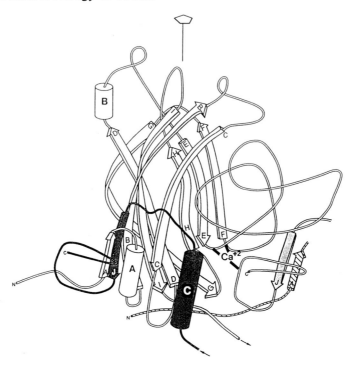

FIGURE 7.5 Richardson diagram of the VP1 monomer subunit from an SV40 pentameric capsomere. Shown are two "invading" C-terminal arms from VP1 molecules in neighboring pentamers. One is shown as contributing part of the calcium binding site, linking the "body" of one pentamer with the C-terminus of another ("C-loop"). The other invading arm shows a C-helix domain that interacts in an amphipathic fashion with other C-helices, and the b-clamp formed by the A and B strands from one pentamer clamping the invading J strand ("C-insert") from the invading arm. The external virus surface is at the top. (Reproduced with permission from Liddington et al., 1991.)

bility that allows for the different contacts occurs at the two ends of the C-helix, permitting different helix clusters and different relative orientations of pentamers.

The contacts between a pentavalent pentamer and its five neighbors ("five-around-one") are more extensive than those between hexavalent pentamers: in the five-around-one, the pentamers dock together closely, making extensive contacts across their side faces; the A strand and A helix of each neighboring pentamer makes largely hydrophobic contacts with an A strand and A helix of the central pentamer. At the junction of three pentamers (α, α', and α''), three C-terminal arms are exchanged in a cyclical fashion.

The contacts between five-around-one units involve hexavalent pentamers and are limited to twofold exchanges of C-terminal arms. In one case (β-β'), the C-helices form a short coiled coil; in the other (γ-γ), the pentamers dock more closely and the helices are disordered (SV40) or partially ordered (poly-

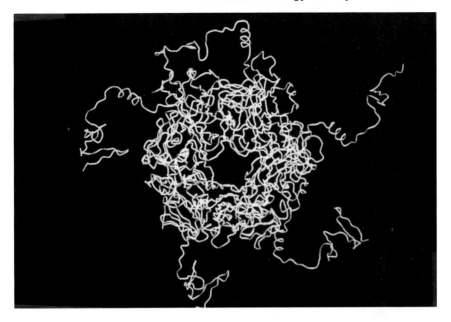

FIGURE 7.6   View down the fivefold axis of a hexavalent pentamer with its full complement of arms as they are configured in the virion. (Reproduced with permission from Liddington et al., 1991.)

oma). The capsid can then be envisaged as an assembly of 12 five-around-one units, arranged on the 12 pentagonal faces of a dodecahedron.

C-Helix Bundles

The structure of SV40 has recently been extended to a resolution of 3.2 Å, allowing a more detailed description of the helix contacts (T. Stehle and S.C. Harrison, personal communication). In the twofold (β-β') contact the helices pack against each other in a symmetric and intimate fashion. They form a short leucine zipper, involving Leu 304, Leu 305, and Leu 308 (see Color Plate 9), which is enclosed by further hydrophobic contacts to Phe 303 and Tyr 298. At the top of the zipper, where the helices diverge, a pair of salt bridges is formed by Asp 307 from each helix and Arg 312 from the other.

In the three-helix bundle, the contacts are less symmetric. One face of the middle part of a helix nestles into the base of its clockwise neighbor; the other face makes hydrophobic contacts with the counterclockwise neighboring helix. At the top of the helices, where they diverge, salt bridges are formed between Asp 307 and Arg 311. These interactions are repeated in a cyclical fashion.

Calcium Binding and Disulfide Bonds

The existence of Ca^{2+} binding sites in the SV40 capsid has been inferred from crystallographic experiments using lanthanide ions (Liddington et al., 1991).

The metal ions form a bridge between a C-terminal arm of one pentamer and an EF loop of another, providing an additional link that fastens an arm into a target pentamer. Two glutamate residues (Glu 157 and Glu 160) from the EF loop and one aspartate residue (Asp 345) from the invading C-terminal arm coordinate the metal. These residues are conserved in all known polyoma VP1 sequences, so the metal binding site is very likely to be common to all members of the polyomavirus family.

The crystal structure of SV40 reveals that there are no disulfides within a subunit or within a pentamer, but there may be one between CD loops of neighboring pentamers (Cys 104), which approach each other just beneath the twofold and threefold clusters of C-helices. Polyoma capsids contain disulfide bonds, but in this case they link VP1 subunits within a pentamer (T. Stehle and S.C. Harrison, personal communication). The bond forms between Cys 20 in the N-terminal arm and Cys 115 in the CD loop of the clockwise neighbor, and appears to act as an additional clamp holding the C-terminal arm of an invading pentamer in place. The disulfide bonds may form while the virus is still in the nucleus, because they are within the interior of the virion, shielded from the intracellular reducing environment. After cell lysis, these bonds certainly must be formed, and stabilize the virion.

These observations provide a rationale for earlier experiments on virion disruption. Walter and Deppert (1974) used 0.05% sodium dodecyl sulfate (SDS) to generate a 140S particle (possibly an empty capsid) that could be completely disrupted with the disulfide reducing agent dithiothreitol (DTT). Christiansen et al. (1977) disrupted SV40 with pH 9.8 and 1-mM DTT to generate a 7S "capsomere" and a 60S DNA–nucleosome–VP2/3 complex. Brady et al. (1977) completely disrupted polyomavirus with ethyleneglycol-bis-(β-aminoethyl ether)-N, N'-tetraacetic acid (EGTA) and DTT at neutral pH. It seems likely that the high pH in the second experiment has an effect analogous to that of EGTA in chelating calcium. Several attempts have been made to reassemble disrupted virions. For both polyoma and SV40, conditions (including readdition of calcium and removal of the disulfide reducing agent) have been defined that allow at least partial reconstitution of infectivity along with reassembly of virionlike particles (Friedmann, 1971; Brady et al., 1979; Yuen and Consigli, 1982; Colomar et al., 1993).

Taken together, the crystallographic and biochemical data suggest that the structural integrity of the virion is maintained by hydrophobic interactions, calcium ions, and disulfide bonds, and that disruption of any two of these contacts leads to dissociation into the minichromosome and VP1 pentamers.

Capsid Assembly

Polyoma VP1, purified after expression in *Escherichia coli*, can be induced to self-assemble in vitro into capsidlike structures (Salunke et al., 1986), either by conditions of high ionic strength or by calcium addition at physiologic salt concentrations. This self-assembly phenomenon appears to be general for the

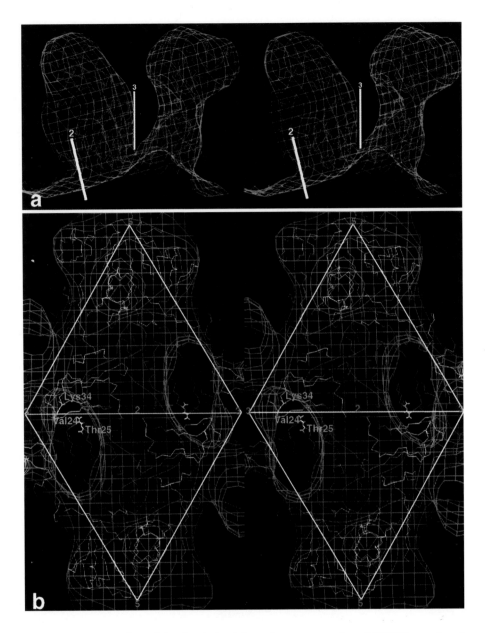

Plate 1. (a) Stereo view of density derived from the reconstruction of CPMV complexed with Fab 10B7 compared with the atomic model of Fab Kol used as a generic model for the Fab. (b) Stereo view of a detailed footprint of the Fab fragment on the surface of CPMV showing the residues in contact with Fab 10B7. Lys 34 (VP2), Val 24 (VP3), and Thr 25 (VP3) are highlighted in the Cα backbone. The electron density of the cryoEM reconstruction is shown as blue contours; the VP2 domain is shown in green; the VP3 domain is shown in red; and VP1 (small subunit) is shown in yellow.

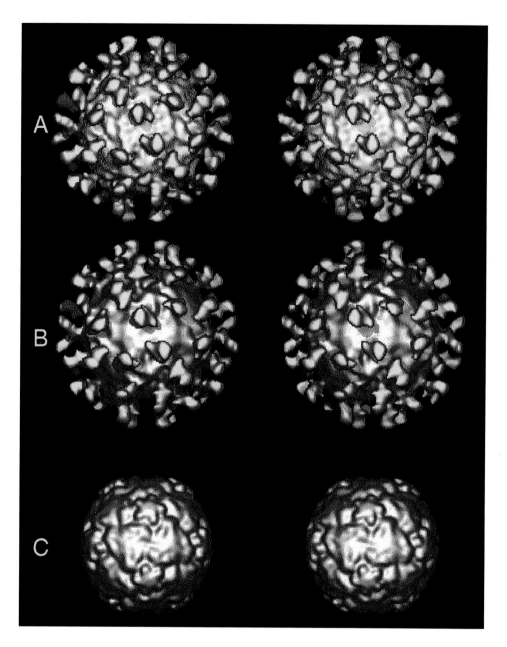

Plate 2. Stereo views of cryo–electron microscopy image reconstructions of (**A**) HRV16 (green)–D1D2 (orange) and (**B**) HRV14 (blue)–D1D2 (orange) complex, viewed along an icosahedral twofold axis in approximately the same orientation as in Figure 4.2. Both **A** and **B** show 60 D1D2 molecules bound to symmetry-equivalent positions in the canyons on the virion surface. (**C**) Shaded-surface view of HRV14 (blue), computed from the known atomic structure (Rossmann et al., 1985—see Chapter 4 references), truncated to 20 Å resolution.

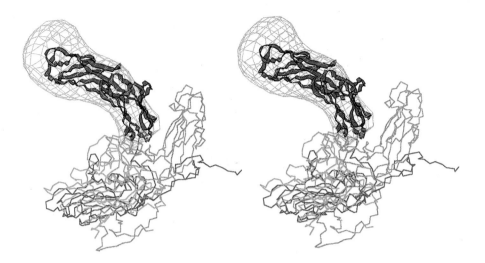

Plate 3. Difference density (orange) between an image reconstruction of HRV16 complexed with D1D2 of ICAM-1 (see Plate 2A), and the crystallographic structure for HRV16. The α-carbon backbones of the HRV16 structural components are shown: VP1 (blue); VP2 (green); VP3 (red). The known crystallographic structures of the D1D2 domains of CD4 (red ribbon), which are homologous to the D1D2 domains of ICAM-1, are placed within the difference density.

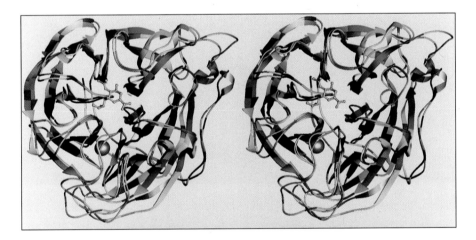

Plate 4. A stereo view of the main chain trace of influenza virus hemagglutinin trimer (H3 subtype) with sialic acid (white) bound in the three receptor binding sites. The trimer is assembled from three identical polypeptides; each polypeptide is later cleaved into the HA1 and HA2 chains (yellow and aqua, green and magenta, red and blue), which remain joined by a disulfide bond. The view is from the side, so the viral membrane would be at the bottom. The figure was made from coordinate file 4HMG in the Brookhaven Protein Data Base.

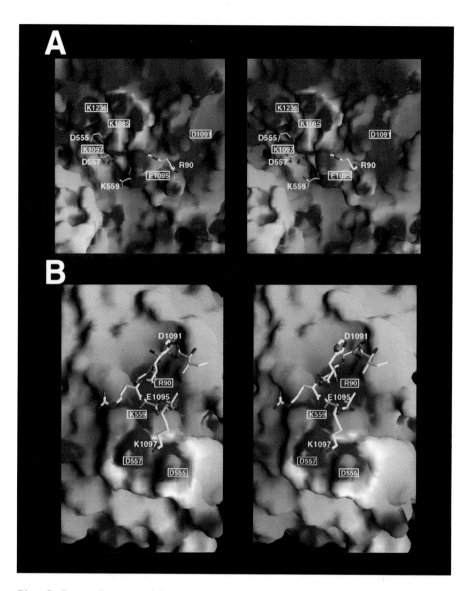

Plate 5. Stereo diagrams of the charge interactions between Fab17-IA and HRV14. (**A**) A surface rendering of the hypervariable region of Fab17-IA. (**B**) The area about the NIm-IA site of HRV14. The surfaces are colored according to their positive (blue) and negative (red) charge character. In **A,** the position of the viral β-structure BC loop, as bound in the complex structure, is shown as a colored stick model. In panel **B,** the BC loop is the large tower in the center right of the diagram, with the canyon directly to the "south." Here, the stick model represents key residues of Fab17-IA proposed to be involved in antibody binding. In both panels, key surface residues are boxed, whereas the stick model labels are not. The number of the Fab residues starts at 1 for the light chain and 501 for the heavy chain. (Reproduced with permission from Liu, H., Smith, T.J., Lee, W.M., Mosser, A.G., Rueckert, R.R., Olson, N.H., Cheng, R.H., and Baker, T.S. 1994. Structure determination of an Fab fragment that neutralizes human rhinovirus 14 and analysis of the Fab-virus complex. J. Mol. Biol. 240:127–37.)

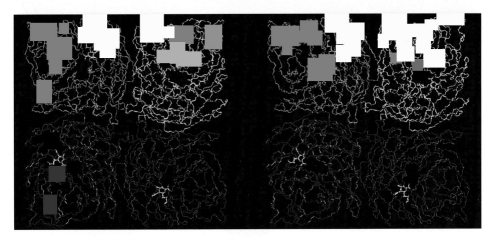

Plate 6. The tetrameric neuraminidase of influenza virus. The stereo view is looking from the top. Each of the four identical chains has a catalytic binding site, which is shown occupied by sialic acid (white).

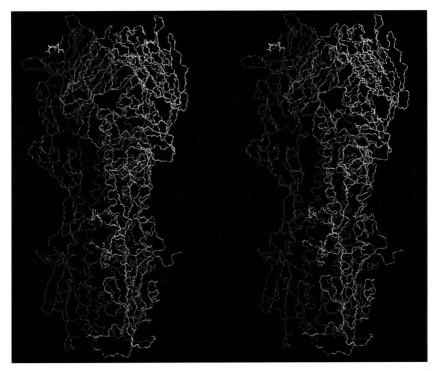

Plate 7. The folding of influenza virus neuraminidases of types A and B. The stereo view shows superimposed α-carbon RIBBON drawings (Carson, 1991) of an influenza type A neuraminidase (A/tern/Australia/G70c/75, N9) with the inhibitor 2-deoxy-2,3-dehydro-*N*-acetylneuraminic acid (DANA) bound in the active site, and a type B neuraminidase (B/-Beijing/1/87). The monomer is viewed from the top, looking toward the viral membrane. The chains are color coded according to the types of amino acid side chains: green, hydrophobic; blue, polar; orange, charged. Note that the chain paths diverge at the surfaces but that they come together in the area immediately surrounding the bound DANA. The coordinates used are 1NNB (Bossart-Whitaker et al., 1993) and 1NSB (Burmeister et al., 1992) from the Brookhaven Protein Data Base. (Courtesy of M. Carson; see Chapter 15 for reference citations.)

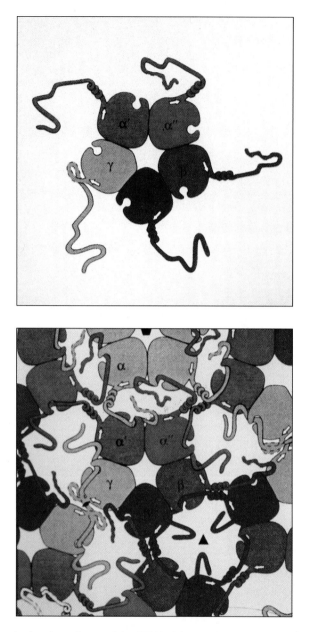

Plate 8. Schematic drawing of part of the 5V40 virion surface, showing the interactions of pentamers around a central hexavalent pentamer. The invading arms are represented by cylinders and lines. Arms emanate from one subunit, form C-helices (small cylinders) and C-inserts (lines running through the invaded subunit), and in three cases also form extended C-loops (hairpins). The labeling of VP1 subunits is as in Figure 7.3. (Reproduced with permission from Caspar, D.L.D. 1992. Virus structure puzzle solved. Curr. Biol. 2:169–171.)

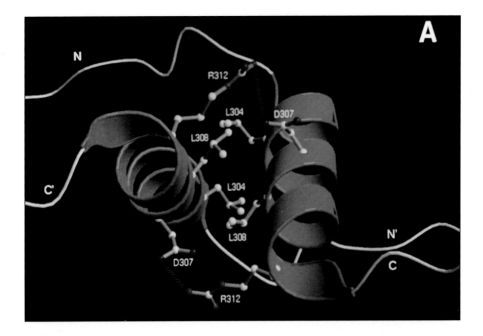

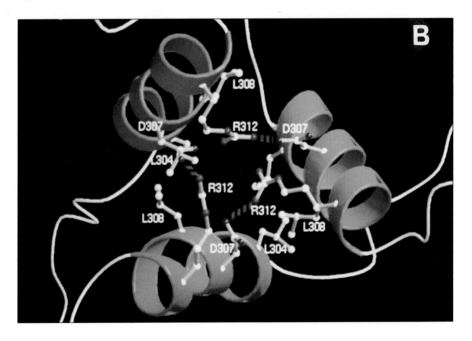

Plate 9. Interaction of C-helices of VP1 subunits of SV40 in (**A**) the two-helix bundle interaction (β-β′ bonds; see Figures 7.2 and 7.6) and (**B**) the three-helix bundle (α, α′, α″; see Figures 7.2 and 7.6).

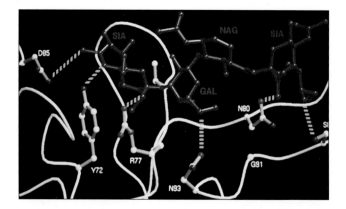

Plate 10. The receptor binding pocket of polyoma VP1 with the receptor analog, 3-sialyl lactose, shown as a ball-and-stick model. Three pockets are evident (numbered 1 to 3). Pockets 1 and 2 accommodate sialic acid and glucose, respectively. Pocket 3 could accommodate the α-2,6-linked sialic acid of a branched-chain receptor. (Reproduced with permission from Stehle, T., Yan, Y., Benjamin, T.L. and Harrison, S.C. 1994. Structure of murine polyomavirus complexed with an oligosaccharide receptor fragment. Nature 369:160–163.)

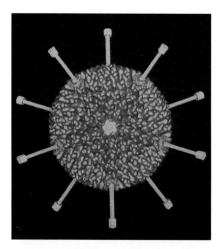

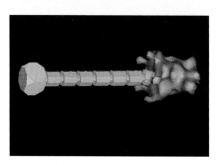

Plate 11. Adenovirus virion, penton base, and fiber. **Top** The penton complex, consisting of the penton base in the capsid (yellow) and the protruding fiber (green), is highlighted on the three-dimensional image reconstruction, which is viewed along an icosahedral fivefold axis. Because the shaft of the fiber (~300 Å) and the knob at its end could not be reconstructed, they are modeled (White, 1993). **Bottom** The penton complex is shown as a composite of the penton base from the reconstruction and a model fiber. The size of the fiber, as well as the distribution of the nine beads in its shaft, are based on EM images of negatively stained single fibers (Ruigrock et al., 1990). The first two beads in the shaft, which are buried within the penton base, and the third bead, which is visible in this view, are visible in the EM difference density. The remaining six beads are modeled, as is the knob at the end of the fiber, whose structure was recently determined (Xia et al., 1994). (Reproduced with permission from Stewart, P.L. and Burnett, R.M. 1995. Adenovirus structure by X-ray crystallography and electron microscopy. Curr. Top. Microbiol. Immunol. 199:25–38. For other citations, see Chapter 8.)

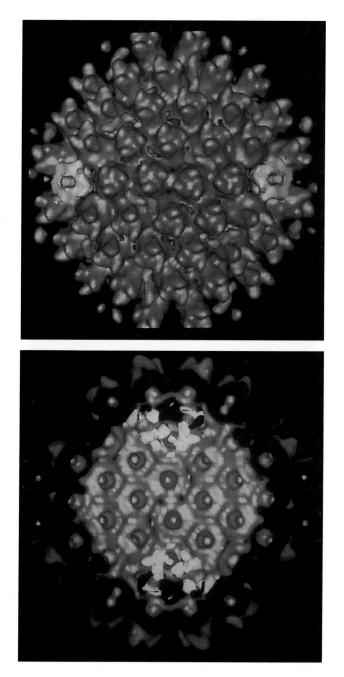

Plate 12. The overall organization of the adenovirus virion. **Top** An external view of the capsid components as identified from the three-dimensional difference map showing hexon (blue), penton base (yellow), fiber (green), polypeptide IIIa (purple), and polypeptide IX (red). **Bottom** An internal view showing the well-ordered density assigned to polypeptide VI (white) under two vertices. (Reproduced with permission from Stewart, P.L., Fuller, S.D. and Burnett, R.M. 1993. Difference imaging of adenovirus: bridging the resolution gap between X-ray crystallography and electron microscopy. EMBO J. 12:2589–2599.)

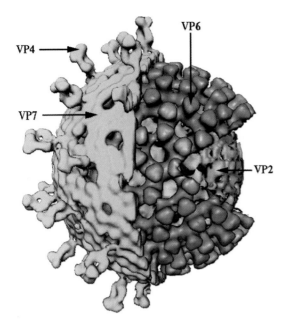

Plate 13. 3-D structure of the rotavirus illustrating the triple-layered architecture. A portion of the outer layer (VP7, yellow) and a portion of the VP6 layer (blue) are removed, showing the VP2 layer (grey). The densities inside the subcore that can be seen through the VP2 layer are shown in red.

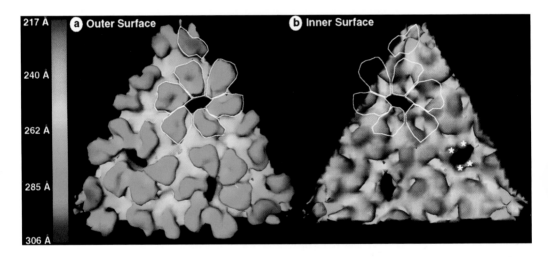

Plate 14. Surface density representation of the outer and inner triangular faces of the P22 procapsid containing scaffolding-proteins, extracted from the 19 Å map (Thurman-Commike et al., 1996). The densities are colored radially as shown in the color bar on the left. The white lines show the seven coat protein subunits in an asymmetric unit in their corresponding positions on both internal and external surfaces. The four asterisks (*) identify the finger-like regions around the hole in one hexon, which have been interpreted as ordered scaffolding protein domains. (Courtesy of Dr. Pam Thuman-Commike and permission for reproduction from Thuman-Commike et al., 1996. Three-dimensional structure of scaffolding-containing phage P22 procapsids by electron cryo-microscopy. J. Mol. Biol. 260:85–98.)

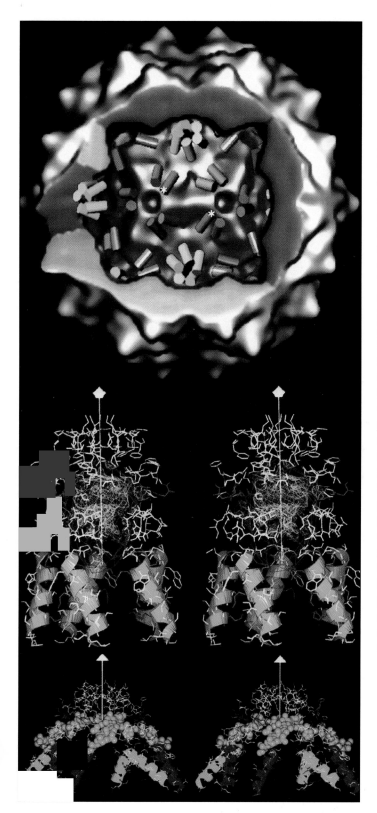

Plate 15.

Top Cutaway view of the whole FHV virion. Electron density derived from the cryoEM reconstruction is blue (Figure 10.2, top), density derived from the difference map is red (see Figure 10.2, bottom). γ_B helices are red cylinders, γ_C helices are green, and γ_A helices are light blue. Although the γ helices all have the same amino acid sequence, quaternary structural interactions create a helical bundle with γ_A peptides (*middle panel*) and helices interspersed with ordered RNA for the γ_B and γ_C peptides (*bottom panel*).

Middle Side view of the pentameric helical bundle (fivefold axis is shown as a white line with a pentagon symbol at the end furthest from the virus center).

Bottom Side view of the γ peptides at the quasi-sixfold axes (true threefold axis is shown as a white line with a triangle at the end).

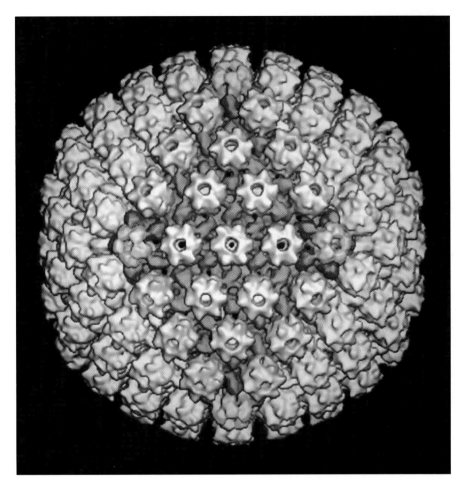

Plate 16. Three-dimensional density map of the capsid of HSV-1 reconstructed to approximately 25 Å resolution from cryo–electron micrographs, as rendered by surface shading (Conway et al., 1995). The capsid is 1250 Å in diameter. The color coding is intended to convey the distribution of capsid proteins and some aspects of quasi-equivalence. The major protein, VP5 (149 kDa), forms both the hexons (depicted as blue on two facets of the icosahedral surface lattice) and the pentons (turquoise) in somewhat different conformations, so that hexons bind six copies of the small capsid protein VP26 (12 kDa) around their outer rims (Booy et al., 1994) but pentons do not (Zhou et al., 1994). Two other capsid proteins—VP19c (50kDa) and VP23 (35 kDa)—form "triplex" complexes at the sites of local threefold symmetry. The triplexes adjacent to pentons (p-P-P triplexes, in the terminology of Figure 12.4c), which are most easily removed by denaturants, are shown as red, and the other kinds of triplexes are orange. (Courtesy of F.P. Booy, J.C. Brown, J.F. Conway, W.W. Newcomb, and B.L. Trus; see Chapter 12 for reference citations.)

Plate 17. Signal transduction model of phage assembly through the exit complex. Binding of the packaging signal, which protrudes from the pV-ssDNA complex, may induce a conformational change in pI that enables its periplasmic domain to bind to the periplasmic domain of the pIV multimer. This, in turn, would stimulate opening of the pIV channel, providing a specific conduit through which the assembling phage can be extruded. OM, outer membrane; PERI, periplasmic domain; IM, inner membrane; CYTO, cytoplasm. Reproduced with permission from Russell, M. 1994. Phage assembly: a paradigm for bacterial virulence factor export? Science 265:612–614.)

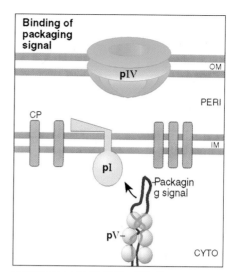

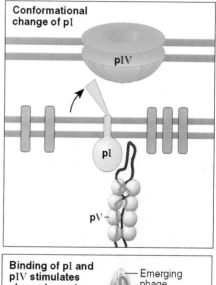

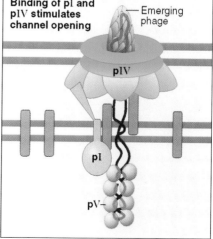

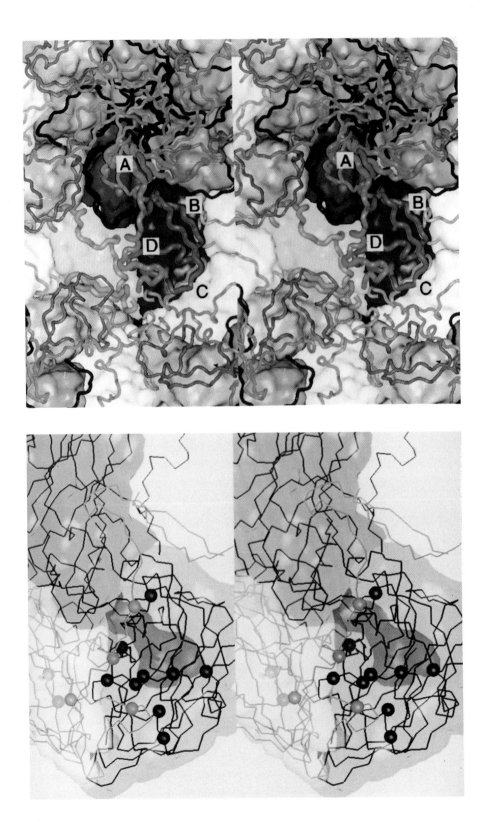

Plate 18.

Top Stereo view of the internal network formed by VP4 and the amino-terminal extensions of VP1, VP2, and VP3 of poliovirus. VP4 (green) and the amino-terminal extensions of VP1 (blue), VP2 (yellow), and VP3 (red) are shown as tubes representing the α-carbon backbone. The remaining portions of VP1, VP2, and VP3 are shown as surfaces. The reference protomer (center) is distinguished from its symmetry-related neighbors by its deeper surface colors and thicker tubes. The fivefold and threefold symmetry axes are located respectively near the top and the bottom of the figure. The labels A, B, C, and D identify locations of interactions that appear to significantly stabilize the virion: (A) The five-stranded β-tube structure (see caption to Figure 6.3A in text). (B) Interactions at the fivefold interface, including contacts between residues 62 to 67 of VP1 in the reference protomer and residues 42 to 52 of VP2 in a fivefold related protomer. (C) The seven-stranded cross-pentamer β-sheet. Depicted is the strand contributed by the amino-terminal extension of VP1 from the reference protomer overlaying two strands formed by the amino-terminal extension of VP2 from a threefold related protomer. The remaining four strands are contributed by the VP3 β-Barrel from the reference protomer (depicted as a surface). See caption to Figure 6.3 for further details and alternate views. (D) The critical VP1 loop contributed by residues 44 to 56, which mediates the intraprotomer interactions between the β-barrels of VP2 and VP3. A number of mutations are located within the loop and the surrounding vicinity.

Bottom Close-up stereo view of the reference protomer depicted in Plate 5, top. The surface representations of VP1 (blue), VP2 (yellow), and VP3 (red) are semi-transparent to reveal the α-carbon backbone of the protomer (thin lines). A number of mutations that effect biological phenotypes of poliovirus map in the VP1 loop or its vicinity. The sites of these mutations are indicated by spheres. See Chapter 6 text for a detailed description of mutant phenotypes.

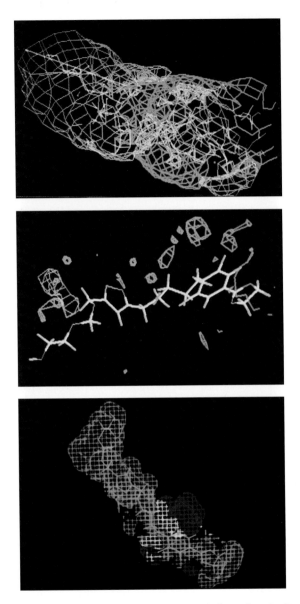

Plate 19. CoMFA contour map representing the correlation of steric (**top**) and electrostatic parameters (**middle**) with biological activity. Blue indicates a negative correlation, and red is indicative of a positive correlation. The data for this grid map were produced from a partial least-squares analysis. These results suggest that the activity of these compounds is influenced to a large extent by steric factors and to a lesser degree by electrostatic factors. **Bottom** A volume map resulting from an overlay of a series of inactive and active structures against HRV14. The conformations of the structures were obtained from X-ray crystallographic studies performed by Dr. Michael G. Rossmann. Volume maps were created by considering the total volume occupied by all of the inactive structures (red) and all of the active structures (blue). One of the most active compounds has been inserted into the map.

papovavirus major capsid proteins; it has been observed for SV40 VP1 (S. Cilmi and S. Harrison, personal communication), budgerigar fledgling disease polyomavirus VP1 (Rodgers et al., 1994), and human papillomavirus-11 L1 protein (Rose et al., 1993). Analysis of the assembly reaction under conditions of variable pH, ionic strength, and calcium concentrations revealed that aggregates other than the $T = 7$ capsids could be formed under nonphysiologic solvent conditions (Salunke et al., 1989). In addition, the structures formed (whether $T = 1$, octahedra, or ribbons) were dependent on the pathway of solvent changes.

In contrast to the variety of assemblies seen in vitro, $T = 7$ capsid assembly is observed with high fidelity in vivo. When either polyoma VP1 or papillomavirus L1 is expressed in a baculovirus system, only unassembled pentamers are observed in the cytoplasm of the insect cells, but near-perfect $T = 7$ capsids are found in the nucleus. The exclusive assembly of virions or capsids in the nucleus is a major puzzle for all DNA viruses.

From current data, several possibilities are suggested for regulating polyoma assembly. First, chaperone proteins could prevent pentamer oligomerization in the cytosol. The SV40 ''agnoprotein'' has been proposed as a possible chaperonin preventing cytoplasmic assembly (see later), and other viruses use heat shock proteins during a phase of their assembly. Second, differential access to calcium in the cytosol versus the nucleus could regulate formation of the crucial calcium-mediated bond between capsomeres. However, no such regulatory mechanism has been identified. Nonetheless, addition of the calcium ionophore ionomycin induces capsid assembly in the cytoplasm for VP1 expressed in insect cells (Montross et al., 1991). Third, a thermodynamic model of capsid assembly has been proposed, suggesting that the assembly of icosahedral viruses is highly concentration dependent (Zlotnick, 1994). In this model, assembly resembles a cooperative reaction, such that the vast majority of observed species are either intact capsids or free pentamers. Furthermore, the concentration of pentamers at which assembly occurs is highly sensitive to small changes in the bond energy between pentamers, so that small changes in calcium concentration could have a dramatic effect on the pentamer–capsid equilibrium (see Chapter 1). Therefore, exclusive assembly in the nucleus may be regulated by the increase in capsid protein concentration that occurs after transport of the pentamers to the smaller volume of the nuclear space, and/or triggered by small changes in calcium concentration.

The property of in vitro VP1 self-assembly can be used to study the effects of mutations on bonding in capsids, because assembly of recombinant mutant proteins is readily screened by electron microscopy. The first informative mutant protein was a truncation of the carboxyl-terminal 63 amino acids of polyoma VP1, ΔNCOVP1, that allowed pentamer formation but not capsid assembly (Garcea et al., 1987). The crystal structure of SV40 revealed that these residues form the critical C-terminal arm. As yet, in vitro self-assembly of purified recombinant VP1 protein has not been possible in the presence of either viral minichromosomes or plasmid DNA (Moreland et al., 1991). Apparently the pentameric VP1 capsomeres, with their high affinity for DNA, bind to the nucleic acid and are unable to interact further. This result suggests that the minor proteins VP2 and VP3 (or other factors) play a role in modulating the interaction between VP1 and the genomic core, such that the intrinsic capsid assembly potential of VP1 can be realized. However, capsids are formed in insect cells expressing VP1, VP2, and VP3, indicating that the minor proteins are not in themselves inhibitory for capsid formation in the absence of the viral minichromosome (Forstova et al., 1993).

Capsid Protein Posttranslational Modification

Phosphorylation of the SV40 and polyoma VP1 proteins was first detected using in vivo labeling followed by virion purification and 1D SDS–polyacrylamide gel electrophoresis (Ponder et al., 1977). The modifications occur rapidly, within 2 min of a pulse label, and at least some of the phosphorylations occur cotranslationally (Garcea et al., 1985; Fattaey and Consigli, 1989). For SV40, no differences in phosphorylated subspecies were detected in VP1 molecules associated with different assembly intermediates (Garber et al., 1980). Temperature-sensitive mutant polyomaviruses are phosphorylated at the restrictive and permissive temperatures, supporting the hypothesis that phosphorylation precedes and is independent of virion assembly, although it may be necessary for forming the final virion (Garcea et al., 1985). Specific VP1 phosphorylation has been implicated in cell receptor recognition, because empty capsids that are underphosphorylated are unable to inhibit infection by virions (Bolen et al., 1981). The isoelectric focusing pattern of VP1 subspecies changes when it is coexpressed with VP2 in a baculovirus system, suggesting that the association of VP1 with VP2 is coupled with VP1 phosphorylation (Forstova et al., 1993). For polyoma, the majority (60 to 70%) of the phosphate is on two threonine residues, Thr 62 and Thr 155 (Anders and Consigli, 1983; Garcea et al., 1985; Li and Garcea, 1994). The crystal structure reveals interesting locations for these residues: Thr 62 is exposed on the upper surface of the pentamer, on the outer edge of the BC1 loop (which forms part of the receptor binding site); Thr 155 is situated just above the FG corner that forms a neck at the top of the inner surface of the pentamer (the presumed location of VP2/3). This region has a high density of threonine residues, including three consecutive threonines at the top of the FG corner.

The polyomavirus VP2 protein is myristylated at the N-terminal glycine (Streuli and Griffin, 1987). Mutant viruses affected at this glycine residue are not myristylated and are efficiently assembled (Krauzewicz et al., 1990; Sahli et al., 1993). However, the resulting virions are 15 to 20-fold less infectious than the wild-type virus in vitro, and have an attenuated in vivo pattern of spread. These results suggest a defect in an early step of infection, at or prior to uncoating.

Intracellular Transport and Virion Assembly

After synthesis in the cytoplasm, the viral capsid proteins must reach the site of nuclear capsid assembly in the correct stoichiometry and without aberrant assembly in the cytosol. The capsid proteins of polyoma and SV40 have nuclear targeting signals that allow them to use cellular functions for their transport (Wychowski et al., 1986, 1987; Clever and Kasamatsu, 1991; Moreland and Garcea, 1991; Chang et al., 1992). From several studies, it appears that stoichiometry in the nucleus is regulated by the formation of a specific complex between VP1 pentamers and VP2/3 proteins in the cytoplasm prior to nuclear transport. Detailed kinetics and cell fractionation analyses of SV40 capsid protein transport have shown that the ratio of VP1 to VP2/3 in the nucleus is constant throughout infection (Lin et al., 1984). In addition, temperature-sensitive mutant viruses affecting SV40 VP1 have defects in both VP1 and VP3 subcellular localization at the nonpermissive temperature (Kasamatsu and Nehorayan, 1979). Protein coexpression experiments, using vaccinia virus vectors in mammalian cells or baculovirus expression in insect cells, have also shown that coexpression of VP2/3 with VP1 facilitates the subcellular localization of the minor capsid proteins to the nucleus (Stamatos et al., 1987; Delos et al., 1993). Indeed, the myristate modification of VP2 appears to override the nuclear targeting signal in the absence of VP1 coexpression, and VP2 is localized exclusively to membranes in the cytoplasm (Delos et al., 1993). Finally, SV40 VP3 has been shown to have a domain near its carboxyl terminus that interacts with VP1 pentamers (Gharakhanian et al., 1988).

Nuclear transport of SV40 VP1 is facilitated in some, but not all, cell types by the viral agnoprotein, a 61-amino-acid protein encoded in the late leader region of the genome (Ng et al., 1985; Carswell and Alwine, 1986; Resnick and Shenk, 1986). Second-site revertants of a VP1 mutant virus have been mapped to the agnogene (Margolskee and Nathans, 1983), and vice versa (Barkan et al., 1987). These results have suggested that the agnoprotein inhibits assembly of VP1 pentamers in the cytoplasm (as described earlier), and that VP1 mutations may compensate for a lack of the agnoprotein by reducing VP1's ability to self-assemble. Without the agnoprotein, (i.e., polyoma, which has a similar open reading frame but is without a detected protein product as yet (Fenton and Basilico, 1982)), host cell chaperone proteins or other mechanisms (as described earlier) may serve to inhibit premature assembly.

Assembly and Cell Transformation

The regulatory genes of all DNA viruses play an integral role in virion production, and many of these regulatory proteins have been identified as oncogenes. The link between the transforming genes of murine polyomavirus and virion assembly has been extensively studied. In particular, the polyoma host-range nontransforming (*hr-t*) mutant viruses have a phenotype that includes an inability to both transform cells in vitro and cause tumors in vivo, as well as a block in virus production during infection of normal cells in vitro (reviewed in Benjamin, 1982). These viruses have mutations in the intron of large T antigen, affecting only the middle and small T antigen proteins. A great deal is now known about the biologic interactions of the T antigens and host cell proteins (reviewed in Fanning, 1992), so that reasonable hypotheses may be formulated that link these interactions with virus reproduction. The large T antigen of both polyoma and SV40 binds the Rb protein, activating the E2F transcription factor. SV40 large T antigen also binds p53. Both of these interactions are thought to lead to activation of cellular DNA synthesis that allows for viral DNA replication. Middle T antigen activates the protein tyrosine kinases $pp60^{c-src}$, $p62^{c-yes}$, and $p60^{c-fyn}$, and also activates the phosphatidylinositol-3-kinase, leading to the accumulation of phosphatidylinositol 3-phosphate (Whitman et al., 1988). Middle and small T antigens associate with both the regulatory and catalytic subunits of protein phosphatase 2A, inhibiting its activity (Pallas et al., 1990; Walter et al., 1990). All of these interactions appear necessary, but not sufficient, for the complete transformation phenotype.

When the block in virus production for *hr-t* mutant nontransforming polyomaviruses was analyzed, two defects were identified: (1) a decrease (3- to 50-fold) in viral DNA accumulation, and (2) inefficient encapsidation of 75S minichromosomes into 240S virions. The viral DNA synthetic defect may be attributed to the failure of T antigen stimulation of cellular DNA replication, despite the interaction of large T antigen with Rb. The inefficient encapsidation has been correlated with decreased to absent phosphorylation of VP1 on Thr 62 and Thr 155 (Li and Garcea, 1994). A mutant virus, Thr 155 to Ala, has a defect in encapsidation similar to *hr-t* mutant viruses, suggesting that phosphorylation of VP1 may be a "proofreading" mechanism such that viral assembly depends on the presence of active middle T antigen.

Two classes of SV40 large T antigen mutant viruses have been identified that affect virion production. The first class are the host-range/adenovirus helper function (*hr/hf*) mutant viruses (Tornow and Cole, 1983; Pipas, 1985; Cole and Stacy, 1987). These mutant viruses, with defects in the carboxyl-terminal 38 amino acids of SV40 T antigen, produce wild-type outputs of virus in BSC monkey cells but 10^{-5} lower titers when grown on CV1P cells, and have defects in late mRNA accumulation (decreased 5- to 15-fold), VP1 and VP3 synthesis (decreased 5- to 15-fold), agnoprotein synthesis (decreased 100-fold), and the addition of VP1 to 75S minichromosomes (Khalili et al., 1988; Stacy et al., 1990; Spence and Pipas, 1994b). One hypothesis is that the primary defect lies in transcriptional regulation, resulting from a failure to activate E2F, and that the effect of the lower levels of the capsid proteins is enhanced by the lack of agnoprotein that would facilitate their nuclear localization. CV1P cells constitutively producing the agnoprotein complement these mutants to some extent (Stacy et al., 1990). A second class of SV40 T antigen mutant viruses with an encapsidation defect is represented by the mutant virus 5002 (Peden et al., 1990; Spence and Pipas, 1994a). This mutant virus synthesizes viral DNA and capsid proteins to near-normal levels, grows poorly in both BSC40 and CV1 cells, and complements the defect of *hr/hf* mutant viruses. Further investigation of the defects of these two classes of assembly mutants, resulting from defects in large T antigen, will be instructive in understanding the recruitment of cellular processes by T antigen during virus assembly.

Murine Polyomavirus Receptor Binding

Unlike SV40, attachment of polyoma to the surface of susceptible cells requires sialic acid. This requirement is reflected in its ability to hemagglutinate erythrocytes (Eddy et al., 1958; Crawford, 1962). The ''large''- and ''small''-plaque virus variants of polyoma have distinctive hemagglutination properties and can be discriminated by the pH dependence of their erythrocyte binding (Diamond and Crawford, 1964). The chemical nature of the linkages recognized by polyoma has been determined as an unbranched or branched (α-2,3)-linked sialyl lactose moiety (Cahan and Paulson, 1980; Fried et al., 1981). Recognition of these moieties differs between small- and large-plaque variants of polyoma: the small-plaque viruses recognize both straight and branched-chain sugars, whereas the large-plaque viruses recognize only the straight chain (Cahan et al., 1983). A single mutation in VP1, Glu 91 to Gly, is sufficient to change the virus phenotype from small to large plaque (Freund et al., 1991), as well as changing the ability of the virus to induce tumors (Dubensky et al., 1991).

The complex between polyoma and the trisaccharide 3'-sialyl lactose has been determined crystallographically (see Color Plate 10) (Stehle et al., 1994). Compared with SV40, polyoma VP1 has an additional 18 residues inserted into the surface loops, which create a shallow oligosaccharide binding pocket on the outer margin of VP1 formed by the larger BC and HI loops. Polyoma has quite low intrinsic affinity for sialic acid ($K_d \sim 1$ mmol/l, similar to influenza hemagglutinin), but when a virus attaches to a cell, many receptors can engage and a strong synergistic interaction may occur. A preliminary structure of a small-plaque strain in complex with the branched oligosaccharide indicates that, in the large-plaque virus, the side chain of Glu 91 would interfere directly with the carboxylate group of the sialic acid. The greater binding specificity of the large-plaque virus may increase its ability to spread rapidly, and so contribute to its virulence.

The attachment of the virus appears to play a significant role in subsequent assembly events. For polyoma, infection of quiescent fibroblasts results in a biphasic transcriptional induction of c-*fos*, c-*myc*, c-*jun*, and *JE* mRNAs (platelet-derived growth factor–inducible gene products) (Zullo et al., 1986; Glenn and Eckhart, 1990). As a consequence, the infected cells transit from G_0 to G_1 and begin DNA synthesis. The first phase of mRNA synthesis can be induced by recombinant VP1 protein alone, suggesting an effect directly caused by attachment to a receptor. Subsequent to the transcription of the viral T antigens, a second, sustained induction of these gene products occurs. Induction of genes involved in DNA synthesis and cell cycle progression is important for replication of these, and most likely other, DNA viruses, because the infected cells must remain in a permissive state for DNA replication. Polyoma *hr-t* mutant viruses, which do not induce the second phase, are markedly defective (3- to 50-fold, dependent on cell type and virus enhancer number) in viral DNA replication, as a likely consequence of this failure to sustain cellular DNA synthesis.

Summary

Based upon current biochemical, cell culture, and structural information, the following scenario can be framed for a polyomavirus infection. The virus attaches to the cell membrane, initiating a signal transduction event that induces gene products that prime the cell for DNA replication. The virus enters the cell by an unknown endocytic pathway. Uncoating is triggered by the low intracellular calcium levels and reducing environment. Once in the nucleus, early viral gene (T antigen) transcription is initiated, and viral DNA synthesis is continually stimulated by the activity of these genes. A transcriptional switch occurs, directing synthesis of the capsid proteins. VP1, synthesized in the cytosol, immediately pentamerizes and then associates with VP2 or VP3; the resulting "capsomere complex" determines the stoichiometry of these proteins as they are transported to the nucleus, utilizing their nuclear targeting signal sequences (Fig. 7.7). Premature capsid assembly in the cytosol is prevented by the low concentrations of pentamers and calcium or by the presence of chaperone proteins. Once in the nucleus, viral genomes are selected for encapsidation by recognition of the nucleosome-free region near the origin of repli-

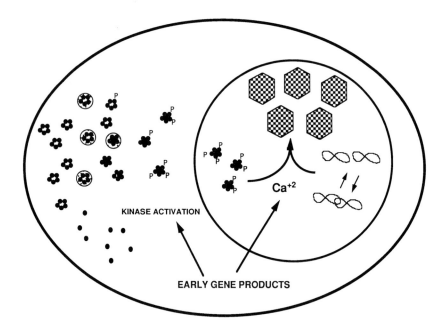

FIGURE 7.7   Schematic representation of a hypothetical pathway for polyomavirus assembly. VP1 capsomeres (rings of five circles) associate with chaperone proteins (open rings around capsomeres) and the VP2/3 proteins (●). Posttranslational modifications such as phosphorylation may occur, and capsomere subunit assemblies are transported to the nucleus. In the nucleus, genomes are selected from the replicating and transcribing pools to be encapsidated. Calcium is required for assembly. The T antigen early gene products may facilitate both capsid protein phosphorylation and minichromosome encapsidation. (Adapted from Hendrix and Garcea, 1994.)

cation. The first capsomere complex attaches to the signal sequence site, followed by a five-around-one polymerization (maximal number of most stable bonding interactions). Capsid polymerization is then rapidly completed, dictated by local bonding constraints as more pentamers are added. Capsomere complexes assemble by interacting with each other and with internucleosomal DNA, displacing histone H1. As the capsid forms, calcium levels may increase (perhaps through T antigen activity) and/or high-affinity Ca^{2+} binding sites are created by apposed VP1 pentamers. The innate ability of pentamers to form $T = 7$ capsids, together with the size of the viral minichromosome, assures accurate assembly. The infection ends as the amount of virions in the nucleus inhibits normal cellular functions and the cell lyses, releasing virus to the environment, where the oxidizing conditions and millimolar Ca^{2+} concentrations assure virion stability.

Obviously much remains to be learned. For example, how does uncoating and endocytic transport to the nucleus occur? What prevents cytosolic capsid protein polymerization? What aspect of the bonding regulates the fidelity of $T = 7$ assembly when other end products are possible? What is the role of VP2/VP3 in the structure? The current knowledge base for the polyomaviruses now enables many of these questions to be addressed. Utilizing site-directed mutant proteins and the in vitro assembly of capsids, the contributions of individual domains of the VP1 C-terminus to interpentamer bonds may be ascertained. The availability of VP2 and VP3 as recombinant proteins will be useful in identifying their interactions with VP1. The possibility of extending both structural and biochemical analyses to papillomavirus proteins may help illuminate general principles in $T = 7$ animal virus construction.

Acknowledgments

We thank Laurie Bankston for a critical reading of the manuscript, and Thilo Stehle and Stephen C. Harrison for providing atomic coordinates prior to publication.

R.L.G. was supported by a grant (CA37667) from the National Cancer Institute, U.S. Department of Health and Human Services.

References

Ambrose, C., Rajadhyaksha, A., Lowman, H. and Bina, M. 1989. Locations of nucleosomes on the regulatory region of simian virus 40 chromatin. J. Mol. Biol. 209: 255–263.

Anders, D.G. and Consigli, R.A. 1983. Comparison of non-phosphorylated and phosphorylated species of polyomavirus major capsid protein VP1 and identification of the major phosphorylation region. J. Virol. 48: 206–217.

Baker, T.S., Drak, J. and Bina, M. 1988. Reconstruction of the three-dimensional structure of simian virus 40 and visualization of the chromatin core. Proc. Natl. Acad. Sci. U.S.A. 85: 422–426.

Barkan, A., Welch, R.C. and Mertz, J.E. 1987. Missense mutations in the VP1 gene of simian virus 40 that compensate for defects caused by deletions in the viral agnogene. J. Virol. 61: 3190–3198.

Barouch, D.H. and Harrison, S.C. 1994. Interactions among the major and minor coat proteins of polyomavirus. J. Virol. 68: 3982–3989.

Baumgartner, I., Kuhn, C. and Fanning, E. 1979. Identification and characterization of fast-sedimenting SV40 nucleoprotein complexes. Virology 96: 54–63.

Benjamin, T.L. 1982. The hr-t gene of polyoma virus. Biochim. Biophys. Acta 695: 69–95.

Blasquez, V., Beecher, S. and Bina, M. 1983. Simian virus 40 morphogenetic pathway: an analysis of assembly-defective tsB201 DNA-protein complexes. J. Biol. Chem. 258: 8477–8484.

Bolen, J.B., Anders, D.G., Trempy, J. and Consigli, R. 1981. Differences in the subpopulations of the structural proteins of polyoma virions and capsids: biological functions of the multiple VP1 species. J. Virol. 37: 80–91.

Brady, J.N., Kendall, J.D. and Consigli, R.A. 1979. In vitro reassembly of infectious polyoma virions. J. Virol. 32: 640–647.

Brady, J.N., Winston, V.D. and Consigli, R.A. 1977. Dissociation of polyomavirus by the chelation of calcium ions found associated with purified virions. J. Virol. 23: 717–724.

Cahan, L.D. and Paulson, J.C. 1980. Polyoma virus adsorbs to specific sialyloligosaccharide receptors on erythrocytes. Virology 103: 505–509.

Cahan, L.D., Singh, R. and Paulson, J.C. 1983. Sialyloligosaccharide receptors of binding variants of polyoma virus. Virology 130: 281–289.

Carswell, S. and Alwine, J.C. 1986. Simian virus 40 agnoprotein facilitates perinuclearnuclear localization of VP1, the major capsid protein. J. Virol. 60: 1055–1061.

Chang, D., Haynes, J.I., Brady, J.N. and Consigli, R.A. 1992. Identification of a nuclear localization sequence in the polyomavirus capsid protein VP2. Virology 191: 978–983.

Chestier, A. and Yaniv, M. 1979. Rapid turnover of acetyl groups in the four core histones of simian virus 40 minichromosomes. Proc. Natl. Acad. Sci. U.S.A. 76: 46–50.

Christiansen, G., Landers, T., Griffith, J. and Berg, P. 1977. Characterization of components released by alkali disruption of simian virus 40. J. Virol. 21: 1079–1084.

Clever, J. and Kasamatsu, H. 1991. Simian virus 40 Vp2/3 small structural proteins harbor their own nuclear transport signal. Virology 181: 78–90.

Clever, J., Yamada, M. and Kasamatsu, H. 1991. Import of simian virus 40 virions through nuclear pore complexes. Proc. Natl. Acad. Sci. U.S.A. 88: 7333–7337.

Coca-Prados, M. and Hsu, M.-T. 1979. Intracellular forms of simian virus 40 nucleoprotein complexes. II. Biochemical and electron microscopic analysis of simian virus 40 virion assembly. J. Virol. 31: 199–208.

Cole, C.N. and Stacy, T.P. 1987. Biological properties of simian virus 40 host range mutants lacking the COOH-terminus of large T antigen. Virology 161: 170–180.

Colomar, M.C., Degoumois-Sahli, C. and Beard, P. 1993. Opening and refolding of simian virus 40 and in vitro packaging of foreign DNA. J. Virol. 67: 2779–2786.

Crawford, L.V. 1962. The adsorption of polyoma virus. Virology 18: 177–181.

Delos, S.E., Montross, L., Moreland, R.B. and Garcea, R.L. 1993. Expression of the polyomavirus VP2 and VP3 proteins in insect cells: coexpression with the major capsid protein VP1 alters VP2/VP3 subcellular localization. Virology 194: 393–398.

Diamond, L. and Crawford, L.V. 1964. Some characteristics of large-plaque and small plaque lines of polyoma virus. Virology 22: 235–244.

Dubensky, T.W., Freund, R., Dawe, C.J. and Benjamin, T.L. 1991. Polyomavirus replication in mice: influences of VP1 type and route of inoculation. J. Virol. 65: 342–349.

Dubochet, J., Adrian, M., Schultz, P. and Oudet, P. 1986. Cryo-electron microscopy of vitrified SV40 minichromosomes: the liquid drop model. EMBO J. 5: 519–528.

Eddy, B.E., Rowe, W.P., Hartley, J.W., Stewart, S.E. and Huebner, R.J. 1958. Hemagglutination with the SE polyoma virus. Virology 6: 290–291.

Fanning, E. 1992. Modulation of Cellular Growth Control by SV40 Large T Antigen: Malignant Transformation by DNA Viruses. Weinheim VCH, New York.

Fanning, E. and Baumgartner, I. 1980. Role of fast-sedimenting SV40 nucleoprotein complexes in virus assembly. Virology 102: 1–12.

Fattaey, A.R. and Consigli, R.A. 1989. Synthesis, posttranslational modification, and nuclear transport of polyomavirus capsid protein VP1. J. Virol. 63: 3168–3175.

Fenton, R.G. and Basilico, C. 1982. Changes in the topography of early region transcription during polyoma virus lytic infection. Proc. Natl. Acad. Sci. U.S.A. 79: 7142–7146.

Fernandez-Munoz, R., Coca-Prados, M. and Hsu, M.-T. 1979. Intracellular forms of simian virus 40 nucleoprotein complexes. I. Methods of isolation and characterization in CV-1 cells. J. Virol. 29: 612–623.

Fields, B.N., Knipe, D.M., Chanock, R.M., Hirsch, M.S., Melnick, J.L., Monath, T.P. and Roizman, B. 1990. Virology. Raven Press, New York.

Forstova, J., Krauzewicz, N., Wallace, S., Street, A.J., Dilworth, S.M., Beard, S. and Griffin, B.E. 1993. Cooperation of structural proteins during late events in the life cycle of polyomavirus. J. Virol. 67: 1405–1413.

Freund, R., Garcea, R.L., Sahli, R. and Benjamin, T.L. 1991. A single amino acid substitution in polyomavirus VP1 correlates with plaque size and hemagglutination behavior. J. Virol. 65: 350–355.

Fried, H., Cahan, L.D. and Paulson, J.C. 1981. Polyoma virus recognizes specific sialyloligosaccharide receptors on host cells. Virology 109: 188.

Friedmann, T. 1971. In vitro reassembly of shell-like particles from disrupted polyoma virus. Proc. Natl. Acad. Sci. U.S.A. 68: 2574–2578.

Garber, E.A., Seidman, M.M. and Levine, A.J. 1978. The detection and characterization of multiple forms of SV40 nucleoprotein complexes. Virology 90: 305–316.

Garber, E.A., Seidman, M.M. and Levine, A.J. 1980. Intracellular SV40 nucleoprotein complexes: synthesis to encapsidation. Virology 107: 389–401.

Garcea, R.L., Ballmer-Hofer, K. and Benjamin, T.L. 1985. Virion assembly defect of polyomavirus hr-t mutants: under phosphorylation of major capsid protein VP1 before viral DNA encapsidation. J. Virol. 54: 311–316.

Garcea, R.L. and Benjamin, T.L. 1983. Isolation and characterization of polyoma nucleoprotein complexes. Virology 130: 65–75.

Garcea, R.L., Salunke, D.M. and Caspar, D.L.D. 1987. Site-directed mutation affecting polyomavirus self-assembly in vitro. Nature 329: 86–87.

Gharakhanian, E., Takahashi, J., Clever, J. and Kasamatsu, H. 1988. In vitro assay for protein-protein interaction: carboxyl-terminal 40 residues of simian virus 40 structural protein VP3 contain a determinant for interaction with VP1. Proc. Natl. Acad. Sci. U.S.A. 85: 6607–6611.

Glenn, G.M. and Eckhart, W. 1990. Transcriptional regulation of early-response genes during polyomavirus infection. J. Virol. 64: 2193–2201.

Griffith, J.P., Griffith, D.L., Rayment, I., Murakami, W.T. and Caspar, D.L.D. 1992. Inside polyomavirus at 25-Å resolution. Nature 355: 652–654.

Harrison, S.C. 1991. What do viruses look like? Harvey Lect. 85: 127–152.

Hendrix, R.W. and Garcea, R.L. 1994. Capsid assembly of dsDNA viruses. Semin. Virol. 5: 15–26.

Jakobovits, E.B. and Aloni, Y. 1980. Isolation and characterization of various forms of simian virus 40 DNA-protein complexes. Virology 102: 107–118.

Jakobovits, E.B., Bratosin, S. and Aloni, Y. 1980. A nucleosome-free region in SV40 minichromosomes. Nature 285: 263–265.

Kasamatsu, H. and Nehorayan, A. 1979. VP1 affects intracellular localization of VP3 polypeptide during simian virus 40 infection. Proc. Natl. Acad. Sci. U.S.A. 76: 2808–2812.

Khalili, K., Brady, J., Pipas, J.M., Spence, S.L., Sadofsky, M. and Khoury, G. 1988. Carboxyl-terminal mutants of the large tumor antigen of simian virus 40: a role for the early protein late in the lytic cycle. Proc. Natl. Acad. Sci. U.S.A. 85: 354–358.

Krauzewicz, N., Strueli, C.H., Stuart-Smith, N., Jones, M.D., Wallace, S. and Griffin, B.E. 1990. Myristylated polyomavirus VP2: role in the life cycle of the virus. J. Virol. 64: 4414–4420.

LaBella, F. and Vesco, C. 1980. Late modifications of simian virus 40 chromatin during the lytic cycle occur in an immature form of virion. J. Virol. 33: 1138–1150.

Li, M. and Garcea, R.L. 1994. Identification of the threonine phosphorylation sites on the polyomavirus major capsid protein VP1: relationship to the activity of middle T antigen. J. Virol. 68: 320–327.

Liddington, R.C., Yan, Y., Moulai, J., Sahli, R., Benjamin, T.L. and Harrison, S.C. 1991. Structure of simian virus 40 at 3.8-Å resolution. Nature 354: 278–284.

Lin, W., Hata, T. and Kasamatsu, H. 1984. Subcellular distribution of viral structural proteins during simian virus 40 infection. J. Virol. 50: 363–371.

Margolskee, R.F. and Nathans, D. 1983. Suppression of a VP1 mutant of simian virus 40 by missense mutations in serine codons of the viral agnogene. J. Virol. 48: 405–409.

Montross, L., Watkins, S., Moreland, R.B., Mamon, H., Caspar, D.L.D. and Garcea, R.L. 1991. Nuclear assembly of polyomavirus capsids in insect cells expressing the major capsid protein VP1. J. Virol. 65: 4991–4998.

Moreland, R.B. and Garcea, R.L. 1991. Characterization of a nuclear localization sequence in the polyomavirus capsid protein VP1. Virology 185: 513–518.

Moreland, R.B., Montross, L. and Garcea, R.L. 1991. Characterization of the DNA-binding properties of the polyomavirus capsid protein VP1. J. Virol. 65: 1168–1176.

Müller, U., Zentgraf, H., Eicken, I. and Keller, W. 1978. Higher order structure of simian virus 40 chromatin. Science 201: 406–415.

Ng, S.-C. and Bina, M. 1984. Temperature-sensitive BC mutants of simian virus 40: block in virion assembly and accumulation of capsid-chromatin complexes. J. Virol. 50: 471–477.

Ng, S.-C., Mertz, J.E., Sanden-Will, S. and Bina, M. 1985. Simian virus 40 maturation in cells harboring mutants deleted in the agnogene. J. Biol. Chem. 260: 1127–1132.

Oppenheim, A., Sandalon, Z., Peleg, A., Shaul, O., Nicolis, S. and Ottolenghi, S. 1992. A cis-acting DNA signal for encapsidation of simian virus 40. J. Virol. 66: 5320–5328.

Pallas, D.C., Shahrik, L.K., Martin, B.L., Jaspers, S., Miller, T.B., Brautigan, D.L. and Roberts, T.M. 1990. Polyoma small and middle T antigens and SV40 small t antigen form stable complexes with protein phosphatase 2A. Cell 60: 167–176.

Peden, K.W.C., Spence, S.L., Tack, L.C., Cartwright, C.A., Srinivasan, A. and Pipas, J.M. 1990. A DNA replication-positive mutant of simian virus 40 that is defective for transformation and the production of infectious virions. J. Virol. 64: 2912–2921.

Pipas, J.M. 1985. Mutations near the carboxyl terminus of the simian virus 40 large tumor antigen alter viral host range. J. Virol. 54: 569–575.

Ponder, B.A.J., Robbins, A.K. and Crawford, L.V. 1977. Phosphorylation of polyoma and SV40 virus proteins. J. Gen. Virol. 37: 75–83.

Rayment, I., Baker, T.S., Caspar, D.L.D. and Murakami, W.T. 1982. Polyoma virus capsid structure at 22.5 Å resolution. Nature 295: 110–115.

Resnick, J. and Shenk, T. 1986. Simian virus 40 agnoprotein facilitates normal nuclear location of the major capsid polypeptide and cell-to-cell spread of virus. J. Virol. 60: 1098–1106.

Rodgers, R.E.D., Chang, D., Cai, X. and Consigli, R.A. 1994. Purification of recombinant budgerigar fledgling disease virus VP1 capsid protein and its ability for in vitro capsid assembly. J. Virol. 68: 3386–3390.

Rose, R.C., Bonnez, W., Reichman, R.C. and Garcea, R.L. 1993. Expression of human papillomavirus type 11 L1 protein in insect cells: in vivo and in vitro assembly of viruslike particles. J. Virol. 67: 1936–1944.

Rossmann, M.G. and Johnson, J.E. 1989. Icosahedral RNA virus structure. Annu. Rev. Biochem. 58: 533–573.

Sahli, R., Freund, R., Dubensky, T., Garcea, R., Bronson, R. and Benjamin, T. 1993. Defect in entry and altered pathogenicity of a polyoma virus mutant blocked in VP2 myristylation. Virology 192: 142–153.

Salunke, D.M., Caspar, D.L.D. and Garcea, R.L. 1986. Self-assembly of purified polyomavirus capsid protein VP1. Cell 46: 895–904.

Salunke, D.M., Caspar, D.L.D. and Garcea, R.L. 1989. Polymorphism in the assembly of polyomavirus capsid protein VP1. Biophys. J. 56: 887–900.

Saragosti, S., Moyne, G. and Yaniv, M. 1980. Absence of nucleosomes in a fraction of SV40 chromatin between the origin of replication and the region coding for the late leader RNA. Cell 20: 65–73.

Spence, S.L. and Pipas, J.M. 1994a. SV40 large T antigen functions at two distinct steps in virion assembly. Virology 204: 200–209.

Spence, S.L. and Pipas, J.M. 1994b. SV40 large T antigen host-range domain functions in virion assembly. J. Virol. 68: 4227–4240.

Stacy, T.P., Chamberlain, M., Carswell, S. and Cole, C.N. 1990. The growth of simian virus 40 (SV40) host range/adenovirus helper function mutants in an African green monkey cell line that constitutively expresses the SV40 agnoprotein. J. Virol. 64: 3522–3526.

Stamatos, N.M., Chakrabarti, S., Moss, B. and Hare, J.D. 1987. Expression of polyomavirus virion proteins by a vaccinia virus vector: association of VP1 and VP2 with the nuclear framework. J. Virol. 61: 516–525.

Stehle, T., Yan, Y., Benjamin, T.L. and Harrison, S.C. 1994. Structure of murine polyomavirus complexed with an oligosaccharide receptor fragment. Nature 369: 160–163.

Streuli, C.H. and Griffin, B.E. 1987. Myristic acid is coupled to a structural protein of polyoma virus and SV40. Nature 326: 619–622.

Tooze, J. 1980. DNA Tumor Viruses. Cold Spring Harbor Laboratory, Cold Spring Harbor, NY.

Tornow, J. and Cole, C.N. 1983. Intracistronic complementation in the simian virus 40 A gene. Proc. Natl. Acad. Sci. U.S.A. 80: 6312–6316.

Varshavsky, A.J., Sundin, O. and Bohn, M. 1979. A stretch of late SV40 viral DNA about 400 bp long which includes the origin of replication is specifically exposed in SV40 minichromosomes. Cell 16: 453–466.

Walter, G. and Deppert, W. 1974. Intermolecular disulphide bonds: an important structural feature of the polyoma virus capsid. Cold Spring Harbor Symp. Quant. Biol. 39: 255–257.

Walter, G., Ruediger, R., Slaughter, C. and Mumby, M. 1990. Association of protein phosphatase 2A with polyoma virus medium tumor antigen. Proc. Natl. Acad. Sci. U.S.A. 87: 2521–2525.

Whitman, M., Downes, C.P., Keeler, M., Keller, T. and Cantley, L. 1988. Type I phosphatidylinositol kinase makes a novel phospholipid, phosphatidylinositol-3-phosphate. Nature 332: 644–646.

Wychowski, C., Benichou, D. and Girard, M. 1986. A domain of SV40 capsid polypeptide VP1 that specified migration into the cell nucleus. EMBO J. 5: 2569–2576.

Wychowski, C., Benichou, D. and Girard, M. 1987. The intranuclear location of simian virus 40 polypeptides VP2 and VP3 depends on a specific amino acid sequence. J. Virol. 61: 3862–3869.

Yuen, L.K.C. and Consigli, R.A. 1982. Improved infectivity of reassembled polyoma virus. J. Virol. 43: 337–341.

Zlotnick, A. 1994. To build a virus capsid: an equilibrium model of the self assembly of polyhedral protein complexes. J. Mol. Biol. 241: 59–67.

Zullo, J., Stiles, C.D. and Garcea, R.L. 1986. Regulation of c-myc and c-fos mRNA levels by polyomavirus: distinct roles for the capsid protein VP1 and the viral early proteins. Proc. Natl. Acad. Sci. U.S.A. 84: 1210–1214.

8.

The Structure of Adenovirus

ROGER M. BURNETT

Adenoviruses were first isolated by Rowe et al. (1953) from human adenoid cells while attempting to isolate the "common cold virus." The *Adenoviridae* family comprises at least 118 species, which have been identified from 18 different animal hosts (Wigand et al., 1982; Hierholzer et al., 1988) as sharing a common chemical composition and architecture (Philipson, 1983; Ginsberg, 1984; Doerfler and Boehm, 1995; Shenk, 1996). The most extensively studied group is that of the human adenoviruses. These are most commonly classified into six subgroups (A to F) by their ability to agglutinate red blood cells, because this property is fairly well correlated with other criteria. They cause a wide variety of diseases, ranging from respiratory infections and conjunctivitis to the severe enteric dysentery that is a leading cause of death in Third World infants (Monto, 1992). Adenoviruses transform cultured rodent cells and so are oncogenic, with some species inducing tumors when injected into newborn rats and hamsters (Shenk, 1996). Most human adults have been exposed to the prototypical type 2 adenovirus (ad2) and the related ad5, which cause only mild respiratory infections.

After their discovery, adenoviruses rapidly became popular as a model system because they are easy to propagate and are good subjects for genetic analysis. They have been important in the study of transcription and replication of the DNA genome, mRNA splicing, expression of proteins, cell cycle control, cellular growth, transformation of cells in vitro, and tumorigenesis. Remarkably, both killed (Hilleman et al., 1955; Huebner et al., 1955) and live (Couch et al., 1963) vaccines were developed within 2 years of adenovirus' discovery.

However, because of their oncogenicity and possible contamination with simian virus 40, their subsequent use was restricted to protecting military recruits against the outbreaks of acute respiratory disease that occur in closed populations such as barracks (Ginsberg, 1984). Most recently, replication-deficient adenoviruses have been used as vectors in human gene therapy (Kozarsky and Wilson, 1993).

Adenovirus was one of the first biological specimens to be imaged using electron microscopy (EM) (Horne et al., 1959), which revealed an icosahedral virion with 252 capsomeres and long fibers projecting from the vertices. Early studies focused on elucidating its architecture by analyzing the virion and capsid dissociation fragments (reviewed by Nermut, 1980). Early biochemical studies were aimed at understanding the complex virion by understanding the properties and roles of its constituent proteins. These studies were facilitated by the 10- to 100-fold excess of structural proteins over intact virions found in infected cells (White et al., 1969). The major coat protein, hexon, is available in soluble form and so can be purified without chemical disruption of the virion. It was the first animal viral protein to be crystallized (Pereira et al., 1968).

This chapter is organized to describe what is currently known about the architecture of the virion and the structures of its component proteins. Reflecting the structural perspective, the biologic properties of adenovirus are described at appropriate points in the text, rather than separately. Space considerations preclude a comprehensive review of all structural work on adenovirus, and the focus will be on a series of studies that arose from an X-ray crystallographic structure investigation of hexon (Burnett, 1984). An initial low-resolution model of hexon revealed several molecular features that could also be seen in EM micrographs of negatively stained capsid dissociation fragments. Subsequent work was driven by the realization that a combination of X-ray crystallography for revealing atomic resolution detail and EM for determining overall organization was the key to obtaining a full understanding of adenovirus and other complex macromolecular assemblies.

Adenovirus Virion

The adenovirus virion (Fig. 8.1A) consists of a nonenveloped icosahedral protein shell, or capsid, enclosing a DNA–protein core complex. The first images from EM showed an icosahedral particle with 252 surface projections, or capsomeres, and long fibers projecting from the 12 vertices (Horne et al., 1959). Because the composition of the capsomeres was unknown, they were named hexons and pentons after the number of their capsid neighbors (Ginsberg et al., 1966). The structural proteins in the virion were originally designated II to XII in order of their decreasing apparent molecular mass on sodium dodecyl sulfate (SDS)–polyacrylamide gels (Maizel et al., 1968a). A ''polypeptide I,'' seen on early SDS gels, arose from incomplete protein denaturation and was eliminated by introducing a boiling step into the method (Maizel et al., 1968a).

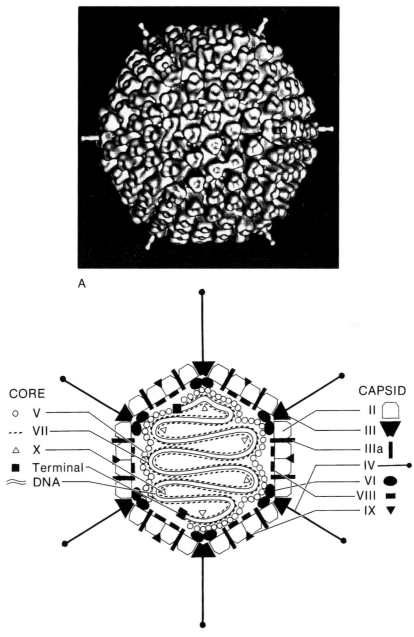

A

CORE

○ V

--- VII

△ X

■ Terminal

≈ DNA

CAPSID

II

III

IIIa

IV

VI

VIII

IX

B

FIGURE 8.1 Adenovirus virion. *A*. The three-dimensional image reconstruction viewed along the threefold axis of the icosahedral capsid. Note that the capsid is rounded to provide a continuous protein shell (cf. the model in Fig. 8.2). Hexon, the major coat protein, has a triangular top with three towers. A fiber protrudes from the penton base at each fivefold vertex. As the individual fibers are bent, their outer parts are not related by icosahedral symmetry and so are washed out when averaged by the reconstruction process. Thus, only the first third of each fiber is seen. (Reproduced with permission from Stewart and Burnett, 1993.) *B*. A stylized section summarizing current structural knowledge of the polypeptide components and the viral DNA. No real planar section through the icosahedral virion would contain all these components. (Reproduced with permission from Stewart and Burnett, 1993; based on an earlier drawing by Ginsberg, 1979.)

TABLE 8.1 Adenovirus Structural Proteins

Polypeptide	Molecular Mass of Monomer (Da)[a]	Number of Residues in Monomer[a]	Biochemical Copy Number of Monomer[a]	Copy Number of Protein in Current Model
II (hexon)	109,077	967	720 ± 7	240 trimers
III (penton base)	63,296	571	56 ± 1	12 pentamers
IIIa				
External portion	41,689	376[c]		
Internal portion	21,846	194[c]		
Total	63,535[b]	570[b]	68 ± 2[b]	60 monomers
IV (fiber)				
N terminus +	13,912	133[c]		
2/8ths of shaft				
6/8ths of shaft	48,048	449[c]		
+ C terminus				
Total	61,960	582	35 ± 1	12 trimers
V (core)	41,631	368	157 ± 1	—[d]
Terminal[e] (core)	~55,000	~500	2	—
VI				
Ordered portion	9,773	91[c]		
Disordered portion	12,345	115[c]		
Total	22,118[b]	206[b]	342 ± 4[b]	60 hexamers
VII (core)	19,412	174	833 ± 19	—
VIII	15,390[b]	140[b]	127 ± 3[b]	—
IX	14,339	139	247 ± 2	80 trimers
μ[f] (core)	~4,000	~36	~104	—

Adapted from Stewart and Burnett (1995).

[a]The molecular masses, residue numbers, and biochemically determined copy numbers for the monomers are taken from van Oostrum and Burnett (1985) but have been updated using the adenovirus protease cleavage sites given by Anderson (1990).

[b]Proteins for which the adenovirus protease cleavage sites of Anderson (1990) were used.

[c]Estimated division.

[d]Dashes indicate proteins that have not so far been imaged.

[e]Estimates from Rekosh (1981).

[f]Estimates from Hosokawa and Sung (1976).

The 11 better characterized structural proteins in the virion are listed in Table 8.1.

The 20 interlocking facets of the capsid are each formed from 12 copies of hexon (Fig. 8.1B). Pentons, a complex of penton base (polypeptide III) and a long fiber (IV), lie at the 12 vertices. With a spherical diameter of 914 Å enclosing the fivefold vertices (Stewart et al., 1991), the virion is about 20 times larger than the spherical RNA plant viruses. The particle mass of ad2 is at least 150×10^6 Da (van Oostrum and Burnett, 1985), of which 22.6×10^6 Da is DNA. The core is formed from a single copy of linear double-stranded DNA covalently linked to two copies of the terminal protein. The DNA is complexed with histonelike viral polypeptides V and VII to form structures termed ''adenosomes'' by analogy with the nucleosomes observed in the nuclei of mammalian cells (Tate and Philipson, 1979). The sequence of the entire ad2 genome (35,937 bp) is known (Roberts et al., 1984).

Disruption of adenovirus capsids under mild conditions results in a progressive loss from the core of the vertex pentons (Fig. 8.2), the adjacent peripentonal hexons, and finally the remaining planar groups-of-nine hexons (GONs) (Smith et al., 1965; Laver et al., 1969; Prage et al., 1970). The characteristic propeller shape of GONs has handedness, and the absolute configuration could only be determined from EM images by knowing from which side the GONs were viewed. Pereira and Wrigley (1974) made this assignment when they reconstituted empty capsids, lacking vertices, from purified GONs. They noted a nonrandom dissociation pattern of GONs, which was later explained by a model (Burnett, 1984; van Oostrum and Burnett, 1985) in which a minor capsid protein, polypeptide IX, lies between hexons and acts as a capsid "cement" (Maizel et al., 1968b; Aleström et al., 1980).

An early theory of viral architecture (Caspar and Klug, 1962; see also Chapter 1) predicted that hexons would be hexameric. Hexons were shown

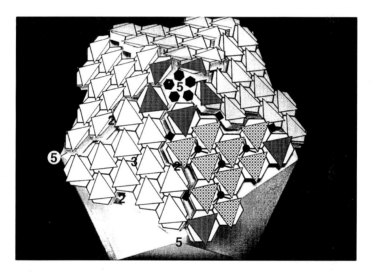

FIGURE 8.2   Model of the icosahedral adenovirus capsid. Hexon (a trimer of polypeptide II) is represented as a triangular top superimposed on a hexagonal base to reflect its morphology. Each of the 20 capsid facets is formed from 12 hexons. Five planar facets are shown superimposed on an icosahedron. The positions of the icosahedral symmetry axes are indicated on the lower left facet. A penton, a complex between fiber (a trimer of polypeptide IV) and penton base (a pentamer of polypeptide III), lies at each of the 12 vertices (complex not shown in the model). The positions of the minor proteins are indicated on the lower right facet. Polypeptide IX (solid circle with three arms) binds as a trimer in four symmetry-related locations in the middle of the facet. It lies in a cavity formed between the towers of three different hexon molecules and cements the central nine hexons in a facet into the GON (highlighted with a pattern). Two copies of the monomeric polypeptide IIIa (solid square) penetrate each edge to rivet two facets together. A ring of five hexamers of polypeptide VI (solid hexagon) lies underneath five peripentonal hexons (shaded gray) to provide stability at each vertex and attach it to the core. Upon dissociation, the penton complex is lost, followed by the peripentonal hexons, which are not cemented into the facet. The GONs then dissociate as stable groups. (Adapted from Burnett et al., 1990.)

later to have threefold, and not sixfold, symmetry when Crowther and Franklin (1972) used rotational filtering (Crowther and Amos, 1971) on electron micrographs of GONs. This result was in accord with a trimer (Horwitz et al., 1970), but the composition remained controversial until Grütter and Franklin (1974) showed that hexon has three identical subunits. Crowther and Amos (1971) found that the individual hexons were arranged on a $p3$ net, rather than the $p6$ net predicted by a hexamer-clustered $T = 25$ surface lattice (Caspar and Klug, 1962; see also Chapter 1). This result, together with the finding that the 12 pentons were chemically distinct from the 240 hexons, had indicated by 1971 that adenovirus was an exception to the Caspar and Klug model of virus architecture.

The 6-Å crystallographic model of hexon (Burnett et al., 1985) was used to develop a model of the capsid (Fig. 8.2) (Burnett, 1985). Hexon has two distinctive molecular features, a triangular top and a hollow pseudo hexagonal base, that were recognizable in published electron micrographs of GONs (Nermut, 1975) and other capsid fragments (Pereira and Wrigley, 1975). The tops are even sometimes barely recognizable in complete virions (Nermut, 1975; Nász and Ádám, 1983). This initial model of the capsid had planar facets with hexons arranged on a $p3$ net and was confirmed by image analysis of electron micrographs of capsid fragments (van Oostrum et al., 1986, 1987a, 1987b).

Because the pseudo hexagonal shape of hexons allows them to form a close-packed capsid, chemical contacts between hexons are conserved in a manner different from that suggested by quasi-equivalence (Caspar and Klug, 1962; see also Chapter 1). If the adenovirus virion had exact $T = 25$ packing, there would be 1500 identical subunits, organized as 240 hexamers and 12 pentamers. These subunits would occupy 25 environments that would differ both topologically and chemically. The real adenovirus capsid has 240 hexons and 12 pentons, and the hexons lie in only four different, but closely related, environments (Burnett, 1985). It can be described for convenience as pseudo $T = 25$ (see Chapter 1), which highlights the distribution of capsomeres, but the nomenclature obscures the fact that these are multimeric proteins. The use of stable multimers as "building blocks" leads to a sixfold reduction in both their overall number and their possible locations. The possibility of an error while incorporating a unit is smaller, leading to greater accuracy in virion assembly (see Chapter 1).

The current model of the virion (Fig. 8.1) is derived from a three-dimensional image reconstruction using cryo-EM, which provides a complete view of the virion at 35-Å resolution (Stewart et al., 1991). The EM reconstruction confirmed that the hexon packing is essentially the same as in the earlier planar model, but showed that the capsid is significantly rounded, with outer radii of 457 Å, 442 Å, and 435 Å along the fivefold, twofold, and threefold axes, respectively. Because the facets are nonplanar, the hexons at the capsid edges are in as close contact as those within the facets. Thus, close packing of hexons occurs throughout the capsid despite the completely different chemical nature of the hexon–hexon interfaces at the edge and within the facet. The actual hexon packing must be subtly different from that predicted by the planar capsid model (Fig. 8.2) (Burnett, 1985) because of the curvature of the facets. Current crystallographic work is aimed at understanding the details of the hexon–hexon interactions.

In other work (Stewart et al., 1993), the 2.9-Å resolution crystallographic hexon model (Athappilly et al., 1994) was used to calculate a simulated capsid by positioning 240 copies of hexon within the EM reconstruction. The simulated capsid then was subtracted from the EM reconstruction to obtain a three-dimensional difference map. This map not only provided a clearer view of the penton base and fiber, but also permitted the assignment of two minor proteins, polypeptides IIIa and VI. The assignment was accomplished by correlating the distribution and volume of difference map density features with the known copy numbers and molecular masses of the minor proteins (van Oostrum and Burnett, 1985). The study demonstrated that X-ray crystallography and EM can be combined quantitatively to investigate complex macromolecular assemblies (Stewart and Burnett, 1993).

Major Coat Proteins

Hexon (Polypeptide II)

Hexon is a homotrimer (Grütter and Franklin, 1974). The polypeptide chain of ad2 contains 967 amino acid residues (Akusjärvi et al., 1984), each with a molecular mass of 109,077 Da (including the acetylated N terminus). The 240 copies of hexon in the capsid account for 63% of the total particle weight (van Oostrum and Burnett, 1985). The 2.9-Å crystal structure (Roberts et al., 1986; Athappilly et al., 1994) reveals that the hexon subunit (Fig. 8.3) has two very similar β-barrel domains in its base. These domains impart the pseudo hexagonal symmetry to the trimeric molecule that is visible at low resolution both by crystallography (Burnett et al., 1985) and by EM (van Oostrum et al., 1987a, 1987b). The β-barrels in hexon have the same topology as in almost all spherical RNA viruses (Roberts and Burnett, 1987; Rossmann and Johnson, 1989) and in tumor necrosis factor (Jones et al. 1989). In adenovirus, the β-barrel axes are parallel to hexon's molecular axis and so normal to the capsid surface. This arrangement is similar to that in tumor necrosis factor but contrasts with the mostly in-plane orientation of the β-barrels in the spherical RNA virus capsids. The topology of the viral β-barrels has been used to derive an evolutionary tree that relates all spherical viruses containing the viral β-barrel to a single precursor (Chelvanayagam et al., 1992).

One of the most distinctive properties of the hexon trimer is its extreme stability. It retains its physical and immunological characteristics even after exposure to 8-M urea (Shortridge and Biddle, 1970) and is highly resistant to proteolysis (Jörnvall and Philipson, 1980). The stability arises from the unusual topology of the interactions between its subunits and their extensive interface (Athappilly et al., 1994), and the abnormal distribution of prolines (Jörnvall and Philipson, 1980).

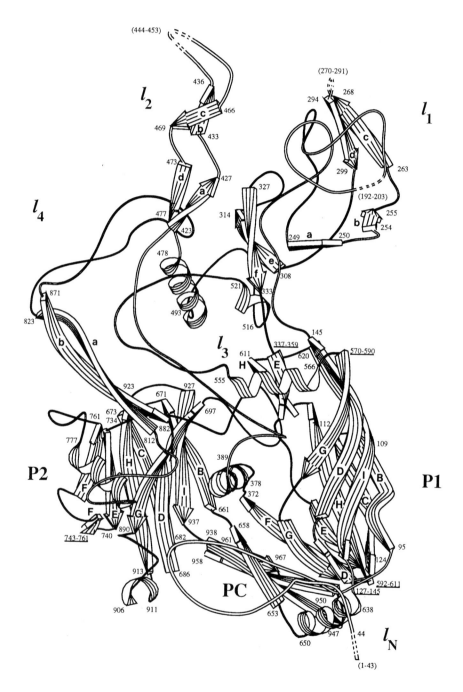

The hexon trimer is locked into a very stable arrangement by the loops (Fig. 8.3) that emanate from the β-barrels in the base to form the towers at the top of the molecule. These intertwine so that the individual subunits of the trimer are clasped so they cannot easily be dissociated from each other. Each tower is formed from one loop from each subunit (l_1, l_2, and l_4), and these have extraordinarily intricate interactions. In addition, the N-terminal arm in the base passes underneath neighboring subunits. Both features clasp the subunits tightly in place. The threefold symmetry is also important in contributing stability because it ensures that a subunit both is clamped by one neighbor and clamps its other neighbor. The cyclic symmetry thus locks the trimer once it is formed so that a subunit cannot be extricated without disrupting both tertiary and quaternary molecular structure.

The unusual ''tubular'' shape of hexon is seemingly at odds with its abnormally high stability. The large cavity in the base and the depression between the towers make the molecule almost hollow, which increases the accessible surface area and reduces the hydrophobic contribution to molecular stability (Kauzmann, 1959). The accessible surface area is 20% larger than that predicted for a multimer with similar mass (Janin et al., 1988). This destabilizing feature of the molecule is outweighed by a large decrease in accessible surface area from 41,004 to 32,868 $Å^2$ when the trimer is formed. Although this 20% drop is within the normal range of 10% to 40% for multimeric proteins (Janin et al., 1988), the magnitude of the 8,136-$Å^2$ decrease is only exceeded in the catalase tetramer. Thus, the very large interface area between the subunits more than compensates for the abnormally large accessible surface area of the complete trimer. In common with other multimeric proteins, each subunit interface surface has scattered hydrophobic patches that are complementary to patches on the neighboring surface (Korn and Burnett, 1991). This feature ensures that the subunit is at least partially soluble and guides its accurate alignment with its neighbor.

The clear instability of the isolated hexon polypeptide resulting from its shape (Fig. 8.3), the complex topology of the trimer, and the hydrophobicity

←——

FIGURE 8.3 Hexon subunit. The view of the ribbon representation (Priestle, 1988) of the molecule is perpendicular to the molecular threefold axis and from inside the molecule. In the virion, the hexon tops form the outer surface of each facet (see Figs. 8.1 and 8.2). The base consists of pedestal domains P1 and P2, which are eight-stranded β-barrels. These are connected and stabilized below by the pedestal connector, PC, and above by loop l_3 from P2. The top is formed from loops l_1 and l_2 from P1 and loop l_4 from P2. In the trimer, the N-terminal loop l_N emerges from P1 in the reference subunit, passes below P2 and then P1 of the closest adjacent subunit, and reaches so far as to touch the P2 domain of the third subunit. The sequence numbers indicate the beginning and end of the α-helices and β-strands. The latter are labeled with capital letters in the base and lower-case letters in the loops. Strands in which breaks in hydrogen bonding occur are indicated by underlining (e.g., $P1^D$: 127–145). The breaks occur at $P1^D$ residues 131 to 132, $P1^E$ (342 to 346 and 350 to 353), $P1^G$ (580 to 584), $P1^H$ (604 to 606), and $P2^F$ (746 to 754). The dashed traces indicate four stretches that were not defined in the crystallographic model at 2.9-Å resolution: the N terminus (residues 1 to 43); loop l_1 (residues 192 to 203 and 270 to 291); and loop l_2 (residues 444 to 453). (Reproduced with permission from Athappilly et al., 1994.)

of the large subunit interface all suggest why transient complex formation with the adenovirus 100K protein is a prerequisite for correct folding of the hexon trimer (Cepko and Sharp, 1982). This system is perhaps the first protein-mediated folding pathway to be identified (Oosterom-Dragon and Ginsberg, 1981), and was known long before chaperones were recognized as a distinct class of proteins (Ellis and Hemmingsen, 1989).

Comparative studies have been made for other hexon species by mapping their sequences onto the known ad2 hexon structure. Figure 8.4 compares six complete sequences, four from humans—ad2 (von Bahr-Lindström et al., 1982), ad5 (Kinloch et al., 1984), ad40 (Toogood et al., 1989), and ad41 (Toogood and Hay, 1988)—and two from animals—bovine Bav3 (Hu et al., 1984) and murine Mav1 (Weber et al., 1994). Overall there is 66% homology, suggesting that other hexon structures cannot be very different from ad2 hexon. Hexons within the individual subgroups C (ad2 and ad5) and F (ad40 and 41) show around 90% homology. The number of known sequences is rapidly growing because the hexon gene provides an excellent clinical probe in clinical assays employing DNA hybridization and polymerase chain reaction techniques to detect adenovirus (Allard et al., 1992; Scott-Taylor and Hammond, 1992). Although these often provide only partial hexon sequences (e.g., human ad1 and ad6; Pring-Åkerblom and Adrian, 1993), the variable regions are usually covered. The expanding hexon database (e.g., ad48; Crawford-Miksza and Schnurr, 1996) promises to be important in understanding adenovirus evolution.

Sequence comparison (Fig. 8.4) (Weber et al., 1994) shows that most of the nonconserved changes occur in the tower regions at the top of the molecule, while the base is conserved. The most striking result is the occurrence of a consistent deletion in the range of residues 139 to 170 in all other serotypes with respect to ad2. In the murine and bovine animal hexons, a long deletion occurs between residues 141 and 170. In human hexons, the corresponding deletion is also long in subgroup F (ad40 and ad41) but is very short in subgroup C (ad1, ad2, ad5, and ad6). It is notable that an unusually long stretch of acidic residues, unique to human subgroup C hexons, is part of this region.

The interaction between an adenovirus virion and a cell is modulated by both a primary receptor, which recognizes the fiber knob upon attachment, and a secondary receptor, which recognizes an Arg-Gly-Asp (RGD) region on the penton base and triggers viral uptake (see later). Wohlfart (1988) has proposed that neutralizing anti–penton base antibodies act by covering the penton base and interfering with receptor-mediated endocytosis. He also suggested that neutralizing anti-hexon antibodies act by preventing a low pH–induced conformational change in the hexon structure and so inhibiting the proper exposure of penton base. It has long been known that pH plays an important role in GON interactions in the capsid (Pereira and Wrigley, 1974). The acidic region in hexon thus may induce a pH-dependent structural change in the molecule, or in hexon-associated proteins in the virion, during cell entry through acidic endosomes (Fig. 8.5) (Greber et al., 1993, 1994).

Because the P1 β-barrel in hexon is in contact with the penton base, a pH-induced structural shift, magnified in the virion by the concerted action of all

240 hexons, could result in the release of the penton base at low pH (Stewart et al., 1993). A similar concerted pH-dependent structural change has been observed in spherical plant viruses. The acidic region may modify the uptake pathway in a subgroup-specific manner, providing a third level of variation beyond those provided by differences in the primary and secondary recognition sites. These differences, and others, probably explain why adenoviruses cause a range of disease. Because the acidic stretch is absent in hexons from subgroup F, which causes enteric dysentery, the cell entry pathway for this subgroup may be different from that for subgroup C, which causes respiratory infections. Further understanding of these cellular recognition events would permit their alteration to modify the uptake of adenovirus by different cell types and organs and make it a more specific vector for human gene therapy.

The long deletion in subgroup F was at first puzzling because it extends into the D strand of the P1 domain. It was not immediately apparent why this would not disrupt the P1 β-barrel in the base of the molecule. However, the residues in the ad41 sequence following the deletion can be accommodated in the D strand without structural disturbance (Toogood et al., 1989). The modeled β-barrel for ad41 has a smaller D strand and an earlier start for loop l_1. Thus, the overall structure of the base is conserved but the top is different. A murine animal hexon has also been modeled (Mav1; Weber et al., 1994). Three major deletions in the murine sequence occur in loops l_1 (residues 141 to 170 and 270 to 284) and l_2 (residues 446 to 455) at the same locations where type-specific antigenic sites are found in the human adenoviruses. Because these changes all occur at the top of hexon, which forms the outer surface of the ad2 virion (Stewart et al., 1991), antigenic sites can evolve with minimal structural changes in the capsid and lead to different adenovirus species. Because deletions are found in all serotypes with respect to ad2, this species may be early in adenovirus evolution.

The immunological properties of adenovirus, such as the positions of the group- and type-specific epitopes, are not nearly as well determined as those for other viruses such as influenza virus (see Chapter 3) and poliovirus (see Chapter 6) that pose more of a problem to human health. Hexon contains both group- and type-specific determinants, but these have not, in general, been correlated with the primary sequence or three-dimensional structure. The best-characterized are type-specific sites in ad2 and ad5 (subgroup C). These were predicted to lie in the region of residues 140 to 290 from biochemical studies based on ad2 and ad5 (Mautner and Willcox, 1974), and were further localized by genetic studies (Boursnell and Mautner, 1981). Once the sequence (Roberts et al., 1984) and structure (Roberts et al., 1986) of hexon were known, the regions of highest variability were seen to lie in the l_1 and l_2 loops, which form the top of the molecule and the outer surface of the virion. The locations in ad2 were further refined by means of antisera raised against peptides corresponding to residues 281 to 292 in loop l_1 and 441 to 455 in loop l_2 (Toogood et al., 1992). This information, and the knowledge from the structure that there are flexible regions in loops l_1 and l_2, suggested sites where a foreign epitope could be inserted into adenovirus (Crompton et al., 1994). An eight-residue

```
        |.....|.....|.....|.....|.....|.....|.....|.....|.....|.....|.....|.....|.....|.....|.....|.....|.....|.....|.....|
Ad2   -ATPSMPQWSYMHISGQDASEYLSPGLVQFARATETYPSLNNKFRNPTVAPTHDVTTDRSQRLTLRFIPVDREDTAYSYKARFTLAVGDNRVLDMASTY   99
Ad5   -A   M  SY  S   E   L  R  ET  SLN              D     FI   DTA  S  A  TA
Ad40  -A   M  SY  A   E   L  R  DT  SLG              D   T FV   ETA  S  V  TA
Ad41  -A   M  SY  A   E   L  R  DT  SLG              D   T FV   DTA  S  V  TA
Bav3  -A   L  SY  A   E   L  Q  ES  NIG              E   Q FV   DTQ  S  T  QA
Mav1  -T   Q  AF  A   Q       V  A   DT  SLG         D   T IV   DSQ  T  Q  S

        |.....|.....|.....|.....|.....|.....|.....|.....|.....|.....|.....|.....|.....|.....|.....|.....|.....|
Ad2   FDIRGVLDRGPTFKPYSGTAYNALAPKGAPNSCEWEQTEDSGRAVAEDEEEEDEDEDEEEEQNARDQATKKTHVYAQAPLSGETITKS-GLQIGSDNAE  198
Ad5          PT    AL KG    PCEWDEAATALEINLEEEDDNEDEVDEQAEQQ---    -KTHVFG  PYSGINITKE-GIQIGVEGQT
Ad40       V PS    SL KG    PSQWTNQN-----                      ---KTNSFG  PYIGQKITNQ-GVQVGLDSNN
Ad41       V PS    SL KT    PCEWKDNN-----                      ---KIKVRG  PFIGTNINKDNGIQTGTDTTN
Bav3       T AS    SF KS    NTQFRQANNG----                     ---HPAQTIA  SYVATIGGANNDLQMGVDERQ
Mav1       R PS    SL RA    NLMFKTGADS----                     ---KTMTIA  NIVGSTID-QDGLKI--NNVA

        |.....|.....|.....|.....|.....|.....|.....|.....|.....|.....|.....|.....|.....|.....|.....|.....|
Ad2   TQAKPVYADPSYQPEPQIGESQWNE----ADANAAGGRVLKKTPMKPCYGSYARPTNPFGGQSVLVPDEKGVPL-PKVDLQFFSNTTSLNDRQGNATKP  293
Ad5   ---PKYADKTFQ EP I ESQ YE----TEINHAAG V KKTTPMK C    EN GQGILVKQQNGKLE-SQ EMQFFSTEATAG-NGDNLT
Ad40  ---RDVFADKTYQ EP V QTQ N-----INPMQNAAG I KQTTPMQ C   R EK GQAKLVKNDDNQTTTN GLNFFTAETAN----FS
Ad41  ---QPIYADKTYQ EP V QTQ NSE-VGAAQKVAG V KDTTPML C    K EK GQASLITNGDQTLTSD NLQFFALPSTPNE-----
Bav3  L--PVYANTTYQ EP L IEG TAG-SMAVIDQAGG V RNFTQ-T C    K EH GITK----------ANTQ EKKYY------RTGDNGN
Mav1  LNPNAG-----E AV E LEN MEELQIGADKFAS V KADQPVL N  Y IN TQTK----------DGTT DKVYMC-------SKGANVQN

        |.....|.....|.....|.....|.....|.....|.....|.....|.....|.....|.....|.....|.....|.....|.....|.....|
Ad2   KVVLYSEDVNMETPDTHLSYKPGKGDE--NSKAMLGQQSMPNRPNYIAFRDNFIGLMYNSTGNMGVLAGQASQLNAVDLQDRNTELSYQLLLDSIGDR   391
Ad5   KV L S DVDIET  ISYMPTIKEG--NSRELMGQQSM   Y A N I   T M   A     LL DSIG
Ad40  KV L S DVNLEA  LVFKPDVN--GTSAELLGQQAA   Y G N I   T M   A     LM DALG
Ad41  KA L A NVSIEA  LVYKPDVAQGTISSADLLTQQAA   Y G N I   T M   A     LM DALG
Bav3  ET F T EADVLT  LV--HAVPAADRAKVEGLSQHAA   F G C V   G L   S     ML ANTT
Mav1  DV L S EVNLQA  LL---DGPGADNPKDRVAFCAA   Y G N I   N Q   A     FL DSLY

        |.....|.....|.....|.....|.....|.....|.....|.....|.....|.....|.....|.....|.....|.....|.....|.....|
Ad2   TRYFSMNQAVDSYDPDVRIIENHGTEDELPNYCFPLGGIGVTDTYQAIKANGNGSGDNGDTTWTKDET-FATRNEIGVGNNFAMEINLNANLWRNFLYS  490
Ad5   T M V  D IEHT ELPNYC P G VINTETLRKVKPKTGQ----ENGWEKDATEFSDKNE RV NF   LN  RN  S
Ad40  S M V  D IEHV ELPNYC P N QGISNSYQGVKTDNG------TNWSQNNTDVSSNNE SI VF   LA  RE  S
Ad41  S M V  D IEHV ELPNYC P S SAATDTYSGIKAN-G------QTWTADDNYADRGAE ES IF   LA  RS  S
Bav3  S M M  E VDVV EMPNYC P S VQIGNRSHEV----QRNQQQWQNVAN-SDNNY GK LP   LA  RS  S
Mav1  S L I  R IEDV DITSFA D R IG-DLPYKVQTHNGN-------QQSANTTDTCY GK MA   IP  LG  A

        |.....|.....|.....|.....|.....|.....|.....|.....|.....|.....|.....|.....|.....|.....|.....|.....|
Ad2   NIALYLPDKLKYNPTNVEISDNPNTYDYMNKRVVAPGLVDCYINLGARWSLDYMDNVNPFNHHRNAGLRYRSMLLGNGRYVPFHIQVPQKFFAIKNLLLL  590
Ad5   I L L KL YS S VKISDNPN D  K VVAPGLVDC I L A   L Y  N     A R   ML  YVP   KN
Ad40  V L L SY IT D ITLPDNKN A  G VAVPSALDT V I A   P P  N     A R   ML  YVP   KN
Ad41  V L L SY IT D ITLPENKN A  G VAVPSALDT V I A   P P  N     A R   ML  YVP   KN
Bav3  V L L NL FT H IQLPPNTN E  G IPVSGLIDT V I T   P V  N     S R   QL  FCD   RN
Mav1  V Q M DF GD S IQMPQDKT A  Q IAPAGLIET V V G   V F  T     E K   QI  FVD   KS
```

220

```
          ·········|····|····|····|····|····|····|····|····|····|····|····|····|····|····|····|····|····|····|····|·
Ad2  PGSYTYEWNFRKDVNMVLQSSLGNDLRVDGASIKFDSICLYATFFPMAHTASTLEAMLRNDTNDQSFNDYLSAANMLYPIPANATNVPISIPSRNWAAF  690
Ad5        S      N         V    S    V   SIKFD IC T   A          D   S N  SA N LY   ANA NV ISI S
Ad40       S      N         I    S    V   SVRFD IN N   A          D   S N  CA N LY   ANA SV ISI S
Ad41       S      N         I    S    V   SVRFD IN N   A          D   S N  CA N LY   SNA SV ISI S
Bav3       T      S         I    T    V   TVNIT VN S   S          D   S N  SA N LY   PNA QL I-- S
Mav1       S      S         V    S    A   RLEIH VN S   A          E   T L  SS T MF   AGQ QV VSI A

          ·|····|····|····|····|····|····|····|····|····|····|····|····|····|····|····*|···|····|····|····|····|·
Ad2  RGWAF TRLKTKETPSLGSGYDPYYTYSGSSIPYLDGTFYLNHTFKKVAITFDSSVSWPGNDRLLTPNEFEIKRSVDGEGYNVAQCNMTKDWFLVQMLANYN  790
Ad5  AF    L TK SL GY YT  SI Y G   N   KK AIT    S                 RSV G  YN A C      V   AN
Ad40 SF    L TK SL GF FT  SV Y G   N   KK SVM    S                 RTV G  YN A C      I   SH
Ad41 SF    L TK SL GF FT  SV Y G   N   KK SIM    S                 RTV G  YN A C      I   SH
Bav3 SL    L QR AL PF FT  TI Y G   S   RK AIQ    T                 ISV G  YN A S      V   AN
Mav1 SF    I QQ NI PY FK  SI F A   T   QR SIM    S                 RHI A  LC G S      V   SN

          ·|····|····|····|····|····|····|····|····|····|····|····|····|····|····|····|····|····|····|····|····|·
Ad2  IGYQGFYIPESYKDRMYSFFRNFQPMSRQVVDDTKYKEYQOVGILHQHNNSGFVGYLAPTM-REGQAYPANVPYPLIGKTAVDSITQKKFLCDRTLWRIP  889
Ad5       FYI ESY MY FR Q MS  VDDTKYKD QQVGILHQHNNSGFVGYLAPTM-   QA ANFPY LIGKTAVDSITQK     C TLW
Ad40      FHV ESY MY FR Q MS  VDTTTYTE QNVTLPFQHNNSGFVGYMGPAI-   QA ANPY  LIGQTAVPSLTQK     C TMW
Ad41      FYV ESY MY FR Q MS  VNTTTYKE QNVTLPFQHNNSGFVGYMGPTM-   QA ANPY  LIGQTAVPSLTQK     C TMW
Bav3      YHL PDY TF LH I MC  PNPAT-EG FGLGIVNHRTTPAVWFRFCRAP-   HP QLALP HWDPRHALRDPER     C TLW
Mav1      FYL ENN TY VR E LA  VDDAAAAN RDVPLSKRFNNSGWRSAGPPIFA   AP ANWPY LCGEAWAAKTQR     V YMY

          ·|····|····|····|····|····|····|····|····|····|····|····|····|····|····|·
Ad2  FSSNFMSMGALTDLGQNLLYANSAHALDMTFEVDPMDEPLLYVLFEVFDVVRVHQPHRGVIETVYLRTPFSAGNATT  967
Ad5          A      L    S A D  EV MD P L V  VV VR V   T T
Ad40         A      M    S A D  EV MD P L V  VV IQ V   A T
Ad41         A      M    S A D  EV MD P L V  VV IQ V   A T
Bav3         S      L    A A D  EM IN P L V  VA VQ V   V T
Mav1         A      L    S S E  NV MQ A F I  CV IQ I   A S
```

FIGURE 8.4 Sequence comparison of hexons. Six different adenovirus hexon amino acid sequences are compared, including four human (ad2, ad5, ad40, ad41), one bovine (Bav3), and one murine (Mav1) sequence. The sequence of ad2 hexon is listed on the top line in full, but regions of identity are blank for other types. A dash indicates an insertion relative to ad2. The initial methionine at position zero is removed and the N-terminal alanine in hexon is acetylated (Laver, 1970; Pettersson, 1971). The tryptic sites at Arg 142, Arg 165, and Arg 286 are indicated (↓), as is the reactive Cys 629 (*). In the current model of ad2 hexon at 2.3-Å resolution (P.R. Kuser and R.M. Burnett, unpublished results), only the N terminus (residues 1 to 43) and one internal stretch in loop l_1 (residues 274 to 291) remain undefined (●). (Updated from Athappilly et al., 1994.)

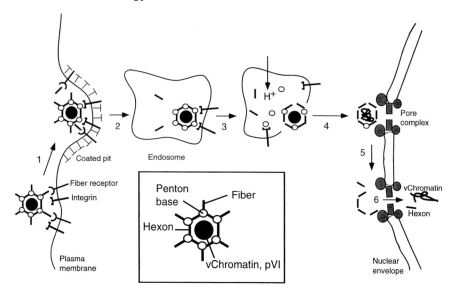

FIGURE 8.5   Stepwise uncoating of adenovirus during cell entry. The virion first binds through its fiber to a so-far unidentified primary receptor at the cell surface (Philipson et al., 1968). The virion is then localized through a cytoskeleton-dependent rearrangement to coated pits on the cell surface, where the penton base protein recognizes a secondary receptor, the fibronectin/vitronectin binding integrin (Wickham et al., 1993). The fibers dissociate from the virion, which is internalized by a coated vesicle (Greber et al., 1993). Shortly thereafter, a drop to acidic pH triggers virus penetration and release into the cytoplasm. The penton base then dissociates from the particle, interactions between the viral chromatin and the capsid are severed, and the minor stabilizing proteins are degraded. The remainder of the virion binds to the nuclear pore complex, the capsid is dismantled, and the DNA core and a small fraction of hexon are imported into the nucleus. (Reproduced by permission from Greber et al., 1994.)

sequence containing an immunodominant antigenic site in the VP1 capsid protein (residues 93 to 100) of poliovirus type 3, P3/Leon/USA/37 strain (Evans et al., 1983), was inserted into l_1 (residues 284 to 291) and l_2 (residues 442 to 449). These locations in hexon are sufficiently mobile that density was not seen in the 2.9-Å-resolution electron density map for residues 270 to 291 and 444 to 453 (Athappilly et al., 1994). Substitution into loop l_1 gave viruses with normal growth, but alterations in loop l_2 produced viruses with impaired growth. Because these viruses are efficiently neutralized by antisera that neutralize poliovirus type 3, the transferred sequence is presumably free to adopt the same three-dimensional structure as the poliovirus epitope.

Penton Base (Polypeptide III)

The penton complex at the vertex is formed from penton base and fiber. Early EM studies indicated that penton base has a spadelike shape, with a polygonal cross section that sometimes revealed a hole (Ginsberg, 1984). A biochemical study (van Oostrum and Burnett, 1985) established that penton base (see Color Plate 11) is a pentamer of polypeptide III, whereas the fiber is a trimer of polypeptide IV (Table 8.1). Thus an intriguing symmetry mismatch occurs within the penton complex. This was visualized by EM image analysis on a capsid fragment containing a vertex (van Oostrum et al., 1987a, 1987b). These results, together with later EM studies on purified penton base and fiber (Ruigrok et al., 1990), indicate that the fiber fills the central approximately 30-Å-diameter core of a hollow penton base. This information was used to define the molecular boundaries for penton in the difference image (Stewart et al., 1993) and separate the two vertex components. A 29-Å-diameter cylinder was used to crop the fiber from the center of the penton base. The boundary between the penton base and the internal ring of polypeptide VI was set to correspond to the previously observed spadelike shape of the penton base. The cropped penton base has a height of 124 Å and maximum and minimum diameters of 112 and 50 Å. There are five small protrusions, roughly 22 Å in diameter, on the top. The molecular volume is consistent with its mass (Table 8.1), but it should be noted that this is very dependent on the exact placement of the molecular boundary. For proteins that are not clearly separated from their neighbors, this assignment is a major difficulty in EM imaging.

During viral infection, the knob at the end of the fiber (see later) attaches to an unidentified cellular receptor and begins the infective process (Fig. 8.5) (Philipson et al., 1968; Defer et al., 1990; Greber et al., 1993, 1994). The penton base carries the RGD recognition sequence for the secondary cellular receptor, which is an integrin receptor that mediates internalization (White 1993; Wickham et al., 1993). Alterations in the RGD region of the penton base modify viral interactions with the cell (Bai et al., 1993). There have been suggestions that the RGD sequence is exposed at the top of the small protuberances on the penton base, but this is inconsistent with the "canyon" hypothesis (Rossmann, 1989; see also Chapter 4). This theory postulates that viral receptor attachment sites lie in surface depressions, as in rhinovirus (see Chapter 4) or influenza virus (see Chapter 3), to protect them from immune surveillance. It is possible that the penton site is protected by lying below the very top of the protuberance, or is in a small depression unseen at the resolution of the EM image.

As the mechanisms of cell entry and disassembly are being unraveled, it is becoming increasingly clear that the process is a series of intricate events, each of which must occur before the next can proceed. This ensures that the process is sequential and error free. It is likely not only that each step is essential for the next step, but that some important steps are preceded by a "trigger" to ensure accuracy. An example is the requirement for two receptors. Separating attachment from uptake increases the odds that the virus is targeted to the appropriate cell. Similar strategies can be seen in assembly, with specific proteolytic steps "locking in" intermediate stages and driving the assembly pathway in one direction.

Fiber (Polypeptide IV)

The fiber (582 residues, M_r 61,960 Da) is the only complete structural protein other than hexon to be crystallized (Mautner and Pereira, 1971) but crystallographic investigations (Devaux et al., 1984, 1990; Stouten et al., 1992) have been plagued by the inability to produce large nontwinned crystals. A recent breakthrough occurred with the crystal structure determination of the C-terminal knob at the end of the fiber from recombinant type 5 protein (Fig. 8.6) (Xia et al., 1994). This revealed the primary receptor binding site and provided unambiguous proof that the fiber is trimeric (van Oostrum and Burnett, 1985). The construction of the knob is similar to that of hexon, with β-barrels forming the core of each subunit and the remainder formed from turns and loops connecting the individual β-strands. Although their β-barrels both have eight strands, the folding topology of that from the fiber is different from that of hexon, and is unlike any other known structure. The primary receptor binding

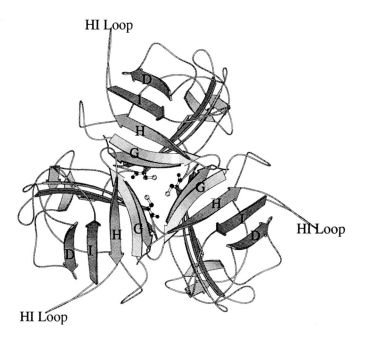

FIGURE 8.6   Fiber knob. The structure of the trimeric ad5 knob at the end of the fiber is shown with the course of the polypeptide chain as a ribbon and the cysteine residues as ball-and-stick models. The view is down the threefold molecular axis toward the virus. About 35% of the subunit is formed from an eight-stranded antiparallel β-sandwich, with the remainder formed from turns and loops connecting the β-strands. Despite the apparent similarity of the knob β-sandwich to the β-barrels in hexon, the folding topology is different. The trimeric knob resembles a three-bladed propeller, with a central depression from which three valleys extend between the individual subunits. The receptor binding site is presumed to lie in these surface depressions because the residues that are conserved between different adenovirus species lie near the symmetry axis, in the β-sheet strands G, H, I, and D and in the HI loop. (Reproduced with permission from Xia et al., 1994.)

site was assigned to the upper surface, the orientation of the molecule being known from the residues on the lower surface leading into the N-terminal shaft. The residues around the threefold symmetry axis in a deep surface depression, and on the floors of the valleys radiating out from the center, are conserved and could form the binding site for the cellular receptor (Xia et al., 1994). This situation is consistent with the canyon hypothesis developed for rhinovirus (see Chapter 4), whereby a conserved binding site is protected from antibody binding by the steric hindrance provided by the surrounding nonconserved residues. Xia et al. (1994) postulate that there are two possible binding modes. The first would be with a cellular receptor binding to the central depression, presumably through trimeric interactions. In the second, up to three dimeric receptors would bind to the valleys. The second mode would explain the knob's very low dissociation constant.

The recombinant knob protein includes 15 residues of the 22nd repeating unit of the shaft domain, but the first 7 are disordered in the crystal structure. This was both surprising, because the shaft is a thin, rigid, noncurved structure suggesting strong interactions between its subunits, and disappointing, because the knob structure provides no information on the validity of the triple-helical shaft model (Stouten et al., 1992).

Ruigrok et al. (1990) showed that the complete fiber protrudes roughly 330 Å from the penton base. However, only a short portion of the thin, 37-Å-diameter fiber protruding from the vertices was visible in the EM reconstruction (Stewart et al., 1991). The fibers are bent and so, because the virion does not obey icosahedral symmetry beyond a radius of 545 Å, the density in the averaged image is "smeared" away. The EM reconstruction shows two "beads" within the penton base and a third roughly spherical region protruding from its top. The spacing between the internal beads is 56 Å, and the outermost pair are 45 Å apart. These observations agree with EM images of negatively stained isolated fibers showing seven to nine small beads along the shaft, each spaced by roughly 35 Å, and a large C-terminal knob (Ruigrok et al., 1990).

Analysis of the primary sequence of the adenovirus type 2 fiber by Green et al. (1983) suggested an N-terminal portion of 44 residues, a shaft of 357 residues with a highly repetitive sequence of 22 segments, and a C-terminal head of 181 residues forming the knob at the distal end of the fiber. This orientation, with the N terminus attached to the penton base, has been confirmed experimentally (Devaux et al., 1987; Weber et al., 1989; Xia et al., 1994). If the shaft corresponds to eight density beads, each composed of two or three sequence repeats and spaced by an average distance of 40 Å to yield the correct overall length, then each bead would be roughly 8% of the total molecular mass. The fiber in the EM reconstruction has roughly the correct volume if the innermost bead is the N-terminal portion and the two outermost density beads are 2/8ths of the shaft (Table 8.1).

The repetitive nature of the sequence led Green et al. (1983) to propose a dimeric model for the shaft, in which a repetitive motif of approximately 15 residues forms the short β-strands in a zigzag dimer. This model was oriented correctly, correlated the variable number of motifs with the fiber lengths observed for different serotypes, and provided a structural explanation for how a long, thin, and rigid fiber shaft could be constructed. A later model (Stouten et al., 1992) incorporated the idea of β-strands into a left-handed triple-helical structure to provide consistency with a trimeric fiber (van Oostrum and Burnett, 1985). The EM reconstruction (Stewart et al., 1993) is in general agreement with this model, but it should be noted that the vertex proteins are fivefold averaged. This process washes out the fine structure of a triple helix but preserves gross features, such as a beading pattern.

A further puzzle to be solved for fiber is the relationship between the discovery of two different types of fiber in ad40 (Kidd et al., 1993) and the long-known existence of two fibers per vertex in the avian adenoviruses (Laver et al., 1971; Hess et al., 1995). It is likely that the solution will explain how the symmetry mismatch between the trimeric fiber and the pentameric penton base is overcome (van Oostrum et al., 1987b).

Minor Coat Proteins and Their Role as Capsid Cement

The adenovirus virion contains several minor proteins, polypeptides IIIa, VI, VIII, and IX (Table 8.1). Polypeptide X, first identified by Maizel et al. (1968a), was later resolved (Anderson et al., 1973) into three very small proteins of 5 to 6.5 kD (X to XII), one of which may correspond to the μ core protein (Hosokawa and Sung, 1976). It has long been a puzzle why adenovirus, and other complex viruses, contain so many proteins apart from those, such as hexon, penton, and fiber, whose structural role is clear. The function of the minor proteins is not so obvious, although Maizel et al. (1968b) suggested that they may act as capsid cement. Colby and Shenk (1981) provided direct evidence that polypeptide IX stabilizes the virion by showing that mutant virions can be assembled without this protein, but that they are more thermolabile than wild-type virions. Further evidence was provided by the finding that the mutant virions cannot package full-length DNA (Ghosh-Choudhury et al., 1987).

Because most minor proteins are small, they are difficult to discern in EM images alone, but they can be revealed by difference imaging, in which X-ray hexon images are subtracted from the EM image (Furcinitti et al., 1989; Stewart et al., 1993). These techniques have revealed the locations of all but polypeptide VIII with reasonable confidence (Figs. 8.1 and 8.2). It is notable that all minor proteins lie at specialized locations in the hexon capsid: polypeptide IX stabilizes hexon−hexon contacts within the center of a facet; polypeptide IIIa spans the capsid to join hexons on adjacent facets; and polypeptide VI anchors the ring of peripentonal hexons inside the capsid and connects the capsid to the core. These positions are consistent with the minor proteins being capsid cement.

The existence of cementing proteins can be rationalized as a mechanism to overcome the conflicting requirements for the accurate assembly and ultimate stability of the completed virion (Burnett, 1985). This leads to a structural paradox: The interactions between the major structural proteins must be weak enough during assembly to allow error correction, but strong enough in the mature virion to protect the genome from the hostile environment outside the cell. Cementing proteins resolve this paradox. In adenovirus, the strategic positions in the capsid occupied by the minor proteins are highly suggestive of their stabilizing function. There is growing experimental evidence for their specialized role in each of their different intermolecular interactions. Polypeptide IX has long been the clearest case, but a particularly pleasing illustration of the paradox has been provided by Matthews and Russell (1995). They have shown that polypeptide VI is assembled as a full-length precursor protein into immature virions. When polypeptide VI is cleaved by the adenovirus protease, which is activated by the 11-residue N-terminal peptide, the binding of polypeptide VI to hexon is dramatically enhanced. Thus, polypeptide VI interacts weakly with hexon during assembly, but once it is in its correct position, a cleavage event locks it into position. The analogy with "fastening" or "cementing" components together is inescapable. There is also a clear analogy between the adenovirus minor proteins and their counterparts in bacteriophage (see Chapters 1 and 11). It is highly likely that cementing proteins are a common, but hitherto overlooked, feature of all large macromolecular assemblies and that their use extends to ribosomes and other cellular organelles.

Polypeptide IIIa

The three-dimensional difference image (Stewart et al., 1993) showed that polypeptide IIIa (570 residues, M_r 63,535 Da) is present as a large, elongated monomeric component. Although its main bulk is on the outer surface of the virion, it spans the capsid to link two facets. It is present as 60 copies, two along each of the 30 icosahedral edges (Fig. 8.7). The molecule penetrates the capsid, tapering to a point in the middle, and appearing on the inside so that polypeptide IIIa appears to be a "rivet" whose role is to fasten capsid facets together. The outer density region has an irregular shape, roughly 30 Å in diameter, that reaches 50 Å above the capsid surface and has contacts with four different hexons. A narrow arm, approximately 15 Å in diameter, contacts the tower of one hexon at a kink in a β-strand (Fig. 8.8). The kink is composed of residues 255 to 261, which are highly conserved among the four human adenovirus hexons that have been sequenced (Toogood et al., 1989). The inner density region, which is roughly 37 Å in diameter, lies at a point where three hexons meet and connects to weaker, disordered density in the core.

Polypeptide VI

Polypeptide VI (206 residues, M_r 23,449 Da) has been assigned (Stewart et al., 1993) to a position on the inner capsid surface, where it anchors the ring of peripentonal hexons as well as connecting the highly ordered capsid to the core region. The molecule has a clear trimeric shape in the difference density, with three 29-Å-diameter lobes separated by 46 Å, and connects the bases of two

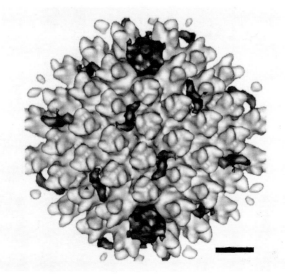

FIGURE 8.7   Penton and polypeptide IIIa. Three-dimensional EM difference density corresponding to penton and polypeptide IIIa (dark gray) is superimposed on a capsid of hexons calculated from their crystallographic coordinates (light gray). Density from the penton base and protruding fiber is seen at the two vertices. Elongated density from polypeptide IIIa is observed along the edges of the hexon capsid where it stabilizes the junction of two facets (cf. Fig. 8.2). Bar, 100 Å. (Reproduced with permission from Stewart and Burnett, 1995.)

adjacent peripentonal hexons (see Color Plate 12). The interior location is supported by enzymatic labeling, suggesting an association between polypeptide VI and core polypeptide V (Everitt et al., 1975) and cross-linking indicating that it binds DNA (Russell and Precious, 1982). Because the volume of each lobe is consistent with a monomer of polypeptide VI, but the stoichiometry indicates that the protein is a hexamer (Table 8.1), it has been suggested that the lobe is a dimer, with half of each polypeptide disordered (Stewart et al., 1993). The middle 115 residues, which are basic (13% arginine and lysine) and highly proline rich (21%), are similar in composition to the 113-residue N-terminal domain of the Sindbis virus core protein that is disordered in its crystallographic structure (Choi et al., 1991). Several other viral capsid proteins have basic N-terminal domains that are crystallographically disordered and are presumed to interact with the internal nucleic acid (Harrison et al., 1978; Abad-Zapatero et al., 1980).

Evidence that polypeptide VI is a capsid cement has emerged from the work of Matthews and Russell (1995). They found that the binding of recombinant polypeptide VI to hexon increased by 40-fold when it was cleaved by the adenovirus protease upon maturation. This processing occurs for several adenovirus structural proteins (IIIa, VI, VII, VIII, terminal protein, μ), that are present as precursors in the immature particle (Anderson et al., 1973). A temperature-sensitive mutant, H2ts1, that is defective in protease function (Weber,

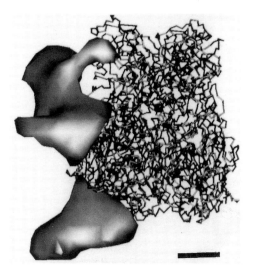

FIGURE 8.8   Interactions of polypeptide IIIa with hexon. Polypeptide IIIa, imaged by EM difference density, is shown adjacent to one of the hexons with which it interacts at the capsid edge. The hexon is represented as a trimer of three identical chains connecting the α-carbon atoms. Because the crystallographic coordinates are known for all atoms, the polypeptide IIIa binding site is localized to within one residue on the crystallographic hexon structure. Polypeptide IIIa has a protruding narrow arm, approximately 15 Å in diameter, that contacts the lower region of one hexon. The orientation is such that the outer surface of the viral capsid is to the top and the inside surface to the bottom. Bar, 25 Å. (Reproduced with permission from Stewart and Burnett, 1995.)

1976) shows that proteolytic processing is an essential step in correct assembly. Virions are formed at nonpermissive temperatures, but they are not infectious.

Polypeptide VIII

Polypeptide VIII could not be identified in the difference maps (Stewart et al., 1993). This is either because its small size (Mr 15,390 Da) makes it difficult to image at 35-Å resolution or because it is disordered. The latter possibility is suggested by its relatively high content of proline (8%) and the basic residues arginine and lysine (11%). Everitt et al. (1975) placed polypeptide VIII on the inside of the capsid. This location is supported by the fact that there was no unassigned external density in the difference map.

Polypeptide IX and Its Role in the Group-of-Nine Hexons

The dissociation fragments known as the GONs have played a key role in developing an architectural model for the virion. GONs provide a well-defined array of hexons aligned so that their threefold molecular axes are perpendicular to the plane. This feature was invaluable in early EM work aimed at determining the orientation of hexons within the capsid and has led to polypeptide

IX (139 residues, M_r 14,339 Da) being the best characterized of the minor proteins.

The origin of the GONs was initially unclear and was attributed to the 60 peripentonal hexons being somehow different from the 180 GON hexons, either chemically (Ginsberg, 1984) or in orientation (Pereira and Wrigley, 1974). Once it was appreciated that all hexons in a facet were chemically identical, and that they lay on a uniform $p3$ net (Burnett, 1984), it was clear that the explanation lay elsewhere. Van Oostrum and Burnett (1985) showed that 12 copies of polypeptide IX bind to each facet. They proposed a model of the GON (Fig. 8.2), in which four trimers of polypeptide IX bind to nine hexons, based on the observed nonrandom dissociation pattern of GONs (Pereira and Wrigley, 1974) and the suggestion by Maizel et al. (1968b) that polypeptide IX could act as a capsid "cement."

Difference imaging of GONs in two dimensions (Furcinitti et al., 1989) at approximately 16-Å resolution revealed four trimers, confirming the model, and defined the overall dimensions of the elongated monomers rather accurately. Three-dimensional difference imaging (Stewart et al., 1993) confirmed that polypeptide IX lies on the outer surface. Each monomer extends from a local threefold axis just above the hexon base along almost the entire length of a hexon–hexon interface. Polypeptide IX is not observed at sites adjacent to each of the three peripentonal hexons.

The location of polypeptide IX in the capsid (Fig. 8.2, see also Color Plate 12) explains several phenomena: how it interacts with hexons to cause the observed dissociation pattern of the virion, which sequentially loses pentons, peripentonal hexons, and then GONs; the nonrandom disintegration pattern of GONs (Pereira and Wrigley, 1974); the lack of reactivity of anti–polypeptide IX antibodies to the intact virion, and the low antibody response to GONs (Cepko et al., 1981). Deletion mutants that lack the gene for polypeptide IX can still form virions, but these are less stable and do not form the GON fragments upon dissociation (Colby and Shenk, 1981). Thus there is excellent evidence supporting the idea that polypeptide IX is a cementing protein that tightly binds the nine central hexons in a facet into a GON to stabilize the capsid.

There are strong indications that the monomers of polypeptide IX are α-helical and that the trimer may have a coiled-coil structure. A preliminary secondary structure prediction of polypeptide IX has been made (P.R. Kuser and R.M. Burnett, unpublished results) using the PHD method (Rost and Sander, 1993, 1994). The PHD method uses evolutionary information contained in multiple sequence alignments as input to neural networks to predict secondary structure. The results show that two long stretches of 22 and 34 residues form α-helices at the C-terminal end of the chain. These would have approximate lengths of 31 and 49 Å and would be approximately 13 Å in diameter. There is also a very high probability that polypeptide IX forms a coiled-coil structure. In a search of all sequences in the database using a coiled-coil prediction algorithm, it had the highest score of any protein (Lupas et al., 1991; A.N. Lupas, personal communication). The shape of polypeptide IX trimers can be explained by a model in which each monomer binds to a neighbor to form a trimer with long coiled-coil arms extending out from the center. This agrees with the observed shape and size of the monomer, which is estimated to be 64 Å long and 18 Å in diameter at its midpoint (Stewart et al., 1993).

Core Structure

Polypeptides V, VII, Terminal Protein, and μ

The core of the adenovirus virion contains a copy of linear double-stranded DNA (35,937 bp in ad2) and four proteins. The DNA is covalently attached to two copies of a 55-kDa protein, the terminal protein. This forms dimers to circularize the DNA, acts as a primer for DNA replication, and attaches the viral genome to the nuclear matrix (see Shenk, 1996). The remaining three proteins resemble the basic arginine-rich histones, and so can neutralize the charged phosphates in DNA. All three bind to DNA (Chatterjee et al., 1986). Polypeptide VII is the major core protein, with over 800 copies (Table 8.1). It complexes tightly with DNA to form compact repeating structures (Black and Center, 1979; Mirza and Weber, 1982; Chatterjee et al., 1986). Polypeptide V can also bind to penton base (Everitt et al., 1975) and so could link the core to the capsid. The function of μ is unknown (Hosokawa and Sung, 1976).

The core structure is poorly understood. Cross-linking studies (Wong and Hsu, 1989), EM of negatively stained preparations of sarcosyl cores (Brown et al., 1975), and ion etching (Newcomb et al., 1984) indicate that the viral DNA is organized into approximately 12 large spherical domains. The EM reconstruction (Stewart et al., 1991) revealed an icosahedrally disordered core, with no evidence of distinct spheres at the vertices. However, it is possible that the disordered domains of polypeptide VI interact with the viral DNA under each vertex. Because the ordered ring of polypeptide VI density on the inner capsid surface has an overall diameter of 160 Å, a loosely ordered region could extend to the spherical diameters of 230 Å (Newcomb et al., 1984) and 216 Å (Brown et al., 1975) observed earlier. These spherical structures may not be visualized by images that have been icosahedrally averaged.

Future Directions

Our picture of the adenovirus virion is much clearer than it was 20 years ago, with atomic-resolution structures for hexon and the fiber knob from X-ray crystallography, an accurate stoichiometry for the structural proteins from biochemistry, and a comprehensive picture for their distribution in the virion from EM and image reconstruction. The secondary cellular receptor has been identified, much more is known about the entry process and uncoating, and the importance of the proteolytic steps in viral maturation are being appreciated.

Nevertheless, much work remains to be accomplished before a full understanding can be reached of the assembly, architecture, and disassembly of this fascinating virus. Important steps will include the identification of the primary cellular receptor; further unraveling of key steps in viral uptake and disassembly, virion formation, and assembly; unambiguous assignments for the minor proteins; insight into the core structure and the role this plays in the expression of the viral genome; and atomic-resolution structures for further components of the virion, including the penton base, the intriguing shaft structure, and the minor proteins.

Adenovirus has not only been a model for molecular biology, but has also provided the impetus for the development of new structural methods such as three-dimensional difference imaging, in which crystallographic and EM images were quantitatively combined for the first time. The next major challenge will be to determine the structure of the nonicosahedrally related parts of the virion. The efforts to establish this new frontier in structural virology will be given impetus by the growing importance of adenovirus as a vector for human gene therapy and vaccine delivery.

Acknowledgments

I am indebted to the many talented people who have contributed to my structural studies on adenovirus over the years. I owe a special debt to Drs. Paula R. Kuser and John J. Rux for many recent contributions, including their help with this chapter. This work would not have been possible without the enormous body of information on adenovirus assembled by many other workers. Their talents and generosity in sharing results and ideas are gratefully acknowledged, while my apologies are offered for the inevitable oversights and omissions in this short chapter. I particularly wish to acknowledge Dr. William C. Russell, to whom this chapter is dedicated on his retirement, and Drs. Johann Deisenhofer, Urs F. Greber, Ari Helenius, and Thomas Shenk for discussions and unpublished material.

This work is currently supported by grants from the National Institute of Allergy and Infectious Diseases (AI 17270), the National Science Foundation (MCB 92-07014), and the Wistar Cancer Center (CA10815).

References

Abad-Zapatero, C., Abdel-Meguid, S.S., Johnson, J.E., Leslie, A.G.W., Rayment, I., Rossmann, M.G., Suck, D. and Tsukihara, T. 1980. Structure of southern bean mosaic virus at 2.8 Å resolution. Nature 286: 33–39.

Akusjärvi, G., Aleström, P., Pettersson, M., Lager, M., Jornvall, H. and Pettersson, U. 1984. The gene for the adenovirus 2 hexon polypeptide. J. Biol. Chem. 259: 13976–13979.

Aleström, P., Akusjärvi, G., Perricaudet, M., Mathews, M.B., Klessig, D.F. and Pettersson, U. 1980. The gene for polypeptide IX of adenovirus type 2 and its unspliced messenger RNA. Cell 19: 671–681.

Allard, A., Albinsson, B. and Wadell, G. 1992. Detection of adenoviruses in stools from healthy persons and patients with diarrhea by two-step polymerase chain reaction. J. Med. Virol. 37: 149–157.

Anderson, C.W. 1990. The proteinase polypeptide of adenovirus serotype 2 virions. Virology 177: 259–272.

Anderson, C.W., Baum, P.R. and Gesteland, R.F. 1973. Processing of adenovirus 2-induced proteins. J. Virol. 12: 241–252.

Athappilly, F.K., Murali, R., Rux, J.J., Cai, Z. and Burnett, R.M. 1994. The refined crystal structure of hexon, the major coat protein of adenovirus type 2, at 2.9 Å resolution. J. Mol. Biol. 242: 430–455.

Bai, M., Harfe, B. and Freimuth, P. 1993. Mutations that alter an arg-gly-asp (RGD) sequence in the adenovirus type 2 penton base protein abolish its cell-rounding activity and delay virus reproduction in flat cells. J. Virol. 67: 5198–5205.

Black, B.C. and Center, M.S. 1979. DNA-binding properties of the major core protein of adenovirus 2. Nucleic Acids Res. 6: 2339–2353.

Boursnell, M.E.G. and Mautner, V. 1981. Recombination in adenovirus: crossover sites in intertypic recombinants are located in regions of homology. Virology 112: 198–209.

Brown, D.T., Westphal, M., Burlingham, B.T., Winterhoff, U. and Doerfler, W. 1975. Structure and composition of adenovirus type 2 core. J. Virol. 16: 366–387.

Burnett, R.M. 1984. Structural investigations on hexon, the major coat protein of adenovirus. In Biological Macromolecules and Assemblies, Vol. 1: Virus Structures (Jurnak, F.A. and McPherson, A., Eds.), pp. 337–385. John Wiley & Sons, New York.

Burnett, R.M. 1985. The structure of the adenovirus capsid. II. The packing symmetry of hexon and its implications for viral architecture. J. Mol. Biol. 185: 125–143.

Burnett, R.M., Athappilly, F.K., Cai, Z., Furcinitti, P.S., Korn, A.P., Murali, R. and van Oostrum, J. 1990. Structure of the adenovirus virion. In Use of X-Ray Crystallography in the Design of Antiviral Agents (Laver, W.G. and Air, G.M., Eds.), pp. 35–48. Academic Press, San Diego.

Burnett, R.M., Grütter, M.G. and White, J.L. 1985. The structure of the adenovirus capsid. I. An envelope model of hexon at 6 Å resolution. J. Mol. Biol. 185: 105–123.

Caspar, D.L.D. and Klug, A. 1962. Physical principles in the construction of regular viruses. Cold Spring Harbor Symp. Quant. Biol. 27: 1–24.

Cepko, C.L., Changelian, P.S. and Sharp, P.A. 1981. Immunoprecipitation with two-dimensional pools as a hybridoma screening technique: production and characterization of monoclonal antibodies against adenovirus 2 proteins. Virology 110: 385–401.

Cepko, C.L. and Sharp, P.A. 1982. Assembly of adenovirus major capsid protein is mediated by a nonvirion protein. Cell 31: 407–415.

Chatterjee, P.K., Vayda, M.E. and Flint, S.J. 1986. Identification of proteins and protein domains that contact DNA within adenovirus nucleoprotein cores by ultraviolet light crosslinking of oligonucleotides ^{32}P-labelled in vivo. J. Mol. Biol. 188: 23–37.

Chelvanayagam, G., Heringa, J. and Argos, P. 1992. Anatomy and evolution of proteins displaying the viral capsid jellyroll topology. J. Mol. Biol. 228: 220–242.

Choi, H.-K., Tong, L., Minor, W., Dumas, P., Boege, U., Rossman, M.G. and Wengler, G. 1991. Structure of Sindbis virus core protein reveals a chymotrypsin-like serine proteinase and the organization of the virion. Nature 354: 37–43.

Colby, W.W. and Shenk, T. 1981. Adenovirus type 5 virions can be assembled in vivo in the absence of detectable polypeptide IX. J. Virol. 39: 977–980.

Couch, R.B., Chanock, R.M., Cate, T.R., Lang, D.J., Knight, V. and Huebner, R.J. 1963. Immunization with types 4 and 7 adenovirus by selective infection of the intestinal tract. Am. Rev. Respir. Dis. 88(Part-2): 394–403.

Crawford-Miksza, L. and Schnurr, D.P. 1996. Analysis of 15 adenovirus hexon proteins reveals the location and structure of seven hypervariable regions containing serotype-specific residues. J. Virol. 70: 1836–1844.

Crompton, J., Toogood, C.I.A., Wallis, N. and Hay, R.T. 1994. Expression of a foreign epitope on the surface of the adenovirus hexon. J. Gen. Virol. 75: 133–139.

Crowther, R.A. and Amos, L.A. 1971. Harmonic analysis of electron microscope images with rotational symmetry. J. Mol. Biol. 60: 123–130.

Crowther, R.A. and Franklin, R.M. 1972. Structure of the groups of nine hexons from adenovirus. J. Mol. Biol. 68: 181–184.

Defer, C., Belin, M.T., Caillet-Boudin, M.-L. and Boulanger, P. 1990. Human adenovirus-host cell interactions: comparative study with members of subgroups B and C. J. Virol. 64: 3661–3673.

Devaux, C., Adrian, M., Berthet-Colominas, C., Cusack, S. and Jacrot, B. 1990. Structure of adenovirus fibre. I. Analysis of crystals of fibre from adenovirus serotypes 2 and 5 by electron microscopy and X-ray crystallography. J. Mol. Biol. 215: 567–588.

Devaux, C., Berthet-Colominas, C., Timmins, P.A., Boulanger, P. and Jacrot, B. 1984. Crystal packing and stoichiometry of the fibre protein of adenovirus type 2. J. Mol. Biol. 174: 729–737.

Devaux, C., Caillet-Boudin, M.L., Jacrot, B. and Boulanger, P. 1987. Crystallization, enzymatic cleavage, and the polarity of the adenovirus type 2 fiber. Virology 161: 121–128.

Doerfler, W. and Boehm, P. (Eds.) 1995. Molecular repertoire of adenoviruses. Curr. Top. Microbiol. Immunol. 199.

Ellis, R.J. and Hemmingsen, S.M. 1989. Molecular chaperones: proteins essential for the biogenesis of some macromolecular structures. Trends Biochem. Sci. 14: 339–342.

Evans, D.M.A., Minor, P.D., Schild, G.C. and Almond, J.W. 1983. Critical role of an eight amino acid sequence of VP1 in the neutralization of poliovirus type 3. Nature 304: 459–462.

Everitt, E., Lutter, L. and Philipson, L. 1975. Structural proteins of adenovirus. XII. Location and neighbor relationship among proteins of adenovirion type 2 as revealed by enzymatic iodination, immunoprecipitation and chemical cross-linking. Virology 67: 197–208.

Furcinitti, P.S., van Oostrum, J. and Burnett, R.M. 1989. Adenovirus polypeptide IX revealed as capsid cement by difference images from electron microscopy and crystallography. EMBO J. 8: 3563–3570.

Ghosh-Choudhury, G., Haj-Ahmad, Y. and Graham, F.L. 1987. Protein IX, a minor component of the human adenovirus capsid, is essential for the packaging of full length genomes. EMBO J. 6: 1733–1739.

Ginsberg, H.S. 1979. Adenovirus structural proteins. In Comprehensive Virology, Vol. 13 (Fraenkel-Conrat, H. and Wagner, R.R., Eds.), pp. 409–457. Plenum, New York.

Ginsberg, H.S. (Ed.) 1984. The Adenoviruses. Plenum, New York.

Ginsberg, H.S., Pereira, H.G., Valentine, R.C. and Wilcox, W.C. 1966. A proposed terminology for the adenovirus antigens and virion morphological subunits. Virology 28: 782–783.

Greber, U.F., Singh, I. and Helenius, A. 1994. Mechanisms of virus uncoating. Trends Microbiol. 2: 52–56.

Greber, U.F., Willetts, M., Webster, P. and Helenius, A. 1993. Stepwise dismantling of adenovirus 2 during entry into cells. Cell 75: 477–486.

Green, N.M., Wrigley, N.G., Russell, W.C., Martin, S.R. and McLachlan, A.D. 1983. Evidence for a repeating cross-β sheet structure in the adenovirus fibre. EMBO J. 2: 1357–1365.

Grütter, M.G. and Franklin, R.M. 1974. Studies on the molecular weight of the adenovirus type 2 hexon and its subunit. J. Mol. Biol. 89: 163–178.

Harrison, S.C., Olson, A.J., Schutt, C.E., Winkler, F.K. and Bricogne, G. 1978. Tomato bushy stunt virus at 2.9 Å resolution. Nature 276: 368–373.

Hess, M., Cuzange, A., Ruigrok, R.W.H., Chroboczek, J. and Jacrot, B. 1995. The avian adenovirus penton: two fibres and one base. J. Mol. Biol. 252: 379–385.

Hierholzer, J.C., Wigand, R., Anderson, L.J., Adrian, T. and Gold, J.W.M. 1988. Adenoviruses from patients with AIDS: a plethora of serotypes and a description of five new serotypes of subgenus D (types 43–47). J. Infect. Dis. 158: 804–813.

Hilleman, M.R., Werner, J.H., Dascomb, H.E. and Butler, R.L. 1955. Epidemiologic investigations with respiratory disease virus RI-67. Am. J. Public Health 45: 203–210.

Horne, R.W., Brenner, S., Waterson, A.P. and Wildy, P. 1959. The icosahedral form of an adenovirus. J. Mol. Biol. 1: 84–86.

Horwitz, M.S., Maizel, J.V. Jr. and Scharff, M.D. 1970. Molecular weight of adenovirus type 2 hexon polypeptide. J. Virol. 6: 569–571.

Hosokawa, K. and Sung, M.T. 1976. Isolation and characterization of an extremely basic protein from adenovirus type 5. J. Virol. 17: 924–934.

Hu, S.-L., Hays, W.W. and Potts, D.E. 1984. Sequence homology between bovine and human adenoviruses. J. Virol. 49: 604–608.

Huebner, R.J., Bell, J.A., Rowe, W.P., Ward, T.G., Suskind, R.G. Jr., Hartley, J.W. and Paffenbarger, R.S. 1955. Studies of adenoidal-pharyngeal-conjunctival vaccines in volunteers. J. Am. Med. Assoc. 159: 986–989.

Janin, J., Miller, S. and Chothia, C. 1988. Surface, subunit interfaces and interior of oligomeric proteins. J. Mol. Biol. 204: 155–164.

Jones, E.Y., Stuart, D.I. and Walker, N.P.C. 1989. Structure of tumour necrosis factor. Nature 338: 225–228.

Jörnvall, H. and Philipson, L. 1980. Limited proteolysis and a reactive cysteine residue define accessible regions in the native conformation of the adenovirus hexon protein. Eur. J. Biochem. 104: 237–247.

Kauzmann, W. 1959. Some factors in the interpretation of protein denaturation. Adv. Protein Chem. 14: 1–63.

Kidd, A.H., Chroboczek, J., Cusack, S. and Ruigrok, R.W.H. 1993. Adenovirus type 40 virions contain 2 distinct fibers. Virology 192: 73–84.

Kinloch, R., Mackay, N. and Mautner, V. 1984. Adenovirus hexon: sequence comparison of subgroup C serotypes 2 and 5. J. Biol. Chem. 259: 6431–6436.

Korn, A.P. and Burnett, R.M. 1991. Distribution and complementarity of hydropathy in multisubunit proteins. Proteins 9: 37–55.

Kozarsky, K.F. and Wilson, J.M. 1993. Gene therapy: adenovirus vectors. Curr. Opin. Genet. Dev. 3: 499–503.

Laver, W.G. 1970. Isolation of an arginine-rich protein from particles of adenoviruses type 2. Virology 41: 488–500.

Laver, W.G., Wrigley, N.G. and Pereira, H.G. 1969. Removal of pentons from particles of adenovirus type 2. Virology 39: 599–605.

Laver, W.G., Younghusband, H.B. and Wrigley, N.G. 1971. Purification and properties of chick embryo lethal orphan virus (an avian adenovirus). Virology 45: 598–614.

Lupas, A., van Dyke, M. and Stock, J. 1991. Predicting coiled coils from protein sequences. Science 252: 1162–1164.

Maizel, J.V. Jr., White, D.O. and Scharff, M.D. 1968a. The polypeptides of adenovirus. I. Evidence for multiple protein components in the virion and a comparison of types 2, 7A, and 12. Virology 36: 115–125.

Maizel, J.V. Jr., White, D.O. and Scharff, M.D. 1968b. The polypeptides of adenovirus. II. Soluble proteins, cores, top components and the structure of the virion. Virology 36: 126–136.

Matthews, D.A. and Russell, W.C. 1995. Adenovirus protein-protein interactions: molecular parameters governing the binding of protein VI to hexon and the activation of the adenovirus 23K protease. J. Gen. Virol. 76: 1959–1969.

Mautner, V. and Pereira, H.G. 1971. Crystallization of a second adenovirus protein (the fibre). Nature 230: 456–457.

Mautner, V. and Willcox, H.N.A. 1974. Adenovirus antigens: a model system in mice for subunit vaccination. J. Gen. Virol. 25: 325–336.

Mirza, M.A. and Weber, J. 1982. Structure of adenovirus chromatin. Biochim. Biophys. Acta 696: 76–86.

Monto, A.S. 1992. Acute respiratory infections. In Maxcy-Rosenau-Last Public Health and Preventive Medicine, 13th ed. (Last, J.M. and Wallace, R.B., Eds.), pp. 125–131. Appleton & Lange, Norwalk, CT.

Nász, I. and Ádám, É. 1983. Arrangement of hexons and polypeptide subunits in the adenovirus capsid. Acta Microbiol. Hung. 30: 169–178.

Nermut, M.V. 1975. Fine structure of adenovirus type 5: I. Virus capsid. Virology 65: 480–495.

Nermut, M.V. 1980. The architecture of adenoviruses: recent views and problems. Arch. Virol. 64: 175–196.

Newcomb, W.W., Boring, J.W. and Brown, J.C. 1984. Ion etching of human adenovirus 2: structure of the core. J. Virol. 51: 52–56.

Oosterom-Dragon, E.A. and Ginsberg, H.S. 1981. Characterization of two temperature-sensitive mutants of type 5 adenovirus with mutations in the 100,000-Dalton protein gene. J. Virol. 40: 491–500.

Pereira, H.G., Valentine, R.C. and Russell, W.C. 1968. Crystallization of an adenovirus protein (the hexon). Nature 219: 946–947.

Pereira, H.G. and Wrigley, N.G. 1974. In vitro reconstruction, hexon bonding and handedness of incomplete adenovirus capsid. J. Mol. Biol. 85: 617–631.

Pettersson, U. 1971. Structural proteins of adenoviruses. VI. On the antigenic determinants of the hexon. Virology 43: 123–136.

Philipson, L. 1983. Structure and assembly of adenoviruses. Curr. Top. Microbiol. Immunol. 109: 1–52.

Philipson, L., Longberg-Holm, K. and Pettersson, U. 1968. Virus-receptor interaction in an adenovirus system. J. Virol. 2: 1064–1075.

Prage, L., Pettersson, U., Höglund, S., Lonberg-Holm, K. and Philipson, L. 1970. Structural proteins of adenoviruses. IV. Sequential degradation of the adenovirus type 2 virion. Virology 42: 341–358.

Priestle, J. P. 1988. RIBBON: a stereo cartoon drawing program for proteins. J. Appl. Crystallogr. 21: 572–576.

Pring-Åkerblom, P. and Adrian, T. 1993. The hexon genes of adenoviruses of subgenus C: comparison of the variable regions. Res. Virol. 144: 117–127.

Rekosh, D. 1981. Analysis of the DNA-terminal protein from different serotypes of human adenovirus. J. Virol. 40: 329–333.

Roberts, M.M. and Burnett, R.M. 1987. Adenovirus hexon: a novel use of the viral beta-barrel. In Biological Organization: Macromolecular Interactions at High Resolution (Burnett, R.M. and Vogel, H.J., Eds.), pp. 113–124. Academic Press, Orlando, FL.

Roberts, M.M., White, J.L., Grütter, M.G. and Burnett, R.M. 1986. Three-dimensional structure of the adenovirus major coat protein hexon. Science 232: 1148–1151.

Roberts, R.J., O'Neill, K.E. and Yen, C.T. 1984. DNA sequences from the adenovirus 2 genome. J. Biol. Chem. 259: 13968–13975.

Rossmann, M.G. 1989. The canyon hypothesis. J. Biol. Chem. 264: 14587–14590.

Rossmann, M.G. and Johnson, J.E. 1989. Icosahedral RNA virus structure. Annu. Rev. Biochem. 58: 533–573.

Rost, B. and Sander, C. 1993. Prediction of protein secondary structure at better than 70% accuracy. J. Mol. Biol. 232: 584–599.

Rost, B. and Sander, C. 1994. Combining evolutionary information and neural networks to predict protein secondary structure. Proteins 19: 55–72.

Rowe, W.P., Huebner, R.J., Gillmore, L.K., Parrott, R.H. and Ward, T.G. 1953. Isolation of a cytopathogenic agent from human adenoids undergoing spontaneous degeneration in tissue culture. Proc. Soc. Exp. Biol. Med. 84: 570–573.

Ruigrok, R.W.H., Barge, A., Albiges-Rizo, C. and Dayan, S. 1990. Structure of adenovirus fibre. II. Morphology of single fibres. J. Mol. Biol. 215: 589–596.

Russell, W.C. and Precious, B. 1982. Nucleic acid-binding properties of adenovirus structural polypeptides. J. Gen. Virol. 63: 69–79.

Scott-Taylor, T.H. and Hammond, G.W. 1992. Conserved sequences of the adenovirus genome for detection of all human adenovirus types by hybridization. J. Clin. Microbiol. 30: 1703–1710.

Shenk, T. 1996. Adenoviridae: and their replication. In Virology (Fields, B., Howley, P. and Knipe, D., Eds.). pp. 2111–2148. Raven Press, New York.

Shortridge, K.F. and Biddle, F. 1970. The proteins of adenovirus type 5. Arch. Ges. Virusforsch. 29: 1–24.

Smith, K.O., Gehle, W.D. and Trousdale, M.D. 1965. Architecture of the adenovirus capsid. J. Bacteriol. 90: 254–261.

Stewart, P.L. and Burnett, R.M. 1993. Adenovirus structure as revealed by X-ray crystallography, electron microscopy, and difference imaging. Jpn. J. Appl. Phys. 32: 1342–1347.

Stewart, P.L. and Burnett, R.M. 1995. Adenovirus structure by X-ray crystallography and electron microscopy. Curr. Top. Microbiol. Immunol. 199: 25–38.

Stewart, P.L., Burnett, R.M., Cyrklaff, M. and Fuller, S.D. 1991. Image reconstruction reveals the complex molecular organization of adenovirus. Cell 67: 145–154.

Stewart, P.L., Fuller, S.D. and Burnett, R.M. 1993. Difference imaging of adenovirus: bridging the resolution gap between X-ray crystallography and electron microscopy. EMBO J. 12: 2589–2599.

Stouten, P.F.W., Sander, C., Ruigrok, R.W.H. and Cusack, S. 1992. New triple-helical model for the shaft of the adenovirus fibre. J. Mol. Biol. 226: 1073–1084.

Tate, V.E. and Philipson, L. 1979. Parental adenovirus DNA accumulates in nucleosome-like structures in infected cells. Nucleic Acids Res. 6: 2769–2785.

Toogood, C.I.A., Crompton, J. and Hay, R.T. 1992. Antipeptide antisera define neutralizing epitopes on the adenovirus hexon. J. Gen. Virol. 73: 1429–1435.

Toogood, C.I.A. and Hay, R.T. 1988. DNA sequence of the adenovirus type 41 hexon gene and predicted structure of the protein. J. Gen. Virol. 69: 2291–2301.

Toogood, C.I.A., Murali, R., Burnett, R.M. and Hay, R.T. 1989. The adenovirus type 40 hexon: sequence, predicted structure and relationship to other adenovirus hexons. J. Gen. Virol. 70: 3203–3214.

van Oostrum, J. and Burnett, R.M. 1985. Molecular composition of the adenovirus type 2 virion. J. Virol. 56: 439–448.

van Oostrum, J., Smith, P.R., Mohraz, M. and Burnett, R.M. 1986. Interpretation of electron micrographs of adenovirus hexon arrays using a crystallographic molecular model. J. Ultrastruct. Mol. Struct. Res. 96: 77–90.

van Oostrum, J., Smith, P.R., Mohraz, M. and Burnett, R.M. 1987a. Morphology of the vertex region of adenovirus. Ann. N. Y. Acad. Sci. 494: 423–426.

van Oostrum, J., Smith, P.R., Mohraz, M. and Burnett, R.M. 1987b. The structure of the adenovirus capsid. III. Hexon packing determined from electron micrographs of capsid fragments. J. Mol. Biol. 198: 73–89.

von Bahr-Lindström, H., Jörnvall, H., Althin, S. and Philipson, L. 1982. Structural differences between hexons from adenovirus types 2 and 5: correlation with differences in size and immunological properties. Virology 118: 353–362.

Weber, J. 1976. Genetic analysis of adenovirus type 2. III. Temperature sensitivity of processing of viral proteins. J. Virol. 17: 462–471.

Weber, J.M., Cai, F., Murali, R. and Burnett, R.M. 1994. Sequence and structural analysis of murine adenovirus type 1 hexon. J. Gen. Virol. 75: 141–147.

Weber, J.M., Talbot, B.G. and Delorme, L. 1989. The orientation of the adenovirus fiber and its anchor domain defined through molecular mimicry. Virology 168: 180–182.

White, D.O., Scharff, M.D. and Maizel, J.V. Jr. 1969. The polypeptides of adenovirus. III. Synthesis in infected cells. Virology 38: 395–406.

White, J.M. 1993. Integrins as virus receptors. Curr. Biol. 3: 596–599.

Wickham, T.J., Mathias, P., Cheresh, D.A. and Nemerow, G.R. 1993. Integrins $\alpha_v\beta_3$ and $\alpha_v\beta_5$ promote adenovirus internalization but not virus attachment. Cell 73: 309–319.

Wigand, R., Bartha, A., Dreizin, R.S., Esche, H., Ginsberg, H.S., Green, M., Hierholzer, J.C., Kalter, S.S., McFerran, J.B., Pettersson, U., Russell, W.C. and Waddell, G. 1982. Adenoviridae: second report. Intervirology 18: 169–176.

Wohlfart, C. 1988. Neutralization of adenoviruses: kinetics, stoichiometry, and mechanisms. J. Virol. 62: 2321–2328.

Wong, M.-L. and Hsu, M.-T. 1989. Linear adenovirus DNA is organized into supercoiled domains in virus particles. Nucleic Acids Res. 17: 3535–3550.

Xia, D., Henry, L.J., Gerard, R.D. and Deisenhofer, J. 1994. Crystal structure of the receptor-binding domain of adenovirus type 5 fiber protein at 1.7 Å resolution. Structure 2: 1259–1270.

9.

Molecular Basis of Rotavirus Replication

Structure-Function Correlations

B.V. VENKATARAM PRASAD & MARY K. ESTES

Rotavirus is the major cause of life-threatening diarrhea in infants and young children worldwide (Blacklow and Greenberg, 1991; Kapikian and Channock, 1995), and an estimated one million children die annually from rotavirus-induced gastroenteritis. Rotavirus is a double-stranded RNA virus belonging to the family *Reoviridae*. Once the rotaviruses were recognized as major human pathogens, they were studied fairly extensively using modern molecular biologic techniques. These studies provided basic information about gene-coding assignments, protein processing, genome expression and replication, viral morphogenesis, and pathogenesis. In addition, molecular epidemiology studies, coupled with the characterization of neutralizing monoclonal antibodies (MAbs) and sequencing of the genes that encode the neutralizing antigens, have begun to provide an understanding at the molecular level of the genetic and antigenic variability of the rotaviruses.

In recent years, rotaviruses also have been studied by three-dimensional (3-D) structural analysis using cryo−electron microscopy and computer image analysis (Prasad et al., 1987, 1988, 1990; Yeager et al., 1990, 1994; Shaw et al., 1993; Prasad and Chiu, 1994). A detailed architectural description of these complex viruses, including the topographic locations of the major structural proteins and their stoichiometric proportions, is being obtained as the resolution of these techniques improves.

Together, the molecular biology and structural studies are permitting a dissection of the molecular mechanisms that underlie the biologic processes in the life cycle of rotavirus. These include cell entry, neutralization, transcription, gene expression, and virus assembly. A more detailed understanding of such structure–function relationships will aid in the development of new and effective antiviral strategies to prevent or cure the infections and disease caused by rotavirus. This chapter reviews our current knowledge of the structure–function relationships for the rotaviruses and their proteins. Because of similarities between rotaviruses and other members of the *Reoviridae*, some of the knowledge obtained for the rotaviruses can be expected to further our understanding of the other viruses in this family. Similarly, structural studies of other viruses in this family, particularly of bluetongue virus and reovirus, enhance understanding of the rotaviruses.

Biology of Rotaviruses

Gene Coding Assignments and Rotaviral Proteins

Rotaviruses are large ($\sim$1000 Å in diameter), spherical viruses. These viruses were classically described as having two shells of protein, referred to as the outer and inner shells, surrounding a central core containing the genome. Recent structural studies have shown that rotavirus has a more complex structure consisting of three protein layers (Prasad and Chiu, 1994) (Fig. 9.1). This discovery has resulted in a new nomenclature for the virus shells that is used in this chapter and elsewhere (Mattion et al., 1994; Estes, 1996). In this nomenclature, the complete virions having all three layers are referred to as triple-layered particles (TLPs, previously called double-shelled particles); particles that lack the outer layer are referred to as double-layered particles (DLPs, previously called single-shelled particles); particles that lack the outer two capsid layers are referred to as single-layered particles (SLPs, previously called core particles); and the central portion of the virion, which contains the genome and the minor proteins, is referred to as the subcore. The genome of rotavirus consists of 11 segments of double-stranded RNA (dsRNA). The protein(s) encoded by each of the rotavirus genes are well established and their properties have been reviewed (Estes and Cohen, 1989; Mattion et al., 1994).

Each of the dsRNA segments codes for at least one protein (Fig. 9.1). Of the 11 proteins encoded by the genomic RNA, 6 are structural and 5 are nonstructural. The structural proteins are designated as viral protein (VP) followed by a number to designate their relative migration when analyzed by electrophoresis on polyacrylamide gels containing the denaturing detergent sodium dodecyl sulfate (SDS) and the reducing agent 2-mercaptoethanol (Mattion et al., 1994). The structural proteins VP7 and VP4, encoded by genes 9 and 4, respectively, of SA11, constitute the outer layer of the virion, with VP7 being the major component. VP6 and VP2, encoded by genes 6 and 2, respectively, each form the intermediate and inner capsid layers. VP1 and VP3, encoded by genes 1 and 3, are present in minor quantities in the subcore of the virion along with the genomic RNA. The five nonstructural proteins (designated NSP followed by a number according to their order of migration on SDS–polyacryl-

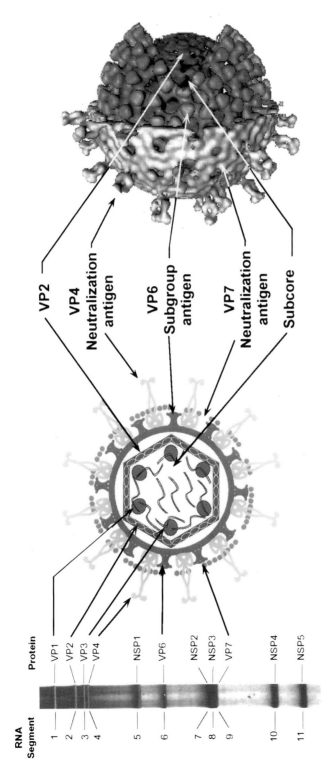

FIGURE 9.1 Rotavirus genes, proteins, and structure. *Left.* A polyacrylamide gel shows the 11 segments of dsRNA that constitute the rotavirus genome and proteins encoded by each of these genes. *Center.* A cartoon representation of the complete rotavirus structure. *Right.* The 3-D structure of a complete virion, in which portions of the VP7 and VP6 layers have been removed to show the VP2 layer. (Reproduced with permission from Estes, 1996.)

amide gel electrophoresis [PAGE]) are NSP1, NSP2, NSP3, NSP4, and NSP5 (Mattion et al., 1994). Although the nonstructural proteins have been detected in various stages of the rotavirus replication cycle, the specific functions of most of these proteins are not yet known. Four of the nonstructural proteins, NSP1 to NSP3 and NSP5, are thought to be involved in RNA replication and/ or assembly of these RNAs into newly made particles. In contrast, NSP4, a tetrameric transmembrane glycoprotein of the endoplasmic reticulum (ER), is thought to act as an intracellular receptor in the unique assembly process that involves budding of the DLPs through the membrane of the rough ER (Au et al., 1989; Meyer et al., 1989).

All the rotaviral genes have been cloned and expressed (Estes and Cohen, 1989; Bellamy and Both, 1990; Mattion et al., 1994). Coexpression of several of the structural proteins has shown that the expressed proteins spontaneously form virus-like particles (Labbé et al., 1991; Chen et al., 1994; Crawford et al., 1994). Coexpression of some of the nonstructural proteins results in the formation of other functional oligomeric complexes (Cohen et al., 1989; Mattion et al., 1992). These results indicate that several of the rotavirus proteins interact with each other.

Another interesting aspect of rotavirus in particular, and segmented viruses in general, is the ability of the virions to assemble a precise set of genome segments into newly made virions. Understanding how this occurs remains a major challenge because this knowledge will lead to the development of reverse genetics systems to allow dissection of the genes of these viruses as well as their use as vectors for delivery of other genes. In the absence of a reverse genetics system, genetic analyses have exploited the fact that the genome segments can undergo reassortment during mixed infections (Greenberg et al., 1981; Ramig and Ward, 1991). Reassortant viruses, or virus-like particles made with wild-type or mutant proteins, which contain capsid proteins of distinct properties, have been useful in mapping specific functions and studying protein–protein interactions because proteins of heterologous origin can be brought together in a single virus particle (Chen and Ramig, 1992; Chen et al., 1989, 1992).

Serogroups and Serotypes

Rotaviruses are classified serologically into both serogroups and serotypes (Estes, 1995). Viruses within each of the six serogroups (A to F) share cross-reacting antigens detectable by serologic tests such as immunofluorescence, enzyme-linked immunoassay, and immunoelectron microscopy. Viruses in groups A, B, and C are found in both humans and animals, whereas viruses in groups D, E, and F have only been found in animals (reviewed in Saif et al., 1994). The group A rotaviruses are the principal pathogens that cause severe dehydrating diarrheal disease in children. Most biochemical and molecular biologic studies have been carried out on group A rotaviruses, but the existing comparative data indicate these different groups of rotaviruses are similar morphologically and biochemically. This review focuses on group A rotaviruses.

Within each serogroup, rotaviruses are further classified into serotypes based on reactivities in neutralization assays using hyperimmune serum prepared in antibody-negative animals. The outer capsid proteins, VP7 and VP4, carry the neutralization epitopes. Rotavirus serotypes defined on the basis of neutralization epitopes in VP4 are called P (for protease-sensitive) serotypes and those defined on the basis of neutralization epitopes in VP7 are called G (for glycoprotein) types. Currently 14 G and 18 P serotypes have been identified (Estes, 1996).

Replication and Pathogenesis

Most information on rotavirus replication has come from studying virus replication in cultured cells. Rotaviruses bind to an unidentified cell receptor and then enter the cell either through direct penetration of the cell membrane or through an endocytotic pathway, depending upon whether the viruses are first trypsinized (reviewed in Estes, 1996) (see Fig. 9.2). Trypsinization results in the cleavage of VP4 into two products, VP5* and VP8*, which remain associated with the virion (Clark et al., 1981; Espejo et al., 1981; Estes et al., 1981). In vivo it is presumed that the proteolysis takes place inside the small intestine by the action of pancreatic proteases (Graham and Estes, 1980; Bass et al., 1992). It has been suggested that productive infection occurs only after virus entry by direct penetration of the cell membrane. It also is thought that, once particles are internalized, the outer capsid shell is removed and the resulting DLPs become transcriptionally active (Cohen et al., 1979). During the process of transcription, the DLPs maintain their structural integrity. Transcription is asymmetric, and all the transcripts are positive strands made from the negative strand of the dsRNA. Cells that rotaviruses infect do not contain the enzymes required for transcription, so the double-layered particle possesses the necessary enzymatic activities, such as transcriptase, nucleotide phosphorylase, guanylyl transferase, and methyl transferase. It is unclear how transcription takes place inside the DLPs, but this process requires energy. The newly made positive-strand RNA transcripts carry out two functions: they are translated into the rotaviral proteins, and they serve as templates for production of negative strands to make the progeny dsRNA. The process by which the correct set of 11 genome segments is encapsidated into each virion remains a mystery. However, it is likely that some, or all, of the nonstructural proteins NSP1, NSP2, NSP3, and NSP4 are involved in this process.

Nine of the proteins encoded by the rotavirus genome are synthesized on free ribosomes (Estes and Cohen, 1989). The glycoproteins VP7 and NSP4 are synthesized on the ribosomes associated with the rough ER (Fig. 9.2). The assembly of DLPs is thought to occur in perinuclear electron-dense structures called viroplasms. Nascent DLPs bud through the rough ER and acquire the outer shell. It is unclear whether the assembly of VP4 takes place prior to the budding event or after. It has been shown that specific interactions between the

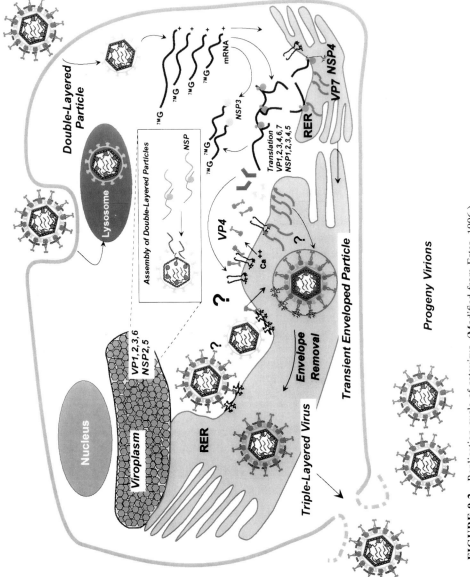

FIGURE 9.2   Replication cycle of rotavirus. (Modified from Estes, 1996.)

244

structural proteins (VP4, VP7, and VP6) and NSP4 are important in the budding process. During the budding process, the viral particles are transiently enveloped. The envelope is lost in the final stages of maturation, and the mature virions are released from the cell by lysis.

Rotavirus replication has also been examined in intestinal biopsies from animals and occasionally from humans (Bishop et al., 1973; Davidson and Barnes, 1979; Hall, 1987; Osborne et al., 1988). Rotavirus infects mature enterocytes via the apical surface of the villi in the small intestine. The infected villous enterocytes are killed and sloughed as the virus replicates in them. Over a period of several days, the infection spreads from the proximal small bowel to the ileum. During infection, the mucosal architecture undergoes remodeling, which includes shortening and atrophy of the villi, crypt hyperplasia, distended cisternae of the ER, and mononuclear cell infiltration in the lamina propria, and results in the stunting of the villi and flattening and loss of enterocytes. The duration of rotavirus infection, in humans and animals, is approximately 1 week. In contrast, in immunocompromised hosts, the infection often persists for months. The precise mechanism by which rotavirus induces diarrhea remain unclear. Studies in animal models have shown that rotavirus infection alters the absorption of selected macromolecules across the intestinal mucosal surface (Stephen, 1989). Malabsorption is one factor that leads to loss of fluid during infection.

Rotavirus Structure

General Architecture

The 3-D structure of rotavirus was first determined by Prasad et al. (1988), using cryo–electron microscopy and computer image process techniques. These techniques have been shown to be particularly useful to study rotaviruses, which are large, complex, nonenveloped isometric particles not currently amenable to crystallographic studies. Figure 9.3 shows electron micrographs of mature rotavirions embedded in vitreous ice. Structural studies have been carried out on two rotavirus strains, the simian rotavirus SA11 (Prasad et al., 1988, 1990) and the rhesus rotavirus RRV (Yeager et al., 1990). The structural features seen in these two strains of rotavirus are very similar, and both have three concentric layers of protein. A surface representation of a 3-D reconstruction of triple-layered (mature) rotavirus along the icosahedral threefold axes is shown in Figure 9.4. The structure has a left-handed $T = 13$ icosahedral symmetry, with distinct structural features including aqueous channels and surface spikes. The overall diameter of the particle, excluding the spikes, is approximately 765 Å.

One of the advantages of structural analysis using electron images of ice-embedded specimens is the ability to retrieve the details of the internal struc-

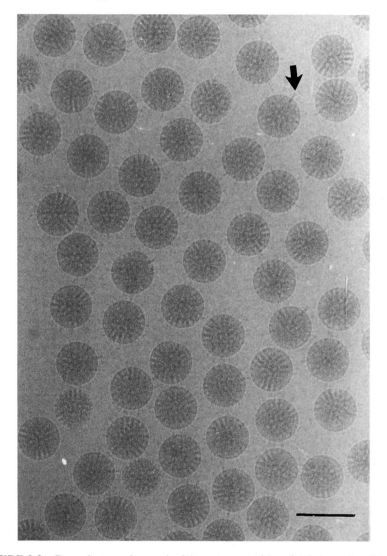

FIGURE 9.3   Cryo–electron micrograph of the mature rotavirions (triple-layered particles). In some of the particles, a characteristic wheel-like appearance is clearly seen. In these images, particles resemble a narrow rim set on short, multiple spokes. Such an appearance is the reason for the name rotavirus (in Latin, *rota* is wheel). One structural feature of the virions that was not observed by conventional negative-stain electron microscopy is that spikes (arrow) emanate from the circular outer margin. The diameter of the particles, excluding the length of the spikes, is 765 Å. Bar, 1000 Å.

ture. This is available because the two-dimensional images are projections that contain complete information of the 3-D density in the particles. This contrasts with structure determination from images of negatively stained specimens, wherein it is possible to obtain only surface features. This concept is demon-

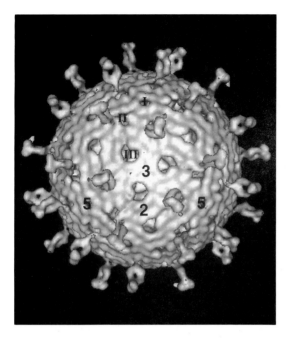

FIGURE 9.4   The 3-D structure of the TLP at a resolution of 28 Å. A set of icosahedral symmetry axes and the locations of the three types of channels are indicated. The surface is a left-handed $T = 13$ icosahedral lattice. In such a lattice, a fivefold axis is reached from its neighboring fivefold axis by stepping over three six-coordinated positions and taking a left turn.

strated in Figure 9.5, which shows a central section through the mature virion, normal to one of the fivefold axes.

Aqueous Channels Are a Distinctive Feature

A distinctive feature of the rotavirus structure is the presence of aqueous channels at all the five- and six-coordinated positions on the $T = 13$ icosahedral lattice (Fig. 9.4). These channels penetrate into the outer two layers of the structure and are about 140 Å deep. They have been classified into three types based on their location with respect to the icosahedral symmetry axes. Type I channels run down the icosahedral fivefold axes, type II are those on the six-coordinated positions surrounding the fivefold axes, and type III are those on the six-coordinated positions neighboring the icosahedral threefold axes. In each virion there are 132 channels: 12 type I, 60 type II, and 60 type III channels. Type II and III channels are about 55 Å wide at the outer surface of the virus. The type I channels, in contrast, have a narrower and more circular opening around 40 Å in diameter. All the channels constrict before widening in the interior. They have their maximum width at the position close to the surface of the inner shell proteins, as seen in the side view of the type III channels in Figure 9.5.

The Virion Outer Layer

Biochemical experiments have shown that the outer shell of the rotavirus contains two proteins, VP7 and VP4 (Estes et al., 1981; Arias et al., 1982). The

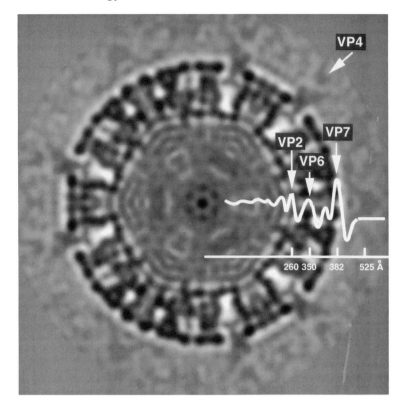

FIGURE 9.5 A central cross section extracted from the 3-D map of the virus. The radial density profile of the structure is overlaid on this section to indicate the radial locations of various layers. The mass density breaks up into three distinct shells between the radii of 210 and 500 Å. These three shells are between the radii 500 and 340 Å, 340 and 270 Å, and 270 and 210 Å. The VP4 spike density is also shown; notice the lower domain of VP4 inside one of the channels.

major component is VP7, a glycoprotein. The protein mass density on the outer shell is distributed uniformly on the local and strict twofold and threefold axes, giving rise to what appears to be a smooth surface. As antibodies specific to VP7 bind to the surface, it is presumed that the VP7 forms the surface (Kapikian and Channock, 1995; B.V.V. Prasad, unpublished results). A close examination of the surface features suggests that VP7 molecules cluster into trimers on the $T = 13$ icosahedral lattice (Fig. 9.4). It is not known whether VP7 molecules aggregate into trimers prior to the assembly of the shell, or if they cluster into trimers upon their assembly into the icosahedral shell. The $T = 13$ icosahedral symmetry of the VP7 layer dictates that each virion has 780 molecules of VP7.

In the immediate vicinity of the channels, the mass density is slightly elevated, giving a craterlike appearance to the channels and depressing the regions between the channels. The depression between type II and type III channels is the most prominent (see Fig. 9.4). This canyonlike feature is about

50 Å wide, 120 Å long, and 20 Å deep. It is not known if there is any functional significance to this structure. This is of interest because the receptor binding domains are located in canyons in other virus systems such as influenza, rhinovirus, and poliovirus (see Chapters 3, 4, and 6).

Surface Spikes

Another distinctive feature of the rotavirus structure is the surface spike. From the surface of the particles, 60 spikes of density extend to a length of 120 Å (Fig. 9.4). The rotavirus spikes are located at the outer edge of the type II channels that surround the icosahedral fivefold axes. The spike has a distinct bilobed structure at the distal end. Each lobe has a diameter of about 25 Å. These lobes are individually connected to rod-shaped densities that are separated from each other by a hole that is approximately 20 Å in diameter. These rod-shaped densities, with a left-handed helical twist, merge together as they approach the surface of the outer layer, making two points of contact.

When the spikes on rotavirus were first discovered, Prasad et al. (1988) predicted that these surface projections were made up of VP4. This prediction was confirmed by immunolabeling studies using MAbs against VP4 (Prasad et al., 1990). Two Fab molecules bound to the sides of the distal bilobes of each spike (Fig. 9.6). The volume of the spike indicates that it is a dimer of VP4, so that each virion contains 120 copies of VP4. Recent higher resolution structural analysis of the virus and in vivo radiolabeling studies have confirmed that the spikes are dimers of VP4 (Shaw et al., 1993). The dimeric nature of the VP4 spike is clearly evident in the higher resolution (~28-Å) structure of rotavirus as shown in Figure 9.4. Gel filtration analyses of expressed VP4 also have provided biochemical evidence that VP4 is an oligomer, likely a

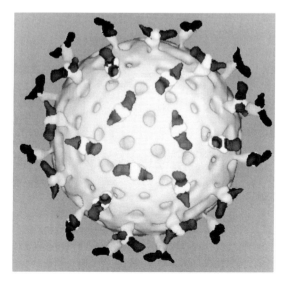

FIGURE 9.6 Surface representation of the 3-D structure of anti-(VP4) Fab-bound triple-layered rotavirus at a resolution of 35 Å, viewed along the icosahedral threefold axis. The Fab molecules are dark and the virus surface is light.

dimer, and interactions between these oligomers are apparently maintained by hydrophobic interactions because these are readily disrupted by detergents (Zhou et al., 1996). This explains why dimers of VP4 have not been detected by simple analysis of VP4 in virus disrupted in SDS and separated by SDS-PAGE.

VP4 Has a Buried Internal Domain

From the 3-D structure of the native virus alone, it was not possible to determine whether there is an inward extension of the VP4 spikes. That is, do the spikes terminate at the surface of the virion or penetrate into the VP7 layer? Based on the volume of VP4 from the 3-D density maps of the native and the Fab-bound virions, Prasad et al. (1990) estimated that about 30 kDa of the VP4 protein is buried inside the virion surface. By computing a difference map between the structures of the native strain and a reassortant strain of rotavirus that lacked spikes, the existence of a large internal domain of VP4 beneath the VP7 surface was confirmed (Shaw et al., 1993). This domain of VP4 is centered on the type II channel in close association with the walls of the channel, made of trimers of VP6 as shown in Figure 9.7 (the structure of the VP6 shell is discussed in The Subcore, later). Similar results were obtained by Yeager et al. (1994) by difference imaging of the virus structure at normal pH and alkaline pH. At alkaline pH, the rotavirus spikes fall off without affecting the integrity of the VP7 layer (Anthony et al., 1991; Yeager et al., 1994). Various views of rotavirus spikes emphasize their asymmetric structure (Fig. 9.8).

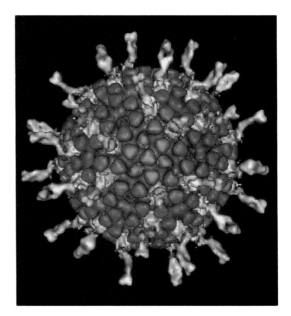

FIGURE 9.7 Interaction of VP4 spikes with the VP6 layer. The VP6 layer of the DLP is dark and the VP4 molecules are light.

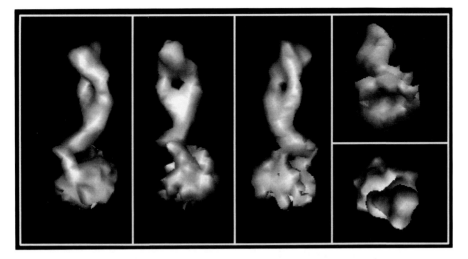

FIGURE 9.8   A close-up view of the isolated VP4 spike as viewed from different angles. *Left.* Side-on views from three different angles. *Right.* Top view, looking down on the spike from the exterior of the virus; Bottom view, looking into the lower globular domain of the spike from the interior of the virus.

The Virion Intermediate Layer

Treatment of intact triple-layered virions with chelating agents (e.g., ethylene diamine tetraacetic acid) removes the outer shell, reduces infectivity, and exposes the inner shell proteins. The resulting DLPs are indistinguishable from those produced in infected cells (Cohen et al., 1979). Electron micrographs of DLPs embedded in vitreous ice (Fig. 9.9) show these particles are 705 Å in diameter, with a bristly surface.

A surface representation of the DLPs (Fig. 9.10) shows the protein mass is mainly concentrated into 260 morphologic units positioned at all the local and strict threefold axes of a $T = 13$ icosahedral lattice. The distribution of protein mass on the surface of the double-layered virion is not as uniform as that observed on the triple-layered virion (Fig. 9.4). The location and the shape of the capsomeres strongly suggest a trimeric clustering of the inner capsid protein VP6. These 260 capsomeric units are arranged in such a way that there are holes or channels at all five- and six-coordinated centers (Fig. 9.10). The 3-D structure of the DLPs has been determined to a resolution of 20 Å using 400-kV electron microscope and spot-scan imaging techniques (B.V.V. Prasad and J. Jakana, unpublished results). The structure of the VP6 molecule in this reconstruction appears to have three domains: the globular upper domain, a lower domain, and a linker domain that connects the upper and the lower domains. The upper globular domain is involved in stabilizing the trimer. The sides of the lower domain seem to be involved in intercapsomeric interactions, and the bottom part of the lower domain interacts with the VP2 layer.

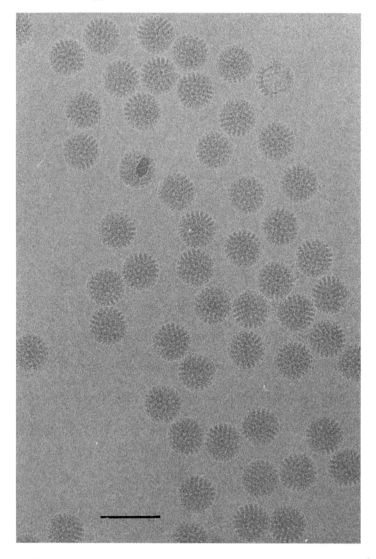

FIGURE 9.9   Cryo–electron micrograph of the rotavirus DLPs. Bar, 1000 Å.

Interactions between Outer and Intermediate Layers

Independent reconstructions of triple- and double-layered virions have made possible the description of the interactions between the outer and inner shell proteins. These interactions between the two shells can be readily seen in Color Plate 13, where the outer shell is partially removed. The interface between the outer and inner shell proteins occurs at the level where the channels begin to widen (Fig. 9.5). The proteins between the outer and inner shells interact in such a way that the channels in these shells are in register. The flat upper

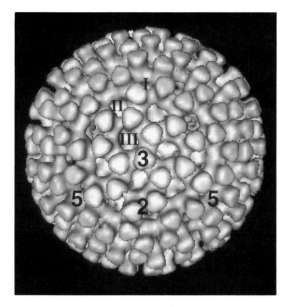

FIGURE 9.10 Surface representation of the 3-D structure of the DLPs at a resolution of 26 Å. A set of icosahedral symmetry axes and the locations of the three types of channels are indicated.

portions of the VP6 trimers protrude into the outer shell such that the close contact between the two shells occurs mainly around the local and strict three-fold axes. The outer shell proteins spread across the spaces between the VP6 trimeric units. Portions of the VP6 trimers are in fact visible through the outer shell channels (see Color Plate 13). It is possible that small molecules or small enzymes may be able to interact with portions of the inner shell (VP6 layer) even in the presence of the outer shell (VP7 layer). Consistent with this idea, at least one MAb to VP6 has been shown to bind to both TLPs and DLPs (Tosser et al., 1994).

The Inner Layer

Although the early protein and electron microscopy analyses defined the composition of the outer two layers of the virion with certainty, the existence and composition of the third shell initially was conjecture. The total mass density in the first shell (i.e., the outer shell) accounts for 780 and 120 molecules of VP7 and VP4, respectively. The mass density in the second shell (i.e., the inner shell) accounts for 780 molecules of VP6. On the basis of the radial density profile, and the fact that VP2 is the most abundant of the remaining three structural proteins, it was proposed that the density between the radii of 230 and 270 Å is due to the shell formed by VP2 molecules (Prasad and Chiu, 1994). The existence of an inner shell was confirmed when SLPs were produced by expression of VP2 alone (Labbé et al., 1991). The volume of this inner shell can accommodate approximately 120 molecules of VP2. This inner shell is a continuous bed of density with small pores surrounding the five-coordinated positions and immediately surrounding the icosahedral threefold axes. Whereas type I and type II channels terminate at this shell, the type III channels continue beyond this shell. The pores around the icosahedral threefold axes are in the proximity of the type III channels, although the two channels are not in register.

The Subcore

Further inside the inner shell, within the radius of 210 Å, we propose that VP1, VP3, and the genome constitute another structural entity called the "subcore." This structure, including the "VP2" shell up to the radius of 270 Å, should be called the SLP to be consistent with the existing terminology (Prasad and Chiu, 1994; Estes, 1995). Further structural studies are required to understand the internal organization of rotavirus, particularly regarding the location of the minor proteins VP1 and VP3. An interesting question is how the 11 segments of the genomic RNA are organized inside the virus. This question can be approached experimentally by studying virus-like particles that contain distinct subsets of the inner proteins. Recent results indicate that a significant portion of the genomic RNA is ordered (Prasad et al., 1996).

Functional Implications of the Rotavirus Structure

Three-dimensional structural studies have been useful in providing a better understanding of the mechanisms underlying various functions of rotaviruses. The following sections discuss functional implications of the virus structure.

Role of VP4 in the Infectivity Process

VP4 is a good candidate for the protein that initiates cell attachment and is involved in viral entry processes, because of its location as a spike that emanates from the surface of virions. Several studies have identified VP4 as the cell attachment protein. For example, the ability of viruses to grow in certain cells is dependent on VP4 (Greenberg et al., 1983; Ramig and Galle, 1990). Binding of virus to cells also is blocked by MAbs to the VP8* region of VP4, whereas MAbs to VP5* and to VP7 do not affect virus binding at concentrations required to neutralize virus infectivity (Ruggeri and Greenberg, 1991). Finally, only those baculovirus-expressed virus-like particles that contain VP4 attach to cells (Crawford et al., 1994).

If VP4 is responsible for viral attachment to cells, does it also play a role in virus entry into cells? The 3-D structural analysis of anti-(VP4) Fab–rotavirus complexes showed that the Fab portion of MAb 2G4 binds to a region close to, but not at, the distal end of the VP4 spike (Prasad et al., 1990). Based on this information, it was proposed that the exposed region of the distal tip of the spike is involved in the initial attachment to cells, and that the region binding to Fab is involved in cell penetration. This prediction has been supported by direct tests of the effects of the neutralizing antibody (2G4) used to prepare the Fabs for the structural studies. This MAb does not block virus binding to cells, but it does inhibit internalization of the virus (Ruggeri and Greenberg, 1991). Thus, it seems likely that if the cell binding domain is on VP4, it is at the distal tip of the spike, and the domain on the sides near the tip is involved in cell penetration. This idea is supported by the similarity of the amino acid sequence of the 2G4 MAb binding domain and the proposed fusion regions in Sindbis and Semliki Forest viruses. Thus this region may be implicated in viral penetration (Mackow et al., 1988). Fusogenic activity of VP4 has yet to be demonstrated, although cleaved VP4 can interact with membranes.

This raises the question: What is the mechanism of protease-enhanced infectivity? First, purified trypsinized particles have been shown to interact with lipid bilayers using liposomes as a model system (Nandi et al., 1992). Rotavirus particles with uncleaved VP4 do not interact with liposomes, but in situ treatment of these particles with trypsin restores their interaction with the liposomes. This interaction is maximum at neutral pH. Virus with cleaved VP4 also interacts with vesicles from intestinal cells (Ruiz et al., 1994). These data support the view that rotavirus enters cells through direct penetration of the plasma membrane, and that cleavage of VP4 is necessary for the interaction of the virus with the membrane (Kaljot et al., 1988). It should be noted that trypsinization does not affect the cell binding of the virus, but facilitates internalization or penetration of the virus. Trypsinized viruses are internalized significantly faster than the nontrypsinized viruses. In addition, the efficiency of rotavirus infection is not significantly affected if trypsinized or nontrypsinized virus is put into cells using lipofectin to bypass the penetration step (Bass et al., 1992).

Is it possible to visualize structural changes in VP4 following trypsinization of virions? Preliminary structural analyses of the trypsinized particles using cryo–electron microscopy, at 24-Å resolution, indicate noticeable structural changes in the spike structure upon trypsinization (A.L. Shaw and B.V.V. Prasad, unpublished results). It is quite possible that trypsinization exposes some regions in VP4 that are critical for virus interactions with cell membranes and consequent internalization.

VP4 Is the Viral Hemagglutinin

VP4 is the viral hemagglutinin based on genetic analyses of reassortants, and the demonstration that VP4, whether expressed alone in insect cells or when bound to virus-like particles, has hemagglutination activity (Kalica et al., 1983; Mackow et al., 1989; Crawford et al., 1994). Monoclonal antibodies to the VP8* region of VP4 also inhibit hemagglutination. Studies mapping the hemagglutination activity of different constructs of VP4 have indicated that the hemagglutinating domain is located between amino acids 92 and 208 (Mackow et al., 1990; Lizano et al., 1991; Zhou et al., 1994; Fuentes-Panana et al., 1995) and is reasonably large (Mackow et al., 1990).

Most of the data on rotavirus hemagglutinin, cell binding, and entry come from studies of animal rotaviruses. Hemagglutination activity is reduced by neuraminidase treatment of red blood cells, indicating a role for sialic acid in virus attachment to red blood cells. However, binding to sialic acid is not an essential step for the infectivity of animal rotaviruses, because it is possible to select sialic–acid resistant mutants that retain full infectivity (Mendez et al., 1994). Knowledge of the sites of amino acid substitutions in these mutants will be important to map the location of the sialic acid binding domain. Possibly this domain will correspond to the hemagglutination domain in VP8* (Mackow et al., 1989; Fiore et al., 1991; Zhou et al., 1994), while cell binding and internalization of rotavirus involve a distinct binding site within VP5*. This explanation would help resolve the question of where the VP8* domain of VP4 is located.

To date, structural studies have not localized VP8*, but MAbs to the VP8* region of VP4 block hemagglutination and cell binding (Mackow et al., 1990; Ruggeri and Greenberg, 1991; Zhou et al., 1994). Because the sides near the tip of the spike are VP5*, it is difficult to envision that VP8* is at the tip of the spike. Instead, it seems likely that VP8* is located near the surface of the virion. This idea is consistent with the following observations. First, although VP8*-specific MAbs clearly aggregate virus particles, it is very difficult to visualize these immunoglobulin molecules, whereas the immunoglobulin in the VP5*-specific Mab 2G4 is visualized easily (Zhou et al., 1994). Perhaps if VP8* is located closer to the surface of the outer layer, the number of antibody molecules that can bind to the particles may be restricted by the VP4 spikes. Second, binding of some VP8*-specific MAbs results in destabilizing the outer capsid structure. Escape mutants to these MAbs alter binding to VP7 MAbs, indicating that these VP8* epitopes may be located near the interface with VP7 (Zhou et al., 1994). Finally, MAbs to the virion surface glycoprotein VP7 block hemagglutination and cell binding. This inhibitory activity has been attributed to steric hindrance, which is reasonable for those MAbs that bind closer to the surface of the virion (Greenberg et al., 1983). It is more difficult to imagine how VP7 MAbs block cell binding if attachment occurs at the tip of the spike, although interference could occur if the surface binding of the VP7 MAbs changes the tip's conformation. In the absence of a high-resolution structure of VP4, additional knowledge of the sites of the amino acid substitutions of the sialic acid–independent mutants and further structural studies using MAbs to distinct regions of VP4 are needed to address these questions.

Role of VP7 in Cell Entry Process

Although there is increasing evidence that VP4 is the main player in the cell binding and entry process, VP7 may still play a role. For example, it is possible that the main role of VP7 is to confer stability to VP4 spikes and facilitate proper presentation of the functional epitopes of VP4. This idea is supported by a reassortant in which the VP4 and VP7 originate from heterologous parental strains, for which the spikes are destabilized and fall off during purification

(Chen et al., 1992; Shaw et al., 1993). It remains possible that VP7 functions as a second molecule in the cell entry process, and some studies have identified VP7 as the cell attachment protein (Matsuno and Inouye, 1983; Sabara et al., 1985; Fukahara et al., 1988). VP4 may function to bind to sialic acid and to a sialic acid–independent receptor, but the latter process may require interactions with VP7.

Immunogenicity of VP7 and VP4, and Neutralization Mechanisms

The two outer layer proteins, VP4 and VP7, each carry epitopes that induce antibodies, and some of these antibodies possess neutralization activity that may be against homotypic or heterotypic viruses. Although both proteins induce neutralizing antibodies, the importance of VP4 in neutralization was not appreciated initially, in part because the majority of high-avidity antibodies in hyperimmune antisera made in animals recognize VP7. Thus, early studies that sought to characterize differences in virus strains based on reactivities with hyperimmune antisera detected differences based on VP7. It is now clear that several distinct epitopes exist on VP4 and VP7 molecules, and classification of rotaviruses into serotypes is based on these epitopes.

Studies with neutralizing and nonneutralizing MAbs specific to VP8* and VP5* regions have been useful to define and map epitopes and to try to understand the mechanism by which neutralization occurs. The VP8* region contains predominantly type-specific sites, whereas cross-reactive sites have been localized on VP5*. Some of the MAbs directed against VP8* neutralize virus infectivity by mechanisms other than simply interfering with the cell entry processes. One such mechanism is to disorder or dislocate the spikes by interfering with the stabilizing interactions between VP4 and VP6 and/or VP4 and VP7. Such a mechanism has been suggested by biochemical studies with the VP8* MAbs 10G6 and 7G6 (Zhou et al., 1994). Structural studies using these MAbs may prove valuable in understanding these novel neutralization mechanisms and regions of interactions between the dimers of VP4. MAbs directed against VP5* have shown little or no effect on cell binding, although some of these MAbs prevent internalization of virus. As already discussed, studies with the animal rotaviruses suggest that the VP8* region is involved in the initial attachment to sialic acid, while the VP5* region may be involved in cell penetration.

Increasing evidence indicates that neutralization by some anti-VP7 MAbs is caused by interfering with the functions of VP4. Thus MAbs bound to VP7 on the virion surface may render some of the functional sites on VP4 inaccessible. Ruggeri and Greenberg (1991) studied several VP7- and VP4-specific antibodies that neutralize RRV, and found that most of the VP7- and VP5*-specific MAbs, in contrast to VP8*-specific MAbs, did not elute cell-bound virus, although they could efficiently neutralize the virus. VP7- and VP5*-specific MAbs interfere with postbinding steps in the virus replication cycle. One general mechanism by which the antibodies may neutralize a virus is by antibody-induced viral aggregation (see Chapter 5). The observation that all

MAbs investigated were able to neutralize RRV after the virus was bound to cells makes this mechanism less likely for rotaviruses (Ruggeri and Greenberg, 1991). Further biochemical and structural studies are necessary to understand the mechanism by which the VP7- and VP5*-specific antibodies neutralize rotavirus.

Uncoating

Following entry into cells, the virus undergoes a process of uncoating. For rotaviruses, the first step in this process is the removal of the outer shell. It is not clear what triggers this process, but it may be initiated by conformational changes in VP4 and VP7 that occurs during the cell entry process. The mechanism used may depend upon the mode of virus entry. If this is by endocytosis, uncoating may be facilitated by lysosomal enzymes. An alternative pathway has been proposed for trypsinized virus (Kaljot et al., 1988), which is presumed to enter by direct penetration as discussed earlier.

It has been hypothesized that a low intracellular calcium level in the cytoplasm of cells is responsible for uncoating of the outer shell. In vitro studies have shown that the stability of the outer shell is calcium dependent (Cohen, 1977; Estes et al., 1979). In vivo experiments have shown that, during the early stages of replication, uncoating of the outer shell can be blocked by calcium ionophores such as A23187 that increase the intracellular calcium levels (Ludert et al., 1986). Uncoating may occur simply by the delivery of the intact virions into the cytoplasm, where the calcium levels are low. Further work is needed to determine where uncoating takes place in the cell and whether specific interactions with cellular proteins are required for this process to proceed.

Particle Integrity Is Necessary for Transcription

Following virus uncoating, the resulting DLPs become transcriptionally active. The exact site of transcription inside the cell is not clear, but it is thought to take place in electron-dense structures, called viroplasms, which are seen adjacent to the nucleus and ER (Fig. 9.2). During transcription, the genomic dsRNA molecules are transcribed and the nascent mRNA molecules emerge from the DLPs, which remain intact and contain all the necessary enzymatic activities required to carry out transcription. This enzymatic complex is quite stable, and milligram quantities of mRNA will be synthesized in vitro providing the precursors are replenished and the particulate complex is occasionally pelleted to remove the buildup of inhibitory metabolites. The transcription process may simply release the templates from structural constraints, allowing them to move past the transcriptase catalytic site.

From the 3-D structure of the TLPs and DLPs, it has been proposed that the channels seen at the five- and six-coordinated positions of the $T = 13$ lattice provide pathways for entry of metabolites and the exit of newly synthesized RNA transcripts (Prasad et al., 1988; Prasad and Chiu, 1994). An interesting question is whether the mRNA exits selectively from one of the three types of channels. If that occurs, a conformational change must occur in the particles because the channels are not in register to provide a continuous path into the core. Three-dimensional structural studies on the in vitro–stimulated transcribing particles, together with the structural information obtained using difference imaging with baculovirus-expressed particles containing different subsets of VP1, VP2, VP3, and VP6, may provide better insight into the structural mechanisms of transcription.

Replication

The transcribed positive-sense RNA molecules act as templates for the synthesis of negative-strand RNA molecules and the viral proteins. The dsRNA molecules are presumed to be synthesized inside replicase particles. Studies of replicase particles from infected cells have indicated that the nonstructural proteins, particularly NSP1, NSP2, and NSP3, play a role during the replication process (Gallegos and Patton, 1989). The inability to purify these replicase particles in large enough amounts has hampered their structural characterization.

The precise role of the nonstructural proteins in the replication process remains unclear. The nonstructural proteins present in the previously studied replicase complexes may perform other functions, such as helping the mRNAs to be transported within cells to the appropriate sites for replication (Poncet et al., 1993; Estes, 1995). Studies have established a template-dependent replication system using only viral core proteins, or baculovirus-expressed virus-like particles, and exogenously added purified mRNAs (Chen et al., 1994). These studies using a cell-free system are exciting because they open the possibility of carrying out structural studies to obtain better insights into the mechanisms underlying the replication process for the rotaviruses.

Assembly

Biochemical and structural studies on baculovirus-expressed virus-like particles have clearly shown that VP2 is required for the proper assembly of the other structural proteins (Rothnagel et al., 1994; Zeng et al., 1994). VP7, which forms the outer layer, or VP6, which forms the intermediate layer, by themselves cannot assemble into the homogeneous structures seen in the native virus. However, together with VP2, these proteins assemble into structures very similar to those in the native virus. These studies show that VP2 provides a proper structural framework for the assembly of VP6 and the consequent assembly of VP4 and VP7. It is thought that the assembly of DLPs in vivo takes place in the electron-dense compartment called the viroplasm. Open questions, not yet answered, concern the role of VP1 and VP3 in the encapsidation of the RNA segments, and whether host cell proteins play specific roles in the assembly process.

One unique and important step in the morphogenesis of rotaviruses is the budding of the progeny DLPs through the rough ER membrane. NSP4, the transmembrane viral glycoprotein, functions as an intracellular ER receptor (Chan et al., 1988; Au et al., 1989; Bergmann et al., 1989; Meyer et al., 1989). Biochemical evidence suggests that NSP4 functions either as a tetramer or as a dimer, though a tetramer is favored. Evidence indicates that the cytoplasmic C terminus of NSP4 has a binding site for both VP6 and VP4 (Meyer et al., 1989; Au et al., 1993), and specific sets of VP6 and NSP4 may be required for efficient virus assembly (Lopez and Arias, 1993). One question about morphogenesis has been whether VP4 is virion bound before particles bud into the ER. Several lines of evidence show that this is the case. First, 3-D structural studies have shown that VP4 has a large globular domain that interacts with VP6 (Shaw et al., 1993). In addition, particles consisting of VP2/4/6 can be isolated from insect cells coexpressing these proteins (Crawford et al., 1994). Further study is required to prove that VP4 is added onto DLPs before these particles bud into the ER. Assembly of VP4 prior to the interaction of DLPs with the ER membrane may have a functional significance. The internal globular domain of VP4, which plugs one of the three types of channels in the DLPs, may arrest any transcriptional activity in the progeny DLPs.

Furthermore, the membrane interacting ability of VP4 may have a role in facilitating the budding of DLPs through the ER membrane or in the removal of transient membrane subsequent to budding. One hypothesis is that VP4 and NSP4, alone or together, participate in the budding process. Further 3-dimensional structural studies may provide answers to the following questions:

1. Where is the binding site for NSP4 on the DLPs?
2. What is the oligomeric state of virion-bound NSP4?
3. Can VP4 stably interact with DLPs?
4. Does NSP4 interact with viral-bound VP4?

These structural analyses seem feasible because both VP4 and NSP4 can be expressed and purified, and expressed VP4 is dimeric (Zhou et al., 1996).

One of the unique features in rotavirus morphogenesis is the formation of transiently enveloped particles inside the ER. How the mature viruses get rid of the envelope is not understood, but this may involve another function of NSP4, which has been shown to have membrane activity based on its ability to release fluorescent dye entrapped in liposomes (Tian et al., 1996). Calcium levels in cells are important during this final stage of virus assembly because this stage is inhibited if cells are put into media lacking calcium. NSP4 again plays an important role in this process because it affects intracellular calcium homeostasis and mobilized calcium from the ER (Tian et al., 1995). It is suspected that imbalances in the intra-ER calcium levels caused by NSP4 ultimately lead to cell death (Tian et al., 1994).

Antiviral Strategies

In principle, antiviral strategies could be developed to block any of the stages in the rotavirus replication cycle. To date, however, relatively few inhibitors have been developed or tested. Several studies have reported that mucin, protease inhibitors, and human milk can reduce infectivity in cultured cells or animals (Ebina et al., 1985; Yolken et al., 1987; Vonderfecht et al., 1988; Ebina and Tsukada, 1991). Most of these inhibitors are thought to inhibit the early stages of virus replication by preventing viral attachment or virus penetration, although the precise mechanisms are not known and their efficacy remains to be clearly established. A key question relating to the effectiveness of any of these inhibitors of the early steps in replication is whether the virus will rapidly mutate and simply evolve resistant mutants.

As is the case for all infectious diseases, the development of an effective vaccine for rotaviruses will have the largest impact on the morbidity and mortality caused by these viruses. Success with this goal has not come easily, yet our new structural knowledge of these viruses is helping these efforts immensely, from the standpoint both of understanding responses in children given vaccines and of developing better vaccines. For example, virus-like particles composed of the structural proteins that make up the three layers in particles (VP2/4/6) and others with spikes (VP2/4/6/7) have been made and shown to be structurally stable. Synthesis of such virus-like particles with a glycoprotein with broadly reactive neutralization epitopes has enabled neutralizing antibody to more than one serotype of virus to be produced (Conner et al., 1995). While vaccines continue to be tested, virus transmission can be minimized by use of inactivating agents such as ethanol and solutions of chelating agents or high and low pH treatments, both of which remove the outer capsid.

Comparison of Rotavirus Structure with Other Viruses of *Reoviridae*

Three-dimensional structural analyses, using cryo−electron microscopy and computer image reconstruction procedures, have been carried out on other members of the *Reoviridae* family. These include bluetongue virus (BTV) in the *Orbivirus* genus (Hewat et al., 1992b; Prasad et al., 1992) and several strains of reovirus of the *Reovirus* genus (Metcalf et al., 1993; Dryden et al., 1993). The overall organization is similar among these viruses; these structures are organized on a left-handed $T = 13$ icosahedral lattice and they all contain multiple protein layers, multiple dsRNA genome segments, and endogenous transcriptase activity, although the numbers of genomic RNA segments and proteins differ between these viruses.

Structurally, the proteins of reovirus are organized into two layers, and there are no morphologic similarities of any layer between reovirus and rotavirus, or reovirus and BTV. However, the structural proteins in BTV are organized into three concentric layers similar to rotavirus. The structural organization of the VP6 and VP2 layers in rotavirus compares well with the organization of the corresponding VP7 and VP3 layers, respectively, in BTV. Not unexpectedly, with these similarities, it has been possible to reconstitute all the protein layers, individually and together, by coexpression of appropriate combinations of structural proteins in both rotavirus and BTV (French and Roy, 1990; French et al., 1990; Labbé et al., 1991; Roy, 1993; Crawford et al., 1994). Structural and biochemical studies involving baculovirus-expressed proteins and their assemblies reinforce the idea that the assembly strategies, at least of the inner layer proteins, and the structural mechanism of endogenous transcription in rotavirus and BTV are likely to be similar (Hewat et al., 1992a, 1992b; Rothnagel et al., 1994).

An interesting question is why these viruses are so complex. The complex nature of the structural organization may have evolved as a direct consequence of the requirement for transcribing multiple dsRNA genome segments within a particulate enzymatic complex. The inner layer proteins provide the necessary architecture to protect and transcribe the genomic RNA, whereas the outer layer proteins facilitate interaction with the cell and internalization. By interacting with the internal proteins and the outer layer proteins, the VP6 layer (or VP7 layer in BTV) integrates the two principal functions, cell entry and transcription, in the intact virus. A high-resolution structure of VP7 of BTV is now available (Grimes et al., 1995) and has been useful in understanding the domain organization of this protein. It is expected that the fitting of the atomic-resolution BTV VP7 structure to the overall structure of the virus will facilitate understanding of how the three capsid layers interact. Because of similarities with rotavirus VP6, it is expected that the BTV VP7 structure will help in understanding the cognate rotavirus protein.

Conclusions and Future Challenges

Structural studies on rotaviruses in particular, and on reoviruses in general, have helped provide a foundation for our understanding of some of the functions of these viruses. The studies described here are just the beginning of the chapter on structure–function relationships in these complex and large viruses. There are several questions that must be answered. Future studies and those in progress are aimed at answering such questions as: Are there specific receptors for rotaviruses? How does the transcription take place inside intact DLPs? How does the virus encapsidate a correct set of the genome segments? What are the roles of the nonstructural proteins in virus replication and self-assembly? What is the molecular mechanism of budding of the progeny DLPs into the ER membrane? What is the role of NSP4 in cell lysis? We anticipate that a more complete picture of how these viruses replicate will emerge as the structural analyses improve in resolution, either by cryo–electron microsopy or by X-ray crystallography, in conjunction with advances in biochemical and molecular biologic studies. An important step will be to establish a reverse genetics system for rotavirus. The goal of all these studies is to obtain detailed molecular and structural information so as to allow the development of more effective strategies to combat, prevent, or treat the clinical outcome of infections with these viruses. Our increasing knowledge of the structure and function of the rotavirus genes and proteins is already being exploited to change the course of infection or disease transmission.

Acknowledgments

Our work is supported in part by grants from the National Institutes of Health (AI 36040 and DK 31044), the W.M. Keck Foundation, and the National Center for Research Resources (RR 02250).

We thank the editors of this book for helpful discussion and critical comments. We thank A.L. Shaw, R. Rothnagel, and J. Lawton for helpful discussions, comments on this chapter, and their help in making the figures.

References

Anthony, I.D., Bullivant, S., Dayal, S., Bellamy, A.R. and Berriman, J.A. 1991. Rotavirus spike structure and polypeptide composition. J. Virol. 65: 4334–4340.

Arias, C.F., Lopez, S. and Espejo, R.T. 1982. Gene protein products of SA11 simian rotavirus genome. J. Virol. 41: 42–50.

Au, K.-S., Chan, W.K., Burns, J.W. and Estes, M.K. 1989. Receptor activity of rotavirus nonstructural glycoprotein NS28. J. Virol. 63: 4553–4562.

Au, K.-S., Mattion, N.M. and Estes, M.K. 1993. A subviral particle binding domain on the rotavirus nonstructural glycoprotein NS28. Virology 194: 665–673.

Bass, D.M., Baylor, M.R., Chen, C., Mackow, E.M., Bremont, M. and Greenberg, H.B. 1992. Liposome-mediated transfection of intact viral particles reveals that plasma membrane penetration determines permissivity of tissue culture cells to rotavirus. J. Clin. Invest. 90: 2313–2320.

Bellamy, A.R. and Both, J. 1990. Molecular biology of rotaviruses. Adv. Virus Res. 38: 1–43.

Bergmann, C.C., Maass, D., Poruchynsky, M.S., Atkinson, P. and Bellamy, A.R. 1989. Topology of the nonstructural rotavirus receptor glycoprotein NS28 in the rough endoplasmic reticulum. EMBO J. 8: 1695–1703.

Bishop, R.F., Davidson, G.P., Holmes, I.H. and Ruck, B.J. 1973. Virus particles in epithelial cells of duodenal mucosa from children with viral gastroenteritis. Lancet 2: 1281–1283.

Blacklow, N.R. and Greenberg, H.B. 1991. Viral gastroenteritis. N. Engl. J. Med. 325: 252–264.

Chan, W.K., Au, K.S. and Estes, M.K. 1988. Topography of the simian rotavirus nonstructural glycoprotein (NS28) in the endoplasmic reticulum membrane. Virology 164: 435–442.

Chen, D., Burns, J.W., Estes, M.K. and Ramig, R.F. 1989. Phenotypes of rotavirus reassortants depend upon the recipient genetic background. Proc. Natl. Acad. Sci. U.S.A. 86: 3743–3747.

Chen, D., Estes, M.K. and Ramig, R.F. 1992. Specific interactions between rotavirus outer capsid proteins VP4 and VP7 determine expression of a cross-reactive, neutralizing VP4-specific epitope. J. Virol. 66: 432–439.

Chen, D. and Ramig, R.F. 1992. Determinants of rotavirus stability and density during CsCl purification. Virology 186: 228–237.

Chen, D., Zeng, C.Q., Wentz, M.J., Gorziglia, M., Estes, M.K. and Ramig, R.F. 1994. Template-dependent, *in vitro* replication of rotavirus RNA. J. Virol. 68: 7030–7039.

Clark, B., Roth, J.R., Clark, M.L., Barnett, B.B. and Spendlove, R.S. 1981. Trypsin enhancement of rotavirus infectivity mechanisms of enhancement. J. Virol. 39: 816–822.

Cohen J. 1977. Ribonucleic acid polymerase activity associated with purified calf rotavirus. J. Gen. Virol. 36: 395–402.

Cohen, J., Charpillienne, A., Chilmonczyk, S. and Estes, M.K. 1989. Nucleotide sequence of bovine rotavirus gene 1 and expression of gene product in baculovirus. Virology 171: 131–140.

Cohen, J., Laporte, J., Charpilienne, A. and Scherrer, R. 1979. Activation of rotavirus RNA polymerase by calcium chelation. Arch. Virol. 60: 177–186.

Conner, M.E., Crawford, S.E., Barone, S.E., Zhou, Y.J. and Estes, M.K. 1995. Induction of heterotypic neutralizing antibodies by serotype G1 VLPs. Presented at the 5th International Symposium on dsRNA Viruses, Jerba, Tunisia.

Crawford, S.E., Labbé, M., Cohen, J., Burroughs, M.H., Zhou, Y. and Estes, M.K. 1994. Characterization of virus-like particles produced by the expression of rotavirus capsid proteins in insect cells. J. Virol. 68: 5945–5952.

Crawford, S.E., Zeng, C.Q-Y., Zhou, Y.-J. and Estes, M.K. 1995b. Analysis of structural and functional domains of rotavirus spike-protein VP4. Presented at the 5th International Symposium on dsRNA Viruses, Jerba, Tunisia.

Davidson, G.P. and Barnes, G.L. 1979. Structural and functional abnormalities of the small intestine in infants and young children with rotavirus enteritis. Acta Paediatr. Scand. 68: 181–186.

Dryden, K.A., Wang, G.J., Yeager, M., Nibert, M.L., Coombs, K.M., Furlong, D.B., Fields, B.N. and Baker, T.S. 1993. Early steps in reovirus infection are associated with dramatic changes in supramolecular structure and protein conformation: analysis of virions and subviral particles by cryoelectron microscopy and image reconstruction. J. Cell Biol. 122: 1023–1041.

Ebina, T., Sato, A., Umezu, K., Ishida, N., Ohyama, S., Oizumi, A., Aikawa, K., Katagiri, S., Katsushima, N., Imai, A. et al. 1985. Prevention of rotavirus infection by oral administration of cow colostrum containing antihuman rotavirus antibody. Med. Microbiol. Immunol. (Berl.) 174: 177–185.

Ebina, T. and Tsukada, K. 1991. Protease inhibitors prevent the development of human rotavirus-induced diarrhea in suckling mice. Microbiol. Immunol. 35: 583–588.

Espejo, R.T., Lopez, S. and Arias, C. 1981. Structural polypeptides of simian rotavirus SA11 and the effect of trypsin. J. Virol. 37: 156–160.

Estes, M.K. 1996. Rotaviruses and their replication. In Virology (Fields, B.N., Knipe, D.M., Chanock, R.M., Melnick, J.L., Roizman, B. and Hope, R.E., Eds.), pp. 1625–1655, Raven Press, New York.

Estes, M.K. and Cohen, J. 1989. Rotavirus gene structure and function. Microbiol. Rev. 53: 410–449.

Estes, M.K., Graham, D.Y. and Mason, B.B. 1981. Proteolytic enhancement of rotavirus infectivity: molecular mechanisms. J. Virol. 39: 879–888.

Estes, M.K., Graham, D.Y., Smith, E.M. and Gerba, C.P. 1979. Rotavirus stability and inactivation. J. Gen. Virol. 43: 403–409.

Fiore, L., Greenberg, H.B. and Mackow, E.R. 1991. The VP8 fragment of VP4 is the rhesus rotavirus hemagglutinin. Virology 181: 553–563.

French, T.J., Marshall, J.J. and Roy, P. 1990. Assembly of double-shelled, viruslike particles of bluetongue virus by the simultaneous expression of four structural proteins. J. Virol. 64: 5695–5700.

French, T.J. and Roy, P. 1990. Synthesis of bluetongue virus (BTV) corelike particles by a recombinant baculovirus expressing the two major structural core proteins of BTV. J. Virol. 64: 1530–1536.

Fuentes-Panana, E.M., Lopez, S., Gorziglia, M. and Arias, C.F. 1995. Mapping the hemagglutination domain of rotaviruses. J. Virol. 69: 2629–2632.

Fukahara, N., Yoshie, O., Kitaoka, S. and Konno, T. 1988. Role of VP3 in human rotavirus after target cell attachment via VP7. J. Virol. 62: 2209–2218.

Gallegos, C.O. and Patton, J.T. 1989. Characterization of rotavirus replication intermediates: a model for the assembly of single-shelled particles. Virology 172: 616–627.

Graham, D.Y. and Estes, M.K. 1980. Proteolytic enhancement of rotavirus infectivity: biologic mechanisms. Virology 101: 432–439.

Greenberg, H.B., Kalica, A.R., Wyatt, R.G., Jones, R.W., Kapikian, A.Z. and Chanock, R.M. 1981. Rescue of noncultivatable human rotavirus by gene reassortment during mixed infection with ts mutants of a cultivatable bovine rotavirus. Proc. Natl. Acad. Sci. U.S.A. 78: 420–424.

Greenberg, H.B., Valdesuso, J., van Wyke, K., Midthun, K., Walsh, M., McAuliffe, V., Wyatt, R.G., Kalica, A.R., Flores, J. and Hoshino, Y. 1983. Production and preliminary characterization of monoclonal antibodies directed at two surface proteins of rhesus rotavirus. J. Virol. 47: 267–275.

Grimes, J., Basak, A.K., Roy, P. and Stuart, D. 1995. The crystal structure of bluetongue virus VP7. Nature 373: 167–170.

Hall, G.A. 1987. Comparative pathology of infection by novel diarrhoea viruses. CIBA Found. Symp. 128: 192–217.

Hewat, E.A., Booth, T.F., Loudon, P.T. and Roy, P. 1992a. 3-D reconstruction of baculovirus-expressed bluetongue virus core-like particles by cryo-electron microscopy. Virology 189: 10–20.

Hewat, E.A., Booth, T.F. and Roy, P. 1992b. Structure of bluetongue virus particles by cryoelectron microscopy. J. Struct. Biol. 109: 61–69.

Kalica, A.R., Flores, J. and Greenberg, H.B. 1983. Identification of the rotaviral gene that codes for hemagglutination and protease-enhanced plaque formation. Virology 125: 194–205.

Kaljot, K.T., Shaw, R.D., Rubin, D.H. and Greenberg, H.B. 1988. Infectious rotavirus enters cells by direct membrane penetration, not endocytosis. J. Virol. 62: 1136–1144.

Kapikian, A.Z. and Chanock, R.M. 1995. Rotaviruses. In Virology (Fields, B.N., Knipe, D.M., Chanock, R.M., Melnick, J.L., Roizman, B. and Hope, R.E., Eds.), pp. 1657–1708. Raven Press, New York.

Labbé, M., Charpilienne, A., Crawford, S.E., Estes, M.K. and Cohen, J. 1991. Expression of rotavirus VP2 produces empty corelike particles. J. Virol. 65: 2946–2952.

Lizano, M., Lopez, S. and Arias, C.F. 1991. The amino-terminal half of rotavirus SA11 4fM VP4 protein contains hemagglutinin domain and primes for neutralizing antibodies to the virus. J. Virol. 65: 1383–1391.

Lopez, S. and Arias, C.F. 1993. Sequence analysis of rotavirus YM VP6 and NS28 proteins. J. Gen. Virol. 74: 1223–1226.

Ludert, J.E., Gil, F., Liprandi, F. and Esparza, J. 1986. The structure of rotavirus inner capsid studied by electron microscopy of chemically disrupted particles. J. Gen. Virol. 67: 1721–1725.

Mackow, E.R., Barnett, J.W., Chan, H. and Greenberg, H.B. 1989. The rhesus rotavirus outer capsid protein VP4 functions as a hemagglutinin and is antigenically conserved when expressed by a baculovirus recombinant. J. Virol. 63: 1661–1665.

Mackow, E.R., Shaw, R.D., Matsui, S.M., Vo, P.T., Dang, M.N. and Greenberg, H.B. 1988. Characterization of the rhesus rotavirus VP3 gene: location of amino acids involved in homologous and heterologous rotavirus neutralization and identification of a putative fusion region. Proc. Natl. Acad. Sci. U.S.A. 85: 645–649.

Mackow, E.R., Yamanaka, M.Y., Dang, M.N. and Greenberg, H.B. 1990. DNA amplification-restricted transcription-translation: rapid analysis of rhesus rotavirus neutralization sites. Proc. Natl. Acad. Sci. U.S.A. 87: 518–522.

Matsuno, S. and Inouye, S. 1983. Purification of an outer capsid glycoprotein in neonatal calf diarrhea virus and preparation of its antisera. Infect. Immun. 39: 155–158.

Mattion, N.M., Cohen, J., Aponte, C. and Estes, M.K. 1992. Characterization of an oligomerization domain and RNA-binding properties on rotavirus nonstructural protein NS34. Virology 190: 68–83.

Mattion, N.M., Cohen, J. and Estes, M.K. 1994. The rotavirus proteins. In Viral Infections of the Gastrointestinal Tract (Kapikian, A., Ed.), pp. 169–249. Marcel Dekker, New York.

Mendez, E., Arias, C.F. and Lopez, S. 1994. Binding of sialic acids is not an essential step for the entry of animal rotaviruses to epithelial cells in culture. J. Virol. 67: 5253–5259.

Metcalf, P., Cyrklaff, M. and Adrian, M. 1991. The 3-D structure of reovirus obtained by cryo-electron microscopy. EMBO J. 10: 3129–3136.

Meyer, J.C., Bergmann, C.C. and Bellamy, A.R. 1989. Interaction of rotavirus cores with the nonstructural glycoprotein NS28. Virology 171: 98–107.

Nandi, P., Charpilienne, A. and Cohen, J. 1992. Interaction of rotavirus particles with liposomes. J. Virol. 66: 3363–3367.

Osborne, M.P., Haddon, S.J., Spencer, A.J., Collins, J., Starkey, W.G., Wallis, T.S., Clarke, G.J., Worton, K.J., Candy, D.C. and Stephen, J. 1988. An electron microscopic investigation of time-related changes in the intestine of neonatal mice infected with murine rotavirus. J. Pediatr. Gastroenterol. Nutr. 7: 236–248.

Poncet, D., Aponte, C. and Cohen. J. 1993. Rotavirus protein NSP3 (NS34) is bound to the 3' end consensus sequence of viral mRNAs in infected cells. J. Virol. 67: 3159–3165.

Prasad, B.V.V., Burns, J.W., Marietta, E., Estes, M.K. and Chiu, W. 1990. Localization of VP4 neutralization sites in rotavirus by 3-D cryo-electron microscopy. Nature 343: 476–478.

Prasad, B.V.V. and Chiu, W. 1994. Structure of rotaviruses. In Rotaviruses (Ramig, R., Ed.), pp. 9–29. Springer-Verlag, Berlin.

Prasad, B.V.V., Rothnagel, R., Zeng, Q., Jakana, J., Lawton, J.A., Chiu, W. and Estes, M.K. 1996. Visualization of ordered genomic RNA and localization of transcriptional complexes in rotavirus. Nature 382: 471–473.

Prasad, B.V.V., Wang, G.J., Clerx, J.P.M and Chiu, W. 1987. Cryo-electron microscopy of spherical viruses: an application to rotavirus. Micron and Microscop. Acta 18: 327–331.

Prasad, B.V.V., Wang, G.J., Clerx, J.P.M. and Chiu, W. 1988. 3-D structure of rotavirus. J. Mol. Biol. 199: 269–275.

Prasad, B.V.V., Yamaguchi, S. and Roy, P. 1992. 3-D structure of single-shelled blue tongue virus. J. Virol. 66: 2135–2142.

Ramig, F.R. and Galle, K.L. 1990. Rotavirus genome segment 4 determines viral replication phenotype in cultured liver cells (HepG2). J. Virol. 64: 1044–1049.

Ramig, R.F. and Ward, R.L. 1991. Genomic segment reassortment in rotavirus and other Reoviridae. Adv. Virus Res. 39: 164–207.

Rothnagel, R., Zeng, Q., Estes, M.K., Chiu, W. and Prasad, B.V.V. 1994. 3-D structural studies on baculovirus-expressed rotavirus sub-assemblies. Presented at the 1994 Biophysical Society Annual Meeting.

Roy, P. 1993. Dissecting the assembly of orbiviruses. Trends Microbiol. 1: 299–305.

Ruggeri, F.M. and Greenberg, H.B. 1991. Antibodies to the trypsin cleavage peptide VP8* neutralize rotavirus by inhibiting binding of virions to target cells in culture. J. Virol. 65: 2211–2219.

Ruiz, M., Alonso-Torre, S.R., Charpilienne, A., Vasseur, M., Michelangeli, F., Cohen, J. and Alvarado, F. 1994. Rotavirus interaction with isolated membrane vesicles. J. Virol. 68: 4009–4016.

Sabara, M., Gilchrist, J.E., Hudson, G.R. and Babiuk, L.A. 1985. Preliminary characterization of an epitope involved in neutralization and cell attachment that is located the major bovine rotavirus glycoprotein. J. Virol. 53: 58–66.

Saif, L.J., Rosen, B.I. and Parwani, A.V. 1994. Animal rotaviruses. In Viral Infections of the Gastrointestinal Tract (Kapikian, A.Z., Ed.), pp. 279–367. Marcel Dekker, New York.

Shaw, A.L., Rothnagel, R., Chen, D., Ramig, R.F., Chiu, W. and Prasad, B.V.V. 1993. Three-dimensional visualization of the rotavirus hemagglutinin structure. Cell 74: 693–701.

Stephen, J. 1989. Functional abnormalities in the intestine. In Viruses and the Gut (Farthing, M.J.G., Ed.), pp. 41–44. Swan Press Ltd., London.

Tian, P., Ball, J.M., Zeng, Q. and Estes, M.K. 1996. The rotavirus non-structural glycoprotein NSP4 possesses membrane destabilization activity. J. Virol. 70: 6973–6981.

Tian, P., Estes, M.K., Hu, Y., Ball, J.M., Zeng, C.Q.-Y. and Schilling, W.P. 1995. The rotavirus nonstructural glycoprotein NSP4 mobilizes Ca^{2+} from the endoplasmic reticulum. J. Virol. 69: 5763–5772.

Tian, P., Hu, Y., Schilling, W.P., Lindsay, D.A., Eiden, J. and Estes, M.K. 1994. The nonstructural glycoprotein of rotavirus affects intracellular calcium levels. J. Virol. 68: 251–257.

Tosser, G., Delaunay, T., Kohli, E., Grosclaude, J., Pothier, P. and Cohen, J. 1994. Topology of bovine rotavirus (RF strain) VP6 epitopes by real-time biospecific interaction analysis. Virology 204: 8–16.

Vonderfecht, S.L., Miskuff, R.L., Wee, S.B., Sato, S., Tidwell, R.R., Geratz, J.D. and Yolken, R.H. 1988. Protease inhibitors suppress the in vitro and in vivo replication of rotavirus. J. Clin. Invest. 82: 2011–2016.

Yeager, M., Berriman, J.A., Baker, T.S. and Bellamy, A.R. 1994. Three-dimensional structure of the rotavirus haemagglutinin VP4 by cryo-electron microscopy and difference map analysis. EMBO J. 13: 1011–1018.

Yeager, M.K., Dryden, K.A., Olson, H.B., Greenberg, H.B. and Baker, T.S. 1990. 3-D structure of rhesus rotavirus by cryoelectron microscopy and image reconstruction. J. Cell. Biol. 110: 2133–2144.

Yolken, R.H., Willoughby, R., Wee, S.B., Miskuff, R. and Vonderfecht, S. 1987. Sialic acid glycoproteins inhibit in vitro and in vivo replication of rotaviruses. J. Clin. Invest. 79: 148–154.

Zeng, Q., Labbé, M., Cohen, J., Prasad, B.V.V., Chen, D., Ramig, R.F. and Estes, M.K. 1994. Characterization of rotavirus VP2 particles. Virology 201: 55–65.

Zhou, Y., Ball, J. and Estes, M.K. 1996. Analysis of structural and functional domains of rotavirus spike-protein VP4. Manuscript in preparation.

Zhou, Y., Burns, J.W., Morita, Y., Tanaka, T. and Estes, M.K. 1994. Localization of rotavirus VP4 neutralization epitopes involved in antibody-induced conformational changes of virus structure. J. Virol. 68: 3955–3964.

10.

Packaging and Release of the Viral Genome

JOHN E. JOHNSON & ROLAND R. RUECKERT

The infectious virus particle must transport a properly packaged genome between cells of a susceptible host and between hosts. Regardless of the complexity of the virus, three components of this particle-mediated genomic transport must be maintained.

1. During assembly the virus must package only its own genome. The mechanism for the protein–nucleic acid recognition event is well understood for some viruses but unclear for many others.
2. The nucleic acid must be condensed and packaged in a highly organized manner to fit within the volume defined by the protein shell in the case of icosahedral particles. At physiologic pH, nucleic acid is negatively charged. To reach levels of condensation required for packaging, the charge must be neutralized either with counter-ions normally present in the cell or by positive charge associated with the capsid subunit or other proteins encoded by the virus.
3. The infectious genome must be packaged so that it can be released in a state in which translation or transcription can occur. The proper structural relationship between the capsid protein components and specific regions of the genome may be required for infectivity.

Data are available for a number of systems that pertain to component 2, and features of condensation in different viruses appear similar regardless of their complexity. Discussion of these data leads naturally to a consideration of specificity of viral gene packaging and its possible dependence on three-dimensional RNA, DNA, and protein structures observed by crystallography. Finally, the productive release of RNA is considered in the context of the "infectosome," with particular emphasis on picornaviruses and nodaviruses, two systems in which enough data have accumulated to warrant informed speculation.

The Packaged Viral Genome

Factors Determining Nucleic Acid Structure

Packaging of double-stranded DNA in bacteriophage has been studied for decades (for a review, see Black, 1989). In phage such as T7, φ29, and P22, DNA packaging is an energy-consuming process, with a molecular pump employed to fill a preformed protein shell. The DNA packaged in phage (Earnshaw et al., 1976) and in herpes simplex virus (Booy et al., 1991) displays 26-Å striations in its structure, which have been interpreted as side-by-side DNA duplexes packed in a latticelike framework to result in a liquid crystalline structure.

In contrast, RNA packaging in rod-shaped (e.g., tobacco mosaic virus [TMV]) and bacilliform-shaped (e.g., alfalfa mosaic virus [ALMV]) viruses is remarkably simple and may be a useful paradigm for examining RNA packaging in simple icosahedral viruses. TMV RNA associates with protein in an equilibrium reaction driven primarily by the increased entropy that is associated with the displacement of water during protein assembly (Lauffer, 1975). Assembly is initiated at a particular sequence in the genome (origin of assembly), which has a high affinity for the coat protein (Zimmern, 1977), and then it proceeds with the addition of oligomeric protein units (one subunit per three oligonucleotides; Butler et al., 1977; Lebeurier et al., 1977). The structure of the packaged RNA is determined entirely by protein–RNA interactions, and it is single stranded (Namba et al., 1989).

Two groups of plant viruses form particles intermediate between the rod-shaped TMV and the more common icosahedral particles. Alfalfa mosaic and isometric labile ringspot (ILAR) viruses coat RNA in a manner reminiscent of TMV, but the intersubunit interactions drive the formation and association of protein hexamers and pentamers rather than simple helices. Associations of this type result in either a cylindrical lattice of hexamers with ends that are closed by half-shells with icosahedral symmetry (Mellema, 1975), or in spheroidal particles probably composed of hexamers and pentamers that do not lie on a quasi-equivalent surface lattice (Johnson and Argos, 1985). Both of these virus groups form particles of a size proportional to the RNA packaged with roughly 13 nucleotides associated with each subunit. In contrast to TMV RNA, there is duplexed RNA in ALMV, and this plays an important role in the initiation of assembly (Zuiedema et al., 1983). Other plant viruses under structural study that are clearly related to ALMV and ILAR in their gene organization, biology, and probably their structures are cucumoviruses and bromoviruses. Both of these virus groups display strict icosahedral symmetry.

Density of Packaged Nucleic Acid in Icosahedral Viruses

In contrast to the helical and bacilliform-shaped viruses, icosahedral viral shells form a rigid container of defined volume for packaging nucleic acid. The relative packing density of RNA or DNA within these shells provides a quantitative means of comparing viruses (see Chapter 1). Crystallographers define a parameter called V_m ($V_{(container)}$ [Å^3]/$M_{r(contents)}$ [daltons]) that describes the volume occupied per dalton of biologic macromolecule (Matthews, 1968). The unit cell of the crystal is the "container" and V_m is an indication of the solvent content in the crystal. A small V_m indicates a small solvent content or high packing density. Protein crystals typically contain approximately 50% solvent and RNA crystals slightly more. Here we use the internal capsid volume as the container and RNA molecular weight as the macromolecular mass. Table 10.1 lists V_m values for some representative viruses and the V_m of a crystalline duplex RNA, $[U(UA)_6A]_2$. For reference, we also compute the V_m of dehydrated RNA. The partial specific volume of dry RNA is 0.55 cm^3/g. Conversion to molecular dimensions with the conversion factor

$$\frac{10^{24} \text{ Å}^3/\text{cm}^3}{6.02 \times 10^{23} \text{ Da/g}}$$

or 1.66 g-Å^3/Da-cm^3 gives $V_m = 0.91$ Å^3/Da, which is the minimum value for RNA.

The minimal volume required for packaging a genome of sufficient size for a replicating virus is readily estimated (see also Chapter 1). A simple RNA

TABLE 10.1 Experimentally Determined Packing Volumes for Virus Particles and RNA Crystals

Virus[a]	RNA M_r ($\times 10^{-6}$)	Volume ($\times 10^{-6}$ Å^3)	V_m (Å^3/Da)
CPMV			
RNA 1	2.02	4.32	2.14
RNA 2	1.22	4.32	3.54
CCMV	1.20	3.59	3.00
SBMV	1.39	4.32	3.11
HRV 14	2.55	4.32	1.69
STNV	0.34	0.97	2.87
U(UA)$_6$A[b]	0.033600	0.073304	2.18

[a]CPMV, cowpea mosaic virus; CCMV, cowpea chlorotic mottle virus; SBMV, southern bean mosaic virus; HRV 14, human rhinovirus 14 (common cold virus); STNV, satellite tobacco necrosis virus.
[b]Crystalline RNA P2$_1$2$_1$2$_1$, $a = 34$ Å, $b = 44$ Å, $c = 49$ Å; four duplex molecules/cell.

virus requires at least two genes, the coat protein gene and a gene to code for a component of a RNA-directed RNA polymerase complex. The former usually encodes a protein of approximately 30 kDa and the latter encodes a protein of approximately 90 kDa. Roughly 4000 nucleotides ($\sim$1.3 $\times$ 10^6 Da) are required to code for these proteins as well as for the generally required 5' and 3' noncoding regions of the genome. Dehydrated RNA of this molecular mass occupies a volume of approximately 1.2 $\times$ 10^6 Å^3, corresponding to a spherical container of radius approximately 65 Å. An icosahedral particle formed of 60 subunits of approximately 30 kDa molecular mass has an interior radius of 60 Å, too small to accommodate the genome of most viruses. To produce a larger cavity, the vast majority of simple RNA viruses have protein shells containing 180 protein subunits arranged either on a quasi-equivalent $T = 3$ lattice or a pseudo quasi-equivalent $P = 3$ (Johnson and Fisher, 1994; see also Chapter 1). These two particle types are described in Figure 10.1. The interior cavity radius of these particles is roughly 100 Å, sufficient to readily accommodate 5000 nucleotides of hydrated RNA and in some cases much higher packing densities. Even assuming that the RNA within the particle is uniformly packed, some of the viruses listed in Table 10.1 have RNA packaged with a density higher than that found in a crystal. Note the exceptionally high packing density of picornaviruses, 1.69 Å^3/Da compared with values greater than 2 Å^3/Da for other viruses.

The size of an animal virus genome is strictly limited by the interior volume of its protein shell. Many plant viruses, however, have multipartite genomes. The genome is split into as many as three parts that are encapsidated in separate particles (all particles are formed of the same subunit type). This strategy allows for much larger genomes; however, it necessitates the cellular entry of particles containing all three genomic segments to initiate an infection. Plant viruses typically cope with this apparent disadvantage by generating an enormous number of particles; thus this requirement is not a disadvantage in practice. Multipartite genomes are known among animal viruses, but they are packaged in the same particle.

Neutralization of Charge in Packaged Nucleic Acid

Neutralization of the polyanionic genome is required for its condensation and packaging. Nucleic acid exists as a sodium or potassium salt in the cell, and these cations are certainly within the virion. Beyond that, many simple virus capsid proteins have a histone-like positively charged amino terminus that is internal and disordered in the structures determined by X-ray crystallography. In many cases this portion of the protein could neutralize up to 50% of the phosphate anions. Such interactions must condense the RNA close to the protein shell and also provide an electrostatic driving force for assembly.

Picornaviruses and plant comoviruses do not have regions of concentrated positive charge in their capsid proteins, but some package substantial quantities of cationic polyamines that can neutralize about half of the RNA. Viruses that do not have a positive charge in the capsid protein often form empty capsids.

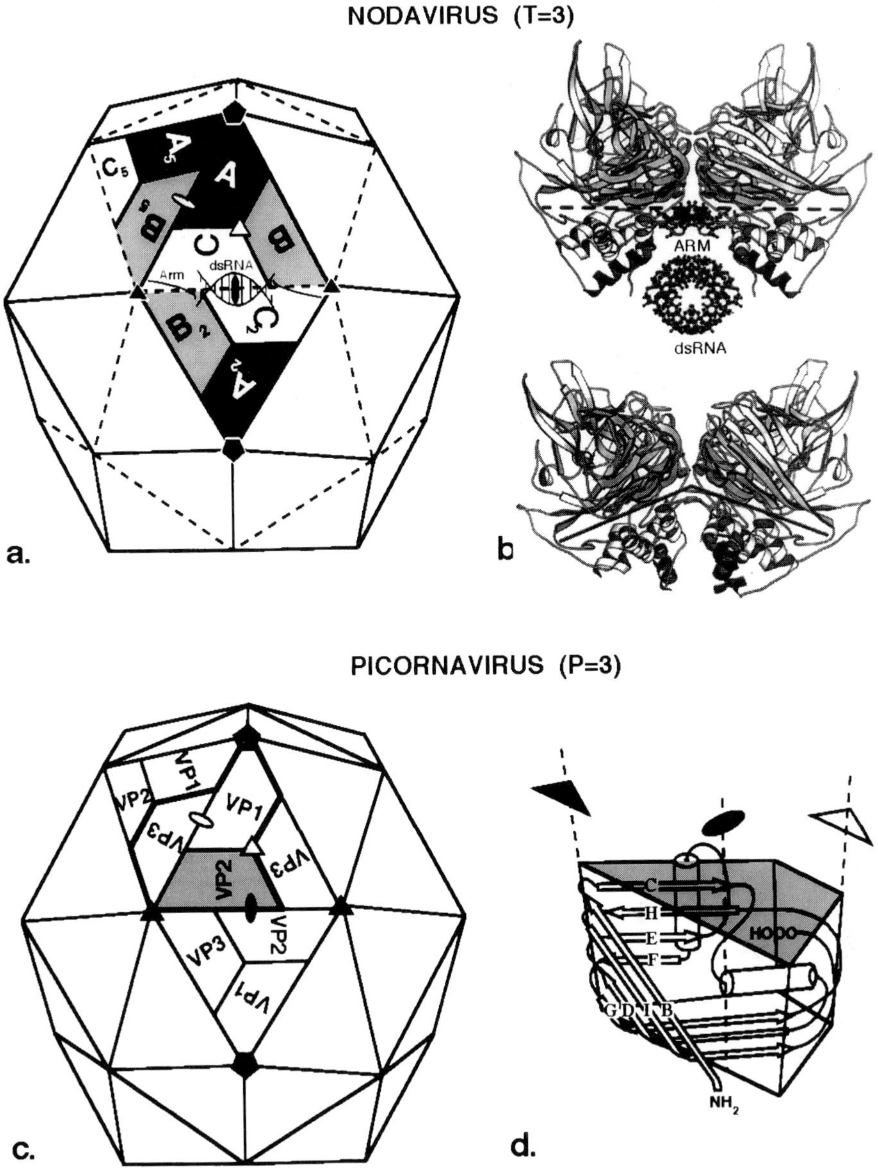

FIGURE 10.1 Comparison of a $T = 3$ nodavirus capsid (a) with a $P = 3$ picornavirus capsid (c). The two capsids are similar in overall shape, size, and organization. The trapezoids labeled A, B, and C in the $T = 3$ shell correspond to the same gene product located in slightly different environments within one icosahedral asymmetric unit (the central triangle of a). The tertiary folds of A, B, and C are virtually identical, but the quasi-equivalent C–C_2 and A–B_5 contacts are flat and bent, respectively (side views shown in b), although the same protein surfaces are juxtaposed. A protein polypeptide (arm) and a portion of duplex RNA (dsRNA) are ordered only at the C–C_2 contact and serve as a wedge to prevent bending. These structures are disordered at the A–B_5 interface, and the contact is bent about an axial hinge that is conserved in both flat and bent contacts. VP1, VP2, and VP3 are different gene products in the picornavirus capsid (c), and therefore the pseudo equivalent VP2–VP2 and VP1–VP3 interfaces have different protein surfaces juxtaposed and there is no need for a molecular switch as observed in the $T = 3$ shell. d. All the trapezoids forming the contiguous shell in both the $T = 3$ and $P = 3$ capsids correspond to an eight stranded β-barrel.

Plant comovirus preparations contain up to 20% empty protein shells, and poliovirus will form empty shells if the cells are exposed to millimolar concentration levels of guanidium hydrochloride during virus growth. The empty particle may serve as a reservoir for excessive coat protein and thereby maintain a favorable subunit:RNA ratio during assembly. It has not been ruled out, however, that an empty particle forms as an intermediate in assembly. The empty and nucleoprotein particles in the cowpea mosaic virus (CPMV) system have been investigated by thermal and high-pressure denaturation (Virudachalam et al., 1985; DaPoian et al., 1994), and it is clear that nucleoprotein components are more stable than the empty capsid.

Bulk Structure of Packaged Nucleic Acid

Solution X-ray scattering at 20-Å resolution reveals packaged viral RNA as a sphere of roughly uniform density (e.g., Schmidt et al., 1983). Spherical symmetry results because of the random orientation of the particles in solution. Protein can be differentiated from RNA for those viruses in which empty capsids exist (e.g., comoviruses) by difference imaging. Cryo–electron microscopy (cryoEM) reconstructions impose icosahedral symmetry on protein and RNA, and the interface between them cannot be readily determined. Recently it has been possible to distinguish between protein and RNA in cryoEM experiments with CPMV (Baker et al., 1992) and flock house virus (FHV; Cheng et al., 1994). Electron density at the same resolution of the cryoEM reconstructions was computed for the protein shell based on the atomic coordinates of the X-ray model (Fisher and Johnson, 1993) of FHV. This density was scaled with the cryoEM data only in the protein region. The correlation coefficient between the two densities was maximized by adjusting radial and density scale factors. The protein density was subtracted from the cryoEM density, leaving density only for the RNA and a feature in the protein shell that was not included in the X-ray model. Figure 10.2 shows a stereo image of the cryoEM and X-ray electron densities and an image of the bulk RNA from the difference map for FHV. The RNA density has an irregular outer surface that follows the protein shell contour closely in some regions, but, in other regions, considerable solvent lies between RNA and the protein. The relationship between bulk RNA and certain regions of the capsid protein identified by this method has suggested the mechanism for release of RNA that is discussed later.

The average secondary structure of packaged RNA can be determined by spectroscopic analysis. Although limited in number, the structures of packaged RNA characterized by different spectroscopic methods provide similar answers for different RNA viruses. Circular dichroism studies of southern bean mosaic virus demonstrate the presence of more than 70% duplex structure in packaged RNA (Odumos et al., 1981), and laser–Raman spectral studies of bean pod mottle virus (BPMV) middle component show that more than 70% of the ribose phosphate backbone of the RNA is in an A-type conformation and nearly all of it is in a base-paired duplex structure (Li et al., 1992). This study also

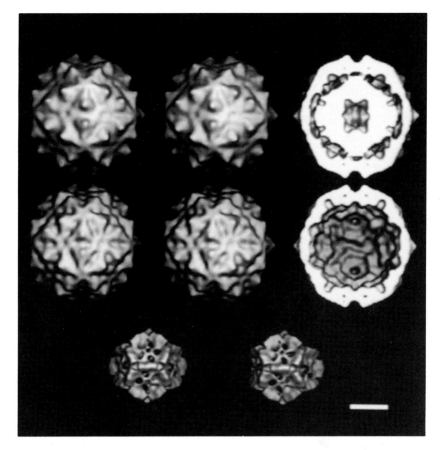

FIGURE 10.2   Electron density distributions displayed as surface-shaded images for the cryo-EM reconstruction of flock house virus (FHV) (top row), the calculated electron density for the X-ray model (middle row), and the difference electron density map (bottom row). *Top left.* A stereo view of the surface-shaded three-dimensional image reconstruction of FHV. The reconstruction was computed to 22-Å resolution and included an explicit correction for the contrast transfer function. *Top right.* An equatorial cut through the reconstruction shown in the left panels, displaying the interior RNA density. High-density values are white and lower density values are dark. *Middle left.* A stereo view of the surface-shaded 22-Å resolution electron density map computed with structure factors based on the 2.8-Å resolution X-ray protein model. *Middle right.* An equatorial cut through the density shown on the left that emphasizes the lack of interior RNA density in the X-ray structure. *Bottom.* A stereo view of the surface-shaded electron density derived from a difference map in which the X-ray model density (middle) was subtracted from the density of the cryo-EM reconstruction (top). The two maps were scaled on the basis of Pearson's correlation coefficient computed in a radial slice (120 to 170 Å) before the subtraction. The map corresponds to electron density not accounted for by the X-ray model, which is primarily due to RNA. Bar, 100 Å.

compared BPMV RNA in the particle and in solution. The amount of secondary structure in the RNA increased by approximately 12% on packaging, and the thermal denaturation temperature for the RNA increased from 43 °C to 53 °C in the packaged form.

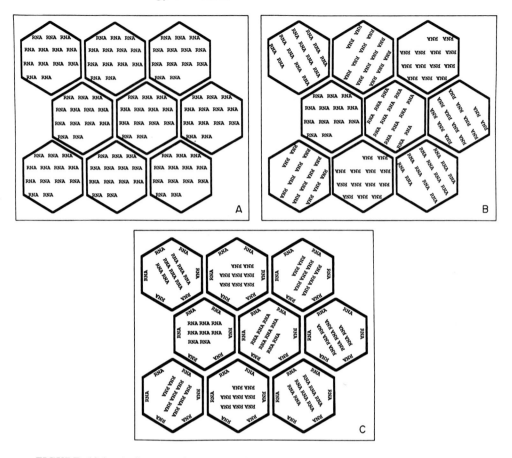

FIGURE 10.3 A diagrammatic representation of possible relationships between packaged RNA and the symmetric capsid in crystal lattices. RNA is represented by the lettering; the hexagonal outline represents the symmetric capsid. *A.* Crystalline arrangement of particles in which the RNA affects the packing of virions. The RNA is ordered and would be visible in the electron density map. Crystallizing virus in a strong magnetic field where the intrinsic dipole of the RNA molecule could affect particle orientation may lead to such an arrangement. *B.* The situation observed in most crystalline spherical viruses. The symmetric capsid packs in six structurally equivalent orientations, which orients the RNA in six different ways, leading to disorder. In an icosahedral shell there are 60 equivalent packing arrangements. *C.* The situation observed in BPMV, wherein a portion of the RNA binds to the symmetric capsid and displays the same symmetry as does the virion. The portion of the RNA bound to the protein is ordered in the crystal lattice, whereas the remaining portion of the RNA is disorded.

RNA packaging stabilized the protein shell as well as the RNA because the naturally occurring empty capsid of BPMV is thermally denatured at 53 °C, whereas the nucleoprotein components are stable to 65 °C. Similar results were found with thermal scanning calorimetry and high-pressure denaturation of CPMV protein and nucleoprotein components (Virudachalam et al., 1985; Da Poian et al., 1994). The thermodynamic stability of the nucleoprotein

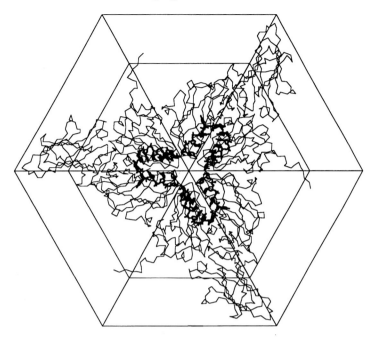

FIGURE 10.4   Diagram showing that the RNA (heavy lines) in BPMV from three icosahedral asymmetric units forms a trefoil shape centered about the threefold particle axes. Three large subunits cluster around the trimer axis, forming a pseudo sixfold unit contacting the RNA trefoil. The small subunits are near fivefold axes and do not interact with the RNA. Thirty-three ribonucleotides form this trefoil-shaped cluster, which appears connected in the electron density map but must have at least one point of entry and exit. Connections between these clusters of 33 ribonucleotides are not visible in the electron density map.

particles demonstrates the remarkable complementarity of protein–protein, protein–RNA and RNA–RNA interactions that are maximized in the assembled particle.

Three-Dimensional Structure of Packaged Nucleic Acid

The bulk properties of packaged RNA structure furnish little insight for specific protein–RNA or RNA–RNA interactions. Crystallographic investigations of icosahedral viruses reported before 1989 provided only information on protein–protein interactions. Four virus structures subsequently determined at high resolution have 20% or more of their packaged RNA visible in electron density maps. Figure 10.3 illustrates the relationship between RNA and protein required for RNA density to be observed in three-dimensional crystal structures. The middle component of BPMV (Chen et al., 1989) contains roughly 20% of its packaged RNA as ordered trefoils surrounding the icosahedral threefold axes. Figure 10.4 illustrates the tertiary structure of BPMV and Figure 10.5 suggests how this structure might relate to the overall BPMV RNA structure.

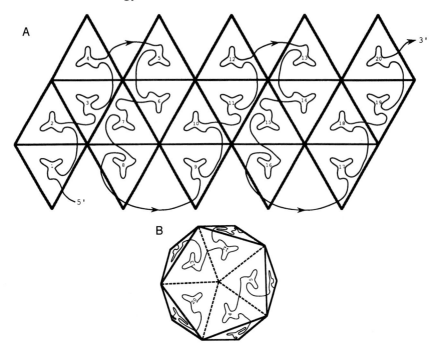

FIGURE 10.5   Model for RNA packaging that is consistent with the crystallographic observations. *A.* Two-dimensional representation of an icosahedron showing the 20 triangular threefold
faces. The RNA molecule is threaded through all the faces, maintaining the same polarity about
each triad. The actual length of the connections between trefoils is not known. *B.* View down a
fivefold axis of the icosahedron shown as a two-dimensional sheet in *A.* The dashed line indicates
the vertex formed by joining the faces labeled 1, 2, 3, 7, and 8 in *A.* The model illustrates the
conserved features at the triad and lack of symmetry in the connecting regions leading to the
observed electron density in the crystal structure.

Satellite tobacco mosaic virus (STMV) packages 44% of its RNA in an ordered
duplex form situated on icosahedral twofold symmetry axes (Larson et al.,
1993), whereas FHV employs segments of ordered duplex RNA as a molecular
switch that differentiates subunit interactions required for the formation of the
$T = 3$ quasi-equivalent lattice (Fisher and Johnson, 1993). Canine parvovirus
(CPV) contains roughly 20% of its DNA genome organized as a 13-nucleotide
loop that is stabilized by a metal ion (Tsao et al., 1991).

The nucleic acid electron density is comparable to that of the protein density in each of the structures described, implying that it has nearly strict icosahedral symmetry. Attempts to identify a periodic sequence in the RNA or
DNA primary structures to explain the repeating tertiary structural patterns have
not been generally successful; however, in the case of BPMV, a consensus
sequence that was roughly consistent with the observed density is repeated 15
times (MacFarlane et al., 1991).

Interactions between the capsid protein and ordered RNA in STMV and
FHV are dominated by electrostatic linkages between basic residues and phos-

phates. The ribose phosphate backbone of the duplex associates with the positive electrostatic potential of the protein in FHV, and the duplex unwinds at its ends to follow a line of lysine residues. The ordered RNA appears to have been incorporated in a sequence-independent fashion to play a structural role in FHV (i.e., nodaviruses) and STMV. In contrast, the nucleic acid structures observed in BPMV and CPV are not dominated by electrostatic linkages with the coat protein, but may have base sequence dependence as well. Whether protein–RNA interactions in BPMV and CPV are fortuitous, or related to assembly and reflective of a specific interaction associated with recognition and packaging of RNA, remains to be determined.

Recognition in RNA Packaging

Nature of Packaging Sites on Viral RNA

Recognition between the coat protein and the genome is a critical step in the assembly of a virion, and it is achieved with remarkable fidelity in the systems that have been analyzed (see Chapter 1). Even very similar viruses (e.g., RNA phages R17 and MS2) infecting the same cell will uniquely package their own RNA (Ling et al., 1970). A number of viral systems have now been well characterized regarding the packaging site on the RNA, but much less is known about the region of the capsid protein that binds the RNA. Figure 10.6 illus-

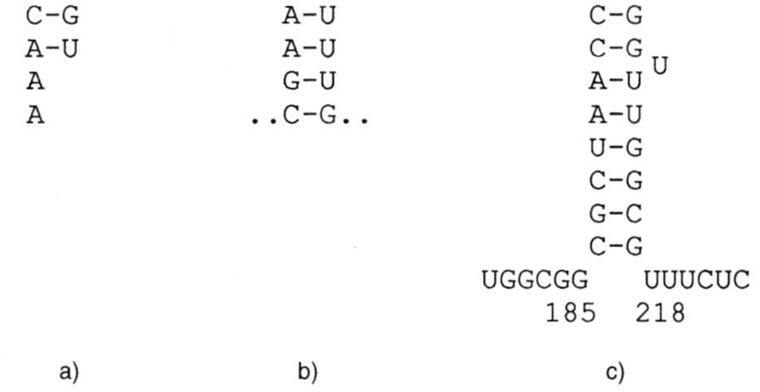

a)   b)   c)

FIGURE 10.6 The primary structures and likely secondary structures of regions of viral genomic RNAs that determine if an RNA molecule is packaged. Such packaging sites have a high affinity for the coat protein and may initiate the assembly process. *a.* Packaging site for the phage R17 (Witherell et al., 1991). *b.* Packaging site for the double-stranded yeast virus L-A (Fujimura et al., 1990). *c.* Packaging site for the insect flock house virus (Zhong et al., 1992).

trates representative sites on viral RNAs that have been investigated (Fujimura et al., 1990; Witherell et al., 1991; Zhong et al., 1992). All of these are composed of stem—loop structures, and two have been investigated to establish precisely which bases are required for packaging.

Specificity of Protein—RNA Interactions in Recognition

The interaction that determines packaging is expected to occur only once during assembly because the site is unique in the RNA primary structure in the cases investigated. It is possible, however, that an interaction with coat protein occurs with exceptionally high affinity at a particular sequence, but also occurs with lower, but significant, affinity in the general sequence of the RNA. The recognition site for RNA binding in TMV includes the sequence AAG, and the high-resolution structure shows that the sequence XXG makes a particularly favorable G-dependent interaction with the capsid protein. The sequence XXG does not occur in phase at a statistically significant level, yet the average base that binds in the third position accommodates to the protein environment that appears optimal for a G (Namba et al., 1989). Although the interaction between protein and RNA in TMV may be different from that found in an icosahedral virus, the concept of RNA structure accommodating to the protein may be general. There is some evidence for this in the phage MS2 (Golmohammadi et al., 1993), where a small region of ordered RNA in the native X-ray structure occurs at the same positions where the recognition sequence binds in an artificial particle assembled with 90 copies of the 19 nucleotides shown in Figure 10.6 (Valegård et al., 1994). The novel structure of the nucleic acid observed in CPV is suggestive of this possibility in a DNA virus.

Infection and RNA Release

The RNA packaging and transport mechanism must deliver the RNA to the appropriate site in the cell in a form ready for translation and replication. Release of RNA from simple icosahedral viruses is an area in which speculation is based on relatively few data. Figure 10.7 illustrates the hypothetical concept of an infectosome, a virion prepared to deliver its RNA in a form that leads to infection. Here we discuss the infectosome on the basis of data and published hypotheses for RNA release from picornavirus and nodavirus capsids. The capsids for these two viruses are compared in Figure 10.1. Although the two viruses have distinctly different capsids, and the processes involved in viral release are not identical, there are parallels that suggest principles that may be more general for icosahedral animal viruses.

Mature Animal Virus Capsid

Both viruses undergo a postassembly cleavage of the capsid protein that is required for infection. The assembled provirion of poliovirus is formed by VP0,

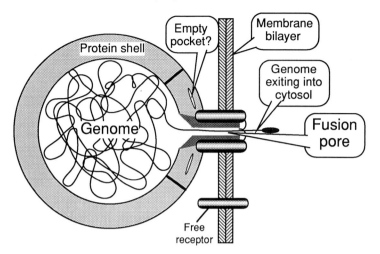

FIGURE 10.7 Hypothetical structure of the infectosome, a transient intermediate that transfers its genome into the cytosol and thereby initiates infection. Receptors mediate infection by positioning the virion near the membrane and then triggering it to form a channel or fusion pore by extruding buried hydrophobic regions of capsid protein into the membrane bilayer. Hydrophobic molecules (pocket factors) may play a role in stabilizing the capsid during transit to the next host cell. If so, the uncoating transition might be facilitated by release of pocket molecules as the capsid interacts with receptor and membrane. Receptors are not yet known to play a role in uncoating of plant viruses. All icosahedral viruses have 12 vertices, each of which must have a unique binding interaction with the asymmetric nucleic acid genome. Thus it is plausible that one of the vertices serves the same role as a kind of invisible tail through which a specific region of the genome exists (cryptic tail hypothesis). Lack of infectivity in provirions of picornaviruses and nodaviruses may result from inability of exiting internal proteins such as VP4 or γ chains to clear the fusion pore unless they have been cleaved free (maturation cleavage) of the anchoring precursor chain. Whether the uncoating transition involved in infectosome formation proceeds through concerted reorientation of all protein subunits, or only of selected vertices as portrayed in the figure, remains to be determined by future work. The "puff" structures in Figure 10.8 may be an example of selective vertex transitions. Application of methods such as crystallography and cryoEM in combination with genetics provides opportunities for unraveling the structures of purified uncoating intermediates to atomic resolution.

VP3, and VP1 (see Chapter 6). After RNA is packaged, VP0 is cleaved to form VP4 and VP2. VP4 is originally the N-terminal 68 residues of VP0 and lies internal, and VP2 is one of the three pseudo equivalent subunits forming the shell (Fig. 10.1). Nodaviruses also form a provirion that matures to an infectious particle by postassembly cleavage of the capsid protein. In this case, the subunits assemble as a 407-amino-acid protein α that is cleaved to β (residues 1 to 363) and γ (residues 364 to 407) postassembly. Like VP4, γ is internal.

Capsid–Cell Surface Interactions and Alterations

The disassembly of rhinovirus 14 and poliovirus is initiated when each virus binds to its respective cellular receptor (see Chapters 4 and 6). The N terminus of VP1, which is normally internal in poliovirus, is exposed at this stage, probably as an amphipathic helix that, together with the myristic acid associated with VP4, interacts with the cellular membrane (Fricks and Hogle, 1990). Poliovirus and rhinovirus both have hydrophobic pockets within the VP1 subunits, the gene product that forms capsid pentamers (Rueckert, 1990). These pockets may be occupied by fatty acids in the wild-type virus (e.g., poliovirus, rhinovirus 16), or the so-called pocket factor may be absent in the purified virus (e.g., rhinovirus 14). A variety of studies have shown that, when this pocket is filled with a tight-binding drug, the virion is stabilized and acquires resistance to RNA uncoating, and therefore becomes noninfectious (reviewed in Rossmann and Johnson, 1989). The endogenous ''pocket factor,'' which stabilizes the isolated virus particle, is probably lost early in the normal infection process, and this permits the necessary flexibility required for the protein shell to disassemble. Although the picornavirus gene products VP2 and VP3 (Fig. 10.1) contain β-barrel domains that are similar to VP1, these subunits do not contain a pocket factor or bind drugs.

Role of Pentamers in RNA Release

The picornavirus studies suggest that one of the roles for the multiple subunit types in $P = 3$ viruses is to differentiate pentamers and hexamers. Such a distinction is not made in a $T = 3$ virus because subunits forming hexamers and pentamers are identical gene products. X-ray and cryoEM analyses of FHV revealed density corresponding to a pocket factor closely associated with the fivefold symmetry axes of the particle. It had an ellipsoidal shape and was located in a hydrophobic environment (Cheng et al., 1994). Similar density on pentamer axes was seen in the X-ray analyses of *Nodaviridae* black beetle virus (Wery et al., 1994) and nodamura virus. The pocket factor in these viruses binds in a position formed by the quaternary structure of the particle rather than the tertiary structure of the subunit, as seen in picornaviruses (see Chapter 4). It occurs only at the fivefold symmetry axes because of a remarkable difference in quasi-equivalent contacts at the fivefold and quasi-sixfold axes. The ellipsoid density is bordered at high radius by five leucine residues and at lower radius by the aliphatic portion of five lysine residues (see Color Plate 15). It is exposed to the interior of the shell at lower radius and it is directly adjacent to the cleaved γ_A polypeptides that form the pentameric helical bundle.

At the quasi-sixfold axes, which are true threefold axes, the corresponding region is occupied not by the pocket factor, but by the side chains of the threefold-related Thr 20 residues that are part of the molecular switch that alters quasi-equivalent contacts between subunits (see Color Plate 15). Thus, the protein portion of the molecular switch (Thr 20–Val 30), which is ordered only at the icosahedral threefold axes, not only contributes to the particle architecture but also determines the chemical composition of the particle by selectively occupying a region on the quasi-sixfold axes that is occupied by the pocket factor on the fivefold symmetry axes.

Another difference in the hexamer and pentamer axes dictated by quaternary interactions supports the role of pentamers in RNA release. The side chains of Met 182 in the A subunits are exposed to the particle exterior (see Color Plate 15). This creates a hydrophobic patch at the pentamer axes and thus a favorable chemical environment for interaction with the lipid of the membrane. The contacts between A subunits near Met 182 are primarily hydrophobic; thus the proposed entry of Met 182 into the membrane might initiate a particle destabilization process by "dissolving" quaternary structure interactions at the pentamer axes in the membrane. Consistent with this hypothesis, Met 182 in the B and C subunits (at the quasi-sixfold symmetry axes) is not as exposed as it is in the A subunits. The difference in quaternary structure at the hexamer axes exposes the side chain of Gln 183 to the viral surface and folds the Met 182 side chains inside the capsid surface, where they interact with each other near the symmetry axis.

The mechanism for the initial interaction of nodaviruses with the cell surface probably includes the release of the pocket factor upon binding the cellular receptor and/or an interaction with the cell membrane. The associated destabilization of the intersubunit contacts along the pentamer axis would then allow the γ_A peptides to emerge along the axis, dragging the RNA with them, via their C-terminal ends. Upon reaching the capsid surface, the hydrophobic portion of the γ_A helix may interact with the membrane and establish a pore or channel. The release of the γ_A peptides as proposed in this model explains the requirement for the maturation cleavage for infectivity.

RNA Release

The possible release of the γ_A peptides associated with RNA was supported by results obtained upon heating FHV. Studies of poliovirus showed that heating it generated the same antigenic site that emerged by the binding and release of the virus from the cellular receptor (see Chapter 6). That antigenic site is now known to be the N terminus of VP1. In a similar study, FHV was heated at 65 °C for 10 min (under these conditions 2 logs of infectivity were lost on the basis of a plaque assay) and electron micrographs of samples stained with uranyl acetate were recorded. Virus particles remained essentially intact after this treatment, but a distinct single "puff" of density appeared that was associated with most of the particles (Fig. 10.8). Each puff may be one bundle of γ_A peptides associated with RNA, consistent with the hypothesis presented earlier. The apparent single-site release of the RNA was not expected, although such specificity might be expected in the interaction of the virus with the cell. These phenomena have at least two explanations. Release of one puff may lead to a configurational relaxation of the virus structure such that further puffs cannot be released. Alternatively, a specific sequence of RNA may prime one pentamer site for release upon perturbation of the particle by destabilizing the pentamer with which it is associated (cryptic tail hypothesis discussed in Figure 10.7).

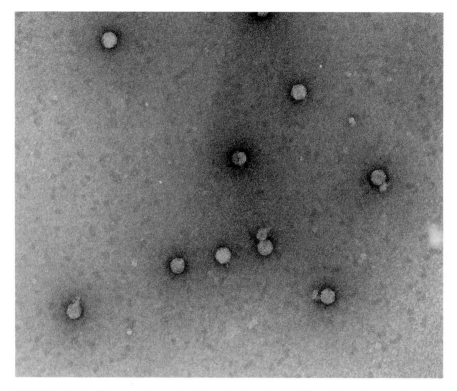

FIGURE 10.8   Electron micrograph of negatively stained FHV particles that were heated to 65 °C. Many of the particles exhibit a ''puff'' of density that appears to protrude from a single site on the capsid. The novel structure seen at each of the FHV pentamer axes makes it a strong candidate for the location of release of a γ-helical bundle and associated RNA. 122,500×

Summary

The protein shell of small icosahedral viruses contains a central cavity into which the genomic RNA is packaged and from which it must later be delivered to infect its next host cell. In most viruses, the building bricks of the shell have a common fold, an eight-stranded antiparallel β-barrel (''shell'' domain) requiring a protein mass on the order of 30 kDa. These domains are packed into an icosahedral surface lattice according to principles that allow 60, 180, 240, ... domains in order to form a closed shell. However, most small viruses contain 180 or more shell domains; this is probably because the cavity generated by sixty 30-kDa proteins is only about 120 Å in diameter, too small to accommodate a nucleic acid encoding the genes required for minimal virus functions (a coat protein of 30 kDa and a replication protein of 90 kDa).

Preliminary studies suggest that genome packaging is initiated by stem–loop regions, but the mechanisms by which small spherical viruses envelop the nucleic acid genome and condense it to densities higher than those observed in crystalline nucleic acids are not clear. In vivo studies indicate that genome packaging is highly specific, but this specificity has not generally been observed in cell-free assembly systems. Most in vitro assembly systems have been based upon reconstitution using protein subunits from dissociated virions. In vivo studies suggest that assembly is coordinated with RNA synthesis and membranes. Studies on poliovirus, showing that cell-free extracts are capable of synthesis and assembly of infectious virus all the way from the RNA genome (Molla et al., 1992), provide a model for studies with many other viruses.

In the past, lattice theory has emphasized the importance of identical bonding environments between protein subunits. This implies that conditions inducing changes in the structure of one protein subunit would produce coordinate changes in all of the other subunits in the shell (''all or nothing'' or ''concerted'' reorientation). This idea has, however, neglected asymmetric interactions of the protein with its genome. This leads to the notion that it may be possible for viruses that are structurally icosahedral on the outer surface to have *functional* asymmetry, such as special regions or cryptic ''tails'' through which the genome is released during infection. Structural studies also point to the fivefold vertex as a common site for exit of nucleic acids. Some questions, such as the nature of virus shell transitions during genome release, are now ripe for analysis by the methods of X-ray crystallography and electron microscopy. Other questions, such as structure of ''infectosomes'' (i.e., virions in the act of productive uncoating), will likely require development of new methods.

Acknowledgments

The authors thank Sanjeev Munshi, Vijay Reddy, Anette Schneemann, and Adam Zlotnick for helpful discussions and suggestions. Work in the authors' laboratories was supported by The U.S. National Institutes of Health.

References

Baker, T.S., Cheng, R.H., Johnson, J.E., Olson, N.H., Wang, G.J. and Schmidt, T.J. 1992. Organized packing of RNA inside viruses as revealed by cryo-electron microscopy and X-ray diffraction analysis. In Proceedings of the 50th Annual Meeting on Electron Microscopy, pp. 454–455. San Francisco Press, San Francisco.

Black, L.W. 1989. DNA packaging in dsDNA bacteriophages. Annu. Rev. Microbiol. 43: 267–292.

Booy, F.P., Newcomb, W.W., Trus, B.L., Brown, J.C., Baker, T.S. and Steven, A.C. 1991. Liquid-crystalline, phage-like packing of encapsilated DNA in herpes simplex virus. Cell 64: 1007–1015.

Butler, P.J.G., Finch, J.T. and Zimmern, D. 1977. Configuration of tobacco mosaic virus RNA during virus assembly. Nature 265: 217–219.

Chen, Z., Stauffacher, C., Li, Y., Schmidt, T., Bomu, W., Kamer, G., Shanks, M., Lomonossoff, G.P. and Johnson, J.E. 1989. Protein-RNA interactions in an icosahedral virus at 3.0 Å resolution. Science 245: 154–159.

Cheng, R.H., Reddy, V.S., Olson, N.H., Fisher, A.J., Baker, T.S. and Johnson, J.E. 1994. Functional implications of quasi-equivalence in a $T = 3$ icosahedral animal virus established by cryo-electron microscopy and X-ray crystallography. Structure 2: 271–282.

DaPoian, A.T., Johnson, J.E. and Silva, J.L. 1994. Differences in pressure stability of the three components of cowpea mosaic virus: implications for virus assembly and disassembly. Biochemistry 33: 8339–8346.

Earnshaw, W., Casjens, S. and Harrison, S.C. 1976. Assembly of the head of bacteriophage P22: X-ray diffraction from heads, proheads, and related structures. J. Mol. Biol. 104: 387–410.

Fisher, A.J. and Johnson, J.E. 1993. Ordered duplex RNA controls the capsid architecture of an icosahedral animal virus. Nature 361: 176–179.

Fricks, C.E. and Hogle, J.M. 1990. Cell-induced conformational change in poliovirus: externalization of the amino terminus of VP1 is responsible for liposome binding. J. Virol. 64: 1934–1945.

Fujimura, T., Esteban, R., Esteban, L.M. and Wickner, R.B. 1990. Portable encapsidation signal of the L-A double stranded virus of S. cerevisiae. Cell 62: 619–628.

Golmohammadi, R., Valegård, K., Fridborg, K. and Liljas, L. 1993. The refined structure of bacteriophage MS2 at 2.8 Å resolution. J. Mol. Biol. 234: 620–639.

Johnson, J.E. and Argos, P. 1985. Virus particle and stability-tricorna virus. In The Virus (Francki, R., Vol. Ed.), pp. 19–56. Plenum Press, New York.

Johnson, J.E. and Fisher, A.J. 1994. Principles of virus structure. In Encyclopedia of Virology (Webster, R.G. and Granoff, A., Eds.), pp. 1573–1586. Academic Press, New York.

Larson, S.B., Koszelak, S., Day, J., Greenwood, A., Dodds, J.A. and McPherson, A. 1993. Double-helical RNA in satellite tobacco mosaic virus. Nature 361: 179–182.

Lauffer, M.A. 1975. Entropy-driven processes in biology: polymerization of tobacco mosaic virus protein and similar reactions. Mol. Biol. Biochem. Biophys, Volume 20.

Lebeurier, G., Nicolaieff, A. and Richards, K.E. 1977. Inside-out model for self-assembly of tobacco mosaic virus. Proc. Natl. Acad. Sci. U.S.A. 74: 149–153.

Li, T., Chen, Z., Johnson, J.E. and Thomas, G.J. Jr. 1992. Conformations, interactions and thermostabilities of RNA and proteins in bean pod mottle virus: investigation of solution and crystal structures by Raman spectroscopy. Biochemistry 31: 6673–6682.

Ling, C.M., Hung, P.P. and Overby, L.R. 1970. Independent assembly of QB and MS2 phages in doubly infected cells of E. coli. Virology 40: 920–929.

MacFarlane, S.A., Shanks, M., Davies, J.W., Zlotnick, A. and Lomonossoff, G.P. 1991. Analysis of the nucleotide sequence of bean pod mottle middle component RNA. Virology 183: 405–409.

Matthews, B.W. 1968. Solvent content of protein crystals. J. Mol. Biol. 33: 491–497.

Mellema, J.E. 1975. Model for the capsid structure of alfalfa mosaic virus. J. Mol. Biol. 94: 643–648.

Molla, A., Paul, A.V. and Wimmer, E. 1992. Cell-free, de novo synthesis of poliovirus. Science 254: 1647–1651.

Namba, K., Pattanayek, R. and Stubbs, G. 1989. Visualization of protein-nucleic acid interactions in a virus: refined structure of intact tobacco mosaic virus at 2.9 Å resolution by X-ray fiber diffraction. J. Mol. Biol. 208: 307–325.

Odumos, A.O., Homer, R.B. and Hull, R. 1981. Circular dichroism studies on southern bean mosaic virus. J. Gen. Virol. 53: 193–196.

Rossmann, M.G. and Johnson, J.E. 1989. Icosahedral RNA virus structure. Annu. Rev. Biochem. 58: 533–573.

Rueckert, R.R. 1990. Picornaviruses and their replication. In Virology (Fields, B.N., Knipe, D.M., Chanock, R.M., Hirsch, M.S., Melnick, J.L., Monath, T.P. and Roizman, B., Eds.), pp. 705–738. Raven Press, New York.

Schmidt, T., Johnson, J.E. and Phillips, W.E. 1983. The spherically averaged structures of cowpea mosaic virus components by X-ray solution scattering. Virology 127: 65–73.

Tsao, J., Chapman, M.S., Agbandje, M., Keller, W., Smith, K., Wu, H., Luo, M., Smith, T.J., Rossmann, M.G., Compans, R.W. and Parrish, C.R. 1991. The three-dimensional structure of canine parvovirus and its functional implications. Science 251: 1456–1464.

Valegård, K., Murray, J.B., Stockley, P.G., Stonehouse, N.J. and Liljas, L. 1994. Crystal structure of an RNA bacteriophage coat protein-operator complex. Nature 371: 623–626.

Virudachalam, R., Harrington, M.A. and Markley, J.L. 1985. Thermal stability of cowpea mosaic virus components: differential scanning colorimetry studies. Virology 146: 138–140.

Wery, J.-P., Reddy, V., Hosur, M.V. and Johnson, J.E. 1994. The refined structure of an insect virus at 2.8 Å resolution. J. Mol. Biol. 235: 565–586.

Witherell, G.W., Gott, J.M. and Uhlenbeck, O.C. 1991. Specific interaction between RNA phage coat proteins and RNA. Prog. Nucleic Acid Res. Mol. Biol. 40: 185–220.

Zhong, W., Dasgupta, R. and Rueckert, R. 1992. Evidence that the packaging signal for nodaviral RNA2 is a bulged stem-loop. Proc. Natl. Acad. Sci. U.S.A. 89: 11146–11150.

Zimmern, D. 1977. The isolation of tobacco mosaic virus RNA fragments containing the origin for viral assembly. Cell 11: 455–462.

Zuiedema, D., Bierhuizen, M.F.A., Cornelissen, B., Bol, J. and Jaspars, E.M.J. 1983. Coat protein binding sites on RNA1 of alfalfa mosaic virus. Virology 125: 361–369.

11.

The Procapsid-to-Capsid Transition in Double-Stranded DNA Bacteriophages

JONATHAN KING & WAH CHIU

The capsids of viruses have generally been regarded as rigid containers designed to protect nucleic acid in the passage from one host cell to another and to deliver the nucleic acid to the appropriate site. This rigid image was reinforced by the first determination of structures of viral capsids at the atomic level (Rossmann, 1984). More recently, deviations from quasi-equivalence (see Chapter 1) have suggested flexibility in subunit–subunit interactions (Liddington et al., 1991; Prasad et al., 1993; see also Chapters 4, 7, and 9). In viruses containing double-stranded (ds) DNA as well as some other viruses, a precursor shell is assembled empty of DNA. These precursor shells participate directly in DNA condensation and are thus a part of the DNA packaging machinery. During this process, the shells transform into the mature, protective capsids capable of DNA delivery. The subunit motions and subunit–subunit reorganization associated with these changes appear to be fundamental features of the dsDNA viruses. This chapter focuses on the structural aspects of the transformation from a DNA packaging apparatus to a DNA protection and delivery vehicle in the best described cases: dsDNA bacteriophages.

Phages as Cloning Vectors and Model Systems

Bacteriophages have long served as test benches for the study of gene expression and macromolecular assembly. As the initial biologic systems used to determine the fine structure of genomes, they have been particularly valuable for the elucidation of the pathways of structure formation and the characteristics

288

of structural intermediates on the way to the final viral particle. Many features of their assembly, such as the procapsid-to-capsid transition and the packaging of DNA through a portal vertex, are only just emerging in animal viruses (Rixon, 1993). In the past two decades they have also served as invaluable cloning vectors for genetic engineering and more recently for the selection of novel protein sequences (Ausubel et al., 1994).

The dsDNA phages represent a large and singularly well-characterized class of phages. They include T4, lambda, T3, T7, P2, and P1 of *Escherichia coli*, P22 of *Salmonella typhimurium*, and φ29 of *Bacillus subtilis*. The systems of amber and temperature-sensitive conditional lethal mutations were originally developed in dsDNA phages (Epstein et al., 1963; Edgar et al., 1964; Edgar and Epstein, 1964). Sodium dodecyl sulfate (SDS) gel electrophoresis was developed as a tool to study the structural proteins of phage T4 (Laemmli, 1970). The coupling of nonsense mutations with SDS gel electrophoresis led to the initial identification of phage structural proteins as the products of known genes (Laemmli, 1970; King and Laemmli, 1971).

Analyses of the structural intermediates that accumulate when individual structural proteins are removed by mutation have provided invaluable insights into the virus assembly process. The general availability of nonsense mutations, still rare for eukaryotic viruses, makes it possible to block assembly pathways by removal of a key protein, accumulating structural intermediates. The transient intermediates are abundant enough in the productive infection to be purified for biochemical and physical studies.

The function of the phages as cloning vectors depends on the assembly of a precursor shell empty of DNA, into which DNA is subsequently packaged. The ability to carry out this reaction in vitro underlies all gene cloning in lambda and other phage vectors (Hohn and Hohn, 1974; Thomas et al., 1974; Murray et al., 1977). The size of the DNA fragment that can be cloned into phage vectors is limited by the volume of the procapsid. The ability to assemble larger procapsids would permit the packaging of larger genome fragments. The mechanism underlying this DNA condensation reaction and the coordinated procapsid-to-capsid transition is the subject of this chapter.

General Features of the Morphogenetic Pathway of
dsDNA Phages

In all dsDNA phages the chromosome is highly condensed within the virion particle, and represents 20% to 50% of its mass (Earnshaw and Harrison, 1977). The mature capsid lattice must be sufficiently robust to maintain the chromosome in the highly condensed form. Unlike many animal viruses, the dsDNA phages deliver their chromosome into the cell through an injection mechanism, rather than through an uncoating process. For all these phages, the initial product of subunit polymerization is not the mature virion but a precursor capsid, into which DNA is packaged, as shown in Figure 11.1. A similar capsid maturation pathway is also found in herpesviruses (Lee et al., 1988; Sherman and

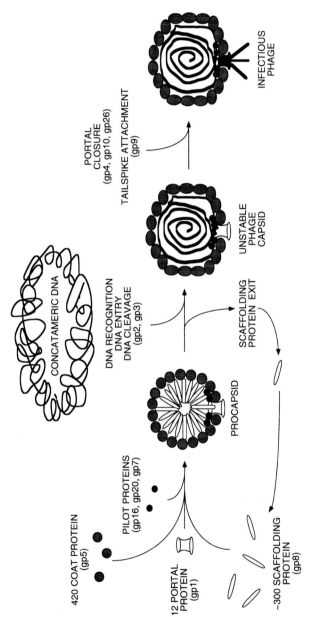

FIGURE 11.1 Pathway for capsid assembly and DNA packaging for dsDNA bacteriophage P22 (Prevelige and King, 1993). The 47-kDa coat protein (gp5) and the 34-kDa scaffolding protein (gp8) copolymerize to form the procapsid. The portal complex is incorporated into or initiates a fivefold vertex and is made up of 11 to 13 copies of the 90-kDa portal protein (gp1) and the pilot proteins (gp16, gp20, and gp7). The latter, required for DNA injection, are also present in small numbers, 5 to 20 copies per virion, and are dispensable for particle assembly. The scaffolding proteins exit from the procapsid and recycle for further rounds of capsid assembly either before or during the DNA packaging. Shell expansion occurs during or soon after scaffolding exit /DNA entry. The DNA encapsidation is carried out through the portal vertex with the assistance of additional DNA packaging proteins gp2 and gp3. The portal vertex is sealed by the portal closure proteins (gp4, gp10, and gp26). Six tailspikes, trimers of the 70-kDa gp9, then assemble to form a stable and infectious capsid.

Bachenheimer, 1988; Rixon, 1993; see also Chapter 12) and in adenoviruses (Edvardsson et al., 1976; D'Halluin et al, 1978).

In all cases the assembly of the procapsid requires scaffolding proteins that must be removed from the precursor shell prior to DNA packaging (Showe and Black, 1973; King and Casjens, 1974; Ray and Murialdo, 1975; Casjens and Hendrix, 1988). The mechanisms of removal are diverse, ranging from proteolytic cleavage of scaffolding subunits by a prehead proteinase that itself is assembled into the procapsid, as in T4 (Showe and Black, 1973), to the release of intact scaffolding proteins followed by their recycling, as in phage P22 (King and Casjens, 1974).

Within infected cells, the initiation of procapsid assembly proceeds from a cyclic portal complex, which sits at or comprises a unique vertex. This portal and the procapsid lattice in which it sits are required for DNA packaging (Murialdo, 1979; Earnshaw and Casjens, 1980; Bazinet and King, 1988). The newly replicated DNA is pumped, spooled, or threaded into the procapsid shells through the portal vertex (Earnshaw and Casjens, 1980; Black, 1989; Turnquist et al., 1992). Associated with this process is the transformation of the procapsid, which expands and changes its character to yield the mature virion. This capsid transformation occurs both for phages in which the capsid subunits undergo significant proteolytic cleavage, such as T4, and those for which there is no proteolytic cleavage, such as P22.

Expansion of viral shells, believed to be associated with release of the RNA upon infection, is well known in plant viruses. In tomato bushy stunt virus, the removal of Ca^{2+} is adequate to trigger the capsid shell expansion (Robinson and Harrison, 1982). However, it is likely that the mechanism of expansion and the extent of conformational change of the capsid proteins are different in different viruses. For example, there is a quaternary but not tertiary structure change of the capsid protein in the expansion of the tomato bushy stunt virus.

In this chapter we focus on the central feature of the morphogenesis of dsDNA phages—the transformation of a procapsid empty of DNA, but competent to package DNA, to a mature virion that protects and delivers the condensed chromosome. The extensive studies on the mechanism of phage DNA packaging have been reviewed by Casjens (1989) and Black (1989). Because the organization of the DNA itself is still obscure, this chapter concentrates on the emerging data on the protein shells. We begin with the mature virion structure and then follow with the earliest shell precursors that have been identified in the assembly pathway.

Structure of Mature Bacterial Virus Capsids

Although there are extensive biochemical and genetic studies of dsDNA bacterial viruses, none of their structures has been determined by X-ray crystallography. Cryo–electron microscopy and computer reconstruction have been used to determine the molecular structures of several bacterial viruses (Table

TABLE 11.1 Phage Capsids Whose Structures Have Been Determined by Cryo–Electron Microscopy

Phage	Gene Product (M_r)	Average Diameter (Å)	T Number	Structure Resolution (Å)
P2	gpN (42 kDa)	600	7	45
P4	gpN	450	4	45
P22 empty procapsid	gp5 (55 kDa)	580	7	28
P22 procapsid	gp5 (55 kDa), gp8 (33 kDa)	580	7	19
P22 mature capsid	gp5	610	7	28
λ procapsid	gpE (38.2 kDa), gpD (11.6 kDa)	540	7	34
λ mature capsid	gpE, gpD	630	7	34
λD⁻ mature capsid	gpE	630	7	45
Hk97 prohead	gp5 (42 kDa)	470	7	25
Hk97 head	gp5* (31 kDa)	550	7	25

11.1). Figure 11.2a shows an example of electron images of ice-embedded particles of the mature P22 capsid. Note that in the same micrograph, some capsid particles appear spherical whereas others appear polygonal. This heterogeneous appearance is due to the particles lying in different orientations with respect to the electron beam. The three-dimensional (3-D) reconstruction from these projection images has been carried out to a resolution of 28 to 45 Å, sufficient to determine the size and shape of the capsid and the capsid triangulation number. In some cases, it has been possible to resolve the molecular interactions among protein subunits in the capsid as well as the extent of symmetric arrangement of capsid proteins around nonicosahedral symmetry axes. The use of a difference map between the wild-type and the mutant capsid with one protein missing should allow one to localize different capsid proteins. Figure 11.3 shows an example of a 3-D reconstruction of the mature P22 capsid (Prasad et al., 1993).

Subunit Organization

The phage capsids so far studied by computer reconstruction exhibit icosahedral symmetry. A number of the dsDNA phages, such as T4 and ϕ29, are not isometric and represent elongated variations on the icosahedral theme. The measured diameter of an icosahedral particle depends on the assignment of the equivalent sphere. The dimensions of the capsid particles shown in Table 11.1 represent the average diameters derived from the reconstructions. The mature capsids of P2, P22, and λ have similar diameters although their capsid proteins do not appear to have any sequence homology (Casjens and Hendrix, 1988), and their molecular masses also differ significantly. Because all phage capsids have a unique vertex marked by the portal complex, only 11 of the 12 vertices can have strict equivalence. Current reconstructions average the portal vertex

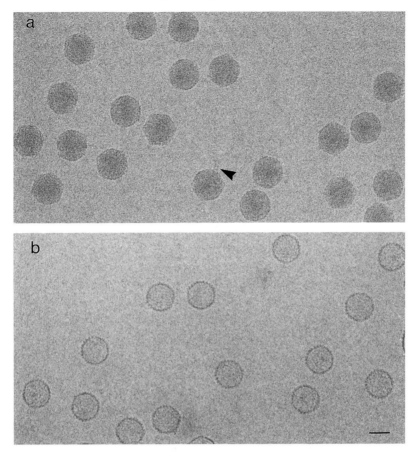

FIGURE 11.2   Electron microscopic images (100 keV) of ice-embedded bacterial viruses of P22 in (*a*) mature form and (*b*) empty procapsids. These images exhibit remarkable contrast. The arrowhead points to a particle for which the portal complex is visible. Bar, 500 Å.

with the other 11 vertices. If portal vertices exist in animal viruses, (such as herpes simplex virus capsid, see Chapter 12) they have also been averaged out in reconstructions.

The bacterial viruses in Table 11.1 all have a triangulation number of $T = 7$ except for the P4 virus (a satellite of P2), which has $T = 4$. In all cases, hexons and pentons are made up of hexameric and pentameric subunits, respectively (Fig. 11.3). These structures provide direct evidence for the arrangement of a monomeric coat protein into 60 hexamers and 12 pentamers on a $T = 7$ icosahedral lattice as previously predicted (Caspar and Klug, 1962). The handedness of all these capsids is left. Because of the $T = 7$ lattice, each asymmetric unit in an icosahedral triangle would contain one penton subunit, six hexon subunits, and three types of connecting arms around the strict and local threefold positions.

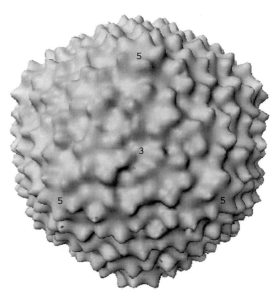

FIGURE 11.3 Three-dimensional density map of P22 mature capsid viewed along a threefold axis. The five, threefold icosahedral symmetry axes in a triangular face in this $T = 7l$ icosahedral lattice are labeled. The densities are rendered 1 standard deviation above the mean density.

Figure 11.4 is a schematic diagram depicting the unique capsomeric subunits and the interactions among the subunits in a $T = 7$ icosahedral triangle. The 420 subunits in the $T = 7$ capsids can be classified into seven types (i.e., A through G) based on their locations with respect to the icosahedral symmetry axes (Fig. 11.3). Type A is peripentonal, whereas types B through G are located around the quasi-sixfold axis (Fig. 11.4). These seven quasi-equivalent subunits constitute the asymmetric unit of the icosahedral structure.

All of the pentameric and hexameric subunits in each of these bacterial viruses have been found to be made up of the same gene product. Because of their chemically distinct environments, the structural conformations of these subunits in each phage do not have to be identical. P4 is a satellite of P2, and the P2 and P4 capsids, which are very different in dimensions, are assembled from the same coat subunit species (Dokland et al., 1992). The difference depends upon a P4-directed scaffolding protein not present in P2-infected cells (Marvik et al., 1994).

In the density map of the λ capsid, one can resolve the capsomeric protein density distribution into two domains. One domain makes up the main body of the capsomer and the other domain mediates the intercapsomeric subunit interactions. The latter domain has flexible conformations in different quasi-equivalent positions (Dokland and Murialdo, 1993). In phage λ, the underlying shell is composed of a single species of the gene E protein, but this E protein is decorated with another protein, the gene D (gpD) protein (Imber et al., 1980). The gpD protein subunits add to the capsid lattice after shell assembly and DNA packaging. The gpD locations have been mapped out from the difference maps of the wild-type λ and λD⁻ mutant, demonstrating that gpD links the intercapsomeric subunits at both strict and local threefold positions.

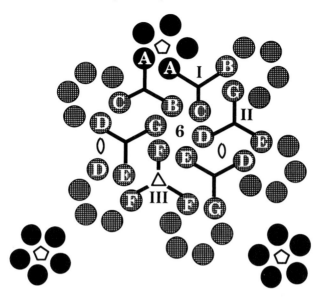

FIGURE 11.4   Schematic diagram of an icosahedral triangular face of a $T = 7$ lattice with three adjacent asymmetric units. The icosahedral five-, three-, and twofold symmetry axes are denoted by different symbols, and one of the quasi-sixfold positions is marked by 6. Each asymmetric unit contains one penton subunit and six hexon subunits, labeled as A through G, respectively. Each of these subunits has a unique chemical environment, and hence can assume different conformations. There are three types of intercapsomeric interactions (I, II, and III) in a $T = 7$ icosahedral lattice. In type I, a pentonal subunit is connected to two other proximal hexon subunits. The type II connects three nonidentical subunits from three separate hexons, while the type III connects three identical subunits from three adjacent hexons related by a strict icosahedral three-fold symmetry axis. (Reproduced with permission from Prasad et al., 1993.)

In phage T4 the pentamers are assembled from a protein other than that used in the hexamers (Onorato and Showe, 1975). The major gp23 subunit of the lattice is only at the hexameric positions. The fivefold vertices are occupied by the gene 24 protein, coded by an adjacent gene and probably evolved by gene duplication. The phage T4 capsid lattice is decorated and stabilized by two outer capsid proteins, Soc and Hoc (Steven et al., 1992). Infectious particles can be formed without these proteins but they are physically unstable. Such proteins may represent a general mechanism for capsid stabilization among animal viruses as well, because both herpesvirus and adenovirus capsid lattices (Newcomb et al., 1993; Stewart et al., 1993; Zhou et al., 1994; see also Chapters 8 and 12) also contain additional proteins at the threefold positions.

Organization of the Condensed DNA

The DNA within phage heads is tightly packed and 250-fold more condensed than the free solution form. The DNA is in the B form, and Raman spectra indicate that its structure is not very perturbed compared to its free solution form (Earnshaw et al., 1978; Reilly and Thomas, 1994). The negative charge is neutralized by Mg^{2+}, spermine or spermidine, or other small counter-ions. Very little protein is associated with the condensed DNA. The DNA is not seen in reconstructions based upon icosahedral symmetry, and is therefore presumably organized with different symmetry, or disorganized. Phages P22 of *Salmonella* and λ and P1 of *E. coli* can package fragments of chromosomal DNA and inject them into a subsequent host cell. Such transducing particles have been essential tools for mapping the genes of *E. coli* and *Salmonella*. Because phages can package DNA independently of the nucleotide sequence, the organization of the DNA within the virus is almost certainly determined by the capsid and the packaging machinery. This unique mechanism of genome packaging is, in fact, one of the reasons for the efficacy of phages as cloning vectors.

Under favorable conditions of negative stain, concentric coiled patterns can be detected within phage particles by electron microscopy (Richards et al., 1973). Low-angle X-ray scattering revealed that the strands are locally close packed, with a spacing of 25 Å (Earnshaw and Harrison, 1977). Earnshaw and coworkers (1978) examined the giant elongated heads that are generated by certain T4 mutants and showed that the DNA was in very long coils, with the axis perpendicular to the long axis of the head. They suggested that the structure of the DNA was that of a spool or solenoid, with the outer coils laid down first, then coils of progressively smaller radius as the DNA was packaged. In such a structure, the strand could be pulled back out from the center without knotting (Earnshaw et al., 1978; Harrison, 1983).

Lepault et al. (1987) have visualized the DNA in the T4 giant heads by cryo−electron microscopy and electron diffraction, confirming that the DNA molecules are in general oriented with the long axis of the head. However, their observations on isometric particles suggested a liquid crystal model, in which the DNA is ordered and parallel in local domains, but the domains are not oriented with respect to each other. Black et al. (1985) found that the last DNA sequences packaged were at the outside of the particle, suggesting that packaging started at the center and proceeded outward. They proposed a spiral-fold model, in which DNA strands are locally packed running back and forth and connected by sharp turns. In contrast Aubrey et al. (1992) argued against sharp turns. The state of the condensed DNA within viruses remains a major unsolved problem in structural virology.

Portal Vertices in Mature Virions

All the dsDNA phages deliver their DNA into the host cell through a portal vertex, to which the phage tail is attached. The portal is referred to as the connector by some authors, emphasizing the head−tail junction role in the mature virion. The initial role of the portal serves as the DNA packaging channel into the procapsid. The portals are cyclic structures that can be released from procapsids or phages by chemical or mechanical means. The portals from T4, λ, P22, and φ29 were originally reported to have 12 subunits. More recently 11 and 13 have been reported for P22, φ29, and other phage (Dube et al., 1993; Kocsis et al., 1993; Tsuprun et al., 1994). The portal complex of φ29 has six molecules of a unique phage-encoded small RNA associated with it (Guo et al., 1987). This RNA has been suggested to function in the assembly of the portal complex or in its association with the DNA packaging proteins (Anderson and Reilly, 1993).

A striking feature of the portal is the symmetry mismatch with the icosahedral fivefold vertex, originally pointed out by Moody (1965) from comparison of the sixfold tail symmetry with the fivefold vertices. The portal–shell mismatch holds whether the portal has 11-fold, 12-fold, or 13-fold symmetry. This asymmetric arrangement prevents the subunits of the portal from having stable equivalent contacts with subunits of the capsid (Hendrix, 1978). Whether the portal displaces the pentamer of capsid subunits or projects through a channel remains to be determined.

During DNA condensation, the DNA packaging enzymes dock against the portal complex. Hendrix (1978) has suggested that the symmetry mismatch might underlie the molecular machinery of the portal: rotation during DNA packaging, driven by the ATP hydrolysis of the complex. In this model, the DNA would be rotating as it was pumped in, with the twists then resolving into the coiled and condensed state in the mature head. More recently, Turnquist et al. (1992) have reported that φ29 DNA wind around the portals. In their model the DNA passes not through the channel of the portal, but around its circumference, as in a winch. Casjens et al. (1993) have described mutants in the portal protein that influence the size of the DNA packaged. This result argues for an active role of the portal.

Procapsid Structure

Procapsid Coat Subunit Lattice

Cells infected with mutant phages blocked in DNA packaging accumulate precursor procapsids. Three-dimensional structures have been determined for the outer capsid lattices of procapsids of *Salmonella* phage P22 (Prasad et al., 1993; Thuman-Commike et al., 1996) and *E. coli* phage λ (Dokland and Murialdo, 1993). Figure 11.2b shows an image of the P22 empty procapsid with the scaffolding subunits removed intentionally during purification. The density is lower because the capsid is empty (Fig. 11.1). Figure 11.5 shows the 3-D reconstruction of the empty procapsid of P22 (Prasad et al., 1993). The organization of the scaffold within procapsids, discussed later, has not yet been solved for any phage except P22 (Thuman-Commike et al., 1996).

Both procapsid lattices of P22 and λ have $T = 7$ icosahedral lattices, consistent with their mature structure. However, there are distinctive differences between precursor and mature lattices. First, the procapsids have smaller diameters than their mature forms (Table 11.1). Second, the subunits in the hexamers appear less hexamerically symmetric. For instance, the λ procapsid capsomeres are skewed into a peanut shape (Dokland and Murialdo, 1993). Third, a significant mass translocation of capsid proteins occurs during the transition

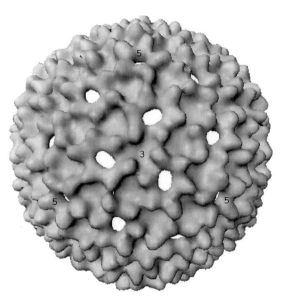

FIGURE 11.5 Three-dimensional density map of the P22 empty procapsid. The subunit around the hexon appears asymmetric. This type of asymmetry is seen in other phage capsids. The fivefold icosahedral symmetry positions are labeled.

from procapsid to mature capsid. Consequently, the capsid shell thickness gets thinner as it matures. Fourth, there are apparent holes in the centers around the hexamers and pentamers in the P22 empty procapsids but not in the mature capsids (Prasad et al., 1993). The diameters of these holes are approximately 25 Å at the former and slightly smaller at the latter positions. These holes have been hypothesized to be the exit points for the gp8 scaffolding protein in the case of P22 phage. They are large enough to allow the scaffolding subunits (33 kDa, with an axial ratio of 9:1) to exit during maturation (Fuller and King, 1981; Greene and King, 1994).

The more recent 19 Å structure of the P22 procapsid containing the scaffolding proteins also shows similar structural characteristics as the empty procapsid (Thuman-Commike et al., 1996). In addition, the higher resolution structure resolves the individual capsid subunits more clearly and exhibits the presence of a local twofold axis in the skewed hexons. This symmetry reduces the number of asymmetric units in the $T = 7$ icosahedron from seven to four coat protein subunits.

Structural Transformation of the Procapsid Lattice to the Mature Capsid Lattice

The need for subunit motion is particularly clear for P22. The empty procapsid is uniformly spherical, with a diameter of 580 Å (Fig. 11.5). In contrast, the mature capsid is polyhedral, with a maximum dimension of 630 Å along the icosahedral fivefold axis and a minimum dimension of 590 Å along the icosahedral threefold axis (Fig. 11.3). The fivefold positions define the apices. The

icosahedral facets, formed by any three neighboring fivefold positions and related by the icosahedral threefold axis, are planar. The most noticeable features
of the P22 structures are their protrusions and depressions. Protrusions appear
around all fivefold and quasi-sixfold axes in both structures. At a slightly lower
radius, the tips of these protrusions connect to form valleys at all local and
strict twofold axes. Both structures also have depressions at the fivefold, quasi-
sixfold, and local and strict threefold axes. Although these depressions are
widest and deepest at the threefold axes, they are rather shallow at the fivefold
axes. In the empty procapsid shell, however, the depressions at the quasi-sixfold
and fivefold positions culminate in prominent holes.

The radial density plots calculated from the respective 3-D maps of these
structures show that the protein mass lies between the radii of 220 and 290 Å
in the empty shell and the radii of 260 and 315 Å in the mature capsid. Thus
there is considerable thinning of the protein shell upon maturation. Despite a
difference of 15 Å in thickness, the volume occupied by the protein mass
remains fairly constant in the reconstructions. The measured shell volumes are
2.71×10^3 Å^3 for the precursor form and 2.65×10^3 Å^3 for the mature form.
Assuming a density of 1.30 g/cm^3, both structures can accommodate 420 molecules of coat protein of 45 kDa, which agrees with previous biochemical
results.

The lattice transformation can be accounted for by a model in which a
domain of the coat protein located at the inner surface swings over toward the
holes at the fivefold and sixfold positions. Although substantial movement of
protein mass would be involved, the secondary structure change could be small,
especially if there was a hinge in the subunit. Analysis by Raman spectroscopy
indicates that the changes in coat secondary structure during the procapsid-to-
capsid lattice transformations are very small, on the order of a few percent
(Prevelige et al., 1993). The seven quasi-equivalent subunits appear to have
different structural conformations in the procapsid compared with those in the
mature capsid (Prasad et al., 1993). Therefore, the molecules of gp5 can assume
different conformations to assemble and maintain the stability of their structures
before and after DNA encapsidation. Their detailed difference awaits future
high-resolution structure determinations.

A lattice expansion and concurrent volume increase is found for all the
dsDNA phages. Bacteriophage T4 proheads are $T = 13$ prolate icosahedrons
with hexamer clustering at a periodicity of 118 Å. The procapsid expands 15%
to 20%, with the lattice constant increasing to 140 Å and the volume expanding
by 65%. Phage HK97 also undergoes a dramatic conformational change between prohead and head forms. The hexons of the prohead exhibit a distinctive
departure from sixfold symmetry while the hexons of the head become more
sixfold symmetrical (Conway et al., 1995).

Organization of Scaffolding Subunits within the Procapsid

The 3-D structure of the organized state of the scaffolding protein within procapsids has not yet been determined for any virus. From electron micrographs

of thin sections of infected cell and of negatively stained particles, it is clear that the scaffolding is organized as an inner shell in P22, λ, and T4 (Lenk et al., 1975). Figure 11.6*A* shows an image of the procapsid, which has similar shape and size as that of the empty procapsid. The radial distribution of electron density (Fig. 11.6*B*) was determined by low-angle X-ray scattering procapsids in solution for P22, λ, and T4 (Earnshaw et al., 1976, 1979). These studies did not reveal the packing of the subunits or the details of the contacts between the scaffolding subunits and the capsid lattice.

Three-dimensional structures also exist for *E.coli* phage P2 and its small-headed variant P4. Because P4 is a satellite, its particles are assembled in cells that are expressing the P2 coat protein. The decision to polymerize into a $T = 4$ rather than $T = 7$ lattice is controlled by a P4-specific scaffolding protein (Marvik et al., 1994). Cryo−electron microscopy and 3-D reconstructions have revealed the surprising result that the scaffolding molecules are on the outside of the procapsid that forms the small $T = 4$ shell (B. Lindquist and T. Dokland, personal communication).

Three-dimensional reconstructions of electron micrographs of complete procapsids embedded in ice have been carried out for P22 (Thuman-Commike et al., 1996). Their reconstruction assumed icosahedral symmetry. Independent reconstructions yield different subunit packing inside the coat protein shell, suggesting either that the scaffolding is not organized with icosahedral symmetry, or that it is organized differently in different particles. However, a careful analysis of the 3-D map at the inner surface of coat protein shell has revealed an additional finger-like mass density around the quasi-sixfold channels extending inward to the scaffolding protein core (see Color Plate 14). These finger-like features have been implemented as domains of the ordered scaffolding proteins in close contact with the coat proteins.

In the case of the T4 head, which has an elongated morphology, the scaffolding cannot be icosahedral. Paulson and Laemmli (1976) examined the structure by optical diffraction of negatively stained images of the scaffolding proteins within the T4 procapsid. They interpreted the results as six helical chains wrapped around a core. In such a structure, the scaffolding and outer capsid, with fivefold symmetry around the long axis, would not match. However, this symmetry mismatch follows that between the dodecameric portal complex and the procapsid lattice (Moody, 1965).

Portal Vertex Organization in the Procapsids

Because the coat lattice of the procapsid has the same general icosahedral geometry as the mature lattice, an 11-, 12-, or 13-fold structure sits at a fivefold vertex. Whether the portal displaces coat subunits or passes between them is unclear. Electron micrographs of procapsids indicate that at least one domain of the portal ring sits between the coat and scaffolding lattices. The images are consistent with models in which the portal is mechanically held within the shell lattice. In that case, the Hendrix model in which the portal rotates during DNA

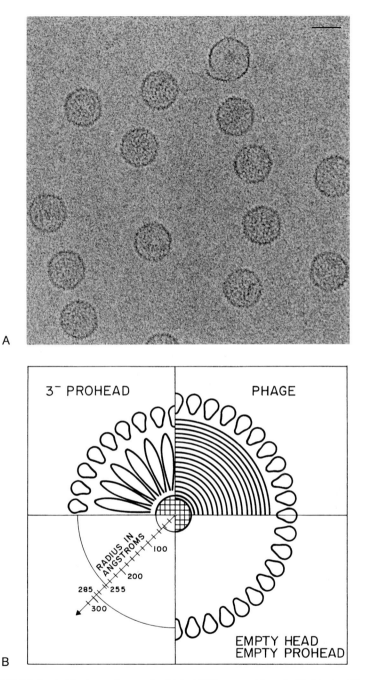

A

B

FIGURE 11.6 *A*. Electron micrograph of intact P22 procapsids embedded in ice. They appear the same size as the empty procapsid except that they have additional density inside. Bar, 500 Å. *B*. Models of procapsids, mature capsids, and empty mature capsids determined from solution X-ray scattering (Reproduced with permission from Earnshaw et al., 1976.) The DNA is shown packed as concentric coils, but other models can account for the density (Earnshaw and Casjens, 1980; Black, 1989). The procapsid samples were purified from cells infected with a mutant in gene 3 that is blocked in DNA packaging. The DNA-containing particles were infectious mature phage. The empty expanded shells were purified from cells infected with a mutant defective in gene 10 that is blocked in closing the portal structure. These particles had packaged DNA within the cells, expanded, and then lost their DNA after packaging (Strauss and King, 1984).

packaging is physically reasonable. If there are specific interactions between the portal subunits and the capsid lattice, rotation would be less likely.

Mechanism of DNA Packaging

In addition to the morphologic changes, the transformation of the procapsid shell to the mature shell involves substantial changes in the properties of both the coat and scaffolding subunits. In all cases, the scaffolding subunits are released. Coat and scaffolding subunits are covalently cleaved in T4, P2/P4, and λ, and capsid proteins in the mature lattices become resistant to further proteolysis. For P22 and T4, the shells become resistant to dissociation by SDS, a technically useful property. For λ and T4, binding sites for the stabilization proteins Soc and Hoc (T4) and gpD (λ) are exposed or created. Steven et al. (1991) have described this as a large-scale allosteric transition, and we consider this a useful formulation.

Initiation and Propagation of the Coat Lattice Transformation

Within cells, the procapsid-to-capsid transition is physiologically coupled to the packaging of the DNA. Because packaging of DNA requires the portal vertex, it seems likely that the procapsid-to-capsid transition is initiated through docking at the portal vertex and propagates throughout the rest of the shell. Because DNA packaging requires hydrolysis of ATP, some of this ATP hydrolysis could be trapped to trigger and propagate the transition. It seems reasonable that the transformation follows a pathway; for example, it could proceed helically around the shell, or could proceed simultaneously up the faces of the icosahedron (see Chapter 1). The formation of polyheads of T4, long cylindrical assemblies of capsid subunits, has made possible the study of the propagation of the transition in these lattices. Upon completion of polymerization of capsid subunits into prolate procapsids, core and shell subunits are processed by prehead proteinase, which removes the N-terminal 66 residues of the 521-residue gp23. Polyheads are formed in cells infected with mutant phages defective in scaffolding assembly, and so lack the inner scaffolding protein layer. The coat protein can still be cleaved by exogenous protease, providing an additional signal with respect to the transition. Upon cleavage, the polyheads expand (Ross et al., 1985; Steven et al., 1992).

The accessory proteins only bind to the cleaved and expanded form of the lattice. In polyheads, the lattice appears coarse, because of the lack of density at the centers of the hexamers and between hexamers. The addition of three molecules of Soc at the triplex positions and one molecule of Hoc at the hexamer centers fills in the surface lattice and converts it to a smooth appearance. This reaction can proceed in vitro, and the resulting polyheads always have a continuous smooth stretch and a continuous coarse stretch. This appearance change suggests that the proteolysis and lattice transformation propagates sequentially along the structure (Steven et al., 1976a, 1976b).

The T4 capsid proteins undergo a substantial conformational change, as shown in the movement of epitopes between the procapsid lattice and mature lattice (Steven et al., 1991). Epitopes found on the inner surface in the mature lattice appear on the outer surface of the procapsid lattice. In the P22 phage, the expansion of the capsid is associated with flattening of the icosahedral faces (Fig. 11.7), and the mass translocation takes place predominantly at the inner surface of the capsid protein shell (Prasad et al., 1993).

Scaffolding Exit and DNA Entry

Procapsids in the productive particle assembly pathways contain hundreds of molecules of scaffolding proteins that are removed near or at the time of capsid expansion. In some cases (T4, λ) the scaffolding molecules are proteolysed into small fragments that then escape from the capsid (Onorato and Showe, 1975; Ray and Murialdo, 1975). In the case of P22 and φ29, the scaffolding molecules exit the procapsid intact, to be recycled in subsequent rounds of procapsid assembly (King and Casjens, 1974; Nelson et al., 1976). The P22 scaffolding probably exits through the channels observed at the sixfold centers of the procapsid, as seen in Figure 11.5 (Prasad et al., 1993). The λ scaffolding proteins are cleaved into much smaller pieces, perhaps explaining why the λ procapsid does not have visible holes (Dokland and Murialdo, 1993). It is not known whether the loss of scaffolding represents a change in the coat subunits prior to the lattice transition, or whether the lattice change also propagates through the scaffolding lattice, and a change in the scaffolding conformation is responsible for release.

The P22 scaffolding protein can be extracted from procapsids with low concentrations of denaturants (Fuller and King, 1981; Prevelige et al., 1988) or heating to 49 °C (Galisteo and King, 1993). The release of scaffolding protein is reversible; upon removal of denaturants, scaffolding proteins can reenter the extracted unexpanded procapsid to regenerate morphologically normal procapsids (Greene and King, 1994). Thus for in vitro extraction, the partial denaturation of a domain of the scaffolding protein is responsible for the release. Although extraction of scaffolding protein leaves the capsid unaltered, it is also possible to purify capsids that have undergone expansion in vivo but are empty of DNA. The scaffolding protein does not interact with these expanded capsids at all (Greene and King, 1994), suggesting that the expansion eliminates the scaffolding binding sites or renders them inaccessible.

Shell expansion, scaffolding protein release, and DNA entry all occur early in the stages of DNA packaging, and their physiologic sequence remains uncertain. Phage T4 procapsids appear competent to encapsidate DNA after expansion (Rao and Black, 1985). In other in vitro systems, shell expansion appears to be dependent on DNA packaging. In T4 and λ, whose scaffolding proteins undergo extensive proteolysis, proteolysis can occur before DNA entry (Black, 1989). For P22, φ29, and T3, whose scaffolding subunits recycle, scaffolding exit and DNA entry have not been separated. In no case have expanded

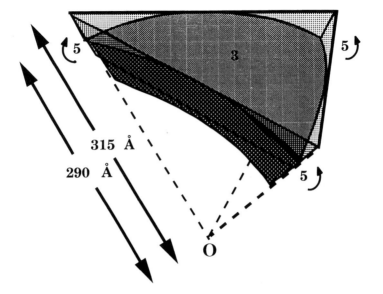

FIGURE 11.7 Schematic diagram illustrating the mechanism of shell expansion in P22, in which the icosahedral triangular face is flattened. The changes start at the fivefold vertex upon the DNA entry and cause the subunits to move closer to each other and outward. These conformational changes propagate throughout the lattice. (Reproduced with permission from Prasad et al., 1993.)

procapsids containing scaffolding protein been recovered, suggesting that protein exit always precedes or accompanies expansion.

What are the signals for scaffolding exit and shell expansion? Short lengths of DNA can enter procapsids in vitro without triggering shell expansion—11.4% of chromosome length for λ (Earnshaw and Casjens, 1980; Hohn, 1983) and less than 25% in T3 (Shibata et al., 1987)—whereas longer lengths are only found in expanded shells. For ɸ29, in vitro packaging of about 30% of the genome resulted in shell transformation (Bjornsti et al., 1983). These results suggest that, in the productive pathway, 10% to 30% of the genome must enter to trigger shell expansion.

Energetics

The energetics of the procapsid-to-capsid transition have been studied directly in P22 and in T4. P22 procapsids are stable for long periods of time in the cold and do not spontaneously dissociate back into subunits, or expand to the mature form. As the temperature is raised, they pass through a series of transitions (Galisteo and King, 1993). The procapsids lose their scaffolding subunits in an endothermic transition centered at 49 °C. The expansion of the shell occurs at 61 °C, and represents an exothermic transition in which energy is released by the shell. Thus the expanded lattice is a lower energy state than the initial unexpanded procapsid shell. A significant kinetics barrier clearly must exist to maintain the procapsid lattice in its higher energy, unexpanded state. The 90-kJ/subunit requirement associated with the shell expansion exotherm represents about 5% to 10% of the energy required to denature at 87 °C. It has been estimated that this represents about 3 ATP molecules per monomer, or 1200 ATP molecules per capsid (Steven, 1993). At 71 °C the expanded shells undergo a localized denaturation of a capsid subunit domain, and finally at 87 °C they undergo the major thermal transition in which the subunits denature and dissociate (Galisteo and King, 1993).

In T4, lattice expansion is coupled to subunit cleavage. Steven et al. (1991) have studied the procapsid-to-capsid transition in T4 using the very long polyheads formed of gp23 capsid protein in the uncleaved conformation. Cleavage of the capsid protein and expansion of the lattice can be induced in vitro. Calorimetry of the cleaved/expanded forms revealed that they were significantly more stable than the uncleaved/unexpanded forms (Ross et al., 1985). It was also possible to obtain polyheads that were uncleaved but expanded, and these were intermediate in stability between the uncleaved/unexpanded and cleaved/expanded forms. Thus a negative enthalpy change is associated with the expansion itself. Steven (1993) proposed that the N-terminal domain when present interfered with the subunit stabilization that occurred during expansion.

Function of Capsid Expansion

Why do the dsDNA phage capsids expand during DNA packaging? A number of proposals have been put forward to account for the lattice expansion and rearrangement that has been observed.

   a. *Cytoplasmic displacement*—Earnshaw and Casjens (1980) proposed that scaffolding might function to prevent the enclosure by the growing shell of the random cytoplasmic proteins that would be encapsidated as the shell formed. Virus-specific scaffolding subunits would be much easier to remove than unrelated host and phage proteins.

   b. *Shell lattice stabilization or locking*—If the packaging process requires subunit–subunit motions, the lattice transition might be a mechanism to lock the subunits of the shell into the mature conformation (Steven et al., 1976b, 1991, 1992). The studies of T4 polyheads (Steven et al., 1992) and P22 phage (Galisteo and King, 1993) showed directly that the subunit rearrangements significantly stabilized the resulting structures. Hendrix has made a good case for such locking by the covalent linking found in Hong Kong 97 phage (R. Duda and R. Hendrix, personal communication).

   c. *Generation of DNA binding sites*—The most general feature shared by procapsids is their precursor role in the packaging of DNA. We develop in additional detail here a model that directly couples scaffolding exit, shell expansion, and DNA packaging. Because the dsDNA phages inject their DNA, leaving their mature capsid intact, the condensed DNA must be free to leave the capsid. Indeed, there is no substantive evidence for direct

binding of packaged DNA by the inner surface of the capsid lattice. However, the initial organization of the DNA within the shell may require binding by the capsid subunits, to organize the first DNA packaged as originally
proposed (Earnshaw et al., 1979). Such DNA binding sites would have to
be transiently exposed as the first DNA entered but then inactivated,
masked, or removed. In this model, the coat protein domains that swing
in to close the scaffolding exit channels would also represent the disruption
of the DNA binding sites.

Given the very high concentrations of coat protein subunits required to
assemble procapsid shells, and the high concentrations of DNA in phage-
infected cells late in infection, it would be problematic if the subunits could
bind DNA. However, if the DNA binding sites were covered by scaffolding
subunits, they would be masked until the point of scaffolding exit. Scaffolding
exit would expose the DNA binding sites, leading to the organization of the
first coils of DNA entering the shell. In P22, the scaffolding protein can be
removed in vitro without triggering shell expansion (Prasad et al., 1993). We
envisage that interaction of the newly entered DNA might then trigger the
expansion and rearrangement reaction, which would lead to both closing the
channels and burying or disrupting the DNA binding site. The remaining DNA
would be pumped in by the packaging enzymes. After the cleavage of the
dsDNA by the packaging complex, the portal would be closed by the addition
of the neck proteins, mechanically fixing the DNA in place (Strauss and King,
1984).

It is striking that phages from different hosts, which lack homology in their
coat protein subunits and have different capsid morphologies, all go through a
related transformation. The most likely connection is DNA packaging, because
the macromolecular properties of the chromosomes packaged are indistinguishable, and because the capsids can package different genomes. We suspect that
capsid expansion is closely coupled to DNA packaging and reflects the need
to organize the entering coils of DNA as they enter the capsid.

Future Prospects

Phage biology has been well advanced for decades. However, the detailed structural mechanisms underlying capsid assembly, DNA condensation, and capsid
maturation remain inadequately defined. Progress using cryo−electron microscopy and biophysical techniques has revealed structural details of the kinetic
intermediates in shell assembly and DNA packaging. The continuing improve-
ment in electron imaging and computational techniques for analysis of particles
with and without icosahedral symmetry (Zhou et al., 1994; Serysheva et al.,
1995) will likely play a significant role in revealing finer structural details of
phage particles at different functional states. With the wealth of genetic and
biochemical information, and the relevance of the phage systems in biotechnology, we look forward to significant progress in elucidating capsid lattice

structure and dynamics at the level of the underlying macromolecular motions and interactions.

Acknowledgments

Support for the preparation of this chapter was provided by National Institutes of Health grants GM17980 to J.K. and AI38469, NCRR (RR02250), and W. M. Keck Foundation Support, to W.C.

We thank Sherwood Casjens, Barrie Greene, and Pam Thuman-Commike for critical reading of the manuscript and many useful suggestions. We thank Pam Thuman-Commike and Z. Hong Zhou for their expert assistance in preparing the figures.

References

Anderson, D.L. and Reilly, B.E. 1993. Morphogenesis of bacteriophage ϕ29. In *Bacillus subtilis* and Other Gram-Positive Bacteria: Physiology, Biochemistry, and Molecular Genetics. (Hoch, J.A., Losick, R. and Sonenshein, A.L., Eds.), pp. 859–867. ASM Publications, Washington, DC.

Aubrey, K.L., Casjens, S.R. and Thomas, G.J., Jr. 1992. Secondary structure and interactions of the packaged dsDNA genome of bacteriophage P22 investigated by Raman difference spectroscopy. Biochemistry 31: 11835–11842.

Ausubel, F.M., Brent, R., Kingston, R.E., Moore, D.D., Seidman, J.G., Smith, J.A. and Struhl, K. 1994. Current protocols in Molecular Biology. Vol. 1. John Wiley & Sons, New York.

Bazinet, C. and King, J. 1988. Initiation of P22 procapsid assembly *in vivo*. J. Mol. Biol. 202: 77–86.

Bjornsti, M.-A., Reilly, B.E. and Anderson, D.L. 1983. Morphogenesis of bacteriophage ϕ29 of *Bacillus subtilus*: Oriented and quantized in vitro packaging of DNA protein gp3. J. Virol. 45: 383–396.

Black, L.W. 1989. DNA packaging in dsDNA bacteriophages. Annu. Rev. Microbiol. 43: 267–292.

Black, L.W., Newcomb, W.W., Boring, J.W. and Brown, J.C. 1985. Ion etching of bacteriophage T4: support for a spiral-fold model of packaged DNA. Proc. Natl. Acad. Sci. USA 82: 7960–7964.

Casjens, S. 1989. Bacteriophage P22 DNA packaging. In Chromosome eukaryotic prokaryotic and viral. (Adolph, K., Ed.), pp. 241–261. CRC Press, Boca Raton.

Casjens, S., Hatfull, G. and Hendrix, R.W. 1993. Evolution of ds DNA tail bacterial phage genome. Semin. Virol. 3: 383–397.

Casjens, S. and Hendrix, R. 1988. Control mechanisms in dsDNA bacteriophage assembly. In The Bacteriophages. (Calender, R., Ed.), pp. 15–91. Plenum, New York.

Caspar, D.L.D. and Klug, A. 1962. Physical principles in the construction of regular viruses. Cold Spring Harbor Symp. Quant. Biol. 27: 1–32.

Conway, J.F., Duda, R.L., Hendrix, R.W. and Steven, A.C. 1995. Proteolytic and conformational control of virus capsid maturation: The bacteriophage HK97 system. J. Mol. Biol. 253: 86–99.

D'Halluin, J.-C.M., Martin, G.R., Torpier, G. and Boulanger, P. 1978. Adenovirus type 2 assembly analyzed by reversible cross-linking of labile intermediates. J. Virol. 26: 357–363.

Dokland, T., Lindquist, B.H. and Fuller, S.D. 1992. Image reconstruction from cryo-electron micrographs reveals the morphopoetic mechanism in the P2-P4 bacteriophage system. EMBO J. 11: 839–846.

Dokland, T. and Murialdo, H. 1993. Structural transitions during maturation of bacteriophage lambda capsids. J. Mol. Biol. 233: 682–694.

Dube, P., Tavares, P., Lurz, R. and van Heel, M. 1993. The portal protein of bacteriophage SPP1: a DNA pump with 13-fold symmetry. EMBO J. 12: 1303–1309.

Earnshaw, W. and Casjens, S. 1980. DNA packaging by double stranded DNA bacteriophages. Cell 21: 319–325.

Earnshaw, W., Casjens, S. and Harrison, S.C. 1976. Assembly of the head of bacteriophage P22: X-ray diffraction from heads, proheads and related structures. J. Mol. Biol. 104: 387–410.

Earnshaw, W. and Harrison, S.C. 1977. DNA arrangement in isometric phage heads. Nature 268: 598–602.

Earnshaw, W.C., Hendrix, R.W. and King, J. 1979. Structural studies of bacteriophage lambda heads and proheads by small angle X-ray diffraction. J. Mol. Biol. 134: 575–594.

Earnshaw, W., King, J., Harrison, S.C. and Eiserling, F.A. 1978. The structural organization of DNA packed within the heads of T4 wild type, isometric, and giant bacteriophages. Cell 14: 559–568.

Edgar, R.S., Denhardt, G.H. and Epstein, R.H. 1964. A comparative genetic study of conditional lethal mutations of bacteriophage T4D. Genetics 49: 635–648.

Edgar, R.S. and Epstein, R.H. 1964. Genetics of bacterial virus. Sci. Am. 212: 71–78.

Edvardsson, B., Everitt, E., Jornvall, H., Prage, L. and Philipson, L. 1976. Intermediates in adenovirus assembly. J. Virol. 19: 533–547.

Epstein, R.H., Bolle, A., Steinberg, C.M., Kellenberger, E., de la Boy, T.E., Chevalley, R., Edgar, R.S., Sussman, M., Denhardt, G.H. and Lielausis, A. 1963. Cold Spring Harbor Symp. Quant. Biol. 28: 375–392.

Fuller, M.T. and King, J. 1981. Purification of the coat and scaffolding protein from procapsids of bacteriophages P22. Virology 112: 529–547.

Galisteo, M.L. and King, J. 1993. Conformational transformations in the protein lattice of phage P22 procapsids. Biophys. J. 65: 227–235.

Greene, B. and King, J. 1994. Binding of scaffolding subunits within the P22 procapsid lattice. Virology 205: 188–197.

Guo, P., Erickson, S. and Anderson, D.L. 1987. A small viral RNA is required for *in vitro* packaging of bacteriophage φ29 DNA. Science 236: 690–694.

Harrison, S.C. 1983. Packaging of DNA into bacteriophage heads: a model. J. Mol. Biol. 171: 577–580.

Hendrix, R. 1978. Symmetry mismatch and DNA packaging in large DNA bacteriophages. Proc. Natl. Acad. Sci. U.S.A. 75: 4779–4783.

Hohn, B. 1983. DNA sequences necessary for packaging of bacteriophage l DNA. Proc. Natl. Acad. Sci. U.S.A. 80: 7456–7460.

Hohn, B. and Hohn, T. 1974. Activity of empty, headlike particles for packaging of DNA of bacteriophage. Proc. Natl. Acad. Sci. U.S.A. 71: 2372–2376.

Imber, R., Tsugita, A., Wurtz, M. and Hohn, T. 1980. Outer capsid protein of bacteriophage lambda. J. Mol. Biol. 139: 277–295.

King, J. and Casjens, S. 1974. Catalytic head assembling protein in virus morphogenesis. Nature 251: 112–119.

King, J. and Laemmli, U.K. 1971. Polypeptides of the tail fibres of bacteriophage T4. J. Mol. Biol. 62: 465–477.

Kocsis, E., Cerritelli, M.E., Trus, B.L. and Steven, A.C. 1993. Determination of rotational symmetry in macromolecules. Biophys. J. 64: 64a.

Laemmli, U.K. 1970. Cleavage of structural proteins during the assembly of the head of bacteriophage T4. Nature 227: 680–685.

Lee, J.Y., Irmiere, A. and Gibson, W. 1988. Primate cytomegalovirus assembly: evidence that DNA packaging occurs subsequent to B capsid assembly. Virology 167: 87–96.

Lenk, E., Casjens, S., Weeks, J. and King, J. 1975. Intracellular visualization of precursor capsids in phage P22 mutant infected cells. Virology 68: 182–199.

Lepault, J., Dubochet, J., Baschong, W. and Kellenberger, E. 1987. Organization of double-stranded DNA in bacteriophages: a study by cryo-electron microscopy of vitrified samples. EMBO J. 6: 1507–1512.

Liddington, R.C., Yan, Y., Moulai, J., Sahli, R., Benjamin, T.L. and Harrison, S.C. 1991. Structure of simian virus 40 at 3.8 Å resolution. Nature 354: 278–284.

Marvik, O.J., Sharma, P., Dokland, T. and Lindqvist, B.H. 1994. Bacteriophage P2 and P4 assembly: alternative scaffolding proteins regulate capsid size. Virology 200: 702–714.

Moody, M. 1965. The shape of the T-even bacteriophage head. Virology 26: 567–576.

Murialdo, H. 1979. Early intermediates in bacteriophage lambda prohead assembly. Virology 96: 341–367.

Murray, N.E., Bruce, D. and Murray, K. 1977. Lambdoid phages that simplify the recovery of *in vitro* recombinants. Mol. Gen. Genet. 150: 53–61.

Nelson, R.A., Reilly, B.E. and Anderson, D.L. 1976. Morphogenesis of bacteriophage ϕ29 of *Bacillus subtilis*: preliminary isolation and characterization of intermediate particles of the assembly pathway. Virology 19: 518–532.

Newcomb, W.W., Trus, B.L., Booy, F.P., Steven, A.C., Wall, J.S. and Brown, J.C. 1993. Structure of the herpes simplex virus capsid: molecular composition of the pentons and the triplexes. J. Mol. Biol. 232: 499–511.

Onorato, L. and Showe, M.K. 1975. Gene *gp21* protein-dependent proteolysis *in vitro* of purified gene *gp22* product of bacteriophage T4. J. Mol. Biol. 92: 395–412.

Orlova, E.V., Serysheva, I.I., van Heel, M., Hamilton, S.L. and Chiu, W. 1996. Two structural configurations of the skeletal muscle calcium-release channel. Nature Struct. Biol. 3: 547–552.

Paulson, J.R. and Laemmli, U.K. 1976. Morphogenetic core of the bacteriophage T4 head: structure of the core in polyheads. J. Mol. Biol. 111: 459–485.

Prasad, B.V.V., Prevelige, P.E., Marietta, E., Chen, R.O., Thomas, D., King, J. and Chiu, W. 1993. Three-dimensional transformation of capsids associated with genome packaging in a bacterial virus. J. Mol. Biol. 231: 65–74.

Prevelige, P.E., Dennis, T.J. and King, J. 1988. Scaffolding protein regulates the polymerization of P22 coat subunits into icosahedral shells *in vitro*. J. Mol. Biol. 202: 743–757.

Prevelige, P.E. Jr. and King, J. 1993. Assembly of bacteriophage P22: a model for dsDNA virus assembly. Prog. Med. Virol. 40: 206–221.

Prevelige, P.E. Jr., Thomas, D., Kelly, A.L., Towse, S.A. and Thomas, G.J. Jr. 1993. Subunit conformational changes accompanying bacteriophage P22 capsid maturation. Biochemistry 32: 537–543.

Rao, V.B. and Black, L.W. 1985. DNA packaging of bacteriophage T4 proheads *in vitro*: evidence that prohead expansion is not coupled to DNA packaging. J. Mol. Biol. 185: 565–578.

Ray, P. and Murialdo, H. 1975. The role of gene Nu3 in bacteriophage lambda head morphogenesis. Virology 64: 247–263.

Reilly, K.E. and Thomas, G.J. 1994. Hydrogen exchange dynamics of the P22 virion determined by time-resolved Raman spectroscopy: effects of chromosome packing on the kinetics of nucleotide exchanges. J. Mol. Biol. 241: 68–82.

Richards, K., Williams, R. and Calendar, R. 1973. Mode of DNA packing within bacteriophage heads. J. Mol. Biol. 78: 255–259.

Rixon, F.J. 1993. Structure and assembly of herpesviruses. Semin. Virol. 4: 135–144.

Robinson, I.K. and Harrison, S.C. 1982. Structure of the expanded state of tomato bushy stunt virus. J. Mol. Biol. 78: 563–568.

Ross, P.D., Black, L.W., Bisher, M.E. and Steven, A.C. 1985. Assembly-dependent conformational changes in a viral capsid protein. J. Mol. Biol. 183: 353–364.

Rossmann, M.G. 1984. Constraints on the assembly of spherical virus particles. Virology 134: 1–11.

Serysheva, I.I., Orlova, E.V., Chiu, W., Sherman, M.B., Hamilton, S.L. and van Heel, M. 1995. Electron cryomicroscopy and angular reconstitution to visualize the skeletal muscle calcium-release channel. Nature Struct. Biol. 2: 18–24.

Sherman, G. and Bachenheimer, S.L. 1988. Characterization of intranuclear capsids made by morphogenetic mutants of HSV-1. Virology 163: 471–480.

Shibata, H., Fujisawa, H. and Minagawa, T. 1987. Characterization of the bacteriophage T3 DNA packaging reaction in vitro in a defined system. J. Mol. Biol. 196: 845–851.

Showe, M.K. and Black, L.W. 1973. Assembly core of bacteriophage T4: an intermediate in head formation. Nature New Biol. 242: 70–75.

Steven, A.C. 1993. Conformational change—an alternative energy source? Exothermic phage transition in phage capsid maturation. Biophys. J. 65: 5–6.

Steven, A.C., Aebi, U. and Showe, M.K. 1976a. Folding and capsomere morphology of the P23 surface shell of bacteriophage T4 polyheads from mutants in five different head genes. J. Mol. Biol. 102: 373–407.

Steven, A.C., Bauer, A.C., Bisher, M.E., Robey, F.A. and Black, L.W. 1991. The maturation dependent conformational change of T4 capsid involves the translocation of specific epitopes between the inner and the outer capsid surfaces. J. Struct. Biol. 106: 236–242.

Steven, A.C., Couture, E., Aebi, U. and Showe, M.K. 1976b. Structure of T4 polyheads II. A pathway of polyhead transformations as a model for T4 capsid maturation. J. Mol. Biol. 106: 187–221.

Steven, A., Heather, C., Greenstone, L., Booy, F.P., Black, L.W. and Ross, P.D. 1992. Conformational changes of a viral capsid protein gp28, thermodynamic rationale for proteolytic regulation of bacteriophage T4 capsid expansion, co-operativity, and super-stabilization by Soc binding. J. Mol. Biol. 228: 870–884.

Stewart, P.L., Fuller, S.D. and Burnett, R.M. 1993. Difference imaging of adenovirus: bridging the resolution gap between X-ray crystallography and electron microscopy. EMBO J. 12: 2589–2599.

Strauss, H. and King, J. 1984. Steps in the stabilization of newly packaged DNA during phage P22 morphogenesis. J. Mol. Biol. 172: 523–543.

Thomas, M., Cameron, J.R. and Davis, R.W. 1974. Viable molecular hybrids of bacteriophage lambda and eukaryotic DNA. Proc. Natl. Acad. Sci. U.S.A. 71: 4579–4583.

Thuman-Commike, P.A., Greene, B., Jakana, J., Prasad, B.V.V., King, J., Prevelige, P.E. Jr. and Chiu, W. 1996. Three-dimensional structure of scaffolding-containing phage P22 procapsids by electron cryo-microscopy. J. Mol. Biol. 260: 85–98.

Tsuprun, V., Anderson, D. and Egelman, E.H. 1994. The bacteriophage φ29 head-tail connector shows 13-fold symmetry in both hexagonally packed arrays and as single particles. Biophys. J. 66: 2139–2150.

Turnquist, S., Simon, M., Egelman, E. and Anderson, D. 1992. Supercoiled DNA wraps around the bacteriophage φ29 head-tail connector. Proc. Natl. Acad. Sci. U.S.A. 89: 10479–10483.

Zhou, Z.H., He, J., Jakana, J., Tatman, J., Rixon, F. and Chiu, W. 1995. Assembly of VP26 in HSV-1 inferred from structures of wild-type and recombinant capsids. Nature Struct. Biol. 2: 1026–1030.

Zhou, Z.H., Prasad, B.V.V., Jakana, J., Rixon, F. and Chiu, W. 1994. Protein subunit structures in the herpes simplex virus A-capsid determined from 400 kV spot-scan electron cryomicroscopy. J. Mol. Biol. 242: 458–469.

12.

Herpesvirus Capsid Assembly and Envelopment

ALASDAIR C. STEVEN & PATRICIA G. SPEAR

Members of the *Herpesviridae* family have linear double-stranded DNA (dsDNA) genomes, ranging in size from about 80,000 to 250,000 bp, and make use of the cell nucleus for genome replication, transcription, and nucleocapsid assembly. The inner nuclear membrane supplies an envelope, which becomes modified or replaced by an altered or new envelope as the virion exits from the cell. Key features of the herpesvirus virion are the DNA-containing core, the icosahedral capsid exhibiting 162 capsomers, the layer of proteinaceous material (tegument) between the capsid outer surface and envelope inner surface, and the membranous envelope containing numerous viral glycoproteins.

Most species of vertebrate animals are susceptible to infection by several distinct, species-specific herpesviruses. These viruses have been classified into three subfamilies based on biologic properties (Roizman et al., 1992). The alphaherpesviruses include the neurotropic viruses, the betaherpesviruses include the cytomegaloviruses, and the gammaherpesviruses, the lymphotropic viruses. This chapter focuses on assembly of the alphaherpesviruses, principally herpes simplex virus (HSV). It seems likely that the main features of morphogenesis will be similar for all the herpesviruses.

Herpes simplex virus type 1 (HSV-1) is the best characterized herpesvirus, with respect to molecular composition and structure. Table 12.1 lists the viral proteins known to be present in its nucleocapsid, tegument, and envelope. Homologous forms of many, but perhaps not all, of these proteins have been or will be found in the virions of other alphaherpesviruses (Mettenleiter, 1994). The total number of HSV-1 virion proteins is currently estimated to be 34. It

TABLE 12.1 Protein Composition of HSV-1 Virions[a]

Nucleocapsid	Tegument	Envelope
—*UL6*	—*UL4*	gL—*UL1*
VP23—*UL18*	—*UL11*	gM—*UL10*
VP5—*UL19*	VP18.8—*UL13*	gH—*UL22*
VP24—*UL26*	—*UL21*	gB—*UL27*
VP26—*UL35*	—*UL25*	—*UL34*
VP19c—*UL38*	VP$^{1/2}$—*UL36*	gC—*UL44*
	—*UL37*	—*UL45*
	VHS—*UL41*	gG—*US4*
	VP11/12—*UL46*	gD—*US6*
	VP13/14—*UL47*	gl—*US7*
	VP16—*UL48*	gE—*US8*
	VP22—*UL49*	
	—*UL56*	
	—*US3*	
	—*US3*	
	—*US9*	
	—*US10*	
	—*US11*	

[a]Names assigned to the proteins, if any, appear before the dash and gene designations according to McGeoch et al. (1988) appear after the dash.

should be noted that other viral (and probably cellular) proteins participate in virion assembly, without remaining or becoming incorporated into the final structure. As described later, there are viral scaffolding proteins present within immature capsid shells that are replaced by the viral DNA in mature nucleocapsids and there are viral proteins that participate in the processing and packaging of viral DNA, probably without becoming incorporated into the virion. Moreover, the tegument and envelope acquired at the inner nuclear membrane may have a different composition than the tegument and envelope of the mature infectious virion released from the infected cell.

Replicative Events Preceding Virion Assembly

Entry of Virus into Cells

Entry of alphaherpesviruses into cells requires attachment of virus to cell surface receptors followed by fusion of the virion envelope with a cell membrane (see reviews by Spear, 1993, and Mettenleiter, 1994). The fusion can occur with the plasma membrane but may perhaps also occur following endocytosis of the virion. The acidification of endosomes is not a requirement for alphaherpesvirus entry because the membrane fusion required for entry can occur at neutral pH.

For several of the alphaherpesviruses (including HSV-1, HSV-2, pseudo-rabies virus [PRV], and bovine herpesvirus type 1), the binding of virions to cells is mediated by interaction of the envelope glycoprotein gC with cell surface glycosaminoglycans (GAGs), preferentially heparan sulfate. These viruses have redundant mechanisms for binding to cells. If gC is absent, the mutant virions can bind with reduced efficiency to at least some cell types. In the case of HSV-1, the binding of gC-negative virions to cells is mediated by gB and is also dependent on the presence of cell surface GAGs (Herold et al., 1994). In the case of PRV, it is not clear how the gC-negative virions bind to cells, but the binding is not dependent on cell surface GAGs (Mettenleiter et al., 1990; Zsak et al., 1990).

At least four other glycoproteins, designated gB, gD, gH, and gL, are required to mediate viral penetration of the cell (virion–cell fusion). For HSV-1, cell surface GAGs may be required to trigger the membrane fusion reaction, at least for virus-induced cell fusion (Shieh and Spear, 1994), and perhaps also for viral penetration. Penetration also requires interactions of virion glycoproteins with one or more cell surface proteins or specific mediators of entry. The cell surface protein that can mediate HSV-1 entry into human T cells is HVEM, a new member of the tumor necrosis factor receptor family (Montgomery et al., 1996). Other cell surface proteins mediate HSV-1 entry into other cell types, indicating the existence of different pathways of entry for different cell types.

The events following fusion between the virion envelope and cell membrane are not well defined but must involve separation of the nucleocapsid from the envelope and release of tegument components as well as the nucleocapsid into the cytoplasm. It is thought that nucleocapsids are transported to nuclear pores, where the DNA is released into the nucleus (Tognon et al., 1981; Batterson et al., 1983). Several of the tegument components are regulatory proteins, some of which must also be targeted to the cell nucleus.

Expression of Viral Genes

In permissive cells, the entry of viral nucleocapsids and tegument proteins into the cell causes an immediate reprogramming of protein synthesis and gene expression, through the action of tegument proteins and cell factors. Two of these tegument proteins, encoded by genes *UL41* and *UL13*, cooperate to mediate by unknown means a degradation of preexisting cytoplasmic mRNA and inhibition of protein synthesis (Read and Frenkel, 1983; Overton et al., 1994). Other tegument proteins influence viral gene transcription. VP16 (α-TIF or Vmw65), encoded by gene *UL48*, interacts with cell transcription factors to upregulate the expression of viral immediate-early or α genes (Ace et al., 1989; Goding and O'Hare, 1989). VP11/12 (*UL46*) and VP13/14 (*UL47*) can modulate this regulatory activity of VP16 (Zhang and McKnight, 1993).

The protein products of the immediate-early genes are themselves regulatory proteins required to turn on or modulate the expression of other viral genes, including the early (β), delayed-early (βγ), and late (γ) genes in temporal sequence (reviewed by Roizman and Sears, 1990). The early genes encode most of the enzymes and factors required for viral DNA replication, and the delayed-early and late genes encode most of the structural proteins of the virion. The transcription of viral genes, replication of the viral genome, and assembly of nucleocapsids take place entirely in the cell nucleus. Thus, the viral proteins required for these activities must carry nuclear localization signals or associate with factors that localize to the nucleus.

Nucleocapsid Structure

Distinct Capsid Species: Interrelationships and Molecular Compositions

A-Capsids, B-Capsids, and C-Capsids

Three main types of intracellular capsids have been distinguished on morphologic grounds in thin sections of infected cells. A-capsids are seemingly empty shells, whereas C-capsids contain darkly staining DNA (Furlong et al., 1972; Puvion-Dutilleul et al., 1987). B-capsids also contain internal material that stains less densely than that in C-capsids, and often has the appearance of an inner ring. Three species of isolated capsids have been separated by density gradient centrifugation: A-capsids (also called ''lights''; Fig. 12.1), B-capsids (''intermediates''; Fig. 12.1) and C-capsids (''heavies''; Fig. 12.2), which correlate with the three species visualized in situ. Pulse-chase experiments done with cells infected by equine herpesvirus type 1 suggest the possibility that B-capsids mature into C-capsids (but see p. 331), whereas A-capsids are abortive products (Perdue et al., 1976), resulting perhaps from failed attempts to package B-capsids with DNA.

Molecular Compositions

Purification of these capsids has allowed their protein compositions to be determined (Gibson and Roizman, 1972; Cohen et al., 1980; Dargan, 1986; Newcomb et al., 1989; Patel and Maclean, 1995). There are seven capsid proteins that each account for more than a few percent (w/w) of the total protein mass. Patel and Maclean (1995) reported that an eighth protein, the product of the *UL6* gene, is also associated with virus capsids. The status of the *UL6* product, and of other minor constituents, as integral components of the capsid has yet to be established. A molecular inventory of highly purified HSV-1 B-capsids (Table 12.2) has been compiled by estimating the relative amounts of the seven major proteins by quantitative sodium dodecyl sulfate–polyacrylamide gel electrophoresis (SDS-PAGE) and translating these mass fractions into copy numbers on the basis of particle masses determined by scanning transmission electron microscopy (STEM) and molecular masses derived from the primary sequences (Newcomb et al., 1993). The B-capsid consists of a shell (~198

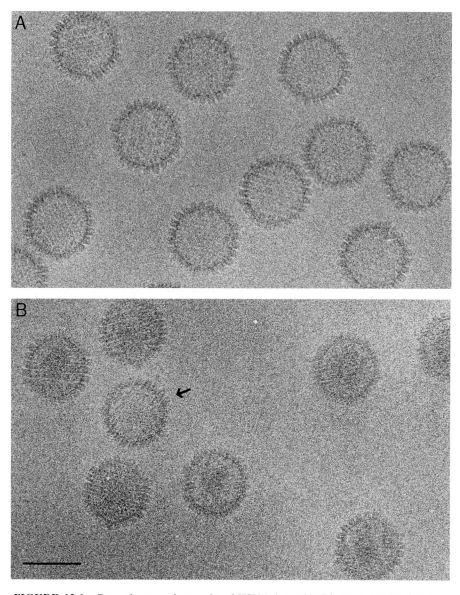

FIGURE 12.1   Cryo–electron micrographs of HSV-1 A-capsids (*A*) and B-capsids (*B*). The arrow in B marks an A-capsid also present in this field. (Courtesy of F.P. Booy and W.W. Newcomb.) Bar, 1000 Å.

MDa) and a proteinaceous core (~47 MDa). The shell composition and structure are common to A-capsids and C-capsids (Booy et al., 1992; Newcomb et al., 1993), whereas most of the B-capsid core proteins are absent from C-capsids and A-capsids.

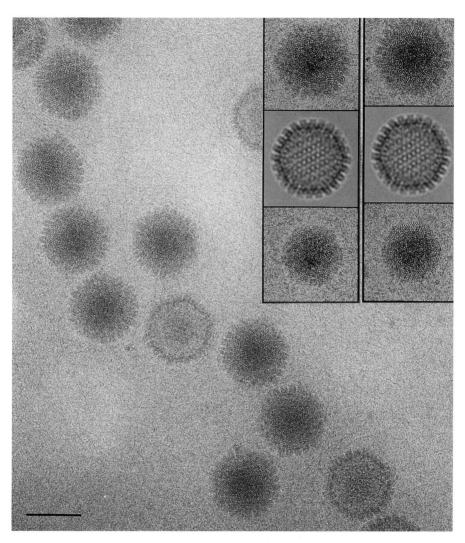

FIGURE 12.2 Cryo–electron micrograph of HSV-1 C-capsids, recorded at a value of approximately 0.7 μm underfocus. Under these conditions, the capsomers and the protein shell are weakly contrasted but the "fingerprint" motif, which represents the projection of the densely packed DNA with a characteristic interduplex spacing of 26 to 28 Å (Booy et al., 1991), is visualized with strong contrast. The ball of DNA fills the capsid, right out to the inner surface of the capsid shell. (Courtesy of F.P. Booy and W.W. Newcomb.) *Inset.* Computational stripping of the capsid shell to reveal the ball of encapsidated DNA. (Courtesy of B.L. Trus.) In these two examples, the capsids are viewed from similar angles, but the packaged genomes look quite different. Bar, 1000 Å.

TABLE 12.2 Protein Composition of HSV-1 B-Capsids

Gene	Protein	M_r (Da)	% Density[a]	Copy No.	Location
UL19	VP5	149,075	54.9 ± 1.4	960	Major shell component
UL38	VP19c	50,260	8.4 ± 0.3	375[b]	Triplexes
UL26	VP21[c]	45,000	1.7 ± 0.8	87	Core
UL26.5	VP22a	33,765	17.2 ± 2.3	1153[d]	Core
UL18	VP23	34,268	8.7 ± 0.9	572[b]	Triplexes
UL26	VP24[c]	26,628	1.7 ± 0.7	147	(Core)
UL35	VP26	12,095	5.1 ± 1.6	952[e]	Hexon tips

Data from Newcomb et al. (1993).
[a]The relative amounts of the proteins were assessed by quantitation of SDS gels of purified capsids stained with Coomassie Brilliant Blue, and so errors in the estimation can arise from nonstoichiometric binding of the dye.
[b]The actual ratio of VP19c to VP23 in triplexes is assumed to be 1:2.
[c]VP21 and VP24 are derived by autoproteolysis of the product of the *UL26* gene (M_r = 62,466 Da) and are therefore expected to be eqimolar.
[d]Earlier work on the molecular composition of equine herpesvirus type 1 (EHV-1) capsids gave a considerably lower copy number (~600) for VP22a (Newcomb et al., 1989). This may represent either a genuine difference between the cores of EHV-1 and HSV-1 or partial loss of VP22a from the EHV-1 capsid.
[e]VP26 is now known to be absent from pentons (see below).

The major capsid protein is VP5. Compared to capsid proteins of most other viruses, it has an exceptionally high molecular mass (149 kDa). VP5 accounts for approximately 72% of the shell mass, with the remainder contributed by VP19, VP23, and VP26 (Table 12.2). There are about 960 copies of VP5, 900 of VP26, and about the same number (960) of VP19c and VP23 combined.

The three major core proteins, VP22a, VP21, and VP24, all derive from two overlapping genes (Fig. 12.3). The number of variants of these proteins (sometimes called ICP35a to f) is large because of proteolytic processing of the two primary gene products, encoded by genes *UL26* and *UL26.5*, respectively (Liu and Roizman, 1991; Preston et al., 1992). VP22a, which is derived from the *UL26.5* product, is the major core protein. It is present in over 1000 copies per B-capsid, and must contribute most of the core density visualized in thin sections. VP24 and VP21 represent, respectively, the amino-terminal and carboxyl-terminal portions of the *UL26* gene product (Davison et al., 1992). VP24 is the viral protease. The carboxyl terminus of VP21 is identical to that of VP22a (see Fig. 12.3a), which is expressed about 10 times more copiously. VP22a and VP21, but not VP24, are expelled from B-capsids upon packaging of viral DNA (Gibson and Roizman, 1974).

Morphology of the Mature Capsid

The large, symmetric, and photogenic capsid appears to be conserved in structure throughout the herpesvirus family. Shortly after the invention of negative staining, the herpesvirus capsid was shown to consist of 150 hexons and 12

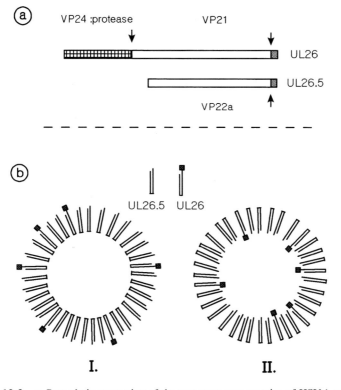

FIGURE 12.3   *a*. Proteolytic processing of the precursor core proteins of HSV-1 capsids, the products of genes *UL26* and *UL26.5*. The protease (VP24) is the amino-terminal domain of *UL26*. The carboxyl-terminal domain (VP21) is identical with that of the *UL26.5* gene product except for the additional peptide between the cleavage point and the start of *UL26.5*. The cleavage that separates VP24 and VP21 is termed the release (R) cleavage. The cleavages near the carboxyl-termini of the translation products are important maturational (M) cleavages. *b*. Two schematic models of the B-capsid core. About 1200 copies of the *UL26.5* product and 150 copies of the *UL26* product are packed in a spherical shell, which is about 200 Å thick (Preston et al., 1983; Rixon et al., 1988). For this to be a monomolecular layer, *UL26.5* should have an elongated conformation and be arranged with its long axis pointing radially. The protease domain (VP24) is shown as a small square. For the B-capsid's separation site and the carboxyl terminal cleavage sites to be accessible to proteolysis on the same side of the core, the molecules are conveyed as having a hairpin shape. Alternatively, the core could become disordered in some way to allow access to cleavage sites. Models I and II correspond to having the protease domains exposed on the outer and the inner surface, respectively.

pentons (Wildy et al., 1960), and therefore conforms to icosahedral symmetry of the triangulation class $T = 16$. The hexons were determined to be sixfold symmetric by image analysis of negatively stained capsids (Furlong, 1978) and capsid fragments (Steven et al., 1986), implying that they are hexamers of the major capsid protein. This finding was not a foregone conclusion because the hexons of two other large dsDNA animal viruses, adenovirus (Nermut, 1980) and iridovirus (Devauchelle et al., 1985), are threefold symmetric (i.e., trimers) (see Chapter 8). That the hexons must be hexamers was further substantiated by STEM mass measurements (Newcomb et al., 1989, 1993). The hexon-as-trimer scenario predicts a capsid mass that is approximately 50% too low.

Cryo–electron microscopy has allowed the three-dimensional structure of the capsid to be defined in a near-native state by digital image reconstruction (Schrag et al., 1989; Baker et al., 1990; Booy et al., 1991; Trus et al., 1992; Conway et al., 1993; Newcomb et al., 1993). These studies have visualized the expected $T = 16$ distribution of capsomers (Fig. 12.4a and d) over an icosahedral lattice that has somewhat rounded facets, and in this respect is inter-

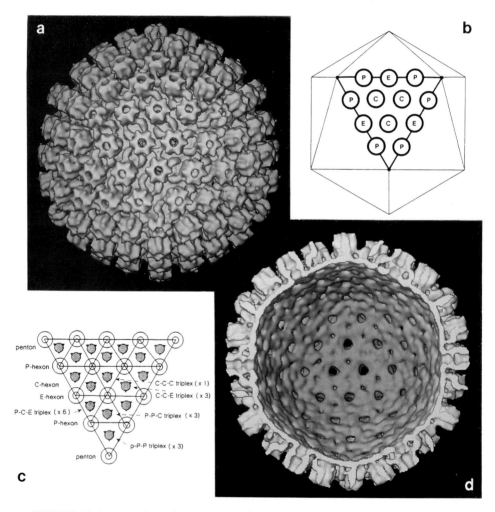

FIGURE 12.4   Three-dimensional reconstruction of the HSV-1 B-capsid calculated to a resolution of 25 Å from 56 particles (Conway et al., 1993). Surface renderings of the outside (a) and inside (d), as viewed along an axis of three-fold symmetry. (Courtesy of J.F. Conway.) The capsid is 1250 Å in diameter. b. Diagram marking the distribution of the three classes of hexons—P, E, and C—on an icosahedral facet. c. Blowup of an icosahedral facet, indicating triplexes.

mediate between a spherical shell and an icosahedron with planar facets. The images also confirmed that the pentons do indeed have fivefold symmetry. In calculating the reconstructions, fivefold symmetry was imposed but, if the pentons were to observe some other rotational symmetry, this symmetrization would result in almost complete cylindrical symmetry. The sharply expressed nature of the penton's fivefold symmetry is strong evidence that this symmetry is genuine.

The shell is approximately 150 Å thick, and the capsomers are held together by interactions between their innermost portions, which pack closely together to form an inner layer, approximately 35 Å thick, called the "floor" (Baker et al., 1990). Each capsomer has a chimneylike protrusion that is approximately 90 Å in outer diameter, is sixfold symmetric in hexons and fivefold symmetric in pentons, and rises 100 Å or so outward from the floor. The capsomers are traversed by axial channels that are nonuniform in diameter and have a pronounced constriction near their midpoints (Fig. 12.5). On the outer surface of the floor, nodules of density occupy the sites of local threefold symmetry (Schrag et al., 1989). On the basis of their association with the trigonal sites, they have been called "triplexes" (Baker et al., 1990). However, the triplexes are not threefold symmetric, with the possible exception of those at the centers of the icosahedral facets (Conway et al., 1993; Zhou et al., 1994) (Fig. 12.4a).

Molecular Anatomy of the HSV-1 Capsid

Several studies have targeted the goal of correlating the morphologic features of the capsid with its constituent proteins. A priori, it is clear that VP5 is the major contributor because it accounts for approximately 72% of the shell mass. However, the limited resolution of the density maps has made it problematic to identify structural features directly with particular minor proteins. To establish correspondences, two approaches have been followed—antibody labeling (Trus et al., 1992), and systematic depletion and reconstitution experiments. In the latter approach, purified capsids were treated with guanidine hydrochloride (Gu-HCl) and urea at concentrations that respect the icosahedral integrity of the capsid but are sufficient to extract certain proteins (Newcomb and Brown, 1991). The proteins lost were detected by SDS-PAGE and correlated with the structural components removed by comparing reconstructions from cryo–electron micrographs (Newcomb et al., 1993). Similarly, depleted capsids have been complemented with purified proteins (Newcomb and Brown, 1991; Booy et al., 1994).

The Pentons Are Pentamers of VP5

The decisive observation in settling the identity of the pentons was that extraction of purified B-capsids with 6-M urea resulted in vacation of the penton sites and solubilization of approximately 6% of the VP5 (Newcomb et al., 1993). This percentage is the amount expected if the pentons are pentamers ($12 \times 5 = 60$ copies) and the hexons are hexamers ($150 \times 6 = 900$ copies) of VP5. Thus, unusually for such a complex virus, HSV conforms to the tenet of

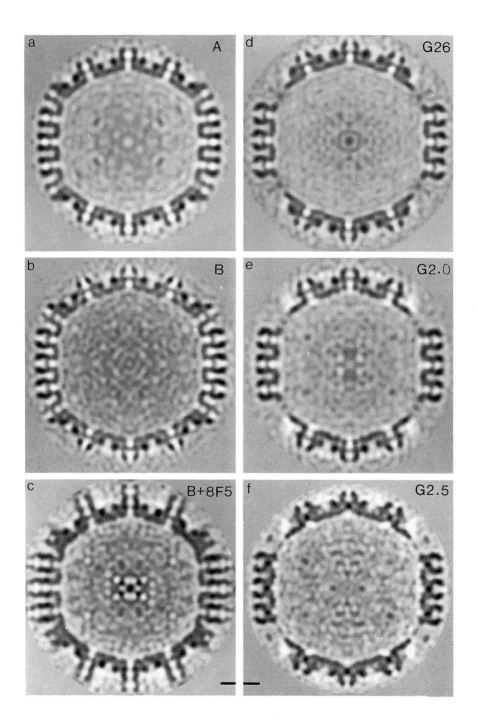

322

quasi-equivalence theory that icosahedral virus shells can be made up of hex-amers and pentamers of the same protein (Caspar and Klug, 1962).

The Triplexes Are Heterotrimers of VP19 and VP23

Extraction of purified capsids with 2-M Gu-HCl was found to reduce their complements of VP19 and VP23 by approximately 25% (Newcomb et al., 1993). Cryo-reconstructions (Fig. 12.5*d*–*f*) showed that the main change, aside from disappearance of the pentons (known to contain VP5), was removal of the two sets of triplexes closest to the pentons. Because no proteolytic degra-dation of VP5 was detected, the missing triplexes could not represent excision of protruding domains of VP5. However, this observation was well explained by the hypothesis that the triplexes are heterotrimers composed of one copy of VP19c and two copies of VP23. There are 320 triplexes and 900 or so copies of the two proteins in a molar ratio of approximately 1:2 (Table 12.2). In principle, triplexes located at different sites (Fig. 12.4*c*) might consist of dif-ferent combinations of VP19c and VP23. However, the fractions of both pro-teins extracted by 2-M Gu-HCl were about the same, suggesting that the tri-plexes removed and those remaining in place have the same stoichiometry (Newcomb et al., 1993).

VP26 Is Located at the Outer Tips of Hexons

VP26, a small (12-kDa) but abundant protein, may be detached from the capsid by either Gu-HCl or urea (Newcomb and Brown, 1991). Purified VP26 rebinds readily to extracted capsids, whereupon the only significant structural alteration is the deposition of density at six sites around the tips of each hexon (Booy et al., 1994). These six structural units (Fig. 12.4*a*) have been inferred to rep-resent monomers of VP26, consistent with the observed stoichiometry (Table

---

←————————————————————————————————————

FIGURE 12.5   Three-dimensional reconstructions calculated from cryo–electron micrographs of HSV-1 A-capsids and B-capsids, and B-capsids modified in vitro in various ways. Central sections (8.3 Å thick) through these reconstructions are compared as viewed along a twofold axis of symmetry. Regions of high protein density are dark. *a*. A-capsid. (Reproduced with permission from Booy et al., 1991.) *b*. B-capsid; note the higher internal density than in the A-capsid. (Reproduced with permission from Trus et al., 1992.) *c*. B-capsids decorated with the monoclonal antibody 8F5, which binds around the hexon tips, presumably between subunits of VP26, and does not bind to pentons. (Reproduced with permission from Trus et al., 1992.) *d*. G26 capsids obtained by complementing G2.0 capsids in vitro with purified VP26, resulting in a lengthening of the hexons as VP26 binds around their tips. (Reproduced with permission from Booy et al., 1994.) *e*. G2.0 capsid obtained by extracting B-capsids with 2-M guanidine hydrochloride—the pentons and the peripentonal triplexes have been removed. (Reproduced with permission from Newcomb et al., 1993.) *f*. G2.5 capsid obtained by extracting B-capsids with 2.5-M guanidine hydrochloride. In addition to removal of the pentons and peripentonal triplexes, changes in density are visible around the inner opening of the transcapsomer channel and the channel, which are attributable to local denaturation effects of the guanidine hydrochloride. (Reproduced with permission from Newcomb et al., 1993.) Bar, 500 Å.

12.2). From a comparison between pentons and hexons visualized in a density map of A-capsids at 26-Å resolution, Zhou et al. (1994) proposed that VP26 is absent from pentons.

VP22a Is the Major Internal Protein of B-Capsids

Several lines of evidence point to VP22a being the major internal protein. First, ion etching studies indicated that VP22a is relatively protected from erosion (Newcomb and Brown, 1989). Second, a cryo–electron microscopic comparison between A-capsids and B-capsids found additional material inside B-capsids, correlated with the presence of VP22a (Baker et al., 1990). Third, earlier immunocytochemistry with anti-VP22a antibodies detected labeling of the interior of B-capsids (Rixon et al., 1988), although these results were considered also to be compatible with earlier studies that placed VP22a on the outside of the capsid on the basis of accessibility to radioiodination (Braun et al., 1984). In retrospect, this reaction may have represented labeling of VP22a released from the capsids, or penetration of the labeling reagent into the capsid interior.

Summary

The major capsid protein, VP5, makes up the basic matrix of the icosahedral HSV capsid, with the other shell proteins attached to its outer surface. VP19 and VP23 form the triplexes, and VP26 binds around the hexon tips. VP22a is the primary constituent of the B-capsid core. VP21 and VP24 have yet to be directly localized, but other evidence—notably, ion etching data (Newcomb and Brown, 1989) and the consideration that they are derived from a joint precursor whose carboxyl terminus is identical to that of VP22a (Fig. 12.3b) —implies that both are also core components.

Expression of Quasi-Equivalence in the HSV Capsid

Most of the results summarized previously were based on maps with resolutions of 30 to 40 Å, and more recent analyses are extending to approximately 24 Å (Conway et al., 1995) and 19 Å (Zhou et al., 1995). Resolutions of this order are sufficient to reveal a wealth of ultrastructural information, but are not sufficient to disclose in detail the adjustments in protein conformation exacted by quasi-equivalence. Nevertheless, some aspects of quasi-equivalence in HSV capsid design are already apparent.

Pentons and Hexons

To a first approximation, pentons resemble fivefold variants of hexons divested of the small horns attributed to VP26 (Fig. 12.4a). The underlying VP5 structure of pentons must be sufficiently different from that of hexons to be ineffective in binding VP26. Moreover, certain monoclonal antibodies specific for

VP5 recognize epitopes that differentiate hexons from pentons (Trus et al., 1992). One antibody (called 3B) bound to penton tips but not to hexons, a second (8F5) bound to hexon tips (Fig. 12.5c) at a site that depends on the presence of VP26 for full antibody binding capability (Booy et al., 1994), and a third (5C) bound to sites on the sides of the hexon protrusions. Finally, structural differentiation between pentons and hexons is reflected in the properties that pentons are more susceptible to proteolytic degradation than hexons (Palmer et al., 1975; Steven et al., 1986) and are more easily detached from the capsid than hexons (Newcomb and Brown, 1991).

The 12 pentons may not all be structurally identical if the phage capsid analogy proves to be correct (see later). As visualized in reconstructed images, the penton structure may represent an average over 11 pentamers of VP5 and one specialized connector/portal. If so, it will be necessary to differentiate between structural differences between penton and hexon subunits from this source and genuine conformational adjustments exacted by quasi-equivalence.

Triplexes

As with VP5, the resistance of triplex components to extraction by Gu-HCl depends on their location on the capsid surface (Figs. 12.4c and 12.5e), and thus on the local curvature of the floor and the underlying VP5 structures. The triplexes closest to the pentons are most easily removed, followed by the next closest set (Newcomb et al., 1993). Conway et al. (1993) and Zhou et al. (1994) have reported discernible structural distinctions between different classes of triplexes. The hexons, pentons, and triplexes and some of the conformational distinctions among them are shown in Color Plate 16.

Nucleocapsid Morphogenesis

Assembly of the Capsid

Viral Proteolytic Activity Required for Assembly

All herpesviruses thus far characterized express a proteolytic enzyme that is incorporated into the assembling capsid. Active protease is required for maturation of precursor capsids (Preston et al., 1983; Rixon et al., 1988; Gao et al., 1994). The protease, also called "assemblin" (Welch et al., 1991), is the amino-terminal domain of the *UL26* gene product (Davison et al., 1992; Deckman et al., 1992; Preston et al., 1992; Liu and Roizman, 1993). Its substrates are itself and the product of the *UL26.5* gene, the precursor of VP22a, which is the principal core or scaffold protein. Both sustain small excisions from their common carboxyl-terminus at the maturational (M) site (Fig. 12.3a). As initially expressed, the full-length protease may be considered as a fusion of the enzyme with a scaffolding protein, these components being separated by cleavage at the release (R) site (Fig. 12.3a). This cleavage event is not required to confer proteolytic activity (Deckman et al., 1992; Liu and Roizman, 1993), but it may facilitate access of the protease to all potential substrates in the maturing capsid. VP24 remains in the C-capsid when DNA is packaged.

Assembly from Proteins Expressed in Insect Cells

Studies (Tatman et al., 1994; Thomsen et al., 1994, 1995; Trus et al., 1995; Zhou et al., 1995) have shown that B-capsids of apparently normal composition and structure can be produced in insect cells infected with baculovirus vectors expressing six HSV-1 gene products: VP5 (*UL19*), VP23 (*UL18*), VP19c (*UL38*), VP26 (*UL35*), VP21/VP24 (*UL26*), and VP22a (*UL26.5*). Omission of VP26 (*UL35*) has little effect on capsid assembly, indicating that this small protein is not essential for assembly. The absence of other capsid shell components (VP5 [*UL19*], VP23 [*UL18*], or VP19c [*UL38*]) prevented the production of capsids of normal size, although smaller closed shells were noted when VP23 (*UL18*) was deleted (Thomsen et al., 1994). Omission of both VP21/VP24 (*UL26*) and VP22a (*UL26.5*) prevented the formation of closed shells; aberrant curved sheets of polymerized capsid proteins were observed instead. Interestingly, capsids were formed when either VP21/VP24 (*UL26*) or VP22a (*UL26.5*) was singly deleted, albeit at reduced efficiency. Absence of VP21/VP24 (*UL26*) alone permitted the formation of capsids containing cores, which appeared to be larger in diameter than those of B-capsids (Tatman et al., 1994). Differences in core conformation might be expected because the protease (VP24) would not have been available to mediate cleavage of the VP22a precursor (Fig. 12.3). Absence of VP22a (*UL26.5*) alone permitted the formation of capsids that appeared to be devoid of cores. Presumably the products of *UL26* (VP21 and VP24) or *UL26.5* (precursor of VP22a) are individually sufficient to serve as scaffold for the formation of closed capsid shells.

Viral Proteins Required in Infected Cells

Results consistent with those described for the baculovirus expression system have been obtained by characterizing the phenotypes of HSV-1 mutants in infected cells (see Table 12.3 for a summary and citations to the literature). Mutants with temperature-sensitive lesions or deletions in genes encoding the capsid shell proteins—VP5 (*UL19*), VP23 (*UL18*), or VP19c (*UL38*)—fail to produce capsids. Mutants that are unable to produce VP26 (*UL35*) have not been described. The results summarized above suggest that such mutants would produce capsids. It is of interest to know what phenotype these VP26 mutants would exhibit (e.g., defects in DNA packaging, envelopment, or virion infectivity). Capsids can be produced in the absence of *UL6*, but they appear to be defective for DNA cleavage and packaging.

Deletions affecting both *UL26* and *UL26.5* products result in the production of aberrant spirals and sheets containing capsid shell proteins, but no closed capsid shells. Mutations in *UL26* that selectively eliminate the proteolytic activity without affecting the expression of *UL26.5* (precursor of VP22a)

TABLE 12.3 HSV-1 Proteins That Influence Capsid Assembly and the Cleavage and Packaging of DNA

Gene	Protein[a]	Location in Virion[b]	Comments
UL6		Nucleocapsid	Required for DNA cleavage/packaging, (Sherman and Bachenheimer, 1987, 1988; Weller et al., 1987); appears to be a constituent of the nucleocapsid (Patel and MacLean, 1995) but is not required for capsid formation
UL12	Alkaline nuclease		May be required to ensure appropriate DNA packaging so that nucleocapsids are stable and can be enveloped (Shao et al., 1993)
UL15			Required for DNA cleavage/packaging (Poon and Roizman, 1993; Baines et al., 1994)
UL18	VP23	Capsid shell	Required for capsid formation (Desai et al., 1993)
UL19	VP5	Capsid shell	Major constituent of hexons and pentons (Newcomb et al., 1993); required for capsid formation (Celluzi and Farber, 1990; Desai et al., 1993)
UL25		Tegument	Required for DNA cleavage/packaging; influences viral penetration (Addison et al., 1984)
UL26	VP21+ VP24 (protease)	(Core)[c]	Required to produce capsids that are functional for DNA cleavage/packaging (Preston et al., 1983; Gao et al., 1994); closed capsid shells not produced in absence of both *UL26* and *UL26.5* (Desai et al., 1994)
UL26.5	VP22a	Core[d]	Major component of B-capsid core; required to produce capsids that are functional for DNA cleavage/packaging (Matusick-Kumar et al., 1994); closed capsid shells not produced in absence of both *UL26* and *UL26.5* (Desai et al., 1994)
UL28	ICP18.5		Required for DNA cleavage/packaging (Addison et al., 1990; Cavalcoli et al., 1993; Tengelson et al., 1993)
UL32			Required for DNA cleavage/packaging (Sherman and Bachenheimer, 1988)
UL33			Required for DNA cleavage/packaging (al-Kobaisi et al., 1991)
UL38	VP19c	Capsid shell	Required for capsid formation (Pertuiset et al., 1989)
UL48	VP16	Tegument	Facilitates DNA cleavage/packaging and envelopment (Weinheimer et al., 1992)

[a]No entry means the protein has not been named.
[b]No entry means that it has not been determined whether the protein is a constituent of the virion.
[c]Location in the mature virion is uncertain.
[d]Present in B-capsids but not in nucleocapsids or virions.

permit the production of capsids. These capsids have distended rather than compact cores and are not functional for the packaging of viral DNA. Moreover, the surface of the capsid shells must be altered because the hexons exhibit altered reactivity with anti-VP5 monoclonal antibodies (Gao et al., 1994). Mutations affecting *UL26.5* products (VP22a), but not *UL26* products (VP21, VP24), permit the formation of capsid shells without obvious core components. Interestingly, a small fraction of the capsids appears to be functional for DNA packaging and virion morphogenesis, and a small amount of infectious virus can be produced. Perhaps VP21 can inefficiently fulfill some of the roles of VP22a. The VP22a mutants revealed that VP22a (or its precursor) is required not only for its role as a core component but also to facilitate the transport of VP5 to the cell nucleus (Matusick-Kumar et al., 1994). Two of the capsid shell proteins, VP5 and VP23, fail to exhibit selective localization to the cell nucleus unless coexpressed with other HSV proteins. Coexpression of VP22a is sufficient for the nuclear localization of VP5 but not of VP23 (Nicholson et al., 1994). The vehicle for nuclear transport of VP23 has not been identified.

Progress in Developing in Vitro Assembly Systems

Using proteins extracted from purified B-capsids, some types of reassembly reactions have been successful. For example, VP26 detached from the hexon tips by Gu-HCl or urea could rebind, after purification and renaturation, to the denuded hexon tips, with deposition of density at six sites around the tips of each hexon (Newcomb and Brown, 1991; Booy et al., 1994). In addition, VP22a extracted from purified B-capsids with 2-M Gu-HCl was found to reassociate into round particles of about 600 Å in diameter upon removal of the denaturant (Newcomb and Brown, 1991). These particles appeared as hollow spherical particles viewed in projection, comparable in dimensions to cores visualized inside intracellular large-cored B-capsids (Rixon et al., 1988).

Using mixed lysates of insect cells producing individual HSV-1 capsid proteins from baculovirus vectors, cell-free assembly of B-capsids has been achieved (Newcomb et al., 1994). The requirements for assembly, in terms of input viral gene products, were much the same as for assembly in insect cells coinfected with multiple baculovirus vectors. Optimal assembly required five viral genes (VP5 [*UL19*], VP19c [*UL38*], VP23 [*UL18*], VP21/VP24 [*UL26*], and VP22a[*UL26.5*]), but not VP26 (*UL35*) or *UL6*. The particles produced appeared to be normal except for a reduced content of VP22a. Reduced numbers of capsids were produced in the absence of *UL26*. Assembly required elevated temperature (>4 °C) and was inhibited by reducing agent.

These in vitro assembly reactions offer many possibilities for further exploration of the requirements for assembly, including environmental and cell factors. They can also facilitate the use of altered viral proteins to explore structure–function relationships pertinent to assembly.

Cleavage and Packaging of Viral DNA

Processing of the DNA

The linear genomic DNA of HSV, consisting of two unique invertible sequences bounded by inverted repeats, circularizes prior to DNA replication. Herpes simplex virus DNA replication depends on seven viral gene products acting in concert with cell factors (see reviews by Challberg, 1991 and Weller, 1995). Replication results in the generation of large concatemeric molecules that are probably branched (Severini et al., 1994; Zhang et al., 1994). It is not clear how these molecules are generated, but it is known that the substrate for packaging of progeny genomes into capsids is concatemeric DNA. Concomitant with packaging, there is cleavage of the concatemers at unique sites within the *a* sequence, which is found at both genomic termini and the internal repeated sequence (Varmuza and Smiley, 1985; Deiss et al., 1986). Nucleocapsid maturation appears to involve site-specific cleavage of DNA as well as a headfull sensing mechanism, because capsids containing less than approximately full-length DNA fail to become enveloped (Vlazny et al., 1982).

Viral Proteins Required

Cleavage of concatemeric viral DNA and packaging appear to be tightly linked in that no viral mutants have been identified in which cleavage can occur, but not packaging. Multiple viral gene products are required for cleavage and packaging, as indicated by the list presented in Table 12.3 along with citations to the relevant literature. These viral gene products include the capsid shell proteins, VP5 (*UL19*), VP23 (*UL18*), and VP19 (*UL38*), inasmuch as formation of the capsid shell appears to be a prerequisite for cleavage as well as packaging. They also include the products of *UL26* and *UL26.5*. Absence of either the protease VP24 (*UL26*) or the major core component VP22a (*UL26.5*) prevents both cleavage and packaging of viral DNA. Finally, several other viral proteins, including but perhaps not limited to the products of the *UL6*, *UL15*, *UL25*, *UL28*, *UL32*, and *UL33* genes, are required for the cleavage of concatemeric DNA and packaging into capsids. One of these proteins (*UL6*) appears to be a minor component of nucleocapsids, whereas another is a tegument protein (*UL25*). It is not yet known whether the remaining proteins are constituents of the virion.

At least two other viral proteins, alkaline nuclease and VP16, appear to influence packaging of viral DNA and subsequent events in virion morphogenesis. HSV-1 mutants deleted for the alkaline nuclease gene (*UL12*) appear to cleave and encapsidate viral DNA normally, but egress of the nucleocapsids to the cytoplasm, and presumably envelopment, are impaired and a higher than normal ratio of A-capsids to B-capsids can be extracted from nuclei (Shao et al., 1993). Those authors suggest that the nuclease may be required to resolve replication or recombination intermediates, and that failure to do so may result in aberrant packaging into nucleocapsids that are conformationally altered and unsuitable for envelopment. HSV-1 mutants deleted for VP16 (*UL48*) can cleave and encapsidate viral DNA, but do so at reduced efficiency, and appear not to undergo envelopment normally at the inner nuclear membrane (Weinheimer et al., 1992). The regulatory role of VP16, enhancement of immediate-early viral gene expression, can be segregated from its structural role in that temperature-sensitive mutations can eliminate the regulatory activity without causing the maturation and morphogenic defects observed on deletion of VP16 (Ace et al., 1989; Weinheimer et al., 1992).

Structure of the Encapsidated Viral Genome

Cryo–electron microscopy has allowed visualization of encapsidated HSV-1 DNA, which takes the form of tightly packed bundles of aligned DNA duplexes with an average spacing of approximately 26 Å (Booy et al., 1991) (Fig. 12.2). Further details of this packaging arrangement—that is, whether the winding is predominantly spool-like (with locally parallel duplexes) or whether it consists of local discrete domains (with parallel or antiparallel packings) and whether the ends are located in specific sites—remain to be settled. Because all of the major proteins of the C-capsid are components of the shell (Booy et al., 1991), it is clear that no histonelike charge-neutralizing protein can be stoichiometrically associated with the encapsidated DNA.

HSV as a Cryptophage

Points of resemblance between herpesvirus nucleocapsid assembly and that of dsDNA bacteriophages attracted comment as early as 1975 (Casjens and King, 1975; Friedmann et al., 1975). Since then, considerably more detailed information has become available for both systems, and these findings have progressively strengthened this analogy in some but not all respects (Lee et al., 1988; Baker et al., 1990; Newcomb et al., 1993). HSV nucleocapsid assembly and DNA phage assembly (see Chapter 11) share the requirements for (1) preformed capsids into which DNA is inserted, (2) assembly of the capsid shell and scaffold with subsequent activation of the encapsidated protease, (3) coupling of DNA concatemer cleavage and packaging, and (4) replacement of scaffolding proteins with the viral DNA. Also, the alignment of DNA duplexes in the HSV nucleocapsid resembles that of bacteriophages T4 and lambda (Lepault et al., 1987) and T7 (Booy et al., 1991). Finally, at least two of the HSV-1 gene products required for cleavage/packaging have some structural/functional similarities to phage proteins also required for cleavage/packaging. The product of the *UL15* gene may be structurally related to one component of the T4 terminase (Davison, 1992). The product of the *UL12* gene, alkaline nuclease, may have a role in morphogenesis (Shao et al., 1993) similar to that of the T4 endonuclease.

There are several points on which HSV nucleocapsid assembly and DNA phage assembly differ, or on which further evidence is required to clarify the similarities or differences. First, phage proheads are usually smaller, rounder, and more fragile than the mature virions. The mature structure results from a conformational change of the capsid (expansion) that generally accompanies DNA packaging. In contrast, isolated B-capsids are indistinguishable in size from A-capsids and C-capsids (Baker et al., 1990; Booy et al., 1991). It remains to be determined whether the HSV capsid passes through a stage equivalent to that of the phage prohead (Fig. 12.6) (in which case isolated B-capsids would correspond to expanded proheads). Second, phage proheads are shortlived and rapidly converted to mature heads in packaging reactions, whereas B-capsids tend to accumulate in the nuclei of infected cells. It may be that the DNA packaging reactions are much slower for HSV, or perhaps not all of the B-capsids visualized in infected cells are maturable particles (Sherman and Bachenheimer, 1988). Third, for phages that have reinforcing triplex proteins, these molecules are not required for formation of the prohead but are added to it after expansion. In contrast, the HSV triplex proteins, VP19 and VP23, are required for capsid formation (see Assembly of the Capsid). Finally, it is not clear whether one of the HSV capsid vertices serves as a unique portal of entry for HSV DNA, analogous to the well-defined phage portal, or whether HSV DNA can enter the capsid through the axial channel of any capsomer (or by some other means).

Envelopment of the Nucleocapsid

Composition of the Tegument and Envelope of the Infectious Virion

Most of the proteins of the virion (Table 12.1) are acquired during the process of envelopment. Integral membrane proteins and glycoproteins are assumed to be constitutents of the envelope (membrane), whereas the other proteins are assumed to form the tegument, the proteinaceous material between the capsid outer surface and membrane inner surface. The organization of the tegument proteins is not well understood. Some of them are difficult to separate from the nucleocapsid (Spear and Roizman, 1972) and must be bound quite tightly to the capsid surface. Other tegument proteins are undoubtedly in intimate contact with the envelope or envelope proteins. The general orientation of the viral membrane glycoproteins in the envelope is known, in most cases. Little is known, however, about the three-dimensional structures of individual envelope proteins or the two-dimensional organization of the envelope proteins in the plane of the membrane.

Tegument Proteins

Roles that have been assigned to tegument proteins include functions related to morphogenesis, uncoating, and regulation of gene expression or protein synthesis. It seems likely that several of the tegument proteins must influence the envelopment of nucleocapsids. To date, however, only two, VP16 (*UL48*) and the product of the *UL11* gene, have been implicated in assembly. VP16 is one

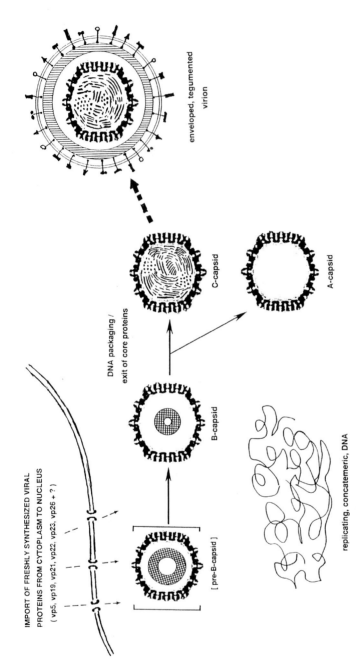

FIGURE 12.6 Diagram of the capsid assembly pathway of HSV-1. By studying particles produced at early times in in vitro assembly, it has recently been possible to characterize a precursor capsid, or procapsid, that subsequently matures spontaneously (Newcomb et al., 1996). This in vitro procapsid appears to be closely related to the pre-B-capsid hypothesized for the in vivo pathway, and to the "large-cored B-capsids" that accumulate at non-permissive temperature with the temperature-sensitive mutant ts1201 (Preston et al., 1983). The structural relationship between the procapsid and the mature capsid (Trus et al., 1996) has many features in common with that linking phage proheads and mature heads: i.e. the procapsid is relatively labile; it is spherical as compared with the polyhedral capsid; its hexons are substantially distorted from six-fold symmetry; and large-scale conformational differences exist between the respective surface lattices, particularly in the 'floor' region. However, unlike phage proheads characterized to date, whose dimensions are generally 10–15% smaller than those of mature heads, the HSV-1 procapsid is about the same size as the mature capsid. It remains to be determined whether the B-capsid is indeed an on-pathway intermediate (as shown), or whether the pre-B-capsid matures directly into the C-capsid, with both the A-capsid and the B-capsid representing abortive by-products.

of the most abundant tegument proteins. Deletion of the gene is lethal (the mutant must be propagated on a cell line that supplies the missing protein) and results in impairment of envelopment at the inner nuclear membrane (Weinheimer et al., 1992). The product of the *UL11* gene is a myristylated protein that is dispensable for the replication of virus in cultured cells but increases the efficiency of envelopment and virion egress (Baines and Roizman, 1992; MacLean et al., 1992).

Following the initial events of viral entry into a cell (binding and initiation of the membrane fusion reaction), the virion envelope must separate from the nucleocapsid, thereby releasing the nucleocapsid and appropriate tegument components into the cell cytoplasm. Although little is known about these events, it seems likely that some tegument proteins may facilitate release of the nucleocapsid from the envelope and its movement toward the cell nucleus. The tegument protein designated VP1/2 (*UL36*) may have a role in facilitating the release of DNA from the capsid. A temperature-sensitive mutant altered in this gene fails to uncoat the viral genome at the nonpermissive temperature (Batterson et al., 1983).

The regulatory activities of tegument proteins have received the most attention to date, as summarized briefly earlier (see Expression of Viral Genes). Most of the tegument proteins are phosphoproteins. At least two are protein kinases, including VP18.8 (*UL13*) and the product of *US3*. VP22 (*UL49*), another tegument protein, is a substrate for VP18.8 (*UL13*) (Purves and Roizman, 1992; Coulter et al., 1993) and the product of *UL34*, probably an integral membrane envelope protein, is a substrate for the product of *US3* (Purves et al., 1991, 1992). It seems likely that phosphorylation–dephosphorylation reactions contribute importantly to modulating the activities of many of the tegument proteins.

Envelope Proteins

The envelope of HSV-1 contains about a dozen integral membrane proteins. The exact number is unknown. Table 12.4 lists known or potential membrane proteins specified by HSV-1 and indicates whether the protein has been detected in virions. The table also summarizes known properties and functions of these membrane proteins, with citations to the relevant literature. All of the proteins given an alphabetic designation preceded by *g* (e.g., gB) are glycoproteins, most of which carry both *N*-linked and *O*-linked glycans (see reviews by Campadelli-Fiume and Serafini-Cessi, 1985, and Olofsson, 1992).

There are formidable technical difficulties with defining the structures and organization of membrane proteins in situ in the viral envelope. Because it seems unlikely that the HSV membrane glycoproteins and tegument components observe icosahedral symmetry, reconstruction from cryo–electron micrographs based on exploiting this symmetry is inapplicable. Moreover, the large dimensions of the intact virion may preclude detailed tomographic reconstructions from frozen-hydrated specimens. Nevertheless, cryo–electron micrographs of intact virions (Schrag et al., 1989; Szilágyi and Berriman, 1994) have established that the virion envelope is approximately spherical with protruding spikes, and this approach would appear to have considerable potential as applied to membrane fragments containing specific viral glycoproteins.

To date, it has been possible to obtain images of proteins protruding from the virion envelope in negatively stained preparations (Wildy et al., 1960; Stannard et al., 1987). The most prominent protrusions or spikes were rigid rods about 140 Å long and were shown by immunogold labeling to be composed of gB (*UL27*). Smaller structures, about 80 to 100 Å long, contained gD (*US6*) and probably other proteins as well. Antibodies specific for gC (*UL44*) bound to structures that must have been about 200 to 240 Å long but could not be resolved (Stannard et al., 1987). In these negatively stained images of damaged and flattened virions, clustering of gB-containing spikes was evident, particularly in regions of the envelope with higher than average curvature (Stannard et al., 1987). This clustering is thought to be artifactual and not indicative of the organization of gB spikes in infectious virions.

Studies of envelope proteins extracted from virions by detergents have revealed certain oligomeric associations that are relatively stable. Glycoprotein B is present in homooligomers, either dimers or trimers that are stable in ionic detergents such as SDS but can be dissociated by heat (Sarmiento and Spear, 1979; Claesson-Welsh and Spear, 1986). Glycoproteins H and L are present in heterooligomers that are stable in nonionic detergent (Hutchinson et al., 1992), as are also gE and gI (Johnson and Feenstra, 1987). The stoichiometry of these heterooligomeric associations has not been fully defined. Cross-linking reagents can be used to induce stable associations between and among various envelope and tegument proteins (Zhu and Courtney, 1988, 1994), but it is not known which of these associations may be functionally significant.

It seems likely that binding of the virion to an appropriate cell surface may trigger changes in protein conformation and new functional interactions among viral envelope proteins and with cell surface components that lead to membrane fusion. It has been shown, for example, that cell surface GAGs or exogenous GAGs are required to trigger HSV-induced cell fusion (Shieh and Spear, 1994). The challenge is not only to identify all the participating molecules in functional interactions required for viral entry but also to investigate the dynamic changes in structure and function required.

Envelopment of the Nucleocapsid at the Inner Nuclear Membrane

Electron microscopic studies have shown that assembly of nucleocapsids in the cell nucleus is followed by budding of the nucleocapsids through the inner nuclear membrane into the perinuclear space (Fig. 12.7), or into invaginations of this space into the nucleus, resulting in acquisition of an envelope (Morgan et al., 1954; Darlington and Moss, 1968). Envelopment at the inner nuclear membrane requires, at a minimum, that the nucleocapsids contain appropriately packaged DNA. Fragments of DNA smaller than genome-sized can be packaged into HSV capsids, provided that appropriate packaging/cleavage signals are present in the DNA, but only nucleocapsids containing genome-sized DNA

TABLE 12.4 HSV-1 Membrane Proteins

Gene	Protein	Present in Virion Envelope	Comments[a]
UL1	gL	Yes	Not an integral membrane protein; can be produced as a secreted glycoprotein (M.J. Novotny and P.G. Spear, unpublished results); oligomerizes with gH (Hutchinson et al., 1992); gH–gL required for viral penetration and cell fusion (Forrester et al., 1992; Roop et al., 1993; Davis-Poynter et al., 1994)
UL3			Has a single hydrophobic domain at the amino terminus; dispensable for viral replication in cultured cells (Baines and Roizman, 1991)
UL10	gM	Yes	Has multiple membrane-spanning domains but no amino-terminal signal sequence; dispensable for viral replication in cultured cells (Baines and Roizman, 1991, 1993; MacLean et al., 1991)
UL20		Yes	Has multiple membrane-spanning domains but no amino-terminal signal sequence; dispensable for viral replication in some cell types; required for virion egress in others (Baines et al., 1991; Ward et al., 1994)
UL22	gH	Yes	Oligomerizes with gL (Hutchinson et al., 1992); gH–gL required for viral penetration and cell fusion (Forrester et al., 1992; Roop et al., 1993; Davis-Poynter et al., 1994)
UL27	gB	Yes	Forms homooligomers and prominent spikes (Sarmiento and Spear, 1979; Stannard et al., 1987); required for viral penetration and cell fusion (Manservigi et al., 1977; Sarmiento et al., 1979; Cai et al., 1988); can mediate virus binding to cell surface GAGs (Herold et al., 1994)
UL34		Yes	Hydrophobic carboxyl terminus but no amino-terminal signal sequence (Marsden et al., 1978; McGeoch et al., 1988); phosphorylated by the protein kinase encoded by *US3* (Purves et al., 1992)
UL43			Has multiple membrane-spanning domains but no amino-terminal signal sequence; dispensable for viral replication in cultured cells (McLean et al., 1991)
UL44	gC	Yes	Mediates binding of virus to cell surface GAGs and enhances infectivity (Herold et al., 1991); dispensable for the infection of some cell types; may be required for entry via the apical surfaces of some polarized cell types (Sears et al., 1991)

Continued

335

TABLE 12.4 *Continued*

Gene	Protein	Present in Virion Envelope	Comments[a]
UL45		Yes	Has a single hydrophobic domain near the amino terminus; dispensable for viral replication in cultured cells but not for cell fusion (Visalli and Brandt, 1991; Haanes et al., 1994)
UL49.5			May not be dispensable (Barker and Roizman, 1992; Barnett et al., 1992)
UL53	gK		Has multiple hydrophobic domains as well as a cleavable amino-terminal signal peptide; missense mutations cause syncytial phenotype (Debroy et al., 1985; Pogue-Geile and Spear, 1987; Dolter et al., 1994)
US4	gG	Yes	Dispensable for viral replication in cultured cells (Longnecker and Roizman, 1987)
US5	gJ		Expression not yet definitively established; dispensable for viral replication in cultured cells (Longnecker and Roizman, 1987)
US6	gD	Yes	Required for viral penetration and cell fusion (Ligas and Johnson, 1988); mediates interference with HSV infection (Campadelli-Fiume et al., 1988; Johnson and Spear, 1989)
US7	gI	Yes	Forms oligomers with gE (Johnson and Feenstra, 1987); gE–gI oligomers facilitate cell-to-cell spread of infection (Dingwell et al., 1994)
US8	gE	Yes	Forms oligomers with gI (Johnson and Feenstra, 1987); has Fc binding activity (Baucke and Spear, 1979; Dubin et al., 1990); gE–gI oligomers facilitate cell-to-cell spread of infection (Dingwell et al., 1994)

[a]Unless noted otherwise, the protein has characteristics of a typical type 1 membrane glycoprotein (see review by Spear, 1993). Many of the proteins listed were first predicted to be membrane proteins by McGeoch et al. (1988).

FIGURE 12.7   Electron micrographs of thin sections of HSV-1–infected HEp-2 cells, showing envelopment at the inner nuclear membrane and various cytoplasmic forms of progeny virus particles. *a.* Budding of nucleocapsids through the inner nuclear membrane into the perinuclear space. *b.* Budding of a nucleocapsid through the inner nuclear membrane, a cytoplasmic nucleocapsid, and four progeny virions in a space between two cells. *c.* Accumulation of enveloped nucleocapsids in the perinuclear space. *d.* Nucleocapsids in the nucleus and cytoplasm and an enveloped particle in the perinuclear space. *e.* Virions inside cytoplasmic vacuoles and cytoplasmic nucleocapsids. *f.* Cytoplasmic nucleocapsids; note that one nucleocapsid appears to be in the process of losing or gaining an envelope. The magnifications shown are different in each panel and can be estimated from the fact that nucleocapsid diameters appear to be about 1000 Å in thin sections. N, nucleus; C, cytoplasm. (Courtesy of B. Roizman.)

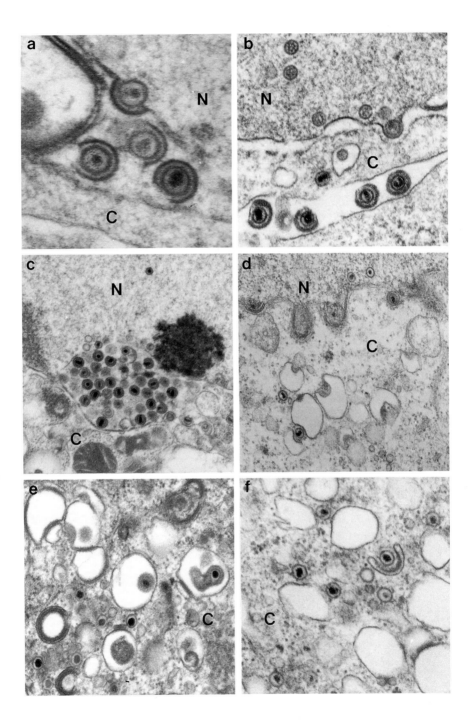

are transported out of the nucleus (Vlazny et al., 1982). Moreover, as mentioned earlier, nucleocapsids produced in the absence of alkaline nuclease are also not transported out of the nucleus efficiently (Shao et al., 1993). Possibly, envelopment of nucleocapsids at the inner nuclear membrane, which is the means of egress from the nucleus, requires that the nucleocapsid surface have an appropriate conformation, which is in turn dependent on presence of a headfull of linear DNA free of recombination and replication intermediates.

Deletion of any of the viral genes listed in Table 12.3 can indirectly block envelopment by preventing the formation of capsids or the packaging of viral genomes into nucleocapsids. At least some tegument proteins appear to have a more direct role in envelopment. Deletion of the gene for VP16 (*UL48*) results in reduced efficiency of DNA packaging and cleavage but also in greatly reduced egress of nucleocapsids from the cell nucleus (Weinheimer et al., 1992). Deletion of the gene *UL11*, which encodes a myristylated tegument protein (MacLean et al., 1992), appears to reduce the efficiency of envelopment at the inner nuclear membrane but not to block it entirely (Baines and Roizman, 1992). In contrast, individual deletion of the genes for a number of other tegument proteins (e.g., *UL4*, *UL13*, *UL21*, *UL41*, *UL46*, *UL47*) has little or no effect on viral replication and presumably none on envelopment (Baines and Roizman, 1991; Zhang et al., 1991; Purves and Roizman, 1992; Smibert et al., 1992; Coulter et al., 1993; Baines et al., 1994a). Interestingly, individual deletions of genes encoding viral envelope glycoproteins also has little if any effect on the envelopment process, although enveloped particles produced may be noninfectious (see review by Spear, 1993).

Subsequent Events in Virion Morphogenesis and Egress

Two Hypotheses about the Source of the Envelope on Mature Virions

There are two schools of thought about morphogenetic events subsequent to envelopment of HSV nucleocapsids at the inner nuclear membrane. One posits that the production of infectious virus requires retention of the envelope acquired at the inner nuclear membrane, with modifications of the viral tegument and envelope proteins as the virion exits the cell via an exocytic pathway (Darlington and Moss, 1968). The other holds that the envelope acquired at the inner nuclear membrane is lost by fusion of the enveloped particle with a cytoplasmic membrane, and that a new envelope is acquired by envelopment of cytoplasmic nucleocapsids at other cytoplasmic membranes, probably those of the Golgi apparatus (Stackpole, 1969; Jones and Grose, 1988). The virion then exits the cell via an exocytic pathway.

The available data can be interpreted to support either hypothesis. Both cytoplasmic nucleocapsids and enveloped nucleocapsids present within cytoplasmic vesicles can be seen in electron micrographs of thin sections of HSV-infected cells during the phase of viral morphogenesis and egress (Fig. 12.7). Images representing the fusion of virions with a cytoplasmic vesicle membrane *or* the budding of cytoplasmic nucleocapsids into cytoplasmic vesicles can also be seen; in static images, it is impossible to say which interpretation is correct. Budding of cytoplasmic nucleocapsids through the plasma membrane or fusion of progeny virions with the plasma membrane is *not* seen despite the fact that progeny virions tend to remain adherent to the plasma membrane of the cell that produced them. Finally, the tegument of perinuclear virions often appears less massive and electron dense than the tegument of extracellular virions.

If the envelope acquired at the inner nuclear membrane must be retained during maturation and egress of the virion, then any cytoplasmic nucleocapsids observed would have resulted from inappropriate fusion of virions with a cytoplasmic membrane during egress and would represent loss of these progeny from the infectious yield. One might predict that the virus has evolved a mechanism to prevent this inappropriate fusion. In fact, it has been shown that expression of gD (*US6*) by transfected cells can block the penetration, but not binding, of HSV (Campadelli-Fiume et al., 1988; Johnson and Spear, 1989). This activity of gD is called interference. The presence of gD in membranes of the infected cell may prevent the fusion of progeny virions with cytoplasmic membranes (Campadelli-Fiume et al., 1991).

Conversely, if the production of infectious virus requires envelopment, deenvelopment, and reenvelopment, then the interference activity of gD is not physiologically relevant to viral morphogenesis and egress. Moreover, the virus must have evolved a mechanism to alter the cell cytoplasm during the early stages of infection so that progeny cytoplasmic nucleocapsids do not lose their DNA or become disassembled.

If viral proteins shown to be present in the tegument or envelope of extracellular virions were found to be absent from perinuclear virions, or vice versa, this would strongly support the envelopment, deenvelopment, and reenvelopment model. The differences in appearance of the teguments of perinuclear and extracellular virions do not necessarily imply differences in composition because of possible artifacts caused by the fixation and staining protocols required for electron microscopy. It has been shown that the phospholipid composition of extracellular virions differed from that of host cell nuclei (van Genderen et al., 1994). The authors concluded that, although the results were consistent with the envelopment, deenvelopment, and reenvelopment hypothesis, they could not rule out the possibilities that viral envelopes acquired at the inner nuclear membrane included lipids unrepresentative of the nuclear envelope or that there could be an exchange of lipids between the virion envelope and cytoplasmic membranes during virion egress. An HSV-1 mutant producing an altered form of gH that remains localized to the endoplasmic reticulum was shown to release gH-negative virions from cells (Browne et al., 1996). If it can also be shown with this mutant that perinuclear virions contain gH, this would constitute the best available evidence in favor of the envelopment, deenvelopment and reenvelopment pathway of virion morphogenesis.

Whichever model is correct, it seems clear that perinuclear virions contain glycoproteins with immature glycans characteristic of endoplasmic reticulum, whereas extracellular virions contain glycoproteins with mature glycans that have been processed through the Golgi apparatus (Johnson and Spear, 1982,

1983; Serafini-Cessi et al., 1983; Banfield and Tufaro, 1990; Torrisi et al., 1992). Either the glycoproteins are processed in situ on the virion envelope during virion egress through the Golgi apparatus or there is an exchange of glycoproteins resulting from exchange of the envelope.

Production of Noninfectious Light Particles

Enveloped particles, similar in size to virions but not quite as regular in size, can be produced by herpesvirus-infected cells along with virions (Szilágyi and Cunningham, 1991). These particles, designated L particles, are devoid of viral DNA and capsid proteins but contain all other known tegument and envelope proteins of virions. They also contain some proteins not normally detected in virions, including five phosphoproteins, one of which is ICP4, a regulatory protein. Production of these particles can occur independently of nucleocapsid envelopment (Rixon et al., 1992), providing the appropriate viral proteins are produced. It has not yet been determined which viral proteins are critically important for the production of L particles. It is also not known where these particles are formed. It seems likely that they bud from cytoplasmic membranes into cytoplasmic vacuoles (Rixon et al., 1992).

L particles lack nucleocapsids and are noninfectious but they do exhibit some biologic activities. Apparently they can bind to and penetrate cells, probably much as virions do. Evidence for this was obtained by showing that the L particles could deliver functional VHS and VP16 to cells (McLauchlan et al., 1992).

Further study of the intracellular distribution of tegument proteins and envelope proteins and of the requirements for production of L particles and virions may help to define the pathway of virion morphogenesis.

Requirements for Virion Egress

Whether the envelope acquired at the inner nuclear membrane is retained or replaced, virions ultimately appear in cytoplasmic transport vesicles, which must fuse with the plasma membrane to release the virions into the extracellular space. It appears that normal cellular membrane transport mechanisms, as well as viral functions, are required for virion egress.

Drugs that interfere with normal membrane transport can also interfere with HSV egress. It is interesting to compare the effects of monensin, which blocks membrane transport from the Golgi apparatus to the cell surface and vice versa, and brefeldin A, which blocks membrane transport from the endoplasmic reticulum to the Golgi apparatus and causes a disappearance of discernible Golgi components, with redistribution of Golgi proteins to the endoplasmic reticulum. Infectious virions are produced in the presence of monensin, but they carry immature glycans (as is usual for glycoproteins produced in the presence of monensin) and accumulate in large cytoplasmic vacuoles believed to be derived from the Golgi apparatus, instead of being exported (Johnson and Spear, 1982; Kousoulas et al., 1983). In contrast, brefeldin A significantly inhibits the production of infectious virus; nucleocapsids are enveloped at the inner nuclear membrane but then appear in the cytoplasm as naked nucleocapsids (Cheung et al., 1991; Chatterjee and Sarkar, 1992). Interpretations consistent with either of the hypotheses about morphogenesis are possible. For example, brefeldin A may block the normal mechanism by which HSV minimizes the loss of envelopes acquired at the inner nuclear membrane and therefore all progeny may rapidly lose infectivity through eclipse. Alternatively, brefeldin A may block the normal mechanism by which deenveloped cytoplasmic nucleocapsids acquire a new cytoplasmic envelope.

At least two mutant cell lines have been shown to be defective for the egress of HSV-1. One is a mutant of baby hamster kidney cells, which was selected to be resistant to ricin and shown to be defective for Golgi enzymes required to add terminal sugars to N-linked glycans. Production of infectious HSV-1 was only somewhat reduced in this mutant cell line, but the virus contained glycoproteins with immature glycans and accumulated intracellularly instead of being exported (Serafini-Cessi et al., 1983). These results suggest a coupling between glycoprotein glycan processing and egress of virions. A second mutant cell line was obtained from mouse L cells by selecting for clones that survived exposure to HSV. This cell line was shown to be defective for the processing of various membrane glycoproteins, marginally deficient in the production of vesicular stomatitis virus, and more markedly deficient in the production and egress of HSV-1. The phenotype was much like that of cells treated with monensin. HSV-1 virions containing glycoproteins with immature glycans accumulated in large cytoplasmic vacuoles (Banfield and Tufaro, 1990).

Finally, mutations in viral proteins also influence HSV egress. Absence of the myristylated tegument protein encoded by *UL11* not only reduces the efficiency of envelopment at the inner nuclear membrane but delays and reduces the export of virus (Baines and Roizman, 1992). Absence of the membrane protein encoded by *UL20* reduces the yield of infectious virus in several cell types and, for some cell types (Vero) but not others, causes a dramatic blockade of egress with accumulation of virions in the perinuclear space (Baines et al., 1991). This suggests that a function required for the movement of perinuclear virions to the cell periphery can be supplied by some cells but not others and can also be supplied by the product of *UL20*. Finally, HSV-1 mutants deleted for gE or gI produce plaques of reduced size, suggesting a deficit in the cell-to-cell transmission of infection (Dingwell et al., 1994). Based on results obtained with similar mutants of PRV (Mettenleiter, 1994), mutations of this kind may influence the egress of progeny virus.

Taken together, these results suggest that HSV egress requires a complex interplay between cell and viral factors, only some of which have been identified.

Anticipation of New Findings

We can anticipate that the structure of the nucleocapsid will be determined at even higher resolution in the future. Perhaps extension of in vitro assembly of nucleocapsids to include cleavage and packaging of viral DNA will prove feasible, facilitating a full definition of the requirements for this assembly. With the development of better molecular probes and the use of evolving imaging techniques, it should be possible to obtain much more complete information about subsequent events in herpesvirus virion morphogenesis, including envelopment and egress.

Acknowledgments

A.C.S. thanks his collaborators, Drs. Jay Brown, Frank Booy, James Conway, Fred Homa, Bill Newcomb, and Benes Trus, for images and ideas. We thank Dr. Bernard Roizman for the electron micrographs presented in Fig. 12–7 and Drs. Steven Bachenheimer and Sandra Weller, and the editors, for helpful comments on the manuscript.

References

Ace, C.I., McKee, T.A., Ryan, J.M., Cameron, J.M. and Preston, C.M. 1989. Construction and characterization of a herpes simplex virus type 1 mutant unable to transinduce immediate-early gene expression. J. Virol. 63: 2260–2269.

Addison, C., Rixon, F.J., Palfreyman, J.W., O'Hara, M. and Preston, V.G. 1984. Characterization of a herpes simplex virus type 1 mutant which has a temperature-sensitive defect in penetration of cells and assembly of capsids. Virology 138: 246–259.

Addison, C., Rixon, F.J. and Preston, V.G. 1990. Herpes simplex virus type 1 UL28 gene product is important for the formation of mature capsids. J. Gen. Virol. 71: 2377–2384.

Al-Kobaisi, M.F., Rixon, F.J., McDougall, I. and Preston, V.G. 1991. The herpes simplex virus UL33 gene product is required for the assembly of full capsids. Virology 180: 380–388.

Baines, J.D., Koyama, A.H., Huang, T. and Roizman, B. 1994a. The U21 gene products of herpes simplex virus 1 are dispensable for growth in cultured cells. J. Virol. 68: 2929–2936.

Baines, J.D., Poon, A.P.W., Rovnak, J. and Roizman, B. 1994b. The herpes simplex virus 1 U15 gene encodes two proteins and is required for cleavage of genomic viral DNA. J. Virol. 68: 8118–8124.

Baines, J.D. and Roizman, B. 1991. The open reading frames U3, U4, U10, and U16 are dispensable for the replication of herpes simplex virus 1 in cell culture. J. Virol. 65: 938–944.

Baines, J.D. and Roizman, B. 1992. The U11 gene of herpes simplex virus 1 encodes a function that facilitates nucleocapsid envelopment and egress from cells. J. Virol. 66: 5168–5174.

Baines, J.D. and Roizman, B. 1993. The U10 gene of herpes simplex virus 1 encodes a novel viral glycoprotein, gM, which is present in the virion and in the plasma membrane of infected cells. J. Virol. 67: 1441–1452.

Baines, J.D., Ward, P.L., Campadelli-Fiume, G. and Roizman, B. 1991. The U20 gene of herpes simplex virus 1 encodes a function necessary for viral egress. J. Virol. 65: 6414–6424.

Baker, T.S., Newcomb, W.W., Booy, F.P., Brown, J.C. and Steven, A.C. 1990. Three-dimensional structures of maturable and abortive capsids of equine herpesvirus 1 from cryoelectron microscopy. J. Virol. 64: 563–573.

Banfield, B.W. and Tufaro, F. 1990. Herpes simplex virus particles are unable to traverse the secretory pathway in the mouse L-cell mutant gro29. J. Virol. 64: 5716–5729.

Barker, D.E. and Roizman, B. 1992. The unique sequence of the herpes simplex virus 1 L component contains an additional translated open reading frame designated U49.5. J. Virol. 66: 562–566.

Barnett, B.C., Dolan, A., Telford, E.A.R., Davison, A.J. and McGeoch, D.J. 1992. A novel herpes simplex virus gene (UL49A) encodes a putative membrane protein with counterparts in other herpesviruses. J. Gen. Virol. 73: 2167–2171.

Batterson, W., Furlong, D. and Roizman, B. 1983. Molecular genetics of herpes simplex virus VIII. Further characterization of a temperature-sensitive mutant defective in release of viral DNA and in other stages of the viral reproductive cycle. J. Virol. 45: 397–407.

Baucke, R.B. and Spear, P.G. 1979. Membrane proteins specified by herpes simplex viruses. V. Identification of an Fc-binding glycoprotein. J. Virol. 32: 779–789.

Booy, F.P., Newcomb, W.W., Trus, B.L., Brown, J.C., Baker, T.S. and Steven, A.C. 1991. Liquid-crystalline, phage-like packing of encapsidated DNA in herpes simplex virus. Cell 64: 1007–1015.

Booy, F.P., Trus, B.L., Newcomb, W.W., Brown, J.C., Conway, J.F. and Steven, A.C. 1994. Finding a needle in a haystack: detection of a small protein (the 12 kDa VP26) in a large complex (the 200 MDa capsid of herpes simplex virus). Proc. Natl. Acad. Sci. U.S.A. 91: 5652–5656.

Booy, F.P., Trus, B.L., Newcomb, W.W., Brown, J.C., Serwer, P. and Steven, A.C. 1992. Organization of dsDNA in icosahedral virus capsids. In Proceedings of the 50th Annual Meeting of the Electron Microscopy Society of America, Vol. 1, pp. 452–453. San Francisco Press, San Francisco.

Braun, D.K., Roizman, B. and Pereira, L. 1984. Characterization of post-translational products of herpes simplex virus gene 35 proteins binding to the surfaces of full capsids but not empty capsids. J. Virol. 49: 142–153.

Browne, H., Bell, S., Minson, T. and Wilson, D.W. 1996. An endoplasmic reticulum-retained herpes simplex virus glycoprotein H is absent from secreted virions: Evidence for reenvelopment during egress. J. Virol. 70: 4311–4316.

Cai, W., Gu, B. and Person, S. 1988. Role of glycoprotein B of herpes simplex virus type 1 in viral entry and cell fusion. J. Virol. 62: 2596–2604.

Campadelli-Fiume, G., Arsenakis, M., Farabegoli, F. and Roizman, B. 1988. Entry of herpes simplex virus 1 in BJ cells that constitutively express viral glycoprotein D is by endocytosis and results in degradation of the virus. J. Virol. 62: 159–167.

Campadelli-Fiume, G., Farabegoli, F., Di Gaeta, S. and Roizman, B. 1991. Origin of unenveloped capsids in the cytoplasm of cells infected with herpes simplex virus 1. J. Virol. 65: 1589–1595.

Campadelli-Fiume, G. and Serafini-Cessi, F. 1985. Processing of the oligosaccharide chains of herpes simplex virus type 1 glycoproteins. In The Herpesviruses, Vol. 3 (Roizman, B., Ed.), pp. 357–382. Plenum Press, New York.

Casjens, S. and King, J. 1975. Virus assembly. Annu. Rev. Biochem. 44: 555–611.

Caspar, D.L. and Klug, A. 1962. Physical principles in the construction of regular viruses. Cold Spring Harbor Symp. Quant. Biol. 27: 1–24.

Cavalcoli, J.D., Baghian, A., Homa, F.L. and Kousoulas, K.G. 1993. Resolution of genotypic and phenotypic properties of herpes simplex virus type 1 temperature-sensitive mutant (KOS) *ts*Z47: evidence for allelic complementation in the UL28 gene. Virology 197: 23–34.

Celluzzi, C.N. and Farber, F.E. 1990. Role of the major capsid protein in herpes simplex virus type-1 capsid assembly. Acta Virol. (Praha) 34: 497–507.

Challberg, M.D. 1991. Functional analysis of the herpes simplex virus gene products involved in gene replication. Semin. Virol. 2: 247–256.

Chatterjee, S. and Sarkar, S. 1992. Studies on endoplasmic reticulum–Golgi complex cycling pathway in herpes simplex virus-infected and brefeldin A-treated human fibroblast cells. Virology 191: 327–337.

Cheung, P., Banfield, B.W. and Tufaro, F. 1991. Brefeldin A arrests the maturation and egress of herpes simplex virus particles during infection. J. Virol. 65: 1893–1904.

Claesson-Welsh, L. and Spear, P.G. 1986. Oligomerization of herpes simplex virus glycoprotein B. J. Virol. 60: 803–806.

Cohen, G.H., Ponce de Leon, M., Diggelmann, H., Lawrence, W.C., Vernon, S.K. and Eisenberg, R.J. 1980. Structural analysis of the capsid polypeptides of herpes simplex virus types 1 and 2. J. Virol. 34: 521–531.

Conway, J.F., Trus, B.L., Booy, F.P., Newcomb, W.W., Brown, J.C. and Steven, A.C. 1993. The effects of radiation damage on the structure of frozen hydrated HSV-1 capsids. J. Struct. Biol. 111: 222–233.

Conway, J.F., Trus, B.L., Booy, F.P., Newcomb, W.W., Brown, J.C. and Steven, A.C. 1996. Visualization of three-dimensional density maps reconstructed from cryo-electron micrographs of viral capsids. J. Struc. Biol. 116: 200–208.

Coulter, L.J., Moss, H.W.M., Lang, J. and McGeoch, D.J. 1993. A mutant of herpes simplex virus type 1 in which the UL13 protein kinase gene is disrupted. J. Gen. Virol. 74: 387–395.

Dargan, D.J. 1986. The structure and assembly of herpesviruses. In Electron Microscopy of Proteins (Harris, J.R. and Horne, R.W., Eds.), pp. 359–437. Academic Press, London.

Darlington, R.W. and Moss, L.H. III. 1968. Herpesvirus envelopment. J. Virol. 2: 48–55.

Davis-Poynter, N., Bell, S., Minson, T. and Browne, H. 1994. Analysis of the contributions of herpes simplex virus type 1 membrane proteins to the induction of cell-cell fusion. J. Virol. 68: 7586–7590.

Davison, A.J. 1992. Channel catfish virus: a new type of herpesvirus. Virology 186: 9–14.

Davison, M.D., Rixon, F.J. and Davison, A.J. 1992. Identification of genes encoding two capsid proteins (VP24 and VP26) of herpes simplex virus type 1. J. Gen. Virol. 73: 2709–2713.

Debroy, C., Pederson, N. and Person, S. 1985. Nucleotide sequence of a herpes simplex virus type 1 gene that causes cell fusion. Virology 145: 36–48.

Deckman, I.C., Hagen, M. and McCann, P.J. 1992. Herpes simplex virus type 1 protease expressed in Escherichia coli exhibits autoprocessing and specific cleavage of the ICP35 assembly protein. J. Virol. 66: 7362–7367.

Deiss, L.P., Chou, J. and Frenkel, N. 1986. Functional domains within the *a* sequence involved in the cleavage-packaging of the herpes simplex virus DNA. J. Virol. 59: 605–618.

Desai, P., DeLuca, N.A., Glorioso, J.C. and Person, S. 1993. Mutations in herpes simplex virus type 1 genes encoding VP5 and VP23 abrogate capsid formation and cleavage of replicated DNA. J. Virol. 67: 1357–1364.

Desai, P., Watkins, S.C. and Person, S. 1994. The size and symmetry of B capsids of herpes simplex virus type 1 are determined by the gene products of the UL26 open reading frame. J. Virol. 68: 5365–5374.

Devauchelle, G., Stoltz, D.B. and Darcy-Tripier, F. 1985. Comparative ultrastructure of iridoviridae. Curr. Top. Microb. Immunol. 116: 1–21.

Dingwell, K.S., Brunetti, C.R., Hendricks, R.L., Tang, Q., Tang, M., Rainbow, A.J. and Johnson, D.C. 1994. Herpes simplex virus glycoproteins E and I facilitate cell-to-cell spread in vivo and across junctions of cultured cells. J. Virol. 68: 834–845.

Dolter, K.E., Ramaswamy, R. and Holland, T.C. 1994. Syncytial mutations in the herpes simplex virus type 1 gK (UL53) gene occur in two distinct domains. J. Virol. 68: 8277–8281.

Dubin, G., Frank, I. and Friedman, H.M. 1990. Herpes simplex virus type 1 encodes two Fc receptors which have different binding characteristics for monomeric immunoglobulin G (IgG) and IgG complexes. J. Gen. Virol. 64: 2725–2731.

Forrester, A., Farrell, H., Wilkinson, G., Kaye, J., Davis-Poynter, N. and Minson, T. 1992. Construction and properties of a mutant of herpes simplex virus type 1 with glycoprotein H coding sequences deleted. J. Virol. 66: 341–348.

Friedmann, A., Coward, J.E., Rosenkranz, H.S. and Morgan, C. 1975. Electron microscopic studies on assembly of herpes simplex virus upon removal of hydroxyurea block. J. Gen. Virol. 26: 171–181.

Furlong, D. 1978. Direct evidence for 6-fold symmetry of the herpesvirus hexon capsomere. Proc. Natl. Acad. Sci. U.S.A. 75: 2764–2766.

Furlong, D., Swift, H. and Roizman, B. 1972. Arrangement of herpesvirus deoxyribonucleic acid in the core. J. Virol. 10: 1071–1074.

Gao, M., Matusick-Kumar, L., Hurlburt, W., DiTusa, S.F., Newcomb, W.W., Brown, J.C., McCann, P.J. III, Deckman, I. and Colonno, R.J. 1994. The protease of herpes simplex virus type 1 is essential for functional capsid formation and viral growth. J. Virol. 68: 3702–3712.

Gibson, W. and Roizman, B. 1972. Proteins specified by herpes simplex virus. VIII. Characterization and composition of multiple capsid forms of subtypes 1 and 2. J. Virol. 10: 1044–1052.

Gibson, W. and Roizman, B. 1974. Proteins specified by herpes simplex virus: staining and radiolabeling properties of B capsid and virion proteins in polyacrylamide gels. J. Virol. 13: 155–165.

Goding, C.R. and O'Hare, P. 1989. Herpes simplex virus Vmw65-octamer binding protein interaction: a paradigm for combinatorial control of transcription. Virology 173: 363–367.

Haanes, E.J., Nelson, C.M., Soule, C.L. and Goodman, J.L. 1994. The UL45 gene product is required for herpes simplex virus type 1 glycoprotein B-induced cell fusion. J. Virol. 68: 5825–5834.

Herold, B.C., Visalli, R.J., Susmarski, N., Brandt, C.R. and Spear, P.G. 1994. Glycoprotein C-independent binding of herpes simplex virus to cells requires cell surface heparan sulphate and glycoprotein B. J. Gen. Virol. 75: 1211–1222.

Herold, B.C., WuDunn, D., Soltys, N. and Spear, P.G. 1991. Glycoprotein C of herpes simplex virus type 1 plays a principal role in the adsorption of virus to cells and in infectivity. J. Virol. 65: 1090–1098.

Hutchinson, L., Browne, H., Wargent, V., Davis-Poynter, N., Primorac, S., Goldsmith, K., Minson, A.C. and Johnson, D.C. 1992. A novel herpes simplex virus glycoprotein, gL, forms a complex with glycoprotein H (gH) and affects normal folding and surface expression of gH. J. Virol. 66: 2240–2250.

Johnson, D.C. and Feenstra, V. 1987. Identification of a novel herpes simplex virus type 1-induced glycoprotein which complexes with gE and binds immunoglobulin. J. Virol. 61: 2208–2216.

Johnson, D.C. and Spear, P.G. 1982. Monensin inhibits the processing of herpes simplex virus glycoproteins, their transport to the cell surface, and the egress of virions from infected cells. J. Virol. 43: 1102–1112.

Johnson, D.C. and Spear, P.G. 1983. O-linked oligosaccharides are acquired by herpes simplex virus glycoproteins in the Golgi apparatus. Cell 32: 987–997.

Johnson, R.M. and Spear, P.G. 1989. Herpes simplex virus glycoprotein D mediates interference with herpes simplex virus infection. J. Virol. 63: 819–827.

Jones, F. and Grose, C. 1988. Role of cytoplasmic vacuoles in varicella-zoster virus glycoprotein trafficking and virion envelopment. J. Virol. 62: 2701–2711.

Kousoulas, K.G., Bzik, D.J. and Person, S. 1983. Effect of the ionophore monensin on herpes simplex virus type 1-induced cell fusion, glycoprotein synthesis, and virion infectivity. Intervirology 20: 56–60.

Lee, J.Y., Irmiere, A. and Gibson, W. 1988. Primate cytomegalovirus assembly: evidence tha DNA packaging occurs subsequent to B-capsid assembly. Virology 167: 87–96.

Lepault, J., Dubochet, J., Baschong, W. and Kellenberger, E. 1987. Organizational of double-stranded DNA in bacteriophages: a study by cryo-electron microscopy of vitrified samples. EMBO J. 6: 1507–1512.

Ligas, M.W. and Johnson, D.C. 1988. A herpes simplex virus mutant in which glycoprotein D sequences are replaced by β-galactosidase sequences binds to but is unable to penetrate into cells. J. Virol. 62: 1486–1494.

Liu, F. and Roizman, B. 1991. The herpes simplex virus 1 gene encoding a protease also contains within its coding domain the gene encoding the more abundant substrate. J. Virol. 65: 5149–5156.

Liu, F. and Roizman, B. 1993. Characterization of the protease and other products of amino-terminus-proximal cleavage of herpes simplex virus 1 UL26 protein. J. Virol. 67: 1300–1309.

Longnecker, R. and Roizman, B. 1987. Clustering of genes dispensible for growth in culture in the S component of the HSV-1 genome. Science 236: 573–576.

MacLean, C.A., Dolan, A., Jamieson, F.E. and McGeoch, D.J. 1992. The myristylated virion proteins of herpes simplex virus type 1: investigation of their role in the virus life cycle. J. Gen. Virol. 73: 539–547.

MacLean, C.A., Efstathiou, S., Elliott, M.L., Jamieson, F.E. and McGeoch, D.J. 1991. Investigation of herpes simplex virus type 1 genes encoding multiply inserted membrane proteins. J. Gen. Virol. 72: 897–906.

Manservigi, R., Spear, P.G. and Buchan, A. 1977. Cell fusion induced by herpes simplex virus is promoted and suppressed by different viral glycoproteins. Proc. Natl. Acad. Sci. U.S.A. 74: 3913–3917.

Marsden, H.S., Stow, N.D., Preston, V.G., Timbury, M.C. and Wilkie, N.M. 1978. Physical mapping of herpes simplex virus-induced polypeptides. J. Virol. 28: 624–642.

Matusick-Kumar, L., Hurlburt, W., Weinheimer, S.P., Newcomb, W.W., Brown, J.C. and Gao, M. 1994. Phenotype of the herpes simplex virus type 1 protease substrate ICP35 mutant virus. J. Virol. 68: 5384–5394.

McGeoch, D.J., Dalrymple, M.A., Davison, A.J., Dolan, A., Frame, M.C., McNab, D., Perry, L.J., Scott, J.E. and Taylor, P. 1988. The complete DNA sequence of the long unique region in the genome of herpes simplex virus type 1. J. Gen. Virol. 69: 1531–1574.

McLauchlan, J., Addison, C., Craigie, M.C. and Rixon, F.J. 1992. Noninfectious L-particles supply functions which can facilitate infection by HSV-1. Virology 190: 682–688.

Mettenleiter, T.C. 1994. Initiation and spread of Alpha-herpesvirus infections. Trends Microbiol. 2: 2–3.

Mettenleiter, T.C., Zsak, L., Zuckermann, F., Sugg, N., Kern, H. and Ben-Porat, T. 1990. Interaction of glycoprotein gIII with a cellular heparin-like substance mediates adsorption of pseudorabies virus. J. Virol. 64: 278–286.

Montgomery, R.I., Warner, M.S., Lum, B.J. and Spear, P.G. 1996. Herpes simplex virus-1 entry into cells mediated by a novel member of the TNF/NGF receptor family. Cell 87: 427–436.

Morgan, C., Rose, H.M., Holden, M. and Jones, E.P. 1954. Electron microscopic observations on the development of herpes simplex virus. J. Exp. Med. 110: 643–656.

Nermut, M.V. 1980. The architecture of adenoviruses: recent views and problems: Brief review. Arch. Virol. 64: 175–196.

Newcomb, W.W. and Brown, J.C. 1989. Use of Ar+ plasma etching to localize structural proteins in the capsid of herpes simplex virus type 1. J. Virol. 63: 4697–4702.

Newcomb, W.W. and Brown, J.C. 1991. Structure of the herpes simplex virus capsid: effects of extraction with guanidine-HCl and partial reconstitution of extracted capsids. J. Virol. 65: 613–620.

Newcomb, W.W., Brown, J.C., Booy, F.P. and Steven, A.C. 1989. Nucleocapsid mass and capsomer protein stoichiometry in equine herpesvirus 1: scanning transmission electron microscopic study. J. Virol. 63: 3777–3783.

Newcomb, W.W., Homa, F.L., Thomsen, D.R., Ye, Z. and Brown, J.C. 1994. Cell-free assembly of the herpes simplex virus capsid. J. Virol. 68: 6059–6063.

Newcomb, W.W., Trus, B.L., Booy, F.P., Steven, A.C., Wall, J.S. and Brown, J.C. 1993. Structure of the herpes simplex virus capsid: molecular composition of the pentons and triplexes. J. Mol. Virol. 232: 499–511.

Newcomb, W.W., Homa, F.L., Thomsen, D.R., Booy, F.P., Trus, B.L., Steven, A.C., Spencer, J.V. and Brown, J.C. 1996. Assembly of the herpes simplex virus capsid. Identification of intermediates in cell-free capsid assembly. J. Mol. Biol. 263: 432–446.

Nicholson, P., Addison, C., Cross, A.M., Kennard, J., Preston, V.G. and Rixon, F.J. 1994. Localization of the herpes simplex virus type 1 major capsid protein VP5 to the cell nucleus requires the abundant scaffolding protein VP22a. J. Gen. Virol. 75: 1091–1099.

Olofsson, S. 1992. Carbohydrates in herpesvirus infections. APMIS Suppl. 100: 84–95.

Overton, H., McMillan, D., Hope, L. and Wong-Kai-In, P. 1994. Production of host shutoff-defective mutants of herpes simplex virus type 1 by inactivation of the UL13 gene. Virology 202: 97–106.

Palmer, E.L., Martin, M.L. and Gary, G.W. Jr. 1975. The ultrastructure of disrupted herpesvirus nucleocapsids. Virology 65: 260–265.

Patel, A.H. and Maclean, J.B. 1995. The product of the UL6 gene of herpes simplex virus type 1 is associated with virus capsids. Virology 206: 465–478.

Perdue, M.L., Cohen, J.C., Randall, C.C. and O'Callaghan, D.J. 1976. Biochemical studies of the maturation of herpesvirus nucleocapsid species. Virology 74: 194–208.

Pertuiset, B., Boccara, M., Cebrian, J., Berthelot, N., Chousterman, S., Puvion-Dutilleul, F., Sisman, J. and Sheldrick, P. 1989. Physical mapping and nucleotide sequence of a herpes simplex virus type 1 gene required for capsid assembly. J. Virol. 63: 2169–2179.

Pogue-Geile, K.L. and Spear, P.G. 1987. The single base pair substitution responsible for the Syn phenotype of herpes simplex virus type 1, strain MP. Virology 157: 67–74.

Poon, A.P. and Roizman, B. 1993. Characterization of a temperature-sensitive mutant of the U15 open reading frame of herpes simplex virus 1. J. Virol. 67: 4497–4503.

Preston, V.G., Coates, J.A.V. and Rixon, F.J. 1983. Identification and characterization of a herpes simplex virus gene product required for encapsidation of virus DNA. J. Virol. 45: 1056–1064.

Preston, V.G., Rixon, F.J., McDougall, I.M., McGregor, M. and Al-Kobaisi, M.F. 1992. Processing of the herpes simplex virus assembly protein-icp35 near its carboxy terminal end requires the product of the whole of the ul26 reading frame. Virology 186: 87–98.

Purves, F.C. and Roizman, B. 1992. The UL13 gene of herpes simplex virus 1 encodes the functions for posttranslational processing associated with phosphorylation of the regulatory protein alpha 22. Proc. Natl. Acad. Sci. U.S.A. 89: 7310–7314.

Purves, F.C., Spector, D. and Roizman, B. 1991. The herpes simplex virus 1 protein kinase encoded by the U3 gene mediates posttranslational modification of the phosphoprotein encoded by the U34 gene. J. Virol. 65: 5757–5764.

Purves, F.C., Spector, D. and Roizman, B. 1992. U34, the target of the herpes simplex virus U3 protein kinase, is a membrane protein which in its unphosphorylated state associates with novel phosphoproteins. J. Virol. 66: 4295–4303.

Puvion-Dutilleul, F., Pichard, E., Laithier, M. and Leduc, E.H. 1987. Effect of dehydrating agents on DNA organization in herpes viruses. J. Histochem. Cytochem. 35: 635–645.

Read, G.S. and Frenkel, N. 1983. Herpes simplex virus mutants defective in the virion-associated shutoff of host polypeptide synthesis and exhibiting abnormal synthesis of Â (immediate early) viral polypeptides. J. Virol. 46: 498–512.

Rixon, F.J., Addison, C. and McLauchlan, J. 1992. Assembly of enveloped tegument structures (L particles) can occur independently of virion maturation in herpes simplex virus type 1-infected cells. J. Gen. Virol. 73: 277–284.

Rixon, F.J., Cross, A.M., Addison, C. and Preston, V.G. 1988. The products of herpes simplex virus type 1 gene UL26 which are involved in DNA packaging are strongly associated with empty but not with full capsids. J. Gen. Virol. 69: 2879–2891.

Roizman, B., Desrosiers, R.C., Fleckenstein, B., Lopez, C., Minson, A.C. and Studdert, M.J. 1992. The family *Herpesviridae*: an update. Arch. Virol. 123: 425–449.

Roizman, B. and Sears, A. 1990. Herpes simplex viruses and their replication. In Virology, 2nd ed. (Fields, B.N., Ed.), pp. 1795–1841. Raven Press, New York.

Roop, C., Hutchinson, L. and Johnson, D.C. 1993. A mutant herpes simplex virus type 1 unable to express glycoprotein L cannot enter cells, and its particles lack glycoprotein H. J. Virol. 67: 2285–2297.

Sarmiento, M., Haffey, M. and Spear, P.G. 1979. Membrane proteins specified by herpes simplex viruses. III. Role of glycoprotein VP7 (B) in virion infectivity. J. Virol. 29: 1149–1158.

Sarmiento, M. and Spear, P.G. 1979. Membrane proteins specified by herpes simplex viruses. IV. Conformation of the virion glycoprotein designated VP7(B). J. Virol. 29: 1159–1167.

Schrag, J.D., Prasad, B.V., Rixon, F.J. and Chiu, W. 1989. Three-dimensional structure of the HSV-1 nucleocapsid. Cell 56: 651–660.

Sears, A.E., McGwire, B.S. and Roizman, B. 1991. Infection of polarized MDCK cells with herpes simplex virus 1: two asymmetrically distributed cell receptors interact with different viral proteins. Proc. Natl. Acad. Sci. U.S.A. 88: 5087–5091.

Serafini-Cessi, F., Dall'Olio, F., Scannavini, M. and Campadelli-Fiume, G. 1983. Processing of herpes simplex virus-1 glycans in cells defective in glycosyl transferases of the Golgi system: relationship to cell fusion and virion egress. Virology 131: 59–70.

Severini, A., Morgan, A.R., Tovell, D.R. and Tyrrell, D.L. 1994. Study of the structure of replicative intermediates of HSV-1 DNA by pulsed-field gel electrophoresis. Virology 200: 428–435.

Shao, L., Rapp, L.M. and Weller, S.K. 1993. Herpes simplex virus 1 alkaline nuclease is required for efficient egress of capsids from the nucleus. Virology 196: 146–162.

Sherman, G. and Bachenheimer, S.L. 1987. DNA processing in temperature-sensitive morphogenic mutants of HSV-1. Virology 158: 427–430.

Sherman, G. and Bachenheimer, S.L. 1988. Characterization of intranuclear capsids made by morphogenic mutants of HSV-1. Virology 163: 471–480.

Shieh, M.-T. and Spear, P.G. 1994. Herpesvirus-induced cell fusion that is dependent on cell surface heparan sulfate or soluble heparin. J. Virol. 68: 1224–1228.

Smibert, C.A., Johnson, D.C. and Smiley, J.R. 1992. Identification and characterization of the virion-induced host shutoff product of herpes simplex virus gene UL41. J. Gen. Virol. 73: 467–470.

Spear, P.G. 1993. Entry of alphaherpesviruses into cells. Semin. Virol. 4: 167–180.

Spear, P.G. and Roizman, B. 1972. Proteins specified by herpes simplex virus. V. Purification and structural proteins of the herpes virion. J. Virol. 9: 143–159.

Stackpole, C.W. 1969. Herpes-type virus of the frog renal adenocarcinoma. I. Virus development in tumor transplants maintained at low termperature. J. Virol. 4: 75–93.

Stannard, L.M., Fuller, A.O. and Spear, P.G. 1987. Herpes simplex virus glycoproteins associated with different morphological entities projecting from the virion envelope. J. Gen. Virol. 68: 715–725.

Steven, A.C., Roberts, C.R., Hay, J., Bisher, M.E., Pun, T. and Trus, B.L. 1986. Hexavalent capsomers of herpes simplex virus type 2: symmetry, shape, dimensions, and oligomeric status. J. Virol. 57: 578–584.

Szilágyi, J.F. and Berriman, J. 1994. Herpes simplex virus L particles contain spherical membrane-enclosed inclusion vesicles. J. Gen. Virol. 75: 1749–1753.

Szilágyi, J.F. and Cunningham, C. 1991. Identification and characterization of a novel non-infectious herpes simplex virus-related particle. J. Gen. Virol. 72: 661–668.

Tatman, J.D., Preston, V.G., Nicholson, P., Elliott, R.M. and Rixon, F.J. 1994. Assembly of herpes simplex virus type 1 capsids using a panel of recombinant baculoviruses. J. Gen. Virol. 75: 1101–1113.

Tengelsen, L.A., Pederson, N.E., Shaver, P.R., Wathen, M.W. and Homa, F.L. 1993. Herpes simplex virus type 1 DNA cleavage and encapsidation require the product of the UL28 gene: isolation and characterization of two UL28 deletion mutants. J. Virol. 67: 3470–3480.

Thomsen, D.R., Newcomb, W.W., Brown, J.C. and Homa, F.L. 1995. Assembly of the herpes simplex virus capsid requirement for the carboxyl-terminal twenty-five amino acids of the proteins encoded by UL26 and UL26.5 genes. J. Virol. 69: 3690–3703.

Thomsen, D.R., Roof, L.L. and Homa, F.L. 1994. Assembly of herpes simplex virus (HSV) intermediate capsids in insect cells infected with recombinant baculoviruses expressing HSV capsid proteins. J. Virol. 68: 2442–2457.

Tognon, M., Furlong, D., Conley, A.J. and Roizman, B. 1981. Molecular genetics of herpes simplex virus V. Characterization of a mutant defective in ability to form plaques at low temperatures and in a viral function which prevents accumulation of coreless capsids at nuclear pores late in infection. J. Virol. 40: 870–880.

Torrisi, M.R., Di Lazzaro, C., Pavan, A., Pereira, L. and Campadelli-Fiume, G. 1992. Herpes simplex virus envelopment and maturation studied by fracture label. J. Virol. 66: 554–561.

Trus, B.L., Booy, F.P., Newcomb, W.W., Brown, J.C., Homa, F.L., Thomsen, D.R. and Steven, A.C. 1996. The herpes simplex virus procapsid: Structure, conformational changes upon maturation, and roles of the triplex proteins VP19c and VP23 in assembly. J. Mol. Biol. 263: 447–462.

Trus, B.L., Homa, F.L., Booy, F.P., Newcomb, W.W., Thomsen, D.R., Cheng, N., Brown, J.C. and Steven, A.C. 1995. Herpes simplex virus capsids assembled in insect cells infected with recombinant baculoviruses: structural authenticity and localization of vp26. J. Virol. 69: 7362–7366.

Trus, B.L., Newcomb, W.W., Booy, F.L., Brown, J.C. and Steven, A.C. 1992. Distinct monoclonal antibodies separately label the hexons or the pentons of herpes simplex virus capsid. Proc. Natl. Acad. Sci. U.S.A. 89: 11508–11512.

van Genderen, I.L., Brandimarti, R., Torrisi, M.R., Campadelli, G. and van Meer, G.1994. The phospholipid composition of extracellular herpes simplex virions differs from that of host cell nuclei. Virology 200: 831–836.

Varmuza, S.L. and Smiley, J.R. 1985. Signals for site-specific cleavage of HSV DNA: maturation involves two separate cleavage events at sites distal to the recognition sequences. Cell 41: 793–802.

Visalli, R.J. and Brandt, C.R. 1991. The HSV-1 UL45 gene product is not required for growth in vero cells. Virology 185: 419–423.

Vlazny, D.A., Kwong, A. and Frenkel, N. 1982. Site-specific cleavage/packaging of herpes simplex virus DNA and the selective maturation of nucleocapsids containing full-length viral DNA. Proc. Natl. Acad. Sci. U.S.A. 79: 1423–1427.

Ward, P.L., Campadelli-Fiume, G., Avitabile, E. and Roizman, B. 1994. Localization and putative function of the U20 membrane protein in cells infected with herpes simplex virus 1. J. Virol. 68: 7406–7417.

Weinheimer, S.P., Boyd, B.A., Durham, S.K., Resnick, J.L. and O'Boyle, D.R. II. 1992. Deletion of the VP16 open reading frame of herpes simplex virus type 1. J. Virol. 66: 258–269.

Welch, A.R., Woods, A.S., McNally, L.M., Cotter, R.J. and Gibson, W. 1991. A herpesvirus maturational protease, assemblin: identification of its gene, putative active site domain, and cleavage site. Proc. Natl. Acad. Sci. U.S.A. 88: 10792–10796.

Weller, S.K. 1995. Herpes simplex virus DNA replication and genome maturation. In The DNA Provirus—Howard Temin's Scientific Legacy (Cooper, G.M., Temin, R.G. and Sugden, B., Eds.), pp. 189–213. American Society for Microbiology, Washington, DC.

Weller, S.K., Carmichael, E.P., Aschman, D.P., Goldstein, D.J. and Schaffer, P.A. 1987. Genetic and phenotypic characterization of mutants in four essential genes that map to the left half of HSV-1 U DNA. Virology 161: 198–210.

Wildy, P., Russell, W.C. and Horne, R.W. 1960. The morphology of herpes virus. Virology 12: 204–222.

Zhang, X., Efstathiou, S. and Simmons, A. 1994. Identification of novel herpes simplex virus replicative intermediates by field inversion gel electrophoresis: implications for viral DNA amplification strategies. Virology 202: 530–539.

Zhang, Y. and McKnight, J.L.C. 1993. Herpes simplex virus type 1 UL46 and UL47 deletion mutants lack VP11 and VP12 or VP13 and VP14, respectively, and exhibit altered viral thymidine kinase expression. J. Virol. 67: 1482–1492.

Zhang, Y., Sirko, D.A. and McKnight, J.L.C. 1991. Role of herpes simplex virus type 1 UL46 and UL47 in ÂTIF-mediated transcriptional induction: characterization of three viral deletion mutants. J. Virol. 65: 829–841.

Zhou, Z.H., He, J., Jakana, J., Tatman, J.D., Rixon, F.J. and Chiu, W. 1995. Assembly of vp26 in herpes simplex virus-1 inferred from structures of wild type and recombinant capsids. Nature Struct. Biol. 2: 1026–1030.

Zhou, Z.H., Prasad, B.V.V., Jakana, J., Rixon, F.J. and Chiu, W. 1994. Protein subunit structures in the herpes simplex virus A-capsid determined from 400 kV spot-scan electron cryomicroscopy. J. Mol. Biol. 242: 456–469.

Zhu, Q. and Courtney, R.J. 1988. Chemical crosslinking of glycoproteins on the envelope of herpes simplex virus. Virology 167: 377–384.

Zhu, Q. and Courtney, R.J. 1994. Chemical cross-linking of virion envelope and tegument proteins of herpes simplex virus type 1. Virology 204: 590–599.

Zsak, L., Mettenleiter, T.C., Sugg, N. and Ben-Porat, T. 1990. Effect of polylysine on the early stages of infection of wild type pseudorabies virus and of mutants defective in gIII. Virology 179: 330–338.

13.

Structure and Assembly of Filamentous Bacteriophages

LEE MAKOWSKI & MARJORIE RUSSEL

The Family of Filamentous Bacteriophages

The filamentous phage constitute a large family of bacterial viruses that infect a variety of gram-negative bacteria. They all have single-stranded (SS), circular DNA genomes. The virus particles are flexible rods about 65 Å in diameter and 0.8 to 2.0 μ in length (0.9 μ for the F-specific phages of *Escherichia coli*). The DNA is extended along the axis of the particle and encapsulated in a tube formed by many copies of the major coat protein. A few copies each of several minor coat proteins are located at the ends of the phage filament. One end contains two proteins that are necessary for efficient particle assembly, and the other end contains two proteins required for particle stability and phage infectivity. Filamentous phage do not kill their host; rather, they are continually extruded or secreted across the bacterial inner and outer membrane without causing cell lysis as the infected cells continue to grow and divide. It is this property, and their filamentous morphology, that distinguishes them from most other bacterial viruses that accomplish their release by lysis of host cells.

The list of filamentous phage that have been identified is quite long, but only a few have been characterized in detail. The best characterized are the very similar phages M13, fd, and f1 (98% identity in DNA sequence) that infect *E. coli* via F pili. Unless otherwise specified, all statements in this chapter refer to these three viruses, designated collectively as Ff. A number of other *E. coli* phage with substantial sequence homology to these F pili–specific phages have been identified, sequenced, and characterized. The class is, however, much broader, and phage specific for *Xanthomonas, Pseudomonas, Salmonella, Vibrio*, and other gram-negative bacteria have been isolated. All appear to be flexible rods, approximately 65 Å in diameter and with lengths that depend both on the size of their genome and on the physical extension of the DNA within the virus particle (axial distance per base). Among the most distant members of the group, some similarities of sequence and genome organization are still discernible.

The Genome

The Ff phage genome encodes 10 proteins, of which five are virion structural proteins, three are required for phage DNA synthesis, and two serve assembly functions (reviewed in Model and Russel, 1988). In addition, there is an intergenic region that does not code for proteins but contains the signals for the initiation of synthesis of both the (+) and (−) strands of DNA, the initiation of capsid formation (packaging signal; PS), and the termination of RNA synthesis. Parts of the intergenic region are dispensable, but all of the phage-encoded proteins are required for synthesis of infectious progeny phage. Figure 13.1 is a diagram of the genome organization.

Viral Assembly

The preeminent feature of filamentous phage assembly is that all steps subsequent to the formation of the protein pV–ssDNA complex occur at or in a

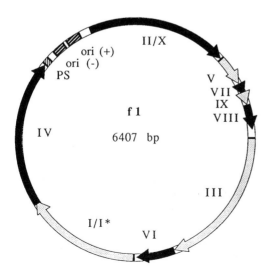

FIGURE 13.1   Genome of filamentous phage of *E. coli* showing genes (Roman numerals) and DNA signals in the intergenic region. PS, packaging signal; ori (−), origin for complementary strand DNA synthesis; ori (+), origin for packaged strand DNA synthesis. Gene X, in-frame with gene II, gives rise to pX, which is identical to the carboxyl-terminal third of pII; similarly, I* is identical to the carboxyl-terminal third of pI. All transcription proceeds clockwise on the map shown.

membrane. This sets these phage apart from most bacterial viruses, which as-semble in the cytoplasm. Because membrane proteins pose difficult problems in biochemical analysis, detailed knowledge of filamentous phage assembly is rudimentary. Instead, most insights are derived from molecular biologic and genetic analysis and from the structure of membrane-associated and particle-associated viral proteins. For these reasons, this chapter is organized to first provide a view of the structures of phage proteins prior to assembly, and then their structures after integration into the completed virus particle. These struc-tures represent the "before" and "after" snapshots of the proteins involved in the assembly process. The dynamic process of assembly is deduced from these snapshots, and from the interactions implied from biochemical, molecular biologic, and genetic analysis.

Structure of the Gene Products Prior to Assembly

Replication Proteins (pII, pX)

Gene II specifies a protein of 410 amino acids required for all phage-specific DNA synthesis other than the formation of the complementary strand of the infecting phage ssDNA. Protein pII is an endonuclease-topoisomerase that in-troduces a specific nick into the (+) strand of supercoiled replicative form (double-stranded) phage DNA. The resulting 3' end serves as primer for the synthesis of a new (+) strand, which is terminated and circularized by pII. The resulting (+) strand either is used as a template for synthesis of the comple-mentary strand by host enzymes or is bound by pV and subsequently assembled into phage. Protein pX is encoded entirely within gene II, starting with an internal ATG (codon 300) that is in phase with the initiating gene II ATG. Hence, pX has the same amino acid sequence as the carboxyl-terminal third of pII. Protein pX is essential in its own right, acting as an inhibitor of pII function (Fulford and Model, 1984).

Single-Stranded DNA Binding Protein (pV)

Gene V encodes an 87-amino-acid protein that binds cooperatively and with high affinity to ssDNA (Alberts and Frey, 1972; Oey and Knippers, 1972). The coating of newly synthesized ssDNA by pV sequesters it and prevents synthesis of the complementary strand. In its absence, ssDNA does not accumulate and few phage are assembled (Pratt and Erdahl, 1968; Salstrom and Pratt, 1971). The crystal structure of the pV dimer has been determined to a resolution of 1.8 Å (Skinner et al., 1994).

The structure of the pV dimer is diagrammed in Figure 13.2. Each mon-omer is largely β-structure (Skinner et al., 1994). Of the 87 residues, 58 are arranged in a five-stranded antiparallel β-sheet and two antiparallel β-ladder loops that protrude out from this sheet. The remainder of the molecule is ar-

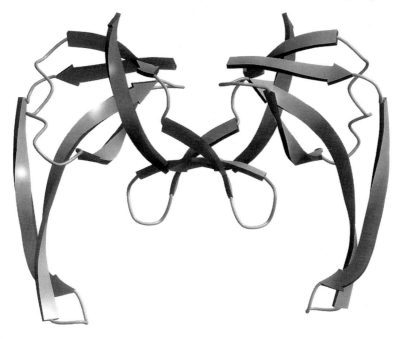

FIGURE 13.2   A ribbon diagram of pV protein dimer structure as viewed down the twofold axis. The saddlelike structure evident in this view reflects the presence of two grooves, one in each monomer, that are postulated to represent the ssDNA binding sites. (Reproduced with permission from Skinner et al., 1994.)

ranged into 3_{10} helical regions (residues 7 to 11 and 65 to 67), β-bends (residues 21 to 24, 50 to 53, and 71 to 74) and one five-residue loop (residues 38 to 42).

The Protein V–ssDNA Complex

The newly synthesized viral ssDNA interacts with pV to form a protein–DNA complex as an obligate first step in the assembly process. Electron microscopy of the pV–ssDNA complex shows that it is a rod, suggesting that the circular single strand has been collapsed by pV dimers (Gray et al., 1981). No minor phage proteins are present in the in vivo complex, although a few molecules of the *E. coli* ssDNA binding protein copurify with it (Grant and Webster, 1984). For phage to assemble, pV, which at physiologic salt concentrations binds DNA quite tightly, must be replaced by the virion structural proteins. Protein pV does not bind to the PS, which is located at one end of this rodlike complex (Bauer and Smith, 1988). In the pV–DNA complex, the monomers making up the pV dimer bind to DNA strands running in opposite directions along the helix. It may be that the only way to construct an organized, helical DNA–protein complex with these properties is to nucleate protein binding at a single site, possibly adjacent to the PS. Multiple nucleation events resulting from random binding of pV along the circular DNA would lead to a topologically complex, chaotic structure with pV "cross-linking" the ssDNA to form loops of random size at random intervals. No simple rearrangement would provide this structure with a pathway for transformation into a regular helical structure.

The complex is a flexible structure, approximately 8800 Å long, and approximately 80 Å in diameter (Gray, 1989; Torbet et al., 1981; Gray et al., 1982a, 1982b). Calculation of mass per unit length deduced from neutron diffraction data (Gray et al., 1982b) suggests that it contains four nucleotides per protein. The radius of gyration of the DNA in the complex is substantially less than that of the protein, indicating that the protein is arranged so that the DNA is, on average, closer to the axis of the structure. The structure appears to be a tightly wound left-handed helix in electron micrographs, although the pitch of the helix may vary from 60 to 120 Å, with the larger values occurring only when turns of the complex appear to part from one another, forming an open helical structure (Gray, 1989). If this stretched form is a result of preparation for electron microscopy, then in vivo the pitch of the complex may be about 75 Å, with eight pV proteins per turn. Consistent with these observations and the recent crystallographic determination of the pV dimer structure, a model of the pV–DNA complex has been constructed (Skinner et al., 1994). Additional modeling will be required to obtain a plausible set of interactions between bound DNA and the pV protein, but the current model is consistent with a strand of ssDNA running approximately from the location of one group of residues (Tyr 26, Leu 28, and Phe 73) to the next in the helix. The distance from one Tyr 26 to the next is about 15 Å in this model, a distance readily spanned by four nucleotides in a partially extended conformation (Skinner et al., 1994). This model places the antiparallel single strands within the grooves formed by the β-strands that loop out from the body of the dimer, as are clearly evident in Figure 13.2.

Major Coat Protein (pVIII)

Gene VIII codes for the major coat protein of the phage. The major coat proteins of all filamentous phages are very small, 44 to 55 amino acids in length, and most are encoded with a signal sequence. In the Ff, the major coat protein is 50 amino acids in length, with a 23-amino-acid signal sequence; that of the *Pseudomonas aeruginosa* phage Pf3 lacks a signal sequence. These coat proteins are organized into three regions: an acidic amino-terminal region that makes up most of the surface of the phage particle, a hydrophobic central region, and a basic carboxyl-terminal region involved in interaction with the DNA. Approximately 2800 copies of pVIII are required to coat the genome in the F-specific viruses. Newly synthesized pVIII is inserted into the host inner membrane, and its signal sequence is removed by host-encoded signal peptidase (Chang et al., 1979). Until it is incorporated into phage particles, the protein remains anchored in the membrane via its hydrophobic central region, with its amino terminus in the periplasm and its carboxyl terminus in the cytoplasm (Ohkawa and Webster, 1981).

In the membrane, pVIII is made up of a pair of α-helices with the amino-terminal helix lying along the surface of the membrane and the transmembrane, carboxyl-terminal helix extending perpendicular to the surface, through the membrane (McDonnell et al., 1993). A model for this structure is shown in Figure 13.3 (left panel). Residues 1 to 5 at the amino terminus are highly mobile and unstructured in both the membrane and the virus particle. Nuclear magnetic resonance (NMR) data indicate that, in the membrane-bound form of the coat protein, the carboxyl-terminal α-helix extends all the way to the carboxyl-terminal residue (McDonnell et al., 1993). If this is true, the carboxyl-terminal helix extends out of the membrane by 16 to 17 Å to accommodate the four basic residues at that end of the coat protein. The bend between the two helices appears to be between Tyr 21 (which is mobile in the membrane-bound form of the protein) and Tyr 24 (McDonnell et al., 1993). The viability of mutants in which residues 23 and 24 are changed to polar residues (Gly 24→Asp, Tyr 24→His, Tyr 24→Asp, and Tyr 24→Asn) suggests that this segment is external to the membrane (Li and Deber, 1991). Protein pVIII of the *E. coli* filamentous phages appears to lack the 7 to 8 amino-acid mobile loop that is present in the membrane-bound form of the major coat protein of

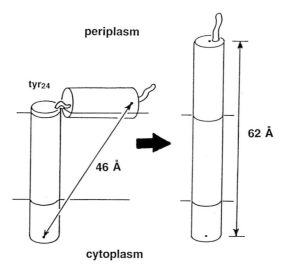

periplasm

tyr24

46 Å

62 Å

cytoplasm

FIGURE 13.3   Model of pVIII in the membrane and in the phage. *Left.* In the membrane, the first few residues at the amino terminus of the protein (top) are disordered and extend out into the periplasm. Residues 6 to 21 form an amphipathic helix that rests on the periplasmic surface of the membrane. Residues 24 to 45 make up a transmembrane α-helix that extends into the cytoplasm. The exact boundary between the two helices is uncertain, but few, if any, residues are in a nonhelical conformation between them. It is also possible that a few residues at the carboxyl terminus may not be in an α-helical conformation. (Modified from McDonnell et al., 1993.) *Right.* In the phage particle, pVIII exists as a single, long helix. The conformational change increases the distance between the two ends of the molecule by about 16 Å, the length increase that the particle undergoes during the integration of each layer of pVIII.

P. aeruginosa phage Pf1 (Shon et al., 1991) and that remains nonhelical in the virion (Nambudripad et al., 1991).

Minor Structural Proteins (pVII, pIX, pIII, and pVI)

Genes VII and IX encode very small proteins (33 and 32 amino acids, respectively) that are located at one end of the virus particle (Lopez and Webster, 1983). Like the major coat protein, these rather hydrophobic proteins are integral inner membrane proteins prior to their assembly (Endemann and Model, 1995). Protein pIX is likely to be anchored in the membrane by its hydrophobic amino-terminal half; the presence of two positively charged residues in the hydrophilic carboxyl-terminal half suggests that this portion of the protein is cytoplasmic. The carboxyl-terminal portion of pVII is highly hydrophobic, but the orientation of this protein is not clear. Although no pVII–pIX complex has been detected (Endemann and Model, 1995), it remains possible that these two small proteins, located at the same tip of the phage particle, associate with one another in the membrane prior to their incorporation into phage. In the absence of either pVII or pIX, almost no phage particles are formed (Webster and Lopez, 1985); genetic evidence suggests that both are involved in the initiation of assembly (Russel and Model, 1989). Approximately five copies each of pVII and pIX are present at the end of the particle that is first to assemble (Webster et al., 1981). If their structure in the virus reflects the fivefold symmetry of the remainder of the particle, there will be exactly five copies of each at the end of the particle.

Gene III encodes a protein located at the other end of the particle, the end assembled last (Webster et al., 1981). This 42-kDa protein has 406 amino acids in the mature form and is synthesized with an 18-amino-acid signal sequence. Prior to assembly, pIII spans the host cell inner membrane via a hydrophobic membrane anchor located five residues from the carboxyl-terminus (Boeke and Model, 1982; Davis et al., 1985). Most of its mass extends into the periplasm, leaving a small carboxyl-terminal domain in the cytoplasm. This protein is required for infectivity; it mediates phage attachment to its receptor (F pili in the case of the Ff) and is also necessary for virus uncoating and penetration of phage DNA into the host cell cytoplasm (Russel et al., 1988). In the absence of pIII, noninfectious, multiple-length ''polyphage'' particles containing several unit-length genomes are produced, indicating that pIII is necessary to terminate the assembly process. In the intact phage there are approximately five copies of pIII, the last structural features to be added to the growing phage particle prior to release from the host. Whether expressed from a phage genome or a plasmid, pIII acts to disrupt the outer membrane, rendering the host cell highly susceptible to detergents and causing leakage of periplasmic proteins from the cell, resistance to infection by F-specific phages, resistance to a variety of colicins, and inability to form mating pairs (Boeke et al., 1982). Removal of sodium dodecyl sulfate (SDS) from SDS-solubilized pIII results in spontaneous oligomerization of the pIII, which, when reconstituted into artificial lipid bilayer membranes, forms aqueous pores with an estimated diameter of 16 Å that remain open for seconds (Glaser-Wuttke et al., 1989). Analysis of deletion mutants indicates that distinct regions of pIII are responsible for these phenotypes (Boeke et al., 1982) (Figure 13.4).

Gene III from filamentous phage that infect *E. coli* via N or P pili (Ike, I2-2) (Peeters et al., 1985) has diverged substantially from that of the F-specific phage, the receptor binding and penetration regions have been rearranged, and a hydrophobic region at the carboxyl terminus appears to be sufficient for association with phage particles (Endemann et al., 1992). Protein pIII from N-specific phage are not normally incorporated into F-specific particles that lack their own pIII (or vice versa); however, some deleted pIII molecules appear to lack this specificity (Endemann et al., 1993).

Gene VI encodes a 113-amino-acid protein, a few copies of which are located at the same end of the phage particle as pIII. It is synthesized without a signal sequence, but contains a hydrophobic segment and is integrated into the inner membrane prior to assembly (Endemann and Model, 1995). Antibodies specific for the carboxyl-terminal end of pVI do not react with phage particles, suggesting that at least a portion of the protein is probably buried (Endemann and Model, 1995). In the absence of pVI, unstable polyphage are produced (Lopez and Webster, 1983).

Morphogenetic Proteins (pI and pIV)

Proteins pI and pIV are required for phage assembly but are not present in the intact virus particle (reviewed by Russel, 1991). Protein pI is a 348-amino-acid protein that lacks an amino-terminal signal sequence. It spans the inner membrane via a hydrophobic membrane anchor, with an amino-terminal cytoplasmic domain of approximately 250 amino acid residues and a carboxyl-terminal periplasmic domain of 75 residues (Guy-Caffey et al., 1992). The anchor functions as an internal signal sequence to direct pI to the membrane, because membrane integration of pI is dependent on the secretion protein SecA (Rapoza and Webster, 1993). Translational initiation from an internal ATG in gene I results in synthesis of pI*, which lacks the large, amino-terminal cytoplasmic domain present in full-length pI and is more abundant in infected cells than pI (Guy-Caffey et al., 1992). This ''eleventh'' protein is also essential for phage assembly in its own right (M.P. Raposa and R.E. Webster, personal communication), but its function is not known. Genetic analysis suggests that pI acts to initiate assembly by recognizing the packaging signal in phage DNA (Russel and Model, 1989), to bind the necessary host factor thioredoxin (Russel and Model, 1983, 1985), and to interact with the major phage coat protein to promote its incorporation into the nascent particle (Russel, 1993). In addition, pI interacts with pIV (Russel, 1993). Protein pI has also been implicated in the formation of phage-specific adhesion zones, regions in which the inner and outer membranes of *E. coli* are in close contact (Lopez and Webster, 1985a). These contact sites, visualized by electron microscopy, may reflect the existence of special assembly sites or exit complexes necessary for phage assembly and extrusion. Protein pI contains a nucleotide binding motif located in the cyto-

pIII

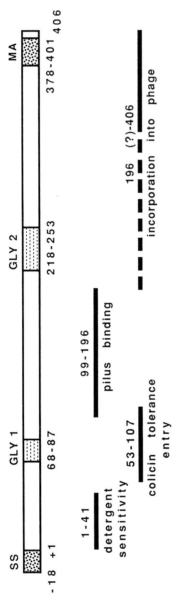

FIGURE 13.4 Structural organization of pIII from F-specific filamentous phage. The regions within the protein that mediate various properties of pIII were defined by characterization of deletion mutants. Thus, residues 1 to 41 are necessary for detergent sensitivity, residues 53 to 107 for tolerance to colicins, and residues 99 to 196 for the ability to bind F pili (Stengele et al., 1990). Protein pIII contains two glycine-rich repetitive regions (residues 68 to 87 and 218 to 253). The first is required both for infectivity and confering tolerance to colicins (Stengele et al., 1990), suggesting that it may interact with the Tol proteins, which are required both for phage penetration and colicin transport into the cell (Sun and Webster, 1986, 1987). The carboxyl-terminal half of pIII is necessary for its attachment to the phage particle; a deletion that lacks the amino-terminal half (to residue 196) is integrated into the phage particle, whereas one deleted to residue 340 is not (Crissman and Smith, 1984). Mutants in which one or a few amino acids in the extreme carboxyl terminus of pIII are changed are nonfunctional (N.G. Davis and P. Model, personal communication); although the nature of the defect has not been analyzed, it seems likely that it has to do with incorporation into the particle. The function for the phage, if any, of the region conferring detergent sensitivity to the host is not known. The host attachment region is required for specific binding to pili, and the colicin tolerance region is presumed to be the domain by which the host Tol proteins mediate internalization of infecting phage DNA. These domains are shuffled in the P pili–specific phage Ike. The region of pIII necessary for its attachment to phage particles has not been precisely defined; it is, however, larger than the hydrophobic domain that anchors pIII in the membrane prior to its incorporation into phage.

360

plasmic domain that is conserved among divergent filamentous phages (Russel, 1991). The role of this motif has not yet been determined, but mutation of either of two consensus residues in the motif in f1 gene I abolishes pI function (J. Feng, P. Model, and M. Russel, unpublished results). When expressed at even moderate levels from a plasmid, pI kills the host; membrane potential is lost and protein synthesis ceases before appreciable pI is synthesized (Horabin and Webster, 1988).

Protein pIV is synthesized with a 21-amino-acid signal peptide. The 43-kDa mature protein (405 amino acids) is secreted across the inner membrane and is transiently soluble in the periplasm; it then becomes an integral protein, presumably of the outer membrane (Brissette and Russel, 1990; Russel and Kazmierczak, 1993). Protein pIV has been hard to localize, with a portion fractionating with the outer membrane and another portion with the inner membrane (Brissette and Russel, 1990). Its amino acid sequence is characteristic of outer membrane proteins. The amino-terminal half of pIV forms a protease-resistant domain exposed in the periplasm (Brissette and Russel, 1990), whereas the carboxyl-terminal half mediates membrane integration and formation of a multimer that appears to be composed of 10 to 12 pIV monomers (Russel and Kazmierczak, 1993; Kazmierczak et al., 1994). A growing family of bacterial proteins (see later) have been identified that share significant sequence homology with the carboxyl-terminal half of pIV. The bacterial homologs also form multimers (Kazmierczak et al., 1994). Mutation of residues in pIV that are absolutely conserved among more than a dozen homologs indicate that Pro 375 of mature pIV is critical for folding and/or multimerization (Russel, 1994a). The morphogenetic proteins pI and pIV, located in different membranes, may interact to form an exit structure through which the assembling phage particle passes (Russel, 1993).

The Packaging Signal

Packaging signals, which earmark viral nucleic acids for encapsidation, have been identified for a number of different RNA- and DNA-containing viruses. In filamentous phage, the PS was initially identified as a short DNA segment from within the intergenic region that enabled heterologous ssDNA molecules to be encapsidated into filamentous phagelike particles (Dotto et al., 1981). In its single-stranded form this segment can be drawn as an imperfect 32-bp hairpin structure with a small bulge. The hairpin below the bulge can be deleted with no loss of function, as can a portion of the longer upper hairpin if the lower hairpin is present. Loop sequences at the tip of the hairpin can be changed and short sequences inserted without loss of function. In addition, the PS of phage Ike, an imperfect hairpin with little sequence similarity to that of f1, can replace that of f1 (Russel and Model, 1989). A perfectly self-complementary segment of DNA functioned poorly as a PS. These results suggest that structure rather than sequence governs PS function in filamentous phage.

The susceptibility of this segment to cross-linking by psoralen treatment of intact phage (Shen et al., 1979) showed that the hairpin exists inside the particle. Webster and colleagues have shown that phage DNA is oriented within the particle, that the PS is located near the end that is first to emerge (the pVII–pIX end) from the infected cell, and that the PS determines the orientation of the DNA (Webster et al., 1981; Lopez and Webster, 1983). Bauer and Smith (1988) subsequently showed that the PS is located at one end of the pV–ssDNA complex. This result suggests that the PS remains exposed in the complex, available to initiate encapsulation.

Russel and Model (1989) constructed phage genomes that lack the PS. When these genomes were introduced into cells, very few phage were produced, but suppressor mutations in the PS-less phage that enabled them to assemble more efficiently arose at high frequency. These mutations altered residues in pI, pVII, and pIX. This result suggests that these proteins normally interact with the PS, directly or indirectly (Fig. 13.5). The mutant proteins function better with genomes that have a PS than those that do not, suggesting

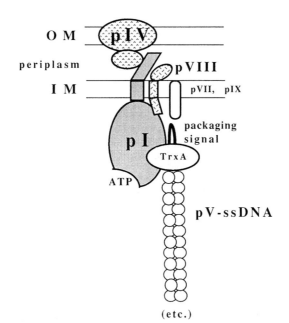

(etc.)

FIGURE 13.5   Diagram of protein–protein interactions predicted from genetic analyses. A series of mutants in gene I compensate for defects in other assembly components. They suggest that a cytoplasmic region of pI, which spans the inner membrane (IM) once with its amino terminus in the cytoplasm and carboxyl terminus in the periplasm, interacts with the PS, which protrudes from the pV–ssDNA complex, thereby initiating encapsidation. The cytoplasmic domain of pI is also proposed to interact with thioredoxin (TrxA) during elongation, which may require ATP hydrolysis by pI. The membrane-spanning region of pVIII, the major coat protein, is postulated to interact with pI. This interaction could mediate transfer of pVIII from the membrane to the growing phage particle. Finally, the periplasmic domains of pI and pIV, an outer membrane (OM) protein, are believed to interact. Formation of these complexes may account for the increased number of adhesion zones seen in phage-infected cells.

that they may have gained the ability to use another small hairpin structure as a "cryptic" PS.

Host Cell Proteins Involved in Assembly

Although several host proteins are required for filamentous phage DNA replication, the only host protein known to be required for assembly is thioredoxin, a ubiquitous protein in nature and a potent reductant of disulfide bonds in proteins (Lim et al., 1985; Russel and Model, 1985). Particle yield is reduced about 10^8-fold in cells from which the thioredoxin gene is deleted. The necessity for thioredoxin was discovered when a mutation that altered a proline residue located between the two active-site cysteines was isolated that rendered *E. coli* unable to support f1 phage production at high temperature (Russel and Model, 1983, 1985). This *trxA* mutant retains full redox activity. In contrast, mutation of either or both cysteine residues (to serine or alanine) destroys the redox activity of thioredoxin yet leaves the mutant proteins competent for phage assembly (Russel and Model, 1986). These active site mutants, which cannot form the oxidized disulfide form of thioredoxin, bypass the requirement for thioredoxin reductase in phage assembly, indicating that it is the thioredoxin in the reduced conformation that is required (Russel and Model, 1986). Thioredoxin is also a subunit of the phage T7 DNA polymerase, where it functions to confer high processivity on the T7-encoded catalytic subunit (Tabor et al., 1987). It may be that, as with the T7 DNA polymerase, thioredoxin confers processivity on the reaction that leads to the displacement of pV from phage DNA and its replacement by virion structural proteins.

Additional host proteins may participate in filamentous phage assembly. Several *E. coli* mutants have been isolated that fail to produce phage particles at a nonpermissive temperature, although phage gene expression occurs. These mutants show additional defects indicative of perturbed membrane structure; thus it is not clear whether phage assembly is affected directly (Model et al., 1981; Lopez and Webster, 1985b).

Structure of the Virion

Structural Organization

Figure 13.6*A* is an electron micrograph of M13, and Figure 13.6*B* is a diagram of its overall structural organization. It is about 65 Å in diameter (Glucksman et al., 1992), with a length dependent on the length of the enclosed genome. A large fraction of the virion mass is made up of many copies of the 50-amino-acid coat protein, pVIII, which forms a 15 to 20-Å thick flexible cylinder about the single-stranded viral genome. There is approximately 0.435 coat protein per nucleotide (Berkowitz and Day, 1976), and the length of the virion is about 3.3 Å per coat protein (or 1.435 Å per nucleotide) plus about 175 Å for the minor proteins (Specthrie et al., 1992).

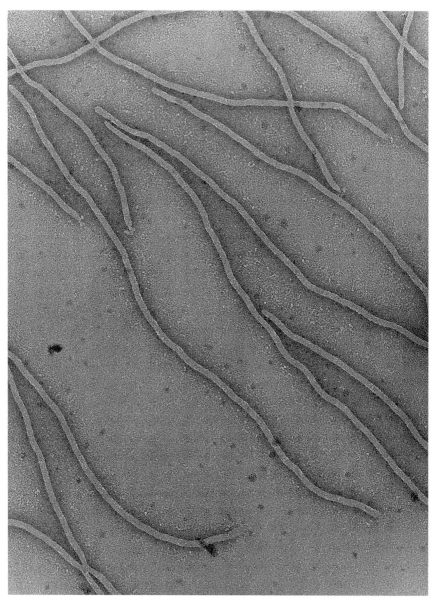

A

FIGURE 13.6 Structural organization of F-specific filamentous phage. *A.* Electron micrograph of negatively stained M13 virus particles. The two ends of the particle are distinctive. One end is relatively blunt with little detail apparent. This is the distal end containing pVII and pIX. The other end is usually pointed, with additional material projecting from the point. This additional material may appear as a long straight projection, an indistinct mass, or occasionally as knobs on the ends of strings. The "knobs" have been identified as the amino-terminal portion of pIII, removed by subtilisin (Gray, Brown and Marvin, 1981). Several phage particles in this field display this appearance for pIII. (Courtesy of I. Davidovich.) Figure continues.

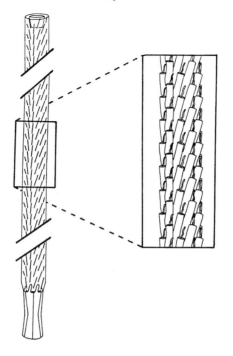

B

FIGURE 13.6 *Continued B.* Diagram of the structural organization of the virion. The virus particle is made up of the circular ssDNA genome, a large number of copies of the major coat protein (pVIII), and five copies each of four minor structural proteins. In wild-type virus, about 2800 copies of pVIII make up about 85% of the virion mass, coating the viral DNA with a 15- to 20-Å-thick flexible, cylindrical shell. The amino-terminal half of pVIII is exposed to the surface of the virus, and the first four to five amino acids at the amino terminus form a flexible arm that appears to extend away from the particle. The remainder of the protein is α-helical, with the possible exception of a few residues at the carboxyl terminus. The distal end (top) is assembled first. This end contains approximately five copies each of pVII and pIX, which may form a plug at the top of the virus particle. The proximal end (bottom) contains approximately five copies of pVI, which are believed to mediate attachment of approximately five copies of pIII to the body of the virion. The pIII molecules are the host cell binding proteins. (Adapted from Bhattacharjee et al., 1992.)

The small, hydrophobic proteins pVII and pIX are located at the end of the virus that is assembled first (Lopez and Webster, 1983), referred to as the "distal" end. Labeling studies (Simons et al., 1981) indicated three to four copies of each of these proteins are present in the virus. If there are five copies per virion, their rotational symmetry would match that of the coat proteins. It has been suggested that they make up a hydrophobic plug at the distal end of the virion (Makowski, 1992).

Protein pIII mediates attachment to the host and penetration of the phage genome into the host cytoplasm. The mature peptide is likely to be attached to the virion through interactions with pVI. There are approximately five copies

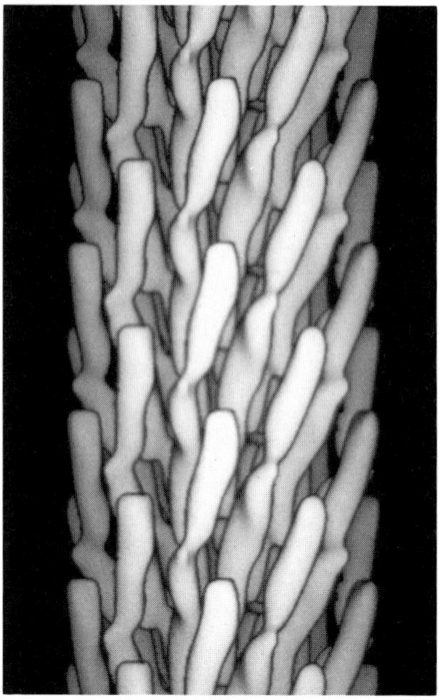

FIGURE 13.7 Computer rendering of the electron density map of M13 calculated from fiber diffraction data to approximately 7 Å resolution. The contour level in this diagram was chosen to be sufficiently high that the interior organization of the particle is apparent beneath the surface layer of protein. In the intact virus, the α-helices on the surface of the phage will pack substantially closer together, creating a solid cylinder of protein around the viral DNA. (Reproduced with permission from Glucksman et al., 1992.)

each of these two proteins at the ''proximal'' end of the virion (Simons et al., 1981).

Protein pVIII Structure

In the virus particle, the pVIII molecules are arranged with a fivefold rotational axis and a twofold screw axis, with a pitch of about 32 Å (Makowski and Caspar, 1981). The coat protein is made up of a single gently curving α-helix extending, probably uninterrupted, from Pro 6 to near the carboxyl terminus (Opella et al., 1987; Glucksman et al., 1992; Marvin et al., 1994). The amino terminus is at the surface of the virion and the carboxyl terminus is at low radius, in close proximity to the viral DNA. The axis of the α-helix is tilted about 20° relative to the virion axis and wraps around the axis in a right-handed helical sense, as shown in Figure 13.7. The amino acid sequence can be divided into four parts: a mobile surface segment (Ala 1 to Asp 4 or Asp 5), an amphipathic α-helix extending from Pro 6 to about Tyr 24, a highly hydrophobic helix extending from Ala 25 to Ala 35, and an amphipathic helix forming the inside wall of the protein coat extending from Thr 36 to Ser 50. Four basic residues near the carboxyl terminus appear to interact with the viral DNA (Hunter et al., 1987; Rowitch et al., 1988).

DNA Structure

The structure of viral DNA in the phage particle is not well characterized. Solid-state NMR studies (Cross et al., 1983) showed that in Pf1 the phosphate backbone appears to have a uniform conformation, whereas in the Ff the phosphates take on a large number of different orientations. Either the DNA is packaged randomly or it has a complex conformation with many different orientations occurring in a nonrandom sequence.

Analysis of the structure of the major coat protein as derived by fiber diffraction (Glucksman et al., 1992; Kishchenko et al., 1994) suggests that the interior space available for the DNA is inadequate for a B–DNA duplex. This would argue that the PS, which apparently forms a hairpin in the intact virus particle (Shen et al., 1979; Ikoku and Hearst, 1981), may require special structural properties to be accommodated within the phage particle. It is possible that the PS in the virion has a somewhat extended conformation relative to B–DNA, or, alternatively, that the coat protein takes on an altered configuration in the vicinity of the PS.

Distal End: pVII and pIX

Analysis of the sequences of pVII and pIX indicates that they have hydrophobic regions and positive charges of length similar to those in pVIII. A model for the distal end of Ff has been constructed (Makowski, 1992) by assuming that these similarities are indicative of structural homologies. Assuming pVII, pIX, and pVIII to be extended α-helical structures results in an energetically unstable model for the end of the virus, as drawn in Figure 13.8A. Analysis of this model suggests that it must be altered, possibly by bending both pVII and the 5-terminal pVIII into L-shaped proteins consisting of two α-helices with an approximate 90° bend. Using sera directed against pVII and pIX, Endemann and Model (1995) have shown that pIX is accessible in the intact phage and at least some epitopes of pVII are not. Although not excluding the model building based on Figure 13.8A, this result is not what would have been predicted on the basis of that model.

Proximal End: pIII and pVI

A partial model for the proximal end of the virion was constructed by assuming that the amino-terminal half of pVI is structurally homologous to two pVIII molecules (Makowski, 1992). A drawing of this model is shown in Figure 13.8B. No attempt was made to model the remainder of pVI nor any of pIII at this level of detail. At this end of the virus particle, the termination of the virion helix affects two layers of pVIII. The terminal pVIIIs are exposed to solvent more than the pVIIIs in the remainder of the virus particle. This potentially unstable situation is relieved by a fold of the pVI, much of which forms an amphipathic surface layer, up and around the last two layers of pVIII. If this model is correct, most of the carboxyl-terminal half of pVI forms an amphipathic girdle about the last two layers of pVIII, holding the virion together and helping to maintain its stability. Antibodies directed against pVI do not interact with intact phage (Endemann and Model, 1995), suggesting that a portion of pIII may fold around pVI, sequestering it from solvent. Contrast-

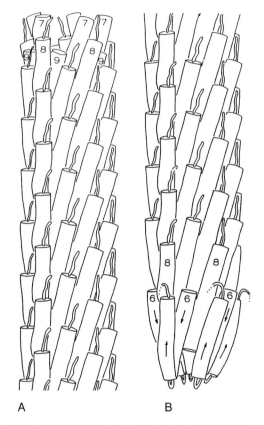

A   B

FIGURE 13.8   Three-dimensional model of the (*A*) distal and proximal (*B*) ends of the virus particle. At the distal end, amino acid similarities between the minor proteins, pVII and pIX, and pVIII suggest a structural arrangement that results in the energetically implausible projection of amphipathic helices from the end of the viral particle. The instability of this arrangement indicates that both pVII and the terminal pVIII probably have structures that are not analogous to the pVIII in the body of the virion (Makowski, 1992). At the proximal end, a similar model-building procedure suggests that the amino-terminal portion of pVI takes the place of two pVIII proteins at the end of the virus particle. It is possible that a portion of pVI or pIII acts to shield some of the exposed hydrophobic portions of the terminal pVIII from the aqueous surroundings. (Reproduced with permission from Makowski, 1992.)

enhanced electron micrographs of negatively stained microphage particles (E. Bullitt and L. Makowski, unpublished results) exhibit a slightly enlarged diameter near the proximal end, consistent with this possibility.

As described earlier, pIII is the largest and most complex structural protein of filamentous phage. In electron micrographs of negatively stained phage (Gray et al., 1979) and 500-Å-long microphage particles (Specthrie et al., 1992), at least a portion of pIII appears flexible. The two glycine-rich tracts in pIII (containing the repetitive motif Ser-Gly-Gly-Gly or Ser-Glu-Gly-Gly-Gly) may contribute to this apparent flexibility. Length analysis of microphage particles indicates that the last 100 Å of the particle cylinder is made up of pVI and pIII (Specthrie et al., 1992). No information about the conformation of pIII within this structure is yet available.

The Assembly Process

Initiation of Virion Assembly

Initiation of phage assembly probably occurs when the PS interacts with the cytoplasmic domain of pI, the morphogenetic protein that spans the inner membrane (Russel and Model, 1989). However, the ability of the cytoplasmic domain of pI to bind the PS (or any DNA) has not been directly determined. Proteins pIX and pVII may also interact with the PS, because mutants with enhanced ability to support assembly of ssDNA that lacks a PS were also isolated in these genes (Russel and Model, 1989). The order of addition of the minor proteins pVII and pIX is not known. Protein pIX has two basic residues that could protrude into the cytoplasm and interact with the DNA, but it is exposed on the virion surface, whereas pVII is not (Endemann and Model, 1995). Efforts to model the phage ends (Figure 13.8A) suggest that the PS may be surrounded by pVIII molecules, rather than by pVII and pIX. If so, the first five molecules of pVIII may have a rather different structure than the pVIII in the remainder of the virus particle. Some pVIII is associated with pVII in the membrane of infected cells (Endemann and Model, 1995), suggesting that it may be a special conformation of this complex that associates with the PS.

Elongation

Genetic evidence suggest that the cytoplasmic domain of pI interacts with host thioredoxin (Russel and Model, 1983). This interaction is likely to promote elongation of the initiated particle. Addition of thioredoxin to crude extracts of infected mutant cells that lack the thioredoxin gene results in the production of a variable but small number of infectious particles; no particles were produced when exogenous ssDNA or pV–ssDNA complexes from another extract were added, suggesting that thioredoxin can only extend previously initiated assembly reactions (P. Model, personal communication). The correspondence in the inability/ability of T7 and filamentous phage to grow on various thioredoxin mutants suggests that thioredoxin might function similarly in both phages. Thioredoxin confers a high degree of processivity on the T7-encoded subunit of T7 DNA polymerase (Tabor et al., 1987); similarly, it might enhance the processivity of pI in the iterative process of removing pV from the DNA and/or replacing it by coat protein(s). This implies that pI acts during elongation as well as at the initiation step. The importance of the consensus nucleotide binding motif in its cytoplasmic domain (see earlier) raises the possibility that pI hydrolyzes ATP to provide energy for phage assembly. The stoichiometry of pI, the number of pI molecules required to assemble a single phage, is not known.

How is the elongating particle translocated from the inner membrane, across the outer membrane, to the cell exterior? The recent observation that pIV forms a large oligomer, probably composed of 10 to 12 monomers, in the outer membrane supports the idea that it may function as an exit channel (Kazmierczak et al., 1994). However, pIV cannot be simply an exit port because a passive hole in the outer membrane large enough to permit passage of the phage (65 Å in diameter) would almost certainly render *E. coli* sensitive to hydrophobic compounds and detergents that do not otherwise penetrate the outer membrane. Expression of substantial amounts of pIV does not have this effect. Interestingly, expression of certain mutant pIVs does render *E. coli* sensitive to hydrophobic compounds and detergents (Russel, 1994a). The glycine residue at position 355 of mature pIV is conserved in all known bacterial homologs (see later); expression of mutant pIVs in which alanine, serine, or any of several other amino acids has been substituted for Gly 355 sensitizes cells to these compounds. It is possible that this conserved glycine is located in a loop that is critical for pIV gating, and that the mutants form a "leaky" gate (Russel, 1994a).

Although pIV from f1 and Ike are not functionally interchangeable, when both pI and pIV are exchanged, some heterologous phage are assembled (Russel, 1993). This suggests that pI and pIV interact to promote assembly. The isolation of a mutation in the periplasmic domain of pI that compensates for a mutation in the periplasmic domain of pIV (Russel, 1993) supports the idea that these proteins interact via their periplasmic domains. Combining these results, Kazmierczak et al. (1994) have proposed that pIV forms a channel that is gated to the initiation complex via an interaction with pI. It is imagined (see Color Plate 17) that binding of the PS to the cytoplasmic domain of pI causes a conformational change in its periplasmic domain, enabling it to interact with and cause the opening of the pIV channel. Protein pIV might also recognize pVII and/or pIX at the tip of the nascent particle, but there is no evidence to support this as yet.

Once assembly has been initiated, pV must be removed from the ssDNA in order for pVIII to be added. This process is likely to involve thioredoxin and pI and may also require ATP hydrolysis. It does not depend on the details of any interaction of pI and pV, because pVs from various phage are interchangeable (Russel, 1992). Because the helical symmetry of DNA in the pV complex is different from its symmetry in the virion, the two structures must rotate relative to one another during the elongation process!

The structure of pVIII in the membrane differs significantly from that in the phage (Figure 13.3). The conformational change, from the two α-helical segments of the membrane-bound form to the single, long α-helix of the virion form, results in the formation of approximately three new intrachain hydrogen bonds and moves the two ends of the protein approximately 16 Å further apart. If coat proteins are added to the growing phage in units of five, as the fivefold rotation axis of phage particles suggests may be the case, approximately 15 new hydrogen bonds would form simultaneously. A 16-Å axial translocation is precisely that required to position the virus particle to accept further additions of pVIII.

Termination of Assembly

Termination of assembly involves the addition of pVI and pIII to the tip of the particle and occurs when the entire DNA molecule has been packaged by pVIII. Termination is a relatively inefficient process, because double-length particles that contain two unit-length genomes are frequent (~5% of the total phage) in a wild-type preparation. The addition of a second genome does not seem to represent a reinitiation event, because ssDNA molecules that lack a PS can be coencapsidated with a PS-containing molecule at about the same frequency (5%) as those that contain one (Russel and Model, 1989).

Proteins pIII and pVI probably do not associate with one another prior to their addition to the particle, because, if they did, the properties of particles formed in the absence of pVI should be the same as those formed in the absence of pIII, which is not the case. Rather, although both are noninfectious polyphage indicating that they lack pIII, the former are less stable, tending to "unravel" at the tip (Lopez and Webster, 1983). These observations indicate that pVI is needed to stabilize the end of the particle and imply that the more stable phage assembled in the absence of pIII contain pVI. This, in turn, suggests that pVI is added before pIII. Protein pVI may bind to the elongating phage in a way that exposes a surface with high affinity for pIII binding. If such a high-affinity binding site was for the stretch of hydrophobic residues that anchor pIII in the inner membrane prior to its incorporation into phage, pVI–pIII interactions could replace lipid–pIII interactions, thereby releasing pIII from the inner membrane.

A certain fraction of pVIII can be coimmunoprecipitated from the membrane of infected cells by pVI antiserum, which suggests that pVI is added to the particle as a pVI–pVIII complex (Endemann and Model, 1995). Reaching the end point of the DNA may alter the affinity of pVIII for the end of the growing phage, so that the pVIII–pVI complex competes successfully for binding to the terminal copies of pVIII.

Detachment from the host cell membrane completes the assembly process. Release of the completed viral particle from the host requires movement of the remainder of the now complete virion through the outer membrane. As pictured schematically in Color Plate 17, this involves movement of the proximal end of the viral particle through the postulated pIV pore in the outer membrane.

Infection

Filamentous phage infection is a two-step process. The first step is binding to its bacterial cell surface receptor, the pilus. This binding is mediated by pIII (Pratt et al., 1969). Treatment of intact phage particles with subtilisin removes a large amino-terminal domain of pIII from the virus particle, renders the particles uninfectious, and removes the globular domains occasionally observed by electron microscopy at the extreme end of the particle (Armstrong et al., 1981; Gray et al., 1981). Analysis of deletion mutants of pIII from Ff indicates that the host binding capacity of the protein is localized in the region between amino acids 99 and 196 (Stengele et al., 1990). This region contains the first of two glycine-rich repeats, but its participation in binding to pili has not been explicitly demonstrated.

Only one or a few pili are present on a cell. Pilus proteins and the proteins required for their assembly are plasmid encoded, and the pili mediate transfer of their genomes into susceptible bacteria. A large number of different pili exist, and pIIIs from different filamentous phages are specific for particular pilus types. Thus the Ff (f1, fd, and M13) bind F pili, whereas Ike and I2-2 infect E. coli strains that specify N or P pili. Assembly of some pili is temperature sensitive; this accounts for the failure of the F-specific phages to infect, or form plaques on, bacteria grown at temperatures below about 32 °C. Upon phage binding, pili appear to "retract," with the pilin subunits believed to depolymerize into the cell inner membrane (Jacobson, 1972; Novotny and Fives-Taylor, 1974). It is not known if phage binding triggers retraction or if phage are brought to the membrane as a consequence of normal cycles of pilus retraction and repolymerization.

During the second step of infection the capsid proteins integrate into the bacterial cell membrane (Trenkner et al., 1967; Smilowitz, 1974) and the phage DNA is uncoated and translocated into the bacterial cytoplasm. Although pVIII depolymerization into the inner membrane occurs in tolA mutants (Smilowitz, 1974), penetration of the DNA requires the host TolQ and R proteins in addition (Sun and Webster, 1987; Levengood and Webster, 1989; Webster, 1991). This penetration step is also mediated by pIII because particles containing defective pIII are not competent for even low-efficiency infection of F^- (tol^+) cells (Russel et al., 1988).

Tol stands for "tolerant," reflecting the initial identification of tolA and tolB mutants that are insensitive, or tolerant, to the lethal action of several colicins, which are proteins produced by some E. coli that kill nonproducing cells (Nagle De Zwaig and Luria, 1967). The tolQ and R genes were identified subsequently and shown to be essential for filamentous phage infection and for susceptibility to some colicins (Sun and Webster, 1986, 1987). Like filamentous phage, colicins obtain access to the cell in a two-step process; one family of colicins binds to the cell surface vitamin B_{12} receptor (Reynolds et al., 1980; Heller et al., 1985) before interacting with the Tol system. The role of the Tol system in macromolecular import has been reviewed by Webster (1991).

Cells infected with filamentous phage that produce functional pIII (Zinder, 1973) or cells expressing pIII from a plasmid show increased tolerance to the effects of E colicins (Boeke et al., 1982). This suggests that pIII and the E colicins may be competing for the same binding sites on the Tol proteins. Both pIII and the E colicins contain glycine-rich repetitive sequences (Yamada et al., 1982; Cole et al., 1985), which may be essential for interaction with one or more of the Tol proteins.

Discussion

The filamentous phage genome is possibly the best characterized genome in the biological world. As such, it can be manipulated to provide answers to large numbers of biological questions. Its contributions are of two kinds, applied and basic. The phage have become workhorses of biotechnology, useful for DNA sequencing and site-directed mutagenesis and, more recently, as "phage display" vectors. Probably because the phage genome is so small and so amenable to genetic and biochemical analysis, filamentous phage have been and continue to be used as an experimental system for elucidating many of the molecular mechanisms involved in each step of the viral life cycle.

Role of Filamentous Phage in Technology

Although the dideoxynucleotide sequencing method was developed using the ssDNA-containing bacteriophage ϕX174, the icosohedral nature of this phage, and the fixed amount of DNA that it contains, made it unsuitable as a cloning vector from which to prepare large quantitites of a foreign gene in single-stranded form. Filamentous phage, in contrast, were ideal for this purpose because additional sequences merely result in production of a longer filament. Because plasmids are often better vectors for producing foreign proteins than filamentous phage–infected cells, plasmid–phage combinations (or phagemids) were developed that combine the advantages of each. Phagemids are plasmids that contain a filamentous phage intergenic region (replication origin and PS). Phagemids normally exist as double-stranded DNA in the cell (they replicate via the plasmid replication origin) and are used in that form to express the products of cloned genes. Upon infection by filamentous helper phage, the phage replication origin is activated, phagelike rolling circle replication generates phagemid ssDNA, and helper phage proteins mediate encapsulation and export of phagemid single strands. ssDNA extracted from these particles can be used as a template for DNA sequencing or oligonucleotide mutagenesis. Detailed descriptions of these techniques and original references can be found in Sambrook et al. (1989).

The development of filamentous phage "display" vectors is opening new areas of utility for these small viruses. Foreign peptides are readily fused with a filamentous phage structural protein by insertion of the corresponding nucleotide sequence into the gene coding for the protein (Smith, 1985, 1991). This insertion results in the "display" of a protein or peptide on the phage surface, provided that the insert does not interfere with the ability of the chimeric protein to be incorporated into the phage particle.

Peptides or whole proteins have been inserted at or near the amino terminus of pIII (Smith, 1985). Phages have been selected that display functional antibodies of defined specificity (Barbas et al., 1991; Clackson et al., 1991); once isolated, display phage can be mutagenized and variants with altered binding properties selected (Lowman and Wells, 1993). Although the chimeric proteins assemble into particles, large insertions often interfere with the infectivity function of pIII (Smith, 1991; Smith and Scott, 1993). This problem can be overcome by expressing the chimeric pIII from a phagemid and infecting it with a helper phage that provides normal pIII. The resulting particles contain both chimeric and normal protein and are infectious. Because there are only five copies of pIII per virion, inserts that mediate high-affinity binding to a particular ligand are preferentially selected. This can be an advantage or disadvantage, depending on the particular application.

Viable inserts of foreign peptides into the amino terminus pVIII have been constructed by many groups (Ilyichev et al., 1989; Castagnoli et al., 1991; Greenwood et al., 1991; Kang et al., 1991). Inserts in pVIII create multivalent displays as a result of the large number of pVIII molecules per phage particle, and as a result phage bearing peptides with low affinity for a particular ligand can be selected. There appears to be an intrinsic limit of 6 to 10 amino acids in the size of foreign peptide that can be inserted into pVIII if no source of normal pVIII is provided. As with pIII, inclusion of normal copies of pVIII allows formation of "hybrid" particles that contain both chimeric pVIII with large inserts and normal pVIII (Greenwood et al., 1991; Markland et al., 1991). The origin of the size limitation is not understood, because these fusion proteins appear capable of carrying out all required steps in the viral life cycle. When wild-type pVIII is present, very large inserts (e.g., immunoglobulin variable genes [Kang et al., 1991]), are tolerated in a small proportion of the pVIII molecules. This means that pVIII with a large insert is properly inserted into the host cell inner membrane and incorporated into the virus particle. It also means that the large insert does not interfere in the infection process. One possible answer is that pVIII molecules at the ends of the virus particle are less tolerant of large inserts than those in the body of the virus particle. A second possibility is that the size of the pIV channel in the outer membrane is insufficient for particles with large inserts in every pVIII protein. In either case, a deeper understanding of the assembly process will be required for the construction of new phage display particles for future technologic applications.

Role of Filamentous Phage in Advancing Fundamental Concepts

Since their isolation in the 1960s, filamentous phage have been used to elucidate a large number of biologic processes and mechanisms that are taken for granted today. (Primary references for most of this body of work can be found in Model and Russel, 1988.) The first recognition site(s) for DNA restriction and modification were recognized genetically in filamentous phage, and characterization of these sites allowed the isolation of the first restriction and modification enzymes. Filamentous phage, and other phages, have been used to elucidate the genetic code, to define promoters, and to investigate aspects of DNA replication, transcription, translation, and translational control. Their intimate association with the host inner membrane has made them useful for various investigations of protein targeting in *E. coli*. Protein pVIII was one of the first procaryotic proteins recognized as containing a signal sequence, and pre-VIII was used as substrate for the purification of signal peptidase from *E. coli*. Detailed analysis of pIII led to the first quantitative description of the parameters necessary for anchoring proteins in membranes. The realization that export of many virulence factors from pathogenic bacteria is a process likely to be mechanistically similar to that required for filamentous phage assembly suggests that these viruses will continue to serve as useful model organisms in the future.

References

Alberts, B. and Frey, L. 1972. Isolation and characterization of gene 5 protein of filamentous bacterial viruses. J. Mol. Biol. 68: 139–152.

Armstrong, J., Perham, R.N. and Walker, J.E. 1981. Domain structure of bacteriophage fd adsorption protein. FEBS Lett. 135: 167–172.

Barbas, C.F. III, Kang, A.S., Lerner, R.A. and Benkovic, S.J. 1991. Assembly of combinatorial antibody libraries on phage surfaces: the gene III site. Proc. Natl. Acad. Sci. U.S.A. 88: 7978–7982.

Bauer, M. and Smith, G.P. 1988. Filamentous phage morphogenetic signal sequence and orientation of DNA in the virion and gene V protein complex. Virology 167: 166–175.

Berkowitz, S.A. and Day, L.A. 1976. Mass, length, composition and structure of the filamentous bacterial virus, fd. J. Mol. Biol. 102: 531–547.

Bhattacharjee, S., Glucksman, M.J. and Makowski, L. 1992. Structural polymorphism correlated to surface charge in filamentous bacteriophages. Biop. J. 61: 725–735.

Boeke, J.D. and Model, P. 1982. A prokaryotic membrane anchor sequence: carboxyl terminus of bacteriophage f1 gene III protein retains it in the membrane. Proc. Natl. Acad. Sci. U.S.A. 79: 5200–5204.

Boeke, J.D., Model, P. and Zinder, N.D. 1982. Effects of bacteriophage f1 gene III protein on the host cell membrane. Mol. Gen. Genet. 186: 185–192.

Brissette, J.L. and Russel, M. 1990. Secretion and membrane integration of a filamentous phage-encoded morphogenetic protein. J. Mol. Biol. 211: 565–580.

Castagnoli, L., Musacchio, A., Jappelli, R. and Cesareni, G. 1991. Selection of antibody ligands from a large library of oligopeptides expressed on a multivalent exposition vector. J. Mol. Biol. 222: 301–310.

Cesareni, G. 1992. Peptide display on filamentous phage capsids: a new powerful tool to study protein-ligand interaction. FEBS Lett. 307: 66–70.

Chang, C.N., Model, P. and Blobel, G. 1979. Membrane biogenesis–cotranslational integration of the bacteriophage f1 coat protein into an *Escherichia coli* membrane-fraction. Proc. Natl. Acad. Sci. U.S.A. 76: 1251–1255.

Clackson, T., Hoogenboom, H.R., Griffiths, A.D. and Winter, G. 1991. Making antibody fragments using phage display libraries. Nature 352: 624–628.

Cole, S.T., Saint-Joanis, B. and Pugsley, A.P. 1985. Molecular characterization of the colicin E2 operon and identification of its products. Mol. Gen. Genet. 198: 465–472.

Colnago, L.A., Valentine, K.G. and Opella, S.J. 1987. Dynamics of fd coat protein in the bacteriophage. Biochemistry 26: 847–854.

Crissman, J.W. and Smith, G.P. 1984. Gene-III protein of filamentous phages: evidence for a carboxyl-terminal domain with a role in morphogenesis. Virology 132: 445–455.

Cross, T.A., Tsang, P. and Opella, S.J. 1983. Comparison of protein and deoxyribonucleic acid backbone structures in fd and Pf1 bacteriophages. Biochemistry 22: 721–726.

Davis, N.G., Boeke, J.D. and Model, P. 1985. Fine structure of a membrane anchor domain. J. Mol. Biol. 181: 111–121.

Dotto, G.P., Enea, V. and Zinder, N.D. 1981. Functional analysis of the bacteriophage f1 intergenic region. Virology 114: 463–473.

Endemann, H., Bross, P. and Rasched, I. 1992. The adsorption protein of phage IKe: localization by deletion mutagenesis of domains involved in infectivity. Mol. Microbiol. 6: 471–478.

Endemann, H., Gailus, V. and Rasched, I. 1993. Interchangeability of the adsorption proteins of bacteriophages Ff and IKe. J. Virol. 67: 3332–3337.

Endemann, H. and Model, P. 1995. Location of filamentous phage minor coat proteins in phage and in infected cells. J. Mol. Biol. 250: 496–506.

Fulford, W. and Model, P. 1984. Gene X of bacteriophage f1 is required for phage DNA synthesis–mutagenesis of in-frame overlapping genes. J. Mol. Biol. 178: 137–153.

Glaser-Wuttke, G., Keppner, J. and Rasched, I. 1989. Pore-forming properties of the adsorption protein of filamentous phage fd. Biochim. Biophys. Acta 985: 239–247.

Glucksman, M.J., Bhattacharjee, S. and Makowski, L. 1992. Three-dimensional structure of a cloning vector; X-ray diffraction studies of filamentous bacteriophage M13 at 7 Å resolution. J. Mol. Biol. 226: 455–470.

Grant, R.A. and Webster, R.E. 1984. Minor protein content of the gene V protein/phage single-stranded DNA complex of the filamentous bacteriophage f1. Virology 133: 315–328.

Gray, C.W. 1989. Three-dimensional structure of complexes of single-stranded DNA-binding proteins with DNA: IKe and fd gene 5 proteins form left-handed helices with single-stranded DNA. J. Mol. Biol. 208: 57–64.

Gray, C.W., Brown, R.S. and Marvin, D.A. 1979. Direct visualization of adsorption protein of fd phage. J. Supramol. Struct. 1979: 91–91.

Gray, C.W., Brown, R.S. and Marvin, D.A. 1981. Adsorption complex of filamentous fd virus. J. Mol. Biol. 146: 621–627.

Gray, C.W., Kneale, G.G., Leonard, K.R., Siegrist, H. and Marvin, D.A. 1982a. A nucleoprotein complex in bacteria infected with pf1 filamentous virus—identification and electron-microscopic analysis. Virology 116: 40–52.

Gray, D.M., Gray, C.W. and Carlson, R.D. 1982b. Neutron-scattering data on reconstituted complexes of fd deoxyribonucleic acid and gene V protein show that the deoxyribonucleic acid is near the center. Biochemistry 21: 2702–2713.

Greenwood, J., Willis, A.E. and Perham, R.N. 1991. Multiple display of foreign peptides on a filamentous bacteriophage: peptides from Plasmodium falciparum circumsporozoite protein as antigens. J. Mol. Biol. 220: 821–827.

Guy-Caffey, J.K., Rapoza, M.P., Jolley, K.A. and Webster, R.E. 1992. Membrane localization and topology of a viral assembly protein. J. Bacteriol. 174: 2460–2465.

Heller, K., Mann, B.J. and Kadner, R.J. 1985. Cloning and expression of the gene for the vitamin B12 receptor protein in the outer membrane of Escherichia coli. J. Bacteriol. 161: 896–903.

Horabin, J.I. and Webster, R.E. 1988. An amino acid sequence which directs membrane insertion causes loss of membrane potential. J. Biol. Chem. 263: 11575–11583.

Hunter, G.J., Rowitch, D.H. and Perham, R.N. 1987. Interactions between DNA and coat protein in the structure and assembly of filamentous bacteriophage fd. Nature 327: 252–254.

Hutchinson, D.L., Barnett, B.L. and Bobst, A.M. 1990. Gene 5 protein-DNA complex: modeling binding interactions. J. Biomol. Struct. Dyn. 8: 1–9.

Ikoku, A.S. and Hearst, J.E. 1981. Identification of a structural hairpin in the filamentous chimeric phage M13gori1. J. Mol. Biol. 151: 245–259.

Ilyichev, A.A., Minenkova, O.O., Tatkov, S.I., Karpyshev, N.N., Eroshkin, A.M., Petrenko, V.A. and Sandakhchiev, L.S. 1989. Production of the m13 phage viable variant with a foreign peptide inserted into the coat basic-protein. Dokl. Akad. Nauk S.S.S.R. 307: 481–483.

Jacobson, A. 1972. Role of F pili in the penetration of bacteriophage f1. J. Virol. 10: 835–843.

Kang, A.S., F, B.C., Janda, K.D., Benkovic, S.J. and Lerner, R.A. 1991. Linkage of recognition and replication functions by assembling combinatorial antibody Fab libraries along phage surfaces. Proc. Natl. Acad. Sci. U.S.A. 88: 4363–4366.

Kazmierczak, B.I., Mielke, D.L., Russel, M. and Model, P. 1994. pIV, a filamentous phage protein that mediates phage export across the bacterial cell envelope, forms a multimer. J. Mol. Biol. 238: 187–198.

King, G.C. and Coleman, J.E. 1988. The Ff gene 5 protein-d(pA)40–60 complex: 1H NMR supports a localized base-binding model. Biochemistry 27: 6947–6953.

Kishchenko, G., Batliwala, H. and Makowski, L. 1994. Structure of a foreign peptide displayed on the surface of bacteriophage M13. J. Mol. Biol. 241: 208–213.

Levengood, S.K. and Webster, R.E. 1989. Nucleotide sequences of the tolA and tolB genes and localization of their products, components of a multistep translocation system in Escherichia coli. J. Bacteriol. 171: 6600–6609.

Li, Z. and Deber, C.M. 1991. Viable transmembrane region mutants of bacteriophage M13 coat protein prepared by site-directed mutagenesis. Biochem. Biophys. Res. Commun. 180: 687–693.

Lim, C.J., Haller, B. and Fuchs, J.A. 1985. Thioredoxin is the bacterial protein encoded by fip that is required for filamentous bacteriophage f1 assembly. J. Bacteriol. 161: 799–802.

Lopez, J. and Webster, R.E. 1983. Morphogenesis of filamentous bacteriophage f1: Orientation of extrusion and production of polyphage. Virology 127: 177–193.

Lopez, J. and Webster, R.E. 1985a. Assembly site of bacteriophage f1 corresponds to adhesion zones between the inner and outer membranes of the host cell. J. Bacteriol. 163: 1270–1274.

Lopez, J. and Webster, R.E. 1985b. FipB and fipC—2 bacterial loci required for morphogenesis of the filamentous bacteriophage f1. J. Bacteriol. 163: 900–905.

Lowman, H.B. and Wells, J.A. 1993. Affinity maturation of human growth hormone by monovalent phage display. J. Mol. Biol. 234: 564–578.

Magusin, P.C. and Hemminga, M.A. 1993a. Analysis of ^{31}P nuclear magnetic resonance lineshapes and transversal relaxation of bacteriophage M13 and tobacco mosaic virus. Biophys. J. 64: 1861–1868.

Magusin, P.C. and Hemminga, M.A. 1993b. A theoretical study of rotational diffusion models for rod-shaped viruses: the influence of motion on ^{31}P nuclear magnetic resonance lineshapes and transversal relaxation. Biophys. J. 64: 1851–1860.

Makowski, L. 1983. Structural diversity in filamentous bacteriophages. In The Viruses, Vol. 1 (McPherson, A., Ed.), pp. 204–253. John Wiley & Sons, New York.

Makowski, L. 1992. Terminating a macromolecular helix: structural model for the minor proteins of bacteriophage M13. J. Mol. Biol. 228: 885–892.

Makowski, L. 1993. Structural constraints on the display of foreign peptides on filamentous bacteriophages. Gene 128: 5–11.

Makowski, L. and Caspar, D.L.D. 1981. The symmetries of filamentous phage particles. J. Mol. Biol. 145: 611–617.

Markland, W., Roberts, B.L., Saxena, M.J., Guterman, S.K. and Ladner, R.C. 1991. Design, construction and function of a multicopy display vector using fusions to the major coat protein of bacteriophage M13. Gene 109: 13–19.

Marvin, D.A., Hale, R.D., Nave, C. and Citterich, M.H. 1994. Molecular models and structural comparisons of native and mutant class I filamentous bacteriophages: Ff (fd, f1, M13), If1 and IKe. J. Mol. Biol. 235: 260–286.

McDonnell, P.A., Shon, K., Kim, Y. and Opella, S.J. 1993. fd coat protein structure in membrane environments. J. Mol. Biol. 233: 447–463.

Model, P. and Russel, M. 1988, Filamentous bacteriophage. In The Bacteriophages Vol. 2 (Calendar, R., Ed.), pp. 375–456. Plenum, New York.

Model, P., Russel, M. and Boeke, J.D. 1981. Filamentous phage assembly: membrane insertion of the major coat protein. In Bacteriophage Assembly (DuBow, M.S., Ed.), pp. 389–400. Alan R. Liss, New York.

Nagle De Zwaig, R. and Luria, S.E. 1967. Genetics and physiology of colicin-tolerant mutants of *Escherichia coli*. J. Bacteriol. 94: 1112–1123.

Nambudripad, R., Start, W., Opella, S.J. and Makowski, L. 1991. Membrane-mediated assembly of filamentous bacteriophage Pf1 coat protein. Science 252: 1305–1308.

Novotny, C.P. and Fives-Taylor, P. 1974. Retraction of F pili. J. Bacteriol. 117: 1306–1311.

Oey, J.L. and Knippers, R. 1972. Properties of the isolated gene 5 protein of bacteriophage fd. J. Mol. Biol. 68: 125–138.

Ohkawa, I. and Webster, R.E. 1981. The orientation of the major coat protein of bacteriophage f1 in the cytoplasmic membrane of *Escherichia coli*. J. Biol. Chem. 256: 9951–9958.

Opella, S.J., Stewart, P.L. and Valentine, K.G. 1987. Protein structure by solid-state NMR spectroscopy. Q. Rev. Biophys. 19: 7–49.

Peeters, B.P., Peters, R.M., Schoenmakers, J.G. and Konings, R.N. 1985. Nucleotide sequence and genetic organization of the genome of the N-specific filamentous bacteriophage IKe: comparison with the genome of the F-specific filamentous phages M13, fd and f1. J. Mol. Biol. 181: 27–39.

Pratt, D. and Erdahl, W.S. 1968. Genetic control of bacteriophage M13 DNA synthesis. J. Mol. Biol. 37: 181–200.

Pratt, D., Tzagoloff, H. and Beaudoin, J. 1969. Conditional lethal mutants of the small filamentous coliphage, II. Two genes for coat proteins. Virology 39: 42–53.

Pratt, D., Tzagoloff, H. and Erdahl, W.S. 1966. Conditional lethal mutants of the small filamentous coliphage M13, I. Isolation, complementation, cell killing, time of cistron action. Virology 30: 397–410.

Rapoza, M.P. and Webster, R.E. 1993. The filamentous bacteriophage assembly proteins require the bacterial SecA protein for correct localization to the membrane. J. Bacteriol. 175: 1856–1859.

Reynolds, P.R., Mottur, G.P. and Bradbeer, C. 1980. Transport of vitamin B_{12} in *Escherichia coli*: some observations on the role of the gene products of *btuC* and *tonB*. J. Biol. Chem. 255: 4313–4319.

Rowitch, D.H., Hunter, G.J. and Perham, R.N. 1988. Variable electrostatic interaction between DNA and coat protein in filamentous bacteriophage assembly. J. Mol. Biol. 204: 663–674.

Russel, M. 1991. Filamentous phage assembly. Mol. Microbiol. 5: 1607–1613.

Russel, M. 1992. Interchangeability of related proteins and autonomy of function: the morphogenetic proteins of filamentous phage f1 and IKe cannot replace one another. J. Mol. Biol. 227: 453–462.

Russel, M. 1993. Protein-protein interactions during filamentous phage assembly. J. Mol. Biol. 231: 689–697.

Russel, M. 1994a. Mutants at conserved positions in gene IV, a gene required for assembly and secretion of filamentous phage. Mol. Microbiol. 14: 357–369.

Russel, M. and Kazmierczak, B. 1993. Analysis of the structure and subcellular location of filamentous phage pIV. J. Bacteriol. 175: 3998–4007.

Russel, M. and Model, P. 1983. A bacterial gene, *fip*, required for filamentous bacteriophage f1 assembly. J. Bacteriol. 154: 1064–1076.

Russel, M. and Model, P. 1985. Thioredoxin is required for filamentous phage assembly. Proc. Natl. Acad. Sci. U.S.A. 82: 29–33.

Russel, M. and Model, P. 1986. The role of thioredoxin in filamentous phage assembly: construction, isolation, and characterization of mutant thioredoxins. J. Biol. Chem. 261: 14997–15005.

Russel, M. and Model, P. 1989. Genetic analysis of the filamentous bacteriophage packaging signal and of the proteins that interact with it. J. Virol. 63: 3284–3295.

Russel, M., Whirlow, H., Sun, T.P. and Webster, R.E. 1988. Low-frequency infection of F-bacteria by transducing particles of filamentous bacteriophages. J. Bacteriol. 170: 5312–5316.

Salstrom, J.S. and Pratt, D. 1971. Role of coliphage M13 gene 5 in single-stranded DNA production. J. Mol. Biol. 61: 489–501.

Sambrook, J., Fritsch, E.F. and Maniatis, T. 1989. Molecular cloning: a laboratory manual. Cold Spring Harbor Laboratory Press, Cold Spring Harbor, NY.

Scott, J.K., Loganathan, D., Easley, R.B., Gong, X. and Goldstein, I.J. 1992. A family of concanavalin A-binding peptides from a hexapeptide epitope library. Proc. Natl. Acad. Sci. U.S.A. 89: 5398–5402.

Scott, J.K. and Smith, G.P. 1990. Searching for peptide ligands with an epitope library. Science 249: 386–390.

Shen, C., Ikoku, A. and Hearst, J.E. 1979. Specific DNA orientation in the filamentous bacteriophage fd as probed by psoralen crosslinking and electron-microscopy. J. Mol. Biol. 127: 163–175.

Shon, K.J., Kim, Y., Colnago, L.A. and Opella, S.J. 1991. NMR studies of the structure and dynamics of membrane-bound bacteriophage Pf1 coat protein. Science 252: 1303–1305.

Simons, G.F.M., Konings, R.N.H. and Schoenmakers, J.G.G. 1981. Genes VI, genes VII, and genes IX of phage m13 code for minor capsid proteins of the virion. Proc. Natl. Acad. Sci. U.S.A. 78: 4194–4198.

Skinner, M.M., Zhang, H., Leschnitzer, D.H., Guan, Y., Bellamy, H., Sweet, R.M., Gray, C.W., Konings, R.N., Wang, A.H. and Terwilliger, T.C. 1994. Structure of the gene V protein of bacteriophage f1 determined by multiwavelength x-ray diffraction on the selenomethionyl protein. Proc. Natl. Acad. Sci. U.S.A. 91: 2071–2075.

Smilowitz, H. 1974. Bacteriophage f1 infection: fate of parental major coat protein. J. Virol. 13: 94–99.

Smith, G.P. 1985. Filamentous fusion phage—novel expression vectors that display cloned antigens on the virion surface. Science 228: 1315–1317.

Smith, G.P. 1991. Presentation of protein epitopes using bacteriophage expression systems. Curr. Opin. Biotechnol. 2: 668–673.

Smith, G.P. and Scott, J.K. 1993. Libraries of peptides and proteins displayed on filamentous phage. Methods Enzymol. 217: 228–257.

Specthrie, L., Bullitt, E., Horiuchi, K., Model, P., Russel, M. and Makowski, L. 1992. Construction of a microphage variant of filamentous bacteriophage. J. Mol. Biol. 228: 720–724.

Stengele, I., Bross, P., Garces, X., Giray, J. and Rasched, I. 1990. Dissection of functional domains in phage fd adsorption protein—discrimination between attachment and penetration sites. J. Mol. Biol. 212: 143–149.

Sun, T. and Webster, R.E. 1986. *fii*, a bacterial locus required for filamentous phage infection and its relation to colicin-tolerant *tolA* and *tolB*. J. Bacteriol. 165: 107–111.

Sun, T. and Webster, R.E. 1987. Nucleotide sequence of a gene cluster involved in the entry of the E colicins and the single-stranded DNA of infecting filamentous phage into *Escherichia coli*. J. Bacteriol. 169: 2667–2674.

Tabor, S., Huber, H.E. and Richardson, C.C. 1987. *Escherichia coli* thioredoxin confers processivity on the DNA polymerase activity of the gene 5 protein of bacteriophage T7*. J. Biol. Chem. 262: 16212–16233.

Torbet, J., Gray, D.M., Gray, C.W., Marvin, D.A. and Siegrist, H. 1981. Structure of the fd DNA-gene 5 protein complex in solution—a neutron small-angle scattering study. J. Mol. Biol. 146: 305–320.

Trenkner, E., Bonhoeffer, F. and Gieres, A. 1967. The fate of the protein components of bacteriophage fd during infection. Biochem. Biophys. Res. Commun. 28: 932–939.

Webster, R.E. 1991. The *tol* gene products and the import of macromolecules into *Escherichia coli*. Mol. Microbiol. 5: 1005–1011.

Webster, R.E., Grant, R.A. and Hamilton, L.A.W. 1981. Orientation of the DNA in the filamentous bacteriophage f1. J. Mol. Biol. 152: 357–374.

Webster, R.E. and Lopez, J. 1985. Structure and assembly of the Class I filamentous bacteriophage. In Virus Structure and Assembly (Casjens, S., Ed.), pp. 235–267. Jones and Bartlett, Boston.

Yamada, M., Ebina, Y., Miyata, T., Nakazawa, T. and Nakazawa, A. 1982. Nucleotide sequence of the structural gene for colicin E1 and predicted structure of the protein. Proc. Natl. Acad. Sci. U.S.A. 79: 2827–2831.

Zinder, N.D. 1973. Resistance to colicins E3 and K induced by infection with bacteriophage f1. Proc. Natl. Acad. Sci. U.S.A. 70: 3160–3164.

14.

Molecular Requirements for Retrovirus Assembly

ROBERT A. WELDON, JR. & ERIC HUNTER

Introduction

Retroviruses

Retroviruses comprise a diverse group of enveloped RNA viruses that require a DNA intermediate for replication. Such viruses include the human immunodeficiency virus (HIV) types 1 and 2, the cause of acquired immunodeficiency syndrome, and human T-cell leukemia virus (HTLV) types 1 and 2, which are responsible for both leukemias and a nervous system degenerative disease. In addition, there are numerous avian and mammalian retroviruses that are the causative agents of cancer and a spectrum of degenerative diseases.

RNA-dependent DNA transcription of the viral genome takes place in the cytoplasm of the infected cell after virus attachment, penetration, and partial uncoating of the viral capsid. Once the completed proviral cDNA enters the nucleus and integrates into the host cell chromosome, its genes behave as cellular genes. As with cellular messenger RNA molecules, the virus-specific transcripts are capped and polyadenylated. Viral protein expression culminates with assembly and release of progeny virus particles from the plasma membrane. These nascent virions are not immediately infectious and must undergo a series of maturation events in order to become so.

Virion Structure

Immature and mature virus particles are similar in size (diameter of 100 to 120 nm) and density (1.16 to 1.18 g/ml) yet morphologically different (Teich, 1984). In general, the immature particle is composed of a host cell–derived lipid bilayer, glycoprotein spikes (Env), and an electron-dense, spherical protein shell that encases the viral RNA genome and that we refer to in this chapter as the immature capsid or procapsid. This nomenclature is based on the fact that the protein shell encompassing the viral genome of most viruses is defined as the viral capsid. Biochemically, the immature capsid is composed of 1) approximately 2000 copies of a 55 to 80-kDa precursor polyprotein, Gag (Davis and Rueckert, 1972); 2) less than 100 copies of a 160 to 200-kDa polyprotein, Gag-Pro-Pol, that contains the viral proteinase (PR), reverse transcriptase (RT), and integrase (IN) enzymes (Stromberg et al., 1974; Panet et al., 1975; Krakower et al., 1977); 3) two copies of single-stranded, plus-sense, genomic RNA; and 4) several host-encoded tRNAs (Linial and Miller, 1990; Schlesinger et al., 1994). A few retroviruses, such as Mason-Pfizer monkey virus (M-PMV), mouse mammary tumor virus (MMTV), HTLV-1, and HTLV-2 also contain a few hundred copies of an additional precursor polyprotein, Gag-Pro, which like the Gag-Pro-Pol polyprotein is made via ribosomal frameshifting (Jacks, 1990).

The mature particle has an envelope that is indistinguishable from that of the immature particle, but its internal structure is quite different. Beneath the envelope are a membrane-associated protein shell and a new structural complex, the core, within which is the ribonucleoprotein complex with the associated replicative enzymes (RT and IN). These structural and enzymatic proteins of the mature core are derived from the Gag and Gag-related precursors during virus maturation via the viral protease (Skalka, 1989; Oroszlan and Luftig, 1990).

Assembly Processes

Final assembly and release of the nascent virion requires that the precursor polyproteins and the viral and host RNAs migrate to the plasma membrane of the cell, where budding occurs. The membrane-spanning glycoproteins are synthesized on membrane-bound polysomes and are transported through the cell's secretory pathway to the plasma membrane, where they colocalize with the immature capsid. In contrast, the Gag and Gag-Pol precursors are synthesized on free polysomes (Eisenman et al., 1974; Purchio et al., 1980). Therefore, they are transported through the cytoplasm to the underside of the plasma membrane. Depending upon the virus, these precursors migrate either individually (or perhaps as small multimers) or as preassembled capsids (procasids). At some point in the assembly process, the genomic-length RNAs and essential cellular tRNAs associate with the Gag precursors. Thus, virion assembly requires stable interactions between Gag–Gag and Gag–Gag-Pol (and perhaps Gag–Env) precursors as well as between Gag and cellular proteins, nucleic acids, and lipids. Ultimately, these intermolecular interactions are themselves dependent upon the three-dimensional structures of the individual components.

Virion Assembly

The Virus Assembly Machine—The Gag Precursor

The structural polyprotein of the immature capsid, encoded by the *gag* gene (Fig. 14.1), contains the necessary information for intracytoplasmic transport, capsid assembly, budding, and release in the absence of the other viral components (Dickson et al., 1984; Wills and Craven, 1991; Hunter, 1994). In most retroviruses, the Gag proteins are modified by cotranslational removal of the amino-terminal methionine and subsequent addition of myristic acid and by phosphorylation (Schultz et al., 1979; Henderson et al., 1983; Schultz and Oroszlan, 1983). Rous sarcoma virus (RSV) and equine infectious anemia virus Gag polyproteins are exceptions and can function properly without myristate, although, in the case of the RSV, its Gag protein is acetylated (Palmiter et al., 1978; Henderson et al., 1987). Shortly after synthesis, such precursors follow one of two morphogenetic pathways during capsid assembly. In most retroviruses, the Gag polyproteins are transported to the plasma membrane, where they first visibly appear as crescent-shaped, electron-dense patches associated with the inner surface of the membrane (Fig. 14.2). As these crescent-shaped structures continue to develop into spherical, immature capsids, they acquire their lipid envelope by budding through the plasma membrane. Viruses that undergo this form of morphogenesis are known as C-type viruses and include the avian and mammalian oncoviruses (i.e., RSV and murine leukemia virus [MuLV]). Lentiviruses, which include HIV, assemble their capsids in a similar fashion. The Gag polyproteins of the second morphogenetic class of viruses preassemble their procapsids in the cytoplasm. These preassembled procapsids are then transported to the plasma membrane, where the envelope is then acquired via viral budding. Viruses that undergo this morphogenic pathway in-

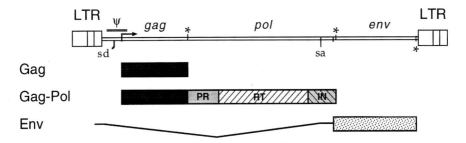

FIGURE 14.1 Sequence organization of a generalized retrovirus genome. The long terminal repeats (LTRs) are located at the 5′ and 3′ ends of the proviral cDNA. The splice donor (sd) and RNA packaging signal (ψ) lie just downstream of the 5′ LTR. The full-length mRNA is initiated in the 5′ LTR and terminates in the 3′ LTR. Translation of this RNA produces the Gag protein (dark box). In general, the viral protease (PR), reverse transcriptase (RT), and integrase (IN) proteins are synthesized as Gag-Pol fusion proteins via either a ribosomal frameshift or suppression of a stop codon at the end of the gag gene. Then Env precursor glycoprotein is translated from a spliced mRNA using a splice acceptor (sa) located in the pol gene. The locations of the stop codons for the *gag*, *pol*, and *env* genes are indicated (*).

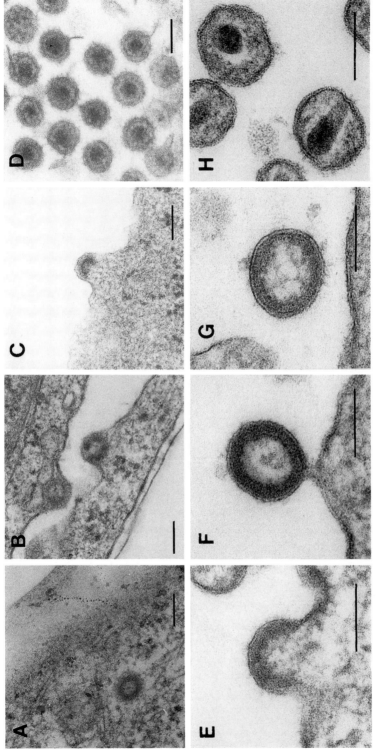

FIGURE 14.2 Electron microscopic analysis of retrovirus assembly and maturation. *A* and *B*. Thin sections of M-PMV–infected cells showing a preassembled intracellular capsid (*A*) and a budding particle (*B*; D-type morphogenesis). *C*. Particle of RSV assembling and simultaneously budding from the plasma membrane (C-type morphogenesis). *D*. Cluster of mature RSV particles. (Panels C and D courtesy of R. Craven and J.W. Wills, Pennsylvania State Medical Center, Hershey, PA.) *E* to *H*. Various stages of HIV morphogenesis. (Courtesy of D. Hockley, NIBSC.) *E*. An early stage of assembly and budding. *F*. A late stage of morphogenesis. *G*. An immature virus particle showing the electron-lucent center. *H*. Mature virus particles with the characteristic cone-shaped capsid shown either along its length (bottom left) or through its center (upper right). Bars, 100 nm.

clude the B-type (MMTV) and D-type (M-PMV and related simian retroviruses [SRV.1 to SRV.5]) viruses, as well as members of the spumavirus family (Teich, 1984). Irrespective of the different morphogenetic pathways, the process by which Gag precursors assemble into capsids is probably similar because a single amino acid change within the Gag protein of M-PMV was shown to divert it to the C-type morphogenetic pathway (Rhee and Hunter, 1990).

All retroviral Gag precursors assemble into structurally similar, immature capsids. Based on thin-section electron microscopy and on cryo−electron microscopy, an immature capsid appears as an electron-opaque ring located be-

A

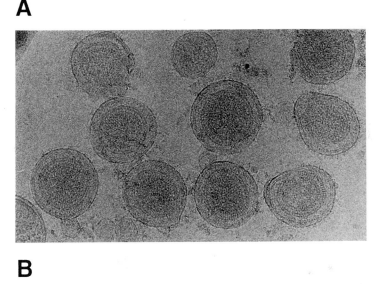

B

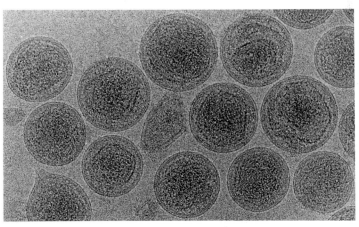

FIGURE 14.3 Cryo−electron micrographs of MuLV. *A.* Immature virus particles resulting from a mutation that inactivates with viral protease. *B.* Mature virus particles. (Micrographs courtesy of E. Kubalek and M. Yeager, Scripps Institute.)

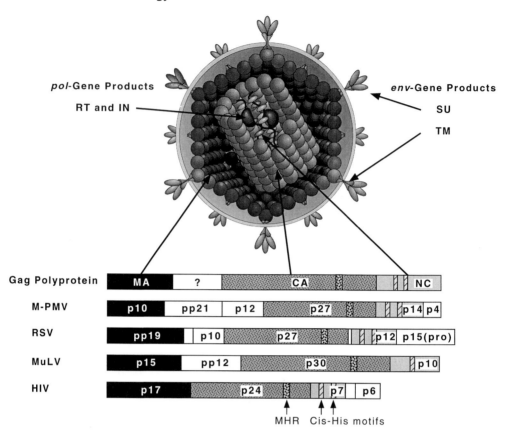

FIGURE 14.4 Organization of the *gag* gene region of a retrovirus and the respective locations of the mature cleavage products in the mature virus particle. *A.* An artistic rendering of a mature retrovirus particle showing the locations of the mature structural and enzymatic proteins. *B.* The Gag precursor polyprotein of all retroviruses contain matrix (MA), capsid (CA), and nucleocapsid (NC) domains linked in that order beginning at the amino-terminal end. The *gag* gene products of M-PMV, RSV, MuLV, and HIV are shown with the approximate locations of their highly conserved major homology regions (MHR) and Cis-His boxes. The unshaded boxes represent regions of the Gag precursor for which no common functions or locations in the mature virion have been established. The specific names associated with the Gag cleavage products are derived from their respective apparent molecular weights ($\times 10^{-3}$).

neath the viral membrane (Figs. 14.2 and 14.3*A*). Around this ring is an electron-lucent space that varies in thickness between retroviruses. In HIV, the electron-dense ring is in tight apposition to the viral membrane. With other retroviruses, the electron-lucent space may be as thick as 100 Å. Although it is generally accepted that the amino-terminal end of Gag associates with the viral membrane and the carboxyl-terminal end extends into the center of the capsid, where it associates with the viral RNA (Dickson et al., 1984; Wills and Craven, 1991), the exact viral components (i.e., regions of Gag and/or RNA) that make up these electron-lucent and -dense regions are currently not known.

For this reason, along with the absence of detailed structural data for the Gag polyprotein, the arrangement of Gag proteins within the immature capsid has not been determined. In fact, cryo–electron micrographs of immature MuLV particles (Fig. 14.3A), as well as HIV and M-PMV (S. Fuller; B.V. Prasad, personal communication), show pleomorphic capsids instead of highly ordered structures (i.e., icosahedral capsids), which have been suggested by others (Nermut et al., 1994). Clearly, resolution of this issue awaits further experimentation.

Structure–Function Relationships of the Mature gag Gene Products

During virus maturation, the Gag precursor polyproteins are cleaved by the viral protease to yield a number of individual proteins found in the mature virion. These include the matrix (MA), capsid (CA), nucleocapsid (NC), and several other proteins of unknown functions (Fig. 14.4). The Gag-Pol precursors are similarly cleaved to release, in addition to those just mentioned, the enzymatic proteins (PR, RT, and IN).

Matrix Protein

In the mature virus, the MA protein forms a protein layer that is closely associated with the viral membrane. In fact, it is the amino-terminal end of the MA protein that provides membrane binding functions. In RSV, this includes the first 35 residues (Pepinsky and Vogt, 1984). Interestingly, the RSV MA protein lacks the amino-terminal myristic acid and any significant stretch of hydrophobic amino acids that might provide a membrane binding function. Therefore, the mechanism by which the MA protein of RSV associates with the membrane is not known. The membrane binding domain of other Gag precursors is thought to involve myristic acid as well as specific amino acid residues in MA (Rhee and Hunter, 1991; Spearman et al., 1994; Zhou et al., 1994). In HIV-1, a fraction of MA molecules are phosphorylated and appear to promote nuclear targeting of a newly synthesized provirus (Farnet and Haseltine, 1991; Sharova and Bukrinskaya, 1991; Bukrinskaya et al., 1992; Gallay et al., 1995).

Recent nuclear magnetic resonance studies of bacterially expressed HIV MA proteins (p17) have provided the first insights into the three-dimensional structure of this normally membrane-associated molecule (Massiah et al., 1994; Matthews et al., 1994). Although the structures determined by these groups appear to adopt similar global folds, there are a few discrepancies. The structure determined by Matthews et al. (residues 15 to 110; Fig. 14.5A) has a compact fold that contains four helices joined by short loops and an amino-terminal, triple-stranded, irregular, mixed β-sheet. Helices A (amino acids 33 to 45), B (amino acids 55 to 69), and D (amino acids 88 to 109) are highly amphipathic and accessible to solvent, in contrast to helix C (amino acids 73 to 88), which runs through the center of the hydrophobic core. The structure determined by

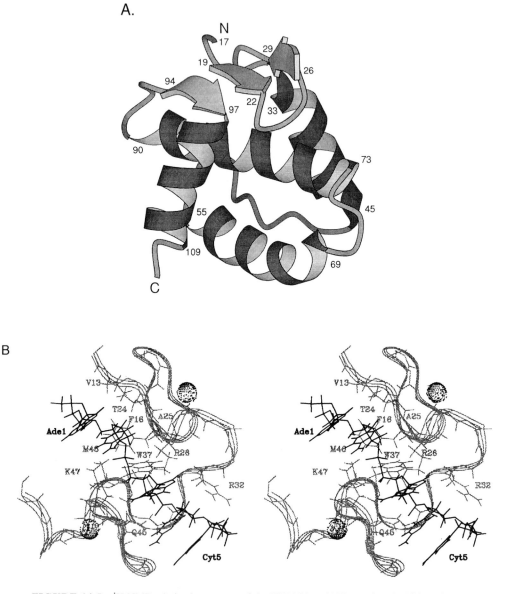

FIGURE 14.5 ¹H NMR–derived structures of the HIV MA and NC proteins. *A*. Ribbon drawing of the HIV MA protein (residues 15 to 110). Arrows and coils represent β-sheets and α-helices, respectively. The remaining regions are random coils. (Reproduced by permission from Matthews et al., 1994.) *B*. Stereo representation of the ¹H NMR–derived structure formed between the single-stranded d[ACGCC] belonging to the virus RNA packaging domain and the NC p7 protein (residues 13 to 64). Residues in contact with the oligonucleotide are labeled. Zinc atoms are represented as their Van der Waals sphere. (Courtesy of H. Déméné and B.P. Roques.)

388

Massiah et al. consists of the entire MA protein (except for the amino-terminal myristic acid) and contains a fifth helix (amino acids 9 to 17). The irregular β-sheet is a prominent feature of p17 (MA), and the solvent-exposed side of this domain provides a surface that could associate with the inner face of the membrane, with several basic side chains (Lys 18, Arg 20, Arg 22, Lys 26 to Lys 28, Lys 30, Lys 32, and Lys 95) available for interaction with phospholipid head groups. Indeed, mutations that alter the charge distribution in this region have significant effects on virus assembly (Gonzalez et al., 1993; Yuan et al., 1993; Zhou et al., 1994).

Capsid Protein

The CA protein forms the shell of the mature core that contains the viral RNA genome, NC proteins, and replicative enzymes (Dickson et al., 1984). While the arrangement of this protein in the core is unknown, recent NMR and crystallographic studies of an N-terminal fragment of the HIV-1 CA have provided some structural information. In contrast to the capsid proteins of the icosahedral animal viruses, which fold into an eight-strand anti-parallel β-barrel, the amino terminal two-thirds of CA is composed of seven α-helices, five of which are arranged in a coil-coil-like structure (Gitti et al., 1996; Momany et al., 1996). In addition to its structural role in the virion, the CA protein of some retroviruses plays an important role in establishing an integrated provirus in newly infected cells. In contrast to that of HIV, the CA protein of MuLV appears to remain with the newly synthesized proviral DNA in the cytoplasm (Bowerman et al., 1989; Varmus and Swanstrom, 1984).

Nucleocapsid Protein

The NC protein is a basic protein that appears to interact with the viral RNA genome in both a sequence-specific and a sequence-independent manner. This protein in all retroviruses except the spumaviruses contains either one or two Cys-His−rich zinc finger motifs (C-X_2-C-X_4-H-X_4-C) that coordinate zinc binding (Fitzgerald and Coleman, 1991; Bess et al., 1992; Chance et al., 1992). In addition to its RNA binding properties, the NC protein has been implicated in mediating packaging of the viral RNA into the virion, linking the two genomic RNA subunits together, and/or facilitating the annealing of the tRNA primer to the viral genome (Meric and Spahr, 1986; Prats et al., 1988; Bieth et al., 1990).

Recent nuclear magnetic resonance (NMR) studies of peptides corresponding to the zinc finger motifs and of intact NC proteins from both HIV and MuLV have yielded new insights into the structural organization of this Gag domain (South et al., 1991; Morellet et al., 1992, 1994; Summers et al., 1992; Déméné et al., 1994). For HIV-1, the two zinc finger motifs, which are separated by a short basic stretch of amino acids (Arg 29-Ala-Pro-Arg-Lys-Lys-Gly 35), fold in a similar manner (Fig. 14.5B) and are brought into spatial proximity by a kink induced by the proline residue in the linker peptide. This structural arrangement shortens the distance between two aromatic residues, Phe 16 and

Trp 37, to approximately 6 Å. Morellet and colleagues (1992, 1994) have speculated that these residues could, through RNA intercalation, direct the specific interaction of NC with the *cis*-acting packaging signal on the genomic RNA of the virus. The Tyr 28 and Trp 35 residues within the NC of MuLV lie on the same face of the single zinc finger present within this protein and may play a similar role (Déméné et al., 1994). Basic residues, which flank the zinc finger domains in both MuLV and HIV, appear to be critical for the stable association of NC with RNA and can function independently from the zinc fingers in this regard, presumably to allow the sequence-independent coating of RNA seen in the virion (De Rocquigny et al., 1992, 1993).

Spatial Organization of the gag *Gene Products*

How are the internal structural proteins arranged in the mature virus? Are they arranged in symmetric arrays (i.e., icosahedral or helical), as are all other viruses (except perhaps members of the *Poxviridae*), or do retroviruses represent a novel class of capsid structure and symmetry? The organization of the MA protein in the outer shell has not been determined because removal of the membrane with detergent, which is needed for visualization of the outer shell, simultaneously destroys this shell (Davis and Rueckert, 1972; Stewart et al., 1990). However, freeze-drying and shadowing as well as computer simulation of thin sections of virions suggested that the MA shells have icosahedral symmetry (Nermut et al., 1972, 1993; Gelderblom, 1990). The precise organization of the CA proteins in the mature capsid is also unknown. However, electron microscopic examination of various retrovirus particles shows the inner core to be cone shaped (HIV), spherical (RSV and MuLV), or cylindrical (M-PMV). Recent in vitro assembly experiments with RSV suggest that its CA–NC domains can assemble on RNA molecules into either rodlike or spherical structures depending on the assembly conditions (Campbell and Vogt, 1995). The HIV CA protein can also oligomerize in vitro into higher order multimers and even long rodlike structures (Ehrlich et al., 1992). Although these structures do not resemble the cone-shaped capsids seen in mature HIV particles, they may, nevertheless, represent assembly intermediates. Additional viral factors (RNA, other viral proteins, and/or membranes) or posttranslational modifications may be required to assemble the cone-shaped capsid shell. Therefore, it is likely that the different core morphologies observed for different retroviruses result from the specific manner in which the monomers assemble.

Genomic Organization of the gag *Gene*

Although the size of the Gag precursor varies between different retroviral families, all Gag precursors contain the MA, CA, and NC domains linked in that order starting from their amino-terminal ends. However, Gag proteins from different retroviruses share little amino acid sequence similarity despite the common organization. Two notable exceptions are a conserved region of approximately 20 amino acids in CA, termed the major homology region (Wills and Craven, 1991; Strambio-de-Castillia and Hunter, 1992; Mammano et al., 1994), and the conserved cysteine-histidine–rich zinc finger–like motifs in the NC domain (Fig. 14.4). Because of the lack of any extensive sequence homology between Gag proteins of different retroviruses, any functional and structural homologies must be reflected in their three-dimensional structures. This is exemplified by the observations that chimeric RSV–HIV and RSV–MuLV Gag proteins can efficiently assemble into viruslike particles (Dupraz and Spahr, 1992; Bennett et al., 1993).

Other Gag Domains

Gag proteins from different retrovirus can contain additional peptide sequences whose positions in the mature virion and functions in the virus life cycle are unknown (Fig. 14.4). These peptide domains may represent spacer peptides that allow the correct folding of the precursor protein for assembly and subsequent processing into the mature virion proteins. Alternatively, because the Gag precursor is the assembly unit for capsid construction, such regions could represent novel domains that function transiently during virus assembly and are then dispensed with during virus maturation. For most oncoviruses, a proline-rich region (i.e., p2 and p10 in RSV and p12 in MuLV; Fig. 14.4) is located between the MA and CA domains (Dickson et al., 1984). It is interesting that, although the HIV Gag protein lacks additional peptide sequences between the MA and CA domains, it does contain a peptide domain (p6) at its carboxyl-terminus that also appears necessary for efficient release of virus particles (Göttlinger et al., 1991; Spearman et al., 1994). A unique feature of the avian leukosis/sarcoma viruses is that the viral protease is contained within the Gag polyprotein and is therefore packaged at high concentrations into the viral particle. The evolutionary basis for this variation is not clear.

The inclusion of the mature structural components within a precursor polyprotein probably serves several roles. First, the arrangement of the structural proteins on the precursor is invariant among all Gag proteins and reflects their position in the mature virion. Second, it ensures that equimolar amounts of each of these proteins are incorporated into the virus. Third, it avoids the need to target each protein to a common assembly site, thus conserving genetic information. Finally, it appears to temporally regulate the assembly functions from those needed during the early phases of infection. Proteolytic processing of Gag after capsid formation not only destroys the assembly functions, which in turn prevents the proteins from reentering the assembly pathway in the cytoplasm of the infected cell, but also activates those functions needed to establish an integrated provirus. For example, proteolysis may uncover nuclear localization signals present in the MA or CA proteins; exposing these signals during virus assembly would certainly be detrimental to virus replication.

Transport of Gag Proteins to the Site of Assembly

The mechanism by which Gag proteins travel from their sites of synthesis to a region within the cell where they assemble into an immature core is unknown. It is thought that Gag proteins arrive at these sites by a specific transport mechanism rather than by passive diffusion (Fig. 14.6). Certain point mutations or deletions within the MA domain of HIV appear to have no effect on capsid assembly but redirect particle formation and budding to the endoplasmic reticulum (Facke et al., 1993; Yuan et al., 1993; Freed et al., 1994). For the B- and D-type viruses, transport involves two distinct stages. The Gag proteins are initially directed to or retained at an intracytoplasmic site where procapsid assembly occurs. From there, the procapsids migrate to the plasma membrane, where envelopment and budding occur. Genetic analyses of M-PMV provide evidence that nascent Gag precursors express a dominant sorting signal that directs the proteins to the assembly site. This signal appears to reside in an 18-amino-acid region of the MA domain that has sequence similarity to the B-type MMTV (Rhee and Hunter, 1990). Transport of the procapsids from the assembly site to the plasma membrane is also dependent upon an active process. A variety of mutations within the MA domain that either block myristic acid addition or alter specific residues other than those required for myristylation result in the accumulation of procapsids deep in the cytoplasm (Rhee and Hunter, 1987, 1991). These procapsids were apparently unable to associate with the transport machinery and thus failed to exit the site of assembly. Thus, myristylation alone is not a sufficient signal for intracytoplasmic transport.

The nature of the transport machinery utilized by Gag proteins is unknown. Gag proteins may require interactions with chaperone or heat shock proteins (e.g., TCP1/cytosolic hsp60, cyclophilin A) for correct protein folding and transport (Luban et al., 1993). It has also been suggested that Gag proteins are transported to the plasma membrane via interactions with secretory vesicles (Hansen et al., 1990; Jones et al., 1990). This scenario, however, seems unlikely for the B-/D-type viruses because there is no evidence of capsids associating with membranes other than the plasma membrane. Finally, transport may utilize elements of cytoskeleton (Damsky et al., 1977; Satake and Luftig, 1982; Edbauer and Naso, 1984; Maldarelli et al., 1987). Studies on the process of Gag/capsid transport using mutants like those described earlier along with evolving technologies are likely to provide new insights into the cell biology of cytoplasmic transport.

Capsid Assembly

Production of a nascent particle with a defined size, density, and morphology requires that Gag proteins (1) find each other; (2) interact in a regular and stable manner to form the spherical, immature core; (3) associate with the plasma membrane; and (4) drive the budding process. The amino acid sequences of Gag that are involved in these processes as well as those that might have other functions in the virus replication cycle are being ascertained through mutational analyses. This approach, which has been explored in a variety of retroviruses, has been reviewed in detail by Wills and Craven (1991). As Wills and Craven pointed out, it is possible that the assembly domains within the Gag precursor may not necessarily reside within the boundaries of the mature cleavage products of Gag but may span the cleavage sites. Thus, PR-mediated processing of the Gag precursor destroys these assembly functions and defines the transition from an assembly function of Gag to an entry/infection function in which there is a requirement for efficient disassembly and release of a transcriptionally active preintegration complex upon infection of a new cell.

All Gag proteins appear to require their amino termini for membrane associations. In RSV, this amino-terminal assembly domain appears to include the first half of the MA (Wills et al., 1991). The membrane binding domains of other Gag proteins are near the amino terminis (Rhee and Hunter, 1987, 1991; Yu et al., 1992; Spearman et al., 1994; Zhou et al., 1994). This domain in HIV, which lies within the first 31 residues, has been shown to bind acidic phospholipids (Zhou et al., 1994). However, it is not known whether these phospholipids are sufficient for membrane associations or if a cellular protein(s) (i.e., a docking protein) is initially required to mediate this interaction (Wills and Craven, 1991).

A second assembly domain has been identified for RSV that appears to function during the late stages of the budding process. This domain has been mapped to the carboxyl-terminus of the p2 "spacer peptide" (Wills et al., 1994). In fact, a "mini" Gag protein of RSV that consists of MA and p2 can direct the budding process (Weldon and Wills, 1993). Likewise, the MA protein from simian immunodeficiency virus (SIV) can drive the budding process, which suggests that it may contain an analogous assembly function (Gonzalez et al., 1993). In HIV, the carboxyl-terminal p6 domain of Gag appears to provide an analogous late-budding function (Göttlinger et al., 1991; J.W. Wills, personal communication, 1995).

For those Gag proteins that have been examined in detail, there appear to be specific domains that are essential for the production of particles with the correct density and size. In RSV, assembly domain 3 is located between the carboxyl-terminal end of the CA domain and the first half of the NC domain and is essential for the production of particles with the correct density (Weldon and Wills, 1993). HIV and MuLV Gag proteins also require analogous regions for the production of particles with the correct density (Jones et al., 1990; Jowett et al., 1992; Bennett et al., 1993). Finally, there appears to be a fourth region in RSV, located within the p10 and the amino-terminal two thirds of the CA domains, that influences particle size (Weldon and Wills, 1993). The mechanisms by which these domains influence particle density and size are not known.

In some retroviruses, additional domains of Gag or other gene products may influence the efficiency of particle assembly without being essential to the process itself. In M-PMV, the p12 domain can be removed without affecting assembly if the Gag protein is expressed at high levels. In contrast, procapsids fail to assemble if such mutants are expressed at levels that mimic that of

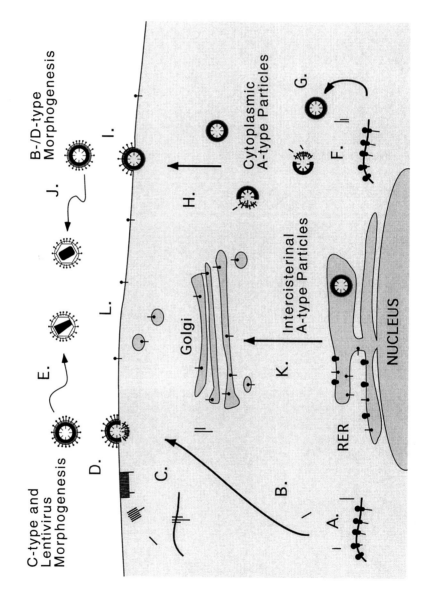

C-type and
Lentivirus
Morphogenesis

B-/D-type
Morphogenesis

Cytoplasmic
A-type Particles

Intercisterinal
A-type Particles

Golgi

RER

NUCLEUS

A. B. C. D. E. F. G. H. I. J. K. L.

FIGURE 14.6 Schematic representation of retrovirus morphogenesis. The assembly pathways of both C-type oncoviruses/lentiviruses (A to E) and B-/D-type oncoviruses (F to J) are shown. The envelope glycoproteins of these viruses are translated on membrane-bound polysomes and are transported to the cell surface through the cell's secretory pathway (K to L). For all morphogenic classes, the Gag and Gag-Pol proteins are synthesized on free polysomes (A and F). In the case of the C-type oncoviruses and lentiviruses (RSV and HIV, respectively), the Gag and Gag-Pol proteins migrate either individually or in small multimers (B) to the plasma membrane, where the immature capsid and envelope develop concurrently (D). At some unknown point in this pathway, the viral RNAs associate with the Gag and Gag-Pol precursors (C) and are incorporated into the developing capsid. In the case of the B-/D-type viruses, the Gag and Gag-Pol precursors are first transported to an intracytoplasmic assembly site (G), where they coalesce into procapsids. These immature capsids are then transported (H) to the plasma membrane, where they associate with the envelope glycoproteins and induce viral budding (I). For all classes of retroviruses, the capsids of the nascent immature particles appear as doughnut-shaped structures and contain unprocessed Gag and Gag-Pol precursors. The mature virus particles contain electron-dense cores with morphologies characteristic of the virus (E and J). The maturation step is required for infectivity and is the result of the activation of the viral protease that cleaves the Gag and Gag-Pol precursors into the internal structural and enzymatic proteins of the virus. RER, rough endoplasmic reticulum.

395

normal infected cells (Sommerfelt et al., 1992). Thus, this domain, which is unique to the B- and D-type retroviruses, may facilitate either the intracellular transport of the Gag precursor protein to the assembly sites or the initial interactions of precursors present at low concentrations in an infected cell. Similarly, the *vpu* gene product of HIV, which is a small membrane-spanning protein, appears to facilitate the intracellular transport of Gag and/or the assembly/release of particles from an infected cell (Terwilliger et al., 1989; Klimkait et al., 1990). This protein may also play an important role in assembly of infectious virus particles by dissociating glycoprotein–receptor complexes that have formed in the endoplasmic reticulum (Willey et al., 1992a, 1992b).

RNA Incorporation

Packaging of the genomic RNA into virions requires both *cis*-acting sequences and *trans*-acting factors. Because retroviruses do not shut off host cell RNA synthesis, nascent viral RNA represents only a small fraction of the total RNA in infected cells. Nevertheless, the vast majority of particles released from cells statistically contain two copies of the genomic-length RNAs rather than a heterogeneous population of cellular RNAs or subgenomic viral RNAs. This implies that a *cis*-acting RNA signal(s) on the genomic RNAs and the Gag precursor govern RNA packaging (Linial and Miller, 1990; Schlesinger et al., 1994).

In most retroviruses, the genomic RNAs contain several *cis*-acting encapsidation elements. The most prominent of these is the Ψ sequence (Fig. 14.1), which is usually located downstream of the major 5′ splice site and is thus absent from the subgenomic *env* RNAs (reviewed in Katz et al., 1986; Linial and Miller, 1990). However, other *cis*-acting elements may also be important because their removal directly affects RNA packaging and viral replication. Depending on the virus, these elements can reside in the long terminal repeats (LTRs), in the untranslated region between the 5′ LTR and *gag*, in the *gag* gene itself, or, in the case of RSV, in the 115-bp direct repeat flanking the *src* gene (Pugatsch and Stacey, 1983; Sorge et al., 1983; Bender et al., 1987; Berkowitz et al., 1993).

The existence of specific *cis*-acting elements for RNA packaging implies that a specific *trans*-acting factor(s) is important as well. Because viruslike particles containing viral RNA can be assembled by coexpressing Gag precursors and Ψ-containing RNA (Oertle and Spahr, 1990; Sakalian et al., 1994), the *trans*-acting element(s) must lie within Gag. In the case of HIV, the Gag precursor has been shown to specifically bind genomic RNA *in vitro* (Luban and Goff, 1991). Genetic analyses suggest that at least part of the *trans*-acting elements include the Cis-His boxes in the NC domain (review in Linial and Miller, 1990; Schlesinger et al., 1994) and perhaps sequences near the major homology region in the CA domain (Sakalian et al., 1994). Precisely how this selection occurs, and when during assembly and where in the cell it transpires, are not known. Nevertheless, understanding this critical step in the production of an infectious virus may lead to the development of antiviral agents that interfere with RNA packaging.

Viral Budding and Release

How do Gag proteins induce viral budding? Clearly, the three-dimensional shape of the monomeric Gag polyprotein dictates its own spatial arrangement in the developing capsid, and the alignment of such proteins is clearly fixed: amino-terminal ends juxtaposed to the viral membrane and carboxyl-terminal ends extending into the center of the virion. Because Gag polyproteins coalesce into spherical capsids, one can speculate that Gag monomers are rod or some what cone shaped such that their carboxyl termini occupy less space in the capsid than their amino-terminal ends (Nermut et al., 1994). It is possible that, as Gag proteins form lateral contacts with each other along their lengths and associate with the plasma membrane through their MA domains, the spherical capsid would form and simultaneously wrap the lipid bilayer around itself. According to this model, it is primarily the forces between Gag proteins and between Gag and the plasma membrane that drive the budding process. Indeed, mutations that apparently alter the tertiary structure of the Gag precursor without preventing the membrane extrusion process or the ability to interact with other precursors often result in malformed and sometimes bizarre budding structures (Göttlinger et al., 1989; Strambio-de-Castillia and Hunter, 1992).

It has also been suggested that conformational changes in the Gag protein occur during the budding process (Wills and Craven, 1991; Hunter, 1994). For example, it is likely that the hydrophobic myristic acid is buried within the tertiary structure of Gag (for the C-type viruses) or the procapsid (for the B-/D-type viruses). Thus, according to this model, as the Gag protein or procapsids associate with the inner surface of the plasma membrane (perhaps through the positive charges clustered near the amino terminus of MA interacting with the polar head groups), a conformational change exposes the hydrophilic fatty acid on the surface, allowing it to insert into the lipid bilayer and provide a more stable Gag–lipid interaction. Clearly, development of an in vitro assembly/budding system would help resolve this poorly understood process.

The final step in virus release requires a membrane fusion event to release the immature particle from the plasma membrane. What facilitates this process is not known, but in HIV the carboxyl-terminal peptide sequence p6 appears to play a role. Truncations or deletions of this domain result in the accumulation of immature particles still attached to the plasma membrane by a thin stalk (Göttlinger et al., 1991). Activation of the viral protease might also be important because similar structures have been seen with HIV-infected cells treated with protease inhibitors (Kaplan et al., 1993).

Virus Glycoprotein Biosynthesis and Incorporation

Whereas the capsid precursor proteins are transported through the cytoplasm by ill-defined pathways, the surface components of the virion, the envelope glycoproteins, are directed to the site of assembly through the relatively well-characterized ''secretory pathway'' of the cell (reviewed in Hunter and Swanstrom, 1990). The envelope glycoprotein complex of replication-competent retroviruses is composed of two polypeptides (Figs. 14.1 and 14.4)—an external, glycosylated, hydrophilic protein (SU) and a membrane-spanning protein (TM)—that form a knob or knobbed spike on the surface of the virion (Dickson et al., 1984). Both polypeptides are encoded by the *env* gene in the order $-NH_2-SU-TM-COOH-$ and are synthesized in the form of a glycosylated polyprotein precursor that oligomerizes in the rough endoplasmic reticulum (Einfeld and Hunter, 1988; Earl et al., 1991).

There is no three-dimensional structural information on any retroviral glycoproteins at the present time, and so attempts have been made to model the tertiary structure of these proteins based on the known crystal structure of the hemagglutinin of influenza virus (Gallaher et al., 1989; Callebaut et al., 1994; see also Chapter 3). According to these models, the TM protein could assume a folded state analogous to that of the HA2 protein of influenza virus (Gallaher et al., 1989), while the receptor binding domain of the SU protein of bovine leukemia virus has been postulated to fold with an overall topology analogous to that of an immunoglobulin constant domain (Callebaut et al., 1994). Gradient centrifugation and chemical cross-linking analyses of oligomeric forms of the RSV, MuLV, and M-PMV glycoproteins point to these complexes being trimeric structures (Einfeld and Hunter, 1988; Kamps et al., 1991), analogous to that of the hemagglutinin of influenza virus. In contrast, similar studies with the HIV-1, HIV-2, and SIV glycoproteins suggest that the most stable glycoprotein complex is a dimer with varying amounts in a higher order structure that appears to be a tetramer (Doms et al., 1990, 1991). Scanning transmission electron microscopy studies have supported this interpretation (Thomas et al., 1991). Because no three-dimensional structural information is available on retroviral glycoproteins, the protein–protein contacts that initiate and stabilize the oligomeric form are not known. Nevertheless, mutational analyses of both the RSV and HIV glycoproteins point to a region in the external domain of the TM protein as being important for this process (Einfeld and Hunter, 1988; Doms et al., 1990). Indeed, expression of the RSV TM coding sequences fused to a signal peptide sequence yielded a protein capable of efficient oligomerization and intracellular transport (Einfeld and Hunter, 1994). The formation of oligomers appears to be required for newly synthesized Env proteins to be exported from the endoplasmic reticulum to the Golgi complex. However, as is the case for the influenza virus hemagglutinin (see also Chapter 3), additional conformational and biochemical changes are probably also necessary for the protein to become transport competent (Einfeld and Hunter, 1994).

The assembled complex is transported, through transport vesicles, to the Golgi apparatus, where the attached oligosaccharides are processed and the precursor is proteolytically cleaved to its mature subunits. Cleavage occurs at a sequence (Arg-X-Lys/Arg-Arg) at the carboxyl-terminus of the SU protein that is highly conserved among divergent retroviruses, and mutational analyses have demonstrated that this processing event is critical for biologic activity of the Env protein (McCune et al., 1988; Dong et al., 1992a). The host cell–encoded enzyme that carries out the cleavage appears to be a member of the furin family of subtilisinlike proteases (Hosaka et al., 1991; Hallenberger et al.,

1992). Following cleavage of the Env precursor, the SU and TM proteins remain associated either through a covalent linkage, as in the case of RSV, or through noncovalent interactions, as is observed with HIV and the majority of other retroviruses.

The cleaved glycoprotein oligomers must then be transported from the Golgi apparatus to the surface of the cell in order to participate in the assembly process. Although the glycoproteins are not required for the assembly of enveloped virus particles, they do play a critical role in the virus replication cycle by recognizing and binding to specific receptors and by mediating the fusion of viral and cell membranes; virus particles lacking envelope glycoproteins are thus noninfectious.

What directs these nascent glycoproteins into the budding virions? It has been generally accepted that they are preferentially incorporated into virions while the bulk of the host cell membrane proteins are excluded from the assembling particle. Support for this concept has evolved from a variety of studies. Electron micrographs of virions released from cells infected with an *env*-deleted RSV provirus showed a lack of surface projections on the "bald" particles, and protein analyses did not demonstrate any significant incorporation of cell membrane components (Kawai and Hanafusa, 1973). Similar results have been obtained following expression of the *gag* gene alone in a variety of retroviral systems. The fact that the TM protein of RSV could be chemically cross-linked to the MA protein in virions supported the concept that an interaction between the glycoprotein and capsid directed the incorporation of glycoproteins into virus (Gebhardt et al., 1984). Moreover, mutations within both the MA protein and the TM protein coding domains of HIV-1 and M-PMV have been shown to reduce the efficiency of glycoprotein incorporation into virions (Rhee and Hunter, 1990; Dubay et al., 1992; Yu et al., 1992, 1993; Dorfman et al., 1994; Lodge et al., 1994), indicating that interactions between these components of the virus are important during the assembly process. Indeed, experiments with HIV-1 have shown that in polarized epithelial cells, the viral glycoproteins can specifically target capsid assembly and release to the basolateral plasma membrane (Owens et al., 1991). Studies have shown that mutations within the carboxyl terminus of the gp41 cytoplasmic domain and those in the matrix protein that block MA−gp41 interactions abolish this specific intracellular targeting event (Lodge et al., 1994). The sum of these experiments would thus point to a specific interaction between the assembling capsid and the viral glycoprotein, so that the latter is positively selected from the mixture of glycoproteins in the plasma membrane; a corollary of such a mechanism would be that, in the absence of this interaction, incorporation of viral glycoproteins would be inefficient.

Although such a positive selection model is attractive, it does not explain the facility with which foreign glycoproteins can be incorporated into retrovirus particles, or the lack of effect that mutations in the cytoplasmic domain of the TM protein have on glycoprotein incorporation in other retroviral systems. In the case of RSV, a truncation of the TM cytoplasmic domain did not block glycoprotein incorporation (Perez et al., 1987); with SIV, in contrast to HIV-1, truncations of the long cytoplasmic domain result in enhanced incorporation of viral glycoproteins into virions (Johnston et al., 1993; Zingler and Littman, 1993). Furthermore, glycolipid-anchored forms of the HIV-1 and MuLV glycoprotein complexes were also efficiently incorporated into virus particles (Salzwedel et al., 1993; Ragheb and Anderson, 1994), suggesting that the protein anchor and cytoplasmic domain were not critical signals for incorporation.

Retrovirus particles, resistant to homologous neutralizing antibody and with an extended host range, are generated when chronically infected cells are superinfected with a virus such as vesicular stomatitis virus (VSV). The formation of these pseudo-type virions demonstrates that foreign glycoproteins can be incorporated into retrovirus particles. In the murine retroviruses, experiments have demonstrated that the host range of a MuLV vector could be extended by coexpressing either the VSV-G protein or Gibbon ape leukemia virus envelope protein with the Gag-Pol proteins of MuLV (Emi et al., 1991; Miller et al., 1991). Similar results have been obtained with HIV, wherein biologically active glycoproteins of MuLV, HTLV-1, and SIV were efficiently incorporated into virus (Spector et al., 1990; Landau et al., 1991). Moreover, in RSV, the insertion of the membrane-spanning and cytoplasmic domains of the RSV TM into a chimeric influenza hemagglutinin protein did not enhance the already efficient incorporation of this foreign protein (Dong et al., 1992b). The incorporation of foreign proteins is not limited to viral glycoproteins; Young et al. (1990) demonstrated that the human lymphocyte membrane protein CD4 could be incorporated efficiently into RSV virions when it was expressed in avian cells producing virus, and a biochemical analysis of HIV virions indicated that certain HL-A species and other cell membrane proteins were present at high levels in virus (Arthur et al., 1992).

The results summarized here show that a variety of nonretroviral glycoproteins can be assembled efficiently and in a biologically active form into retrovirus particles and that for some viruses removal of the domain most likely to interact with capsids has no negative effect on glycoprotein incorporation. It is difficult to reconcile these results with a specific requirement for strong capsid–glycoprotein interactions. How, then, might viral proteins be preferentially incorporated into virions? Given the intimate MA–membrane association, it is possible that, as virus assembly occurs, the associated areas of the plasma membrane are cleared of cellular glycoproteins that have connections with cytoplasmic counterparts (e.g., cytoskeletal elements or associated kinases). This active exclusion of cellular proteins might then allow mobile viral (and cellular) membrane proteins to freely diffuse into this region and be passively incorporated into virions (Fig. 14.7). Such an active exclusion/passive inclusion mechanism would not preclude interactions between the capsid and glycoprotein, but neither would it require this interaction to take place for the glycoprotein complex to be incorporated into virus.

Indeed, it may be possible to reconcile the apparent contradiction of a requirement for specific capsid–glycoprotein interactions during assembly of HIV and the ''passive incorporation'' of foreign glycoproteins. Peptide se-

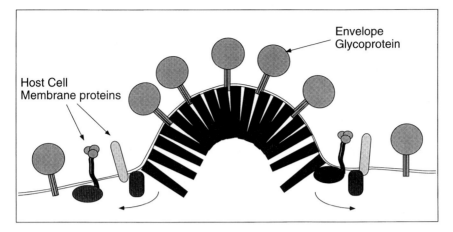

FIGURE 14.7 Possible mechanism for the active exclusion/passive inclusion of glycoproteins into virions. The association of the amino-terminal MA domain of the Gag precursor and the plasma membrane pushes cellular membrane proteins that are associated with cytoplasmic structures away from the budding particle. Viral glycoproteins with no such cellular counterparts can then diffuse into the assembly site and be incorporated into the virus. This model does not exclude specific interactions between the Gag precursor and the viral glycoproteins that might retain those glycoproteins in the assembly site or compete for the binding of cellular proteins involved in endocytosis.

quences similar to the internalization signals of cell membrane receptor molecules (Collawn et al., 1990; Matter et al., 1993) have been found in the cytoplasmic domains of TM proteins from all retroviruses analyzed to date (E. Hunter and W. Gallaher, unpublished results; J. Hoxie, personal communication). Such sequences might play a role in reducing the level of surface expression and the host's immune response to retroviral glycoproteins. Recognition of these sequences in the TM protein by components of the cell's endocytic machinery might effectively result in the exclusion of the viral glycoprotein from the assembling virion in the same manner as that proposed for host cell proteins (Fig. 14.7). For those viruses in which this process is efficient, a competing interaction with the capsid precursor proteins might be required to prevent internalization and to allow incorporation of viral glycoproteins into virions. Thus, mutations that interfere with the capsid–viral glycoprotein interaction (Yu et al., 1992; Dorfman et al., 1994; Lodge et al., 1994) would be expected to reduce the efficiency with which glycoprotein is incorporated into virus. Conversely, mutations that remove internalization signals from the TM cytoplasmic domain could result in higher levels of virion-associated glycoprotein, as is seen in SIV (Johnston et al., 1993; Zingler and Littman, 1993). In this extended model, therefore, the efficiency of glycoprotein incorporation would reflect an equilibrium between the competing interactions of viral glycoprotein endocytosis and viral glycoprotein–capsid association.

Conclusions

The importance of HIV as a human pathogen continues to drive the quest to define at the molecular level the structure of retroviral proteins and the molecular mechanisms by which new infectious viruses are assembled. Such information, it is hoped, will provide new insights into approaches and therapeutic compounds that can be used to interfere with this critical step in the virus life cycle. High-resolution NMR has already provided three-dimensional structural information on some of the smaller structural proteins of HIV, and crystallographic studies of both the viral proteinase and the reverse transcriptase have yielded structures for these viral enzymes at resolutions suitable for rational drug design (see Chapter 17). Similar approaches are being followed for the viral structural proteins, and it seems likely that within the next few years information on the structure of the mature capsid proteins will be available. For an understanding of the protein–protein interactions involved in the assembly process itself, however, it will be necessary to have structural information on the capsid precursor proteins and the immature capsids themselves. The recent description of *in vitro* assembly of retroviral capsids (Klikova et al., 1995), as well as high-level production of such structures in recombinant expression systems, should allow both cryo–electron microscopic and virus crystallographic studies to be initiated. Three-dimensional structural information on retroviral glycoproteins may prove to be more difficult to obtain because extensive glycosylation of these molecules leads to heterogeneity and high solubility. Nevertheless, a variety of approaches are being taken to address these problems, such as reducing the number of carbohydrate addition sites and potential loops on the surface of the molecule that are not required for at least the initial interactions with the viral receptor. Clearly much remains to be learned about the structure and assembly of retroviruses, but the rewards from such knowledge promise to be great.

Acknowledgments

We thank Drs. D. Hockley and M. Nermut for providing electron micrographs of HIV; Drs. R. Craven and J. Wills for those of RSV; and Drs. E. Kubalek and M. Yeager for those of MuLV. We also thank Drs. S. Matthews and P. Roques for the NMR structures of the HIV MA and NC proteins (respectively), and Drs. V. Vogt, M. Sakalian, M. Nermut, and S. Weldon for their helpful comments.

The work of the authors is supported by research grants from the National Cancer Institute and the National Institute of Allergy and Infectious Diseases.

References

Arthur, L., Bess, J. Jr., Sowder, R. II, Benveniste, R., Mann, D., Chermann, J. and Henderson, L. 1992. Cellular proteins bound to immunodeficiency viruses: implications for pathogenesis and vaccines. Science 258: 1935–1938.

Bender, M.A., Palmer, T.D., Gelinas, R.E. and Miller, A.D. 1987. Evidence that the packaging signal of Moloney murine leukemia virus extends into the gag region. J. Virol. 61: 1639–1646.

Bennett, R.P., Nelle, T.D. and Wills, J.W. 1993. Functional chimeras of the Rous sarcoma virus and human immunodeficiency virus gag proteins. J. Virol. 67: 6487–6498.

Berkowitz, R.D., Luban, J. and Goff, S.P. 1993. Specific binding of human immunodeficiency virus type 1 gag polyprotein and nucleocapsid protein to viral RNAs detected by RNA mobility shift assays. J. Virol. 67: 7190–7200.

Bess J.W. Jr., Powell, P.J., Issaq, H.J., Schumack, L.J., Grimes, M.K., Henderson, L.E. and Arthur, L.O. 1992. Tightly bound zinc in human immunodeficiency virus type 1, human T-cell leukemia virus type I, and other retroviruses. J. Virol. 66: 840–847.

Bieth, E., Gabus, C. and Darlix, J.-L. 1990. A study of the dimer formation of Rous sarcoma virus RNA and of its effect on viral protein synthesis in vitro. Nucleic Acids Res. 18: 119–127.

Bowerman, B., Brown, P.O., Bishop, J.M. and Varmus, H.E. 1989. A nucleoprotein complex mediates the integration of retroviral DNA. Genes Dev. 3: 469–478.

Bukrinskaya, A.G., Vorkunova, G.K and Tentsov, Y.Y. 1992. HIV-1 matrix protein p17 resides in cell nuclei in association with genomic RNA. AIDS Res. Hum. Retroviruses 8: 1795–1801.

Callebaut, I., Portetelle, D., Burny, A. and Mornon, J.P. 1994. Identification of functional sites on bovine leukemia virus envelope glycoproteins using structural and immunological data. Eur. J. Biochem. 222: 405–414.

Campbell, S. and Vogt, V.M. 1995. Self assembly in vitro of purified CA-NC proteins from Rous sarcoma virus and human immunodeficiency virus type 1. J. Virol. 69: 6487–6497.

Chance, M.R., Sagi, I., Wirt, M.D., Frisbie, S.M., Scheuring, E., Chen, E., Bess, J.W. Jr., Henderson, L.E. Arthur, L.O., South, T.L., Perez-Alvarado, G. and Summer, M.F. 1992. Extended x-ray absorption fine structure studies of a retrovirus: equine infectious anemia virus cysteine arrays are coordinated to zinc. Proc. Natl. Acad. Sci. U.S.A. 89: 10041–10045.

Collawn, J.F., Stangel, M., Kuhn, L.A., Esekogwu, V., Jing, S.Q., Trowbridge, I.S. and Tainer, J.A. 1990. Transferrin receptor internalization sequence YXRF implicates a tight turn as the structural recognition motif for endocytosis. Cell 63: 1061–1072.

Damsky, C.H., Sheffield, J.B., Tusznski, G.P. and Warren, L. 1977. Is there a role for actin in virus budding. J. Cell. Biol. 75: 593–605.

Davis, N.L. and Rueckert, R.R. 1972. Properties of a ribonucleoprotein particle isolated from Nonidet P-40-treated Rous sarcoma virus. J. Virol. 10: 1010–1020.

De Rocquigny, H., Ficheux, D., Gabus, C., Allain, B., Fournie-Zaluski, M.C., Darlix, J.L. and Roques, B.P. 1993. Two short basic sequences surrounding the zinc finger of nucleocapsid protein NCp10 of Moloney murine leukemia virus are critical for RNA annealing activity. Nucleic Acids Res. 21: 823–829.

De Rocquigny, H., Gabus, C., Vincent, A., Fournie-Zaluski, M.C., Roques, B. and Darlix, J.L. 1992. Viral RNA annealing activities of human immunodeficiency virus type 1 nucleocapsid protein require only peptide domains outside the zinc fingers. Proc. Natl. Acad. Sci. U.S.A. 89: 6472–6476.

Déméné, H., Jullian, N., Morellet, N., de Rocquigny, H., Cornille, F., Maigret, B. and Roques, B.P. 1994. Three-dimensional ^{1}H NMR structure of the nucleocapsid protein NCp10 of Moloney murine leukemia virus. J. Biomol. NMR 4: 153–170.

Dickson, C., Eisenman, R., Fan, H., Hunter, E. and Teich, N. 1984. Protein biosynthesis and assembly. In RNA Tumor Viruses, 2nd ed., Vol. 1 (Weiss, R., Teich, N., Varmus, H. and Coffin, J., Eds.), pp. 513–648. Cold Spring Harbor Laboratory Press, Cold Spring Harbor, NY.

Doms, R.W., Earl, P.L., Chakrabarti, S. and Moss, B. 1990. Human immunodeficiency virus types 1 and 2 and simian immunodeficiency virus env proteins possess a functionally conserved assembly domain. J. Virol. 64: 3537–3540.

Doms, R.W., Earl, P.L. and Moss, B. 1991. The assembly of the HIV-1 env glycoprotein into dimers and tetramers. Adv. Exp. Med. Biol. 300: 203–219.

Dong, J., Dubay, J.W., Perez, L.G. and Hunter, E. 1992a. Mutations within the proteolytic cleavage site of the Rous sarcoma virus glycoprotein define a requirement for dibasic residues for intracellular cleavage. J. Virol. 66: 865–874.

Dong, J., Roth, M.G. and Hunter, E. 1992b. A chimeric avain retrovirus containing the influenza virus hemagglutinin gene has an expanded host range. J. Virol. 66: 7374–7382.

Dorfman, T., Mammano, F., Haseltine, W.A. and Göttlinger, H.G. 1994. Role of the matrix protein in the virion association of the human immunodeficiency virus type 1 envelope glycoprotein. J. Virol. 68: 1689–1696.

Dubay, J.W., Roberts, S.J., Hahn, B.H and Hunter, E. 1992. Truncation of the human immunodeficiency virus type 1 transmembrane glycoprotein cytoplasmic domain blocks virus infectivity. J. Virol. 66: 6616–6625.

Dupraz, P. and Spahr, P.-F. 1992. Specificity of Rous sarcoma virus nucleocapsid protein in genomic RNA packaging. J. Virol. 66: 4662–4670.

Earl, P.L., Moss, B. and Doms, R.W. 1991. Folding, interaction with GRP78-BiP, assembly and transport of the human immunodeficiency virus type 1 envelope protein. J. Virol. 65: 2047–2055.

Edbauer, C.A. and Naso, R.B. 1984. Cytoskeleton-associated Pr65^{gag} and assembly of retrovirus temperature-sensitive mutants in chronically infected cells. Virology 134: 389–397.

Ehrlich, L.S., Agresta, B.E. and Carter, C.A. 1992. Assembly of recombinant human immunodeficiency virus type 1 capsid protein in vitro. J. Virol. 66: 4874–4883.

Einfeld, D. and Hunter, E. 1988. Oligomeric structure of a prototype retrovirus glycoprotein. Proc. Natl. Acad. Sci. U.S.A. 85: 8688–8692.

Einfeld, D.A. and Hunter, E. 1994. Expression of the TM protein of Rous sarcoma virus in the absence of SU shows that this domain is capable of oligomerization and intracellular transport. J. Virol. 68: 2513–2520.

Eisenman, R.N., Vogt, V.M. and Diggelmann, H. 1974. Synthesis of avian RNA tumor virus structural proteins. Cold Spring Harbor Symp. Quant. Biol. 39: 1067–1075.

Emi, N., Friedmann, T. and Yee, J.-K. 1991. Psuedotype formation of murine leukemia virus with the G protein of vesicular stomatitis virus. J. Virol. 65: 1202–1207.

Facke, M., Janetzko, A., Shoeman, R.L. and Krausslich, H.G. 1993. A large deletion in the matrix domain of the human immunodeficiency virus gag gene redirects virus particle assembly from the plasma membrane to the endoplasmic reticulum. J. Virol. 67: 4972–4980.

Farnet, C.M. and Haseltine, W.A. 1991. Determination of viral proteins present in the human immunodeficiency virus type 1 preintegration complex. J. Virol. 65: 1910–1915.

Fitzgerald, D.W. and Coleman, J.E. 1991. Physicochemical properties of cloned nucleocapsid protein from HIV: interactions with metal ions. Biochemistry 30: 5195–5201.

Freed, E.O., Orenstein, J.M., Buckler, W.A. and Martin, M.A. 1994. Single amino acid changes in the human immunodeficiency virus type 1 matrix protein block virus particle production. J. Virol. 68: 5311–5320.

Gallaher, W.R., Ball, J.M., Garry, R.F., Griffin, M.C. and Montelaro, R.C. 1989. A general model for the transmembrane proteins of HIV and other retroviruses. AIDS Res. Hum. Retroviruses 5: 431–440.

Gallay, P., Swingler, S., Aiken, C. and Trono, D. 1995. HIV-1 infection of nondividing cells: C-terminal tyrosine phosphorylation of the viral matrix protein is a key regulator. Cell 80: 379–388.

Gebhardt, A., Bosch, J.V., Zremiccki, A. and Fris, R.R. 1984. Rous sarcoma virus p19 and gp35 can be chemically crosslinked to high molecular weight complexes: an insight into virus assembly. J. Mol. Biol. 174: 297–317.

Gelderblom, H. 1990. Morphogenesis, maturation, and fine structure of lentiviruses. In Retroviral Proteases: Control of Maturation and Morphogenesis (Pearl, L., Ed.), pp. 159–180. Stockton Press, New York.

Gitti, R.K., Lee, B.M., Walker, J., Summers, M.F., Yoo, S. and Sundquist, W.J. 1996. Structure of the amino-terminal core domain of the HIV-1 capsid protein. Science 273: 231–235.

Gonzalez, S.A., Affranchino, J.L., Gelderblom, H.R. and Burny, A. 1993. Assembly of the matrix protein of simian immunodeficiency virus into virus-like particles. Virology 194: 548–556.

Göttlinger, H.G., Dorfman, T., Sodroski, J.G. and Haseltine, W.A. 1991. Effect of mutations affecting the p6 *gag* protein on human immunodeficiency virus particle release. Proc. Natl. Acad. Sci. U.S.A. 88: 3195–3199.

Göttlinger, H.G., Sodroski, J.G. and Haseltine, W.A. 1989. Role of capsid precursor processing and myristoylation in morphogenesis and infectivity of human immunodeficiency virus type 1. Proc. Natl. Acad. Sci. U.S.A. 86: 5781–5785.

Hallenberger, S., Bosch, V., Angliker, H., Shaw, E., Klenk, H.-D. and Garten, W. 1992. Inhibition of furin-mediated cleavage activation of HIV-1 glycoprotein gp160. Nature 360: 358–361.

Hansen, M., Jelinek, L., Whiting, S. and Barklis, E. 1990. Transport and assembly of *gag* proteins into Moloney murine leukemia virus. J. Virol. 64: 5306–5316.

Henderson, L.E., Krutzsch, H.C. and Oroszlan, S. 1983. Myristyl amino-terminal acylation of murine retrovirus proteins: an unusual post-translational protein modification. Proc. Natl. Acad. Sci. U.S.A. 80: 339–343.

Henderson, L.E., Sowder, R.C., Smythers, G.W. and Oroszlan, S. 1987. Chemical and immunological characterizations of equine infectious anemia virus *gag*-encoded proteins. J. Virol. 61: 1116–1124.

Hosaka, M., Nagahama, M., Kim, W.S., Watanabe, T., Hatsuzawa, K., Ikemizu, J., Murakami, K. and Nakayama, K. 1991. Arg-X-Lys/Arg-Arg motif as a signal for precursor cleavage catalyzed by furin within the constitutive secretory pathway. J. Biol. Chem. 266: 12127–12130.

Hunter, E. 1994. Macromolecular interactions in the assembly of HIV and other retroviruses. Semin. Virol. 5: 71–83.

Hunter, E. and Swanstrom, R. 1990. Retrovirus envelope glycoproteins. Curr. Top. Microbiol. Immunol. 157: 187–254.

Jacks, T. 1990. Translational suppression in gene expression in retroviruses and retrotransposons. Curr. Top. Microbiol. Immunol. 157: 93–124.

Johnston, P., Dubay, J. and Hunter, E. 1993. Truncations of SIV transmembrane protein confer expanded virus host range by removing a block to virus entry into cells. J. Virol. 67: 3077–3086.

Jones, T.A., Blang, G., Hansen, M. and Barklis, E. 1990. Assembly of *gag*-beta-galactosidase proteins into retrovirus paricles. J. Virol. 64: 2229–2265.

Jowett, J.B.M., Hockley, D.J., Nermut, M.V. and Jones, I.M. 1992. Distinct signals in human immunodeficiency virus type 1 Pr55 necessary for RNA binding and particle formation. J. Gen. Virol. 73: 3079–3086.

Kamps, C.A., Lin, Y.C. and Wong, P.K. 1991. Oligomerization and transport of the envelope protein of Moloney murine leukemia virus-TB and of ts1, a neurovirulent temperature-sensitive mutant of MoMuLV-TB. Virology 184: 687–694.

Kaplan, A.H., Zack, J.A., Knigge, M., Paul, D.A., Kempf, D.J., Norbeck, D.W. and Swanstrom, R. 1993. Partial inhibition of the human immunodeficiency virus type 1 protease results in aberrant virus assembly and the formation of noninfectious particles. J. Virol. 67: 4050–4055.

Katz, R.A., Terry, R.W. and Skalka, A.M. 1986. A conserved cis-acting sequence in the 5' leader of avian sarcoma virus RNA is required for packaging. J. Virol. 59: 163–167.

Kawai, S. and Hanafusa, H. 1973. Isolation of defective mutant of avian sarcoma virus. Proc. Natl. Acad. Sci. U.S.A. 70: 3493–3497.

Klikova, M., Rhee, S.S., Hunter, E. and Ruml, T. 1995. Efficient in vivo and in vitro assembly of retroviral capsids from Gag precursor proteins expressed in bacteria. J. Virol. 69: 1093–1098.

Klimkait, T., Strebel, K., Hoggan, M.D., Martin, M.A. and Orenstein, J.M. 1990. The human immunodeficiency virus type-1-specific protein vpu is required for efficient virus maturation and release. J. Virol. 64: 621–629.

Krakower, J.M., Barbacid, M. and Aaronson, S.A. 1977. Radioimmunoassay for mammalian type C viral reverse transcriptase. J. Virol. 22: 331–339.

Landau, N.R., Page, K.A. and Littman, D.R. 1991. Pseudotyping with human T-cell leukemia virus type I broadens the human immunodeficiency virus host range. J. Virol. 65: 162–169.

Linial, M.L. and Miller, A.D. 1990. Retroviral RNA packaging: sequence requirements and implications. Curr. Top. Microbiol. Immunol. 157: 125–152.

Lodge, R., Göttlinger, H. Gabuzda, D., Cohen, E.A. and Lemay, G. 1994. The intracytoplasmic domain of gp41 mediates polarized budding of human immunodeficiency virus type 1 in MDCK cells. J. Virol. 68: 4857–4861.

Luban, J., Bossolt, K.L., Franke, E.K., Kalpana, G.V. and Goff, S.P. 1993. Human immunodeficiency virus type 1 Gag protein binds to cyclophilins A and B. Cell 73: 1067–1078.

Luban, J. and Goff, S.P. 1991. Binding of human immunodeficiency virus type 1 (HIV-1) RNA to recombinant HIV-1 *gag* polyprotein. J. Virol. 65: 3203–3212.

Maldarelli, F., King, N.J. and Yagi, M.J. 1987. Effects of cytoskeletal disrupting agents on mouse mammary tumor virus replication. Virus Res. 7: 281–295.

Mammano, F., Ohagen, A., Hoglund, S. and Göttlinger, H.G. 1994. Role of the major homology region of human immunodeficiency virus type 1 in virion morphogenesis. J. Virol. 68: 4927–4936.

Massiah, M.A., Starich, M.R., Paschall, C., Summers, M.F., Christensen, A.M. and Sundquist, W.I. 1994. Three-dimensional structure of the human immunodeficiency virus type 1 matrix protein. J. Mol. Biol. 244: 198–223.

Matter, K., Whitney, J.A., Yamamoto, E.M. and Mellman, I. 1993. Common signals control low density lipoprotein receptor sorting in endosomes and the Golgi complex of MDCK cells. Cell 74: 1053–1064.

Matthews, S., Barlow, P., Boyd, J., Barton, G., Russell, R., Mills, H., Cunningham, M., Meyers, N., Burns, N., Kingsman, S., Kingsman, A. and Cambell, I. 1994. Structural similarity between the p17 matrix protein of HIV-1 and interferon-γ. Nature 370: 666–668.

McCune, J.M., Rabin, L.B., Feinberg, M.B., Lieberman, M., Kosek, J.C., Reyes, G.R. and Weissman, I.L. 1988. Endoproteolytic cleavage of gp160 is required for the activation of human immunodeficiency virus. Cell 53: 55–67.

Meric, C. and Spahr, P.-F. 1986. Rous sarcoma virus nucleic acid-binding protein p12 is neccessary for viral 70S RNA dimer formation and packaging. J. Virol. 60: 450–459.

Miller, A.D., Garcia, J.V., von Suhr, N., Lynch, C.M., Wilson, C. and Eiden, M.V. 1991. Construction and properties of retrovirus packaging cells based on gibbon ape leukemia virus. J. Virol. 65: 2220–2224.

Momany, C., Kovari, L.C., Prongay, A.J., Keller, W., Gitti, R.K., Lee, B.M., Gorbalenva, A.E., Tong, L., McClure, J., Ehrlich, L.S., Summers, M.F., Carter, C. and Rossman, M.G. 1996. Crystal structure of dimeric HIV-1 capsid protein. Nature Struct. Biol. 3: 763–770.

Morellet, N., de Rocquigny, H., Mely, Y., Jullian, N., Demene, H., Ottmann, M., Gerard, D., Darlix, J.L., Fournie-Zaluski, M.C. and Roques, B.P. 1994. Conformational behaviour of the active and inactive forms of the nucleocapsid NCp7 of HIV-1 studied by ^{1}H NMR. J. Mol. Biol. 235: 287–301.

Morellet, N., Jullian, N., De Rocquigny, H., Maigret, B., Darlix, J.L. and Roques, B.P. 1992. Determination of the structure of the nucleocapsid protein NCp7 from the human immunodeficiency virus type 1 by ^{1}H NMR. EMBO J. 11: 3059–3065.

Nermut, M.V., Frank, H. and Schäfer, W. 1972. Properties of mouse leukemia virus. III. Electron microscopic appearance as revealed after conventional preparative techniques as well as freeze-drying and freeze etching. Virology 49: 345–358.

Nermut, M.V., Grief, C., Hashmi, S. and Hockley, D.J. 1993. Further evidence of icosahedral symmetry in human and simian immunodeficiency virus. AIDS Res. Hum. Retroviruses 9: 929–38.

Nermut, M.V., Hockley, D.J., Jowett, J.B., Jones, I.M., Garreau, M. and Thomas, D. 1994. Fullerene-like organization of HIV gag-protein shell in virus-like particles produced by recombinant baculovirus. Virology 198: 288–296.

Oertle, S. and Spahr, P.-F. 1990. Role of the gag polyprotein in packaging and maturation of Rous sarcoma virus genomic RNA. J. Virol. 64: 5757–5763.

Oroszlan, S. and Luftig, R.B. 1990. Retroviral proteinases. Curr. Top. Microbiol. Immunol. 157: 153–185.

Owens, R.J., Dubay, J.W., Hunter, E. and Compans, R.W. 1991. Human immunodeficiency virus envelope protein determines the site of virus release in polarized epithelial cells. Proc. Natl. Acad. Sci. U.S.A. 88: 3987–3991.

Palmiter, R.D., Gagnon, J., Vogt, V.M., Ripley, S. and Eisenman, R.N. 1978. The NH_2-terminal sequence of the avian oncovirus gag precursor polyprotein (Pr76^{gag}). Virology 91: 423–433.

Panet, A., Baltimore, D. and Hanafusa, T. 1975. Quantitation of avian RNA tumor virus reverse transcriptase by radioimmunoassay. J. Virol. 16: 146–152.

Pepinsky, R.B. and Vogt, V.M. 1984. Fine-structure analyses of lipid-protein and protein-protein interactions of gag Protein p19 of the avian sarcoma and leukemia viruses by cyanogen bromide mapping. J. Virol. 52: 145–153.

Perez, L.G., Davis, G.L. and Hunter, E. 1987. Mutants of the Rous sarcoma virus envelope glycoprotein that lack the transmembrane anchor and cytoplasmic domains: analysis of intracellular transport and assembly into virions. J. Virol. 61: 2981–2988.

Prats, A.C., Sarih, L., Gabus, C., Litvak, S., Keith, G. and Darlix, J.L. 1988. Small finger protein of avian and murine retroviruses has nucleic acid annealing activity and positions the replication primer tRNA onto genomic RNA. EMBO J. 7: 1777–1783.

Pugatsch, T. and Stacey, D.W. 1983. Identification of a sequence likely to be required for avian retroviral packaging. Virology 128: 505–511.

Purchio, A.F., Jovanovich, S. and Erikson, R.L. 1980. Sites of synthesis of viral proteins in avian sarcoma virus-infected chicken cells. J. Virol. 35: 629–636.

Ragheb, J.A. and Anderson, W.F. 1994. Uncoupled expression of Moloney murine leukemia virus envelope polypeptides SU and TM: a functional analysis of the role of TM domains in viral entry. J. Virol. 68: 3207–3219.

Rhee, S.S. and Hunter, E. 1987. Myristylation is required for intracellular transport but not for assembly of D-type retrovirus capsids. J. Virol. 61: 1045–1053.

Rhee, S.S. and Hunter, E. 1990. A single amino acid substitution within the matrix protein of a type D retrovirus converts its morphogenesis to that of a type C retrovirus. Cell 63: 77–86.

Rhee, S.S. and Hunter, E. 1991. Amino acid substitutions within the matrix protein of type D retroviruses affect assembly, transport, and membrane-association of a capsid. EMBO J. 10: 535–546.

Sakalian, M., Wills, J.W. and Vogt, V.M. 1994. Efficiency and selectivity of RNA packaging by Rous sarcoma virus Gag deletion mutants. J. Virol. 68: 5969–5981.

Salzwedel, K., Johnston, P.B., Roberts, S.J., Dubay, J.W. and Hunter, E. 1993. Expression and characterization of glycophospholipid-anchored human immunodeficiency virus type 1 envelope glycoproteins. J. Virol. 67: 5279–5288.

Satake, M. and Luftig, R.B. 1982. Microtubule-depolymerizing agents inhibit Moloney murine leukaemia virus production. J. Gen. Virol. 58: 239–249.

Schlesinger, S., Makino, S. and Linial, M.L. 1994. *Cis*-acting genomic elements and *trans*-acting proteins involved in the assembly of RNA viruses. Semin. Virol. 5: 39–49.

Schultz, A.M. and Oroszlan, S. 1983. In vivo modification of retroviral *gag* gene-encoded polyproteins by myristic acid. J. Virol. 46: 355–361.

Schultz, A.M., Rabin, E.H. and Oroszlan, S. 1979. Post-translational modification of Rauscher leukemia virus precursor polyproteins encoded by the *gag* gene. J. Virol. 30: 255–266.

Sharova, N. and Bukrinskaya, A. 1991. p17 and p17-containing gag precursor of input human immunodefiency virus are transported into the nuclei of infected cells. AIDS Res. Hum. Retroviruses 7: 303–306.

Skalka, A.M. 1989. Retroviral proteases: first glimpses at the anatomy of a processing machine. Cell 56: 911–913.

Sommerfelt, M.A., Rhee, S.S. and Hunter, E. 1992. Importance of the p12 protein in Mason-Pfizer monkey virus assembly and infectivity. J. Virol. 66: 7005–7011.

Sorge, J.D., Ricci, W. and Hughes, S.H. 1983. *Cis*-acting RNA packaging locus in the 115-nucleotide direct repeat of Rous sarcoma virus. J. Virol. 48: 667–675.

South, T.L., Blake, P.R., Hare, D.R. and Summers, M.F. 1991. C-terminal retroviral-type zinc finger domain from the HIV-1 nucleocapsid protein is structurally similar to the N-terminal zinc finger domain. Biochemistry 30: 6342–6349.

Spearman, P., Wang, J.J., Vander, H.N. and Ratner, L. 1994. Identification of human immunodeficiency virus type 1 Gag protein domains essential to membrane binding and particle assembly. J. Virol. 68: 3232–3242.

Spector, D.H., Wade, E., Wright, D.A., Koval, V., Clarck, C., Jaquish, D. and Spector, S.A. 1990. Human immunodeficiency virus pseudotypes with expanded cellular and species tropism. J. Virol. 64: 2298–2308.

Stewart, L., Schatz, G. and Vogt, V.M. 1990. Properties of avian retoviral particles defective in viral protease. J. Virol. 64: 5076–5092.

Strambio-de-Castillia, C. and Hunter, E. 1992. Mutational analysis of the major homology region of Mason-Pfizer monkey virus by use of saturation mutagenesis. J. Virol. 66: 7021–7032.

Stromberg, K., Hurley, N.E., Cavis, N.L., Rueckert, R.R. and Fleissner, E. 1974. Structural studies of avian myeloblastosis virus: comparison of polypeptides in virion and core component by dodecyl sulfate-polyacrylamide gel electrophoresis. J. Virol. 13: 513–528.

Summers, M.F., Henderson, L.E., Chance, M.R., Bess, J.W. Jr., South, T.L., Blake, P.R., Sagi, I., Perez-Alvarado, G., Sowder, R.C. III, Hare, D.R., Arthur L.O. 1992. Nucleocapsid zinc fingers detected in retroviruses: EXAFS studies of intact viruses and the solution-state structure of the nucleocapsid protein from HIV-1. Protein Sci. 1: 563–574.

Teich, N. 1984. Taxonomy of retroviruses. In RNA Tumor Viruses, 2nd ed., Vol. 1 (Weiss, R., Teich, N., Varmus, H. and Coffin, J., Eds.), pp. 25–207. Cold Spring Harbor Laboratory Press, Cold Spring Harbor, NY.

Terwilliger, E.F., Cohen, E.A., Lu, Y., Sodroski, J.G. and Haseltine, W.A. 1989. Functional role of human immunodeficiency virus type 1 *vpu*. Proc. Natl. Acad. Sci. U.S.A. 86: 5163–5167.

Thomas, D.J., Wall, J.S., Hainfeld, J.F., Kaczorek, M., Booy, F.P., Trus, B.L., Eiserling, F.A. and Steven, A.C. 1991. gp160, the envelope glycoprotein of human immunodeficiency virus type 1, is a dimer of 125-kilodalton subunits stabilized through interactions between their gp41 domains. J. Virol. 65: 3797–3803.

Varmus, H.E. and Swanstrom, R. 1984. Replication of retroviruses. In RNA Tumor Viruses, 2nd ed., Vol. 1 (Weiss, R., Teich, N., Varmus, H. and Coffin, J., Eds.), pp. 369–512. Cold Spring Harbor Laboratory Press, Cold Spring Harbor, NY.

Weldon, R.A. Jr. and Hunter E. 1997. Molecular requirements for retrovirus assembly. In: Structural Biology of Viruses. (Chiu, W., Burnett, R.M. and Garcea, R.L., Eds.), pp 381–410, Oxford University Press, New York.

Weldon, R.A. Jr. and Wills, J.W. 1993. Characterization of a small (25-kilodalton) derivative of the Rous sarcoma virus Gag protein competent for particle release. J. Virol. 67: 5550–5561.

Willey, R., Maldarelli, F., Martin, M. and Strebel, K. 1992a. Human immunodeficiency virus type 1 Vpu protein induces rapid degradation of CD4. J. Virol. 66: 7193–7200.

Willey, R., Maldarelli, F., Martin, M. and Strebel, K. 1992b. Human immunodeficiency virus type 1 Vpu protein regulates the formation of intracellular gp160-CD4 complexes. J. Virol. 66: 226–234.

Wills, J.W., Cameron, C.E., Wilson, C.B., Xiang, Y., Bennett, R.P. and Leis, J. 1994. An assembly domain of the Rous sarcoma virus Gag protein required late in budding. J. Virol. 68: 6605–6618.

Wills, J.W. and Craven, R.C. 1991. Form, function, and use of retroviral Gag proteins. AIDS 5: 639–654.

Wills, J.W., Craven, R.C., Weldon, R.A. Jr., Nelle, T.D. and Erdie, C.R. 1991. Suppression of retroviral MA deletions by the amino-terminal membrane-binding domain of p60^{src}. J. Virol. 65: 3804–3812.

Young, J.A., Bates, P., Willert, K. and Varmus, H.E. 1990. Efficient incorporation of human CD4 protein into avian leukosis virus particles. Science 250: 1421–1423.

Yu, X., Yuan, X., Matsuda, Z., Lee, T.-H. and Essex, M. 1992. The matrix protein of human immunodeficiency virus type 1 is required for incorporation of viral envelope protein into mature virus. J. Virol. 66: 4966–4971.

Yu, X., Yuan, X., McLane, M.F., Lee, T.H. and Essex, M. 1993. Mutations in the cytoplasmic domain of human immunodeficiency virus type 1 transmembrane protein impair the incorporation of Env proteins into mature virions. J. Virol. 67: 213–221.

Yuan, X., Yu, X., Lee, T.H. and Essex, M. 1993. Mutations in the N-terminal region of human immunodeficiency virus type 1 matrix protein block intracellular transport of the Gag precursor. J. Virol. 67: 6387–6394.

Zhou, W., Parent, L.J., Wills, J.W. and Resh, M.D. 1994. Identification of a membrane-binding domain within the amino-terminal region of human immunodeficiency virus type 1 Gag protein which interacts with acidic phospholipids. J. Virol. 68: 2556–2569.

Zingler, K. and Littman, D.R. 1993. Truncation of the cytoplasmic domain of the simian immunodeficiency virus envelope glycoprotein increases *env* incorporation into particles and fusogenicity and infectivity. J. Virol. 67: 2824–2831.

15.

Structure-Guided Therapeutic Strategies

GILLIAN M. AIR & MING LUO

The basic concepts of structure-based rational drug design were described 100 years ago. In 1894, Emil Fischer reported that glycolytic enzymes distinguish between stereoisomers of sugars and explained the difference as a requirement for complementary shapes—the often-quoted ''lock-and-key'' metaphor. Paul Ehrlich (1897) in his paper on the quantitation of activity in diphtheria antiserum, described how the specificity and affinity of the antigen–antibody interaction could be explained by the laws of structural chemistry (Silverstein, 1989). Ehrlich later turned his attention from immunology to pharmacology, articulating in 1909 the concept that chemical treatment of disease relies on eliminating the invading microbe without damage to the host (Cohen, 1979).

From these great insights, developed in the absence of any knowledge of macromolecular structure, comes the straightforward idea that it should be possible to design molecules that interact with other molecules. The first high-resolution structure of a protein to be determined was that of myoglobin (Kendrew et al., 1960), which was accompanied by the 5.5-Å-resolution structure of hemoglobin (Perutz et al., 1960). The first successful examples of a priori design of small molecules that could interact with a known protein were hemoglobin ligands, designed to fit in the diphosphoglycerate binding site (Beddell et al., 1976). The first designed drug to reach the market was captopril (Cushman et al., 1977). This inhibitor of angiotensin-converting enzyme was developed using knowledge of a related structure, carboxypeptidase, from which the active site of the target enzyme could be modeled. More complicated designs have been attempted recently for drugs targeted to thymidylate synthase

411

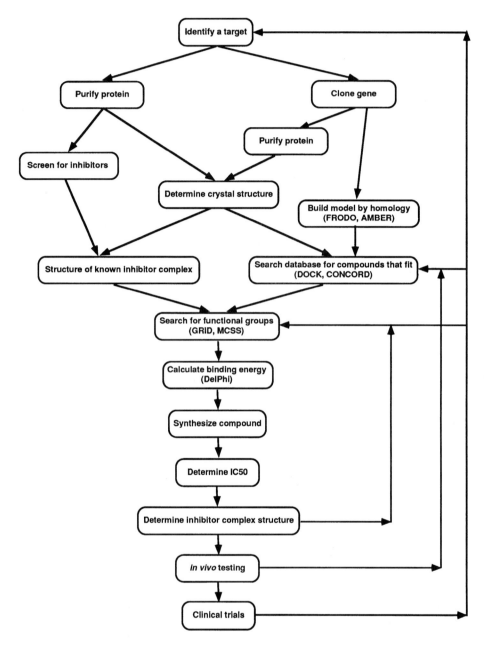

FIGURE 15.1 An example of an iterative drug design cycle. Note that various pathways can be used to arrive at a starting model of candidate inhibitor bound to the protein target. The selection of iterative cycles is determined by experimental results. Computer programs FRODO, AMBER, DOCK, CONCORD, GRID, MCSS, and DelPhi are described in the text. IC50 is the concentration of drug that gives 50% inhibition of activity.

as anticancer therapy (Appelt et al., 1991) and to purine nucleoside phosphorylase as therapy for psoriasis and other immune disorders (Montgomery et al., 1993) . Some of the designed inhibitors have reached the stage of clinical trials. The entire process is now developed as an iterative cycle, as illustrated in Figure 15.1.

The database of known protein structures has expanded exponentially over the past few years, but there are still relatively few atomic structures of human viruses or viral proteins, and in 1994 there are no antiviral drugs licensed in the United States that were designed from the structure of the target molecule. Several, however, are now undergoing clinical testing.

The Need for Antiviral Drugs

The most important protective response in mammals against invading foreign organisms is immunologic. Many viruses have only one opportunity to replicate in a given host, because a single infection of a young child confers lifelong immunity. Although most vaccines are not as effective as a natural infection in stimulating the various branches of the immune system, vaccination has been successful in eliminating smallpox from the world and poliovirus from the Americas. Other viruses have proved more difficult to control by vaccination. The first reason is that many vaccines have limited stability and must be stored frozen or refrigerated and used within a certain time period. This lability of vaccine components is why diseases such as measles are well controlled in wealthier parts of the world but remain at high incidence in the less developed regions. The second reason is that several viruses have developed ways to escape from the immune system. Escape by changing the structures of antigens was one of the first mechanisms to be described, for influenza, in the 1940s. Since then, it has been recognized that many viruses exist in a multitude of serotypes, and some of these, such as influenza (see Chapter 3), are constantly changing. Human immunodeficiency virus (HIV) is one example of these extraordinarily variable viruses (see Chapter 14). HIV has another effective defense; in addition to escaping from existing antibodies by genetic variation, HIV infects those T cells that are necessary to make antibodies of new specificity. Thus the acquired immunodeficiency syndrome (AIDS) patient has only a limited capacity to make antibodies, either against the viral infection or in response to vaccination.

There is obviously a limit to how much variation can be tolerated by a virus. The integrity of the outer surface that is the target of neutralizing antibodies must be maintained, and mutation to escape antibodies must preserve the functions of this surface (see Chapters 4, 5, 6, and 7). Thus an ideal vaccine would be directed against essential regions where any mutation would result in loss of viability. Unfortunately, the crystal structures show that these essential regions are usually in depressions on the surface of the virus or molecule, and these ''canyons'' are inaccessible to antibodies (Rossmann and Rueckert, 1987). They are, however, accessible to smaller molecules, and this is the reason much effort is being directed toward the development of antiviral drugs that bind to viral components and inhibit viral functions (see Chapters 16 and 17).

Identification of Targets

Viruses differ greatly in size and complexity. The smallest consist only of genetic material inside a protein capsid, but even these offer many virus-specific targets for intervention by antivirals. In no case does a viral genome code only for the coat protein. All animal viruses code for several proteins involved in replication, often a DNA or RNA polymerase itself, alternatively or together with ancillary proteins required for replication. Considerable advances have been made in understanding virus genetics as a result of the biotechnology explosion. Cloning and polymerase chain reaction techniques have allowed viral genomes to be sequenced even when the virus itself has not been characterized (Spiropoulou et al., 1994). However, the functions of many virus-coded proteins are not understood.

The virus life cycle begins with recognition of receptors on the surface of the host cell. The virus coat contains one or more proteins that recognize and bind to the receptors. It is becoming clear that many viruses use more than one type of molecule to gain entry into the cell, and the concept of a single receptor species that allows infection to be completed is an oversimplification of the events involved (Haywood, 1994). Identification of viral receptors is often difficult for this reason, and also because the interaction is typically weak, requiring multivalent attachment to stabilize the binding. The first target for antiviral design could be to block attachment by using a ligand that binds either to the receptor binding site on the virus, or to the virus binding receptor on the cell. Because receptors have not evolved to bind virus but have a function in the normal cell, the most promising target is the site on the virus. So far there are no therapies in use that block the binding, but many efforts are underway to find ways to block the CD4 binding site on HIV, the intercellular adhesion molecule-1 binding site on the major group of human rhinoviruses (see Chapter 4), or the sialic acid binding site on influenza virus (see Chapter 3).

The entry processes of human viruses are not well characterized. Many viruses enter the endosomal pathway; others enter directly through the plasma membrane. Viruses that possess a lipid envelope have at least one structural protein that brings about fusion of viral membrane with host cell membrane. The mechanisms are not understood, but in several cases a hydrophobic fusion peptide has been identified, and mimics of this peptide have been shown to reduce infectivity (Richardson et al., 1980). In the case of influenza, the low pH of the endosome causes a massive conformational change in the receptor binding protein (hemagglutinin) that exposes the fusion peptide (see Chapter 3). However, the resulting release of internal virus components into the cytoplasm is not enough for viral replication to occur. The nucleocapsids must also be exposed to low pH to be freed from the structural matrix protein, and this is mediated by a virus-coded ion channel protein (M2) that is present in the virion at low levels. The only currently licensed anti-influenza drugs (amantadine and rimantadine) act to block this ion channel, but resistant strains frequently develop.

Entry of nonenveloped viruses is even less well characterized. Poliovirus capsids undergo a conformational change after attaching to the receptor, a necessary destabilizing step for a virus that must remain resistant to stomach acid, intestinal detergents, and proteases yet release its genomic material into the cell for replication. Several picornaviruses are sensitive to the ''WIN'' (Sterling Winthrop) and Janssen inhibitors (see Chapter 16). These compounds bind in a pocket deep in the capsid and prevent the conformational change that leads to uncoating; thus they are effective inhibitors of the viruses.

Once the viral genome is released into the cell, it begins a replication cycle, usually being first transcribed and then translated into the viral-specific polymerases. Viral polymerases are attractive targets for inhibitor design, and most of the currently marketed antivirals are nucleoside analogs that have been empirically modified to have higher specificity for the viral enzyme than for the equivalent protein in the host replication system. In some cases the drug is inactive until processed by a virus-specific enzyme; thus, although the active form is not selective for the viral polymerase, it will only operate in virus-infected cells. Design of polymerase inhibitors, however, is in its infancy. There is only one X-ray crystal structure of an animal virus polymerase, that of HIV reverse transcriptase. Although the structure is known (Kohlstaedt et al., 1992), currently only α-carbon coordinates have been released, so inhibitor design cannot yet be widely performed. There are inhibitors of HIV reverse transcriptase in use and several more are under test, but all were found by random screening and all suffer from rapid development of resistance. A high-resolution structure of HIV would allow design of inhibitors that contact essential amino acids, so that mutations in these residues to escape the inhibitor would lead to self-destruction of the virus by inactivation of the polymerase.

Many viruses code for transcription factors that ensure preferential transcription of the viral mRNAs or allow the maximum use of a limited amount of genetic material. All viruses use at least some of the host translation machinery, but again there may be virus-encoded factors to increase selective translation of viral mRNAs or to shut down host mRNA translation. All of these are potential targets for inhibitors, but so far no structures are known and therefore inhibitor design cannot yet be initiated. Some antivirals identified by random screening methods appear to act at these levels, but in no case is the actual target known or the mechanism understood.

Assembly of new virus particles is an ordered process that must be well regulated to coordinate the packing of genome and required proteins within a sealed protein shell. One method used by several groups of viruses to regulate assembly is to synthesize polyprotein precursors that are cleaved by specific proteases at the appropriate times (see Chapter 14). Some viruses use cellular proteases to cleave membrane-anchored spike proteins to release a fusion peptide ready for the next round of infection. Many viruses, however, encode their own specific proteases. The HIV protease has been crystallized and structures of recombinant and synthetic forms reported by several groups (Lapatto et al., 1989; Navia et al., 1989; Wlodawer et al., 1989; Spinelli et al., 1991). HIV protease is being intensively studied as a target for rational drug design (see Chapter 17). The first compounds were peptide analogs (DesJarlais et al., 1990), but the later and more successful inhibitors resemble a peptide only in their disposition in the active site. One problem with the HIV protease inhibitors has been that the enzyme cleaves at multiple recognition sites, so the active site is rather plastic and mutation to resistance often occurs.

The exit of newly formed virus particles from the cell is even less understood than the entry. Enveloped viruses gather their external glycoproteins in the cell membrane and bud out nucleocapsids with a matrix shell through that membrane. Unlike bacteriophage, animal viruses are not known to code for lysis factors. There are probably potential targets for antivirals—proteins or domains that are required for virus exit or budding—but none has been characterized. There is, however, a further stage that has shown surprising promise. Influenza virus requires its second surface glycoprotein, neuraminidase, to cleave sialic acid receptors from the glycoproteins of the virus and on the cell surface so that the virus can spread (see Chapter 3). In the absence of neuraminidase, large aggregates of virus are formed and the infection stops. The neuraminidase structure is known, and inhibitors derived from a sialic acid analog have been developed based on the structure.

Antivirals from Random Screens

Until 1996, all antivirals on the market were developed from compounds identified by random screening methods. In many cases the screen was against viral growth, no specific functions were targeted, and the mechanism of action was worked out later (or in some cases not at all). In other programs there was some targeting, (e.g., the screen was for compounds that inhibited the viral polymerase). The screening efforts have been enormous, and rather remarkably unproductive. For viruses other than HIV, the cause of AIDS, there are a total of six classes of drugs licensed in the United States (Table 15.1). Even with the accelerated testing procedures for antivirals against HIV, only four drugs were approved to 1995 (zidovudine, didanozine, celcitobine, and stavudine). During 1996, the first structure-based antivirals were licensed, the HIV protease inhibitors retonavir, indinavir and saquinavir.

Precedents for Drug Design and the Tools Available

The first requirement for a structure-based drug design project is, of course, the molecular structure. Great advances have been made over the past few years in structure determination by X-ray crystallography and nuclear magnetic resonance (NMR). Methods to overexpress proteins of interest from cloned genes

TABLE 15.1 Antiviral Agents Currently Licensed (September, 1996)

Drug	Compound	Virus	Target	Mechanism
Acyclovir	9-(2-Hydroxyethoxy)methyl guanine	Herpes simplex (HSV), varicella zoster (VZV)	DNA polymerase	Activated by viral thymidine kinase; causes chain termination
Ganciclovir*	9-(1,3-Dihydroxy-2-propoxy)methyl guanine	Cytomegalovirus, acyclovir-resistant HSV, VZV	DNA polymerase	Same as acyclovir
Vidarabine	Adenine arabinoside (araA)	HSV, VZV	DNA polymerase	Activated by cellular kinases; inhibits viral polymerase
Idoxuridine, Trifluorothymidine	5'-Iodo-2'-deoxyuridine	HSV conjunctivitis	DNA polymerase	Incorporated into viral DNA, causing abnormal transcription/translation
Foscarnet	Phosphonoformate	Cytomegalovirus, HSV, VZV	DNA polymerase	Noncompetitive inhibitor
Amantadine	1-Adamantanamine HCl	Influenza	M2 protein	Blocks ion channel
Rimantadine	α-Methyl amantadine	Influenza	M2 Protein	Blocks ion channel
Ribavirin	1-β-D-Ribofuranosyl-1,2,4-triazole-3-carboxamide	Respiratory syncytial virus	?	Activated by cellular kinases
Zidovudine (AZT)	Azidothymidine	HIV	Reverse transcriptase	Activated by kinases; causes chain termination
Didanosine	2',3'-Dideoxyinosine	HIV	Reverse transcriptase	Same as AZT
Stavudine (d4T)	2'-3'-Dideoxy-2',3'-didehydrothymidine	HIV	Reverse transcriptase	Same as AZT
Zalcitabine	2',3'-Dideoxycytidine	HIV	Reverse transcriptase	Same as AZT
Nevirapine	Not a nucleoside analog	HIV	Reverse transcriptase	Unclear
Retonavir	Complex	HIV	Protease	Blocks active site
Indinavir	Complex	HIV	Protease	Blocks active site
Saquinavir	Complex	HIV	Protease	Blocks active site
Interferon-α		Hepatitis B, C virus, papillomavirus	Complex	Complex

*Recently, several other acyclic guanine analogs have been added (famciclovir, valaciclovir, penciclovir), offering better oral availability or to combat development of resistance in immunocompromised patients.

have been developed, and synchrotron sources enable data collection from crystals so small that they would have been useless a few years ago. Nevertheless, there are not large numbers of virus or viral protein structures available. Crystallization is often difficult and is the point at which many endeavors fail. Many viruses are difficult and/or hazardous to grow in sufficiently high amounts, and many of the proteins of interest are posttranslationally modified and are not active when expressed in bacteria. For viruses of importance to human medicine, atomic-resolution structures are known for the icosahedral viruses rhinovirus (see Chapter 4), poliovirus (see Chapter 6), mengovirus, Theiler's virus, parvovirus, simian virus 40 (see Chapter 7), and adenovirus (see Chapter 8). Viral proteins of known structure are influenza hemagglutinin (see Chapter 3) and several neuraminidases, HIV and Rous sarcoma virus protease (see Chapter 17), hepatitis A virus protease, HIV reverse transcriptase, Sindbis capsid, and the DNA binding domain of the E2 protein of bovine papillomavirus. Cryo−electron microscope image reconstruction is an emerging tool in virus structure determination (see Chapter 2), but so far the resolution, although visually exciting, is not high enough to use for drug design. However, once the structure is known, there are several tools available to assist the drug designer.

The first requirement is a display and modeling program such as FRODO (Jones, 1978) or CHAIN. It is obvious that manual fitting of even a known inhibitor is not reliable, and computational methods must be used to determine how a prospective inhibitor might bind to the protein (see Chapters 16 and 17).

If the three-dimensional structure of a protein has not been determined, it is sometimes possible to model the structure from a homologous protein using libraries of known substructures and side chain rotamers (Ramachandran et al., 1963; Jones and Thirup, 1986; Ponder and Richards, 1987; Chothia and Finkelstein, 1990). This homology modeling has been done for several proteases (Murphy et al., 1988; Greer, 1990), and was indeed the method used for the first "designed" drug, captopril (Cushman et al., 1977; Ondetti et al., 1977). Nowadays the geometric model is "refined" to an energetic minimum, taking into account the type of each noncovalent interaction, using programs such as AMBER (Weiner and Kollman, 1981), SYBYL (Tripos Associates), or CHARMM (Brooks et al., 1983). There is considerable potential for error in such models, but on the side of the modeler is the fact that the active site targets for inhibitor design are often spatially conserved to a remarkable extent. The amino acid sequences of the neuraminidase of influenza virus types A and B differ by more than 70%, yet the side chains of the enzyme active sites superimpose (Burmeister et al., 1992; Bossart-Whitaker et al., 1993).

Where does one begin to design an inhibitor? The program DOCK (Kuntz et al., 1982; Meng et al., 1993) was developed to allow screening of a small molecule database "*in compuo*" (Sun and Cohen, 1993). It is based on Connolly's (1983) program that describes the surface of a protein by rolling a water-sized sphere (radius 1.4 Å) over the exterior residues, resulting in a smooth description of the water-accessible surface shape. The database is screened for compounds that (1) fit the shape of the active site and (2) have electrostatic compatibility. The program was originally used to search the Cambridge Small Molecule structure database, but because small molecules are relatively easy to model using programs such as CONCORD (Rusinko et al., 1988) or SYBYL, DOCK is now often used to search the databases of all commercially available organic compounds, the Fine Chemicals Directory, or industrial collections.

To extend the search to novel small molecules, the program GRID was developed (Goodford, 1985). GRID calculates the interaction energy for each of a variety of functional groups on a three-dimensional grid of the active site. The grid is then contoured by energy, with the contours showing where a given functional group is predicted to bind. Model building is then used to connect these functional groups into a chemically feasible molecule. An alternative strategy to GRID is the multicopy simultaneous search (MCSS) method, which combines random placement with energy minimization/quenched dynamics techniques so that not only positions but orientations of functional groups in the binding site are determined (Miranker and Karplus, 1991). MCSS also can provide CH_3—R bond vectors to join the functional groups. This step has recently been extended to link functional groups with molecular skeletons taken from a database to construct molecules containing the identified functional groups in the correct orientation. The program is known as HOOK (Eisen et al., 1994), and novel molecules have been identified that are predicted to bind in the active site of influenza hemagglutinin. However, the method has not yet been tested by synthesizing these compounds.

When a novel molecule is modeled into the target binding site, it is useful to have an idea what the binding constant will be. Using computer programs such as DelPhi (Gilson and Honig, 1988; Gilson et al., 1988), the theoretical binding constant K_i can be estimated by electrostatic calculations. Terms contributing to the total electrostatic energy of a protein system are partitioned by DelPhi into three categories: 1) coulombic interaction of charges (ΔG_c^0); 2) interaction of charges with a polarizable solvent (ΔG_p^0); and 3) interaction of charges with the ionic atmosphere in the solvent (ΔG_a^0). The total electrostatic energy, ΔG^0, can then be written as:

$$\Delta G^0 = \Delta G_c^0 + (\Delta G_p^0 + \Delta G_a^0) = \Delta G_c^0 + \Delta G_s^0 = \sum_i (\Delta G_{ci}^0 + \Delta G_{si}^0)$$

Where $\Delta G_s^0 = \Delta G_p^0 + \Delta G_a^0$ is a change of solvent energy of protein and inhibitor, and i denotes the contribution of each charge. Results from such calculations are informative but are still far from reaching a good agreement with experimental results.

Water molecules bound in the active site must be taken into account, with a decision made as to whether they should be displaced by the ligand or incorporated into the binding pattern. Early design projects replaced water molecules with integral parts of the ligand. However, bound water that becomes buried on binding inhibitor may provide additional stability to the interaction, and it is now recognized that it may not be optimal to displace all water molecules (Verlinde and Hol, 1994). By and large, increase in hydrophobic interactions tends to enhance the binding of an inhibitor to a protein target (Reich et al., 1992).

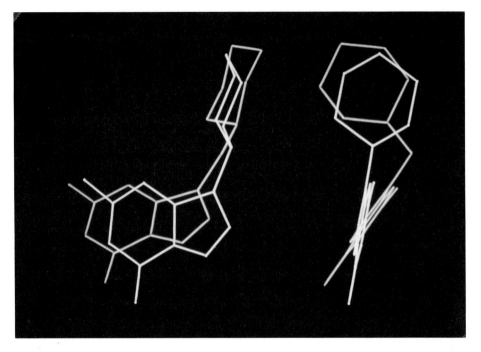

FIGURE 15.2 Designed inhibitors of purine nucleoside phosphorylase did not always bind as expected. End view and side view comparison of the binding of 9-cyclohexyl-9-deaza-guanine (thick lines) with that of 9-cyclohexylmethyl-9-deazaguanine (thin lines). The two cyclohexyl groups occupy approximately the same space in the active site. The additional methylene group was expected to optimize the geometry between the cyclohexane and purine rings, but instead it pulled the purine ring from its bound position. (Courtesy of Dr. Y.S. Babu.)

It should be emphasized that these programs are based on a fixed conformation of the protein, and sometimes of the ligand. In many enzyme active sites there are dramatic changes in conformation on binding substrate or inhibitor, and the ligand is often bound in a conformation that is not the lowest energy of the unbound form. In such cases, the predictive programs fail rather dramatically, although once the changes are established experimentally (by X-ray crystallography or NMR), the design process can proceed. This was the case in the ultimately very successful endeavor to design inhibitors against the human enzyme purine nucleoside phosphorylase (Ealick et al., 1993). A benzene ring added to the inhibitor bound tightly as predicted, but in fact so tightly as to pull the purine ring from its inhibitory disposition, so that the compound was not a good inhibitor (Fig. 15.2). Other reasons for failure include a lack of understanding of how to accurately compute the interaction energy between macromolecule and inhibitor and derive a theoretical K_i. It is still necessary to synthesize at least some of the intermediates in the design process, determine the inhibitory capacity of each, and solve the structure of the complex to determine if the compound is binding as predicted, or to explain why its orientation is not as expected.

Designed Antivirals

Several targets and their designed inhibitors are described in other chapters of this book. They include HIV protease (see Chapter 17), the WIN compounds that inhibit uncoating of human rhinoviruses (see Chapter 16), and two functional sites on the influenza virus hemagglutinin, the fusion site and the sialic acid binding site (see Chapter 3). We will not repeat the discussion of these, but rather describe another target on influenza virus, the neuraminidase.

Influenza continues to be a serious disease. It has a major impact on the health and economy of even the most developed countries, and has eluded the recent advances in vaccine and drug therapy that have eliminated smallpox and brought about a spectacular reduction in the incidence of many other viral diseases. In an average season, influenza causes 10,000 to 20,000 excess deaths, primarily among children and elderly persons. In a more severe outbreak, 40,000 to 70,000 deaths can occur. In the most devastating influenza pandemic of modern times, 20 to 30 million people died worldwide in 1918–1919; in this case, young adults were frequent victims. Estimated costs to the economy of the United States are $1 billion to $3 billion each winter, due largely to morbidity in the workforce (Murphy and Webster, 1990). Vaccines have been available for many years, yet they do not control influenza in the human population. Influenza virus has a remarkable capacity to change its antigenic structure in response to antibodies; thus the vaccines are out-of-date soon after they are made.

There are three types of influenza virus, A, B, and C, distinguished by antigenic properties of internal proteins. Influenza A viruses are further divided into subtypes based on antigenic cross-reactivities of the two surface glycoprotein antigens, hemagglutinin and neuraminidase. Currently 14 subtypes of hemagglutinin (H1 to H14) and 9 subtypes of neuraminidase (N1 to N9) are recognized by the World Health Organization. Two distinct degrees of antigenic variation occur in influenza A viruses, antigenic drift and antigenic shift. Antigenic drift is a gradual distancing of new strains from the original. Point mutations in the genes coding for the hemagglutinin and neuraminidase antigens cause amino acid sequence changes that alter antigenic properties, and these accumulate over time. Antigenic shift is a much more dramatic event, and occurs when a virus with a "new" A subtype of hemagglutinin and sometimes neuraminidase replaces the previously circulating viruses. Viruses that have been isolated in recent epidemics are A subtypes H1N1 and H3N2, and influenza B. In most winters one of these three is dominant, but there is no way to predict which one. Vaccines currently in use contain antigens of a representative of each of these three types and subtypes, and at least one of them is updated each year. It is clearly complex, expensive, and not very effective to make new vaccines each year.

An alternative to vaccination is a specific antiviral drug. Because viral functions are conserved amid the antigenic variation, it is possible to envision an antiviral agent that is effective against several or all influenza viruses.

The only drugs currently approved in the United States for the specific prophylaxis and therapy of influenza virus infections are amantadine and its close relative, rimantadine. Amantadine was discovered empirically; it interferes with virus uncoating by blocking the viral-coded ion channel protein M2. However, mutants resistant to amantadine are readily isolated (Hayden et al., 1989), and amantadine is recommended to be used in conjunction with vaccination but not as a replacement for the vaccine. A more serious deficiency of amantadine is that it is totally ineffective against influenza B infections because a different ion channel is used by the type B virus.

The targets of known structure suitable for a directed drug design against influenza are the two surface proteins, hemagglutinin and neuraminidase (Weis et al., 1988; Varghese et al., 1992). These are embedded in the lipid envelope of the virus and project outward to form surface spikes (Fig. 15.3) The he-

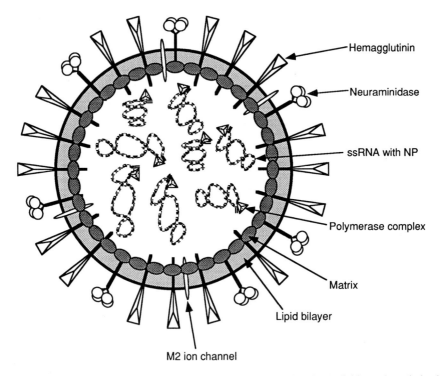

FIGURE 15.3   Components of the influenza virion. The virus has a lipid envelope derived from the host cell. The two surface antigens, hemagglutinin and neuraminidase, are anchored in this membrane, together with smaller numbers of the ion channel protein M2. The major structural protein (matrix) is thought to be associated with the inside surface of the membrane, and within this shell are the nucleocapsid complexes, consisting of the eight segments of the RNA genome complexed with nucleoprotein (NP) and the three polymerase proteins.

magglutinin is a trimer (subunit M_r 60,000) and the neuraminidase is a tetramer (subunit M_r 50,000).

The hemagglutinin has two important functions (see Color Plate 4). It contains a sialic acid binding site that recognizes and binds to specific receptors on the cell surface. The bound virus then enters the cell in an endosome, where, at the low pH found in endosomes, the hemagglutinin mediates fusion of the viral membrane with the cell membrane to allow the nucleocapsid to enter the cytoplasm. Fusion is totally dependent on a proteolytic cleavage within the hemagglutinin to form HA1 and HA2, which releases a hydrophobic sequence at the amino terminus of the HA2 polypeptide. Thus there are three targets for inhibitors of hemagglutinin: the receptor binding site, the cleavage reaction, and the fusion activity. Cleavage involves cellular protease activity, which is presumably used by the cell for normal function and which therefore would be inappropriate as a target. Fusion involves a large conformational change, and an approach to inhibitor design based on preventing this change is discussed in Chapter 3. The receptor binding site is a shallow depression specific for the α-2,3 or α-2,6 linkage between terminal sialic acid and galactose (Weis et al., 1988), and inhibitors are also being designed to fit that site (see Chapter 3).

The neuraminidase has a rather large cavity in which substrate, terminal sialic acid attached to galactose, is bound and cleaved (see Color Plate 4). There is almost 80% difference in sequence between neuraminidases of influenza types A and B, yet the active site pocket of neuraminidase is lined by a large number of conserved basic, acidic, and hydrophobic amino acids. The folding of the two proteins is the same, although the chains do not superimpose (see Color Plate 6). However, in the active site, the side chain positions (Fig. 15.4) are essentially identical (Burmeister et al., 1992; Bossart-Whitaker et al., 1993). Conservative changes of 12 of these amino acids introduced by site-directed mutagenesis abolished or markedly reduced enzyme activity (Lentz et al., 1987).

The function of the influenza neuraminidase is uncertain. In tissue culture, its only role is to facilitate virus release and spread (Palese et al., 1974; Liu and Air, 1993) yet in the absence of active neuraminidase, virus replication is not detected by normal methods. It is generally considered that inhibition of neuraminidase will markedly slow viral replication rather than prevent infection. Therefore, an important advantage in targeting antiviral agents to the neuraminidase is that it is likely that enough virus will replicate to give an immune response, but not enough to cause disease. Thus the antineuraminidase agent will modify an infection so it acts as an attenuated live vaccine (Cohen, 1979).

To effectively design an inhibitor to influenza virus neuraminidase, it is necessary to understand the enzyme mechanism. The only way to avoid the emergence of resistant strains is to have the inhibitor bind tightly to only essential residues of the active site. The interaction of the product of neuraminidase, sialic acid, with the active site is depicted in Figure 15.5. The binding of sialic acid in the active site is dominated by the charge–charge attraction between the carboxylate anion and a pocket of three positively charged arginine

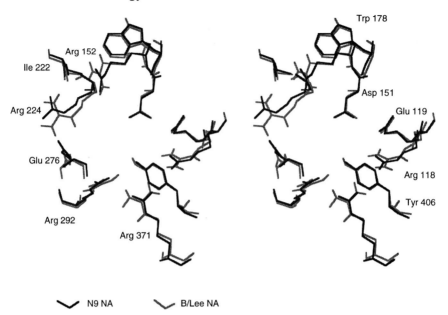

FIGURE 15.4   A more detailed view of the active site in Color Plate 7 showing the positions of side chains lining the active site of influenza type A (solid line) (Bossart-Whitaker et al., 1993) and type B (hatched line) (Janakiraman et al., 1994).

residues (Arg 118, Arg 292, and Arg 371). Opposite the arginine pocket, the methyl group of the acetylamido moiety fits into a small hydrophobic pocket formed by residues Ile 222 and Trp 178, as well as the side chain of Arg 224. The acetylamido carbonyl oxygen forms a hydrogen bond with the guanidinium group of Arg 152. The hydroxyl group at C4 forms hydrogen bonds with Asp 151 and Glu 119, and the last two hydroxyl groups of the glycerol moiety form hydrogen bonds with Glu 276. The Tyr 406 hydroxyl is directly underneath the C2 and O6 sugar ring atoms. The mechanisms proposed (Varghese et al., 1992; Burmeister et al., 1993; Janakiraman et al., 1994) involve common steps, but differ with regard to the cleavage of the glycosidic bond and the force that stabilizes the transition state. As the sialyl group of the substrate binds to the active site, it undergoes a ring distortion probably caused by the strong ionic interactions between the carboxylate of the substrate and the three guanidinium groups of Arg 118, Arg 292, and Arg 371. Varghese et al. (1992) proposed that Asp 151 catalyzes the cleavage of the glycosidic bond at this stage as a general acid catalyst of hydrolysis. However, the pH range for neuraminidase activity (4.5 to 9.0) is inconsistent with the pK_a of the solvent-exposed Asp 151. Alternatively, Burmeister et al. (1993) suggested that an unknown catalyst breaks the glycosidic bond to produce the transition state, oxocarbonium ion, and the hydroxyl group of Tyr 406 is the major residue for stabilization of the transition state. Janakiraman et al. (1994), in contrast, argued that the confor-

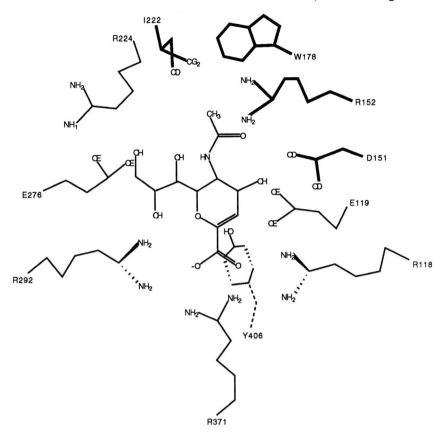

FIGURE 15.5   Schematic showing the interactions between neuraminidase and DANA, a moderate inhibitor that mimics the transition state.

mational changes induced in the sialic acid when it binds to the rigid active site lead to formation of the strained oxocarbonium ion in the active site, thus cleaving the glycosidic bond. The force for the formation and the stabilization of the transition state comes solely from the overall interaction of the substrate with the active site. In the final step, the aglycon moiety leaves the active site with the glycosidic oxygen, which becomes protonated by solvent.

Knowing that the enzyme reaction involves the tight binding between the transition state and 11 conserved side chains of the active site of influenza neuraminidase, novel inhibitors can be derived from the transition state or its analogs. Sialic acid (N-acetylneuraminic acid), a product of the reaction, inhibits the enzyme with millimolar affinity. The derivative 2-deoxy-2,3-dehydro-N-acetylneuraminic acid (DANA; Fig. 15.6), a transition state analog, binds with micromolar affinity. The planar structure of DANA at the C2 position fixes the sugar ring in a conformation that resembles the distorted boat conformation of sialic acid bound in the neuraminidase active site as observed in

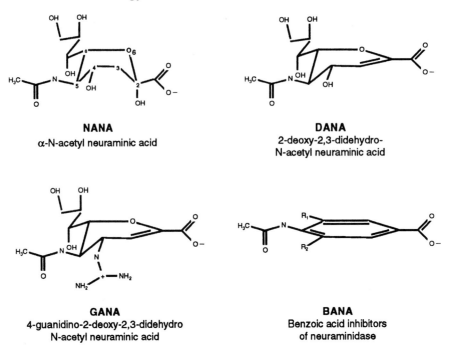

FIGURE 15.6 Structures of inhibitors of influenza virus neuraminidase: NANA, DANA, GANA, and BANA. When NANA is bound to neuraminidase, it assumes a more planar configuration of the sugar ring than that shown, as in the unsaturated derivatives. The planar form appears to be the transition state of the substrate in the enzyme reaction (Janakiraman et al., 1994), and led to the idea of designing inhibitors based on a benzene ring (Jedrzejas et al., 1995).

the crystal structure (Burmeister et al., 1992; Janakiraman et al., 1994). Derivatives of DANA with higher affinities were investigated and shown to block influenza infection in tissue culture, but failed to show an effect in animal models of influenza when given intravenously (Palese and Schulman, 1977). It is now recognized that DANA derivatives have difficulty crossing biologic membranes and are rapidly cleared from the circulation. Further derivatives of DANA have recently been synthesized. The program GRID applied to the neuraminidase–DANA complex showed the possibility of improved binding with a 4-amino group added, and this was shown to be the case. Further inspection of the complex suggested that 4-guanidino-DANA would add a salt bridge: this is the most potent neuraminidase inhibitor described so far. Crystal structures of the new inhibitor–neuraminidase complexes confirmed the designed binding mode of these inhibitors. The compound was effective in ferrets when delivered in a nasal spray, but did not protect against most virulent viruses of chickens, which spread systemically (von Itzstein et al., 1993). To overcome the problems in delivery and duration of DANA derivatives, a new class of inhibitors have been designed. The distortion of the sugar ring when sialic acid is bound to neuraminidase suggested that it might be possible to replace it with a benzene ring, and a series of compounds based on *p*-aminobenzoic acid were designed and synthesized (Fig. 15.6). These compounds (Jedrzejas et al., 1995) have been shown to bind to the neuraminidase active site similarly to DANA, to inhibit neuraminidase enzyme activity in vitro, and to inhibit influenza virus infection in cell culture.

Concluding Remarks

Structure-based drug design is rapidly becoming an important addition to the available methods to control viral disease. As increasing numbers of research teams report their successes and failures along the way to developing high-potency inhibitors of known targets, there is increasing understanding of the interactions that give rise to tight binding. It is anticipated that the computational methods will steadily become more accurate and the predictions more reliable. Further in the future is the prospect of predicting the target protein structure rather than the current necessity of determining its structure by X-ray crystallography or by homology modeling. Many proteins fold spontaneously into their active conformation, implying that eventually it should be possible to deduce the folding pathways and final structure from the amino acid sequence. Other proteins may fold correctly only after association with one or several chaperone proteins, and for the foreseeable future experimental structure determination of these protein targets will be required before inhibitor design can be accomplished.

There are also many steps from obtaining a highly potent inhibitor to a useful therapeutic drug. The final compound must be able to reach the target protein, preferably from oral administration. Thus it must be soluble, stable, and able to cross cell membranes. It must not be too toxic or difficult to synthesize or purify. Although these are complex issues, there is a wealth of experience in the pharmaceutical area, and it should be possible to incorporate the desired characteristics into newly designed inhibitors to make effective antiviral drugs.

Acknowledgments

Work in the authors' laboratories was supported in part by grants AI-18203 and AI-31888 from National Institutes of Health. We thank Dr. Wayne Brouillette for permission to cite unpublished work resulting from his suggestion of using *p*-aminobenzoic acid as the starting point for neuraminidase inhibitor design, Dr. Y.S. Babu for Figure 15.2, Dr. M. Carson for Color Plate 6, and Dr. R. Whitley for information used in Table 15.1.

References

Appelt, K., Bacquet, R.J., Bartlett, C.A., Booth, C.L., Freer, S.T., Fuhry, M.A., Gehring, M.R., Herrmann, S.M., Howland, E.F., Janson, C.A., Jones, T.R., Kan, C.-C.,

Kathardekar, V., Lewis, K.K., Marzoni, G.P., Matthews, D.A., Mohr, C., Moomaw, E.W., Morse, C.A., Oatley, S.J., Ogden, R.C., Reddy, M.R., Reich, S.H., Schoettlin, W.S., Smith, W.W., Varney, M.D., Villafranca, J.E., Ward, R.W., Webber, S., Webber, S.E., Welsh, K.M. and White, J. 1991. Design of enzyme inhibitors using iterative protein crystallographic analysis. J. Med. Chem. 34: 1925–1934.

Beddell, C.R., Goodford, P.J., Norrington, F.E., Wilkinson, S. and Wootton, R. 1976. Compounds designed to fit a site of known structure in human haemoglobin. Br. J. Pharmacol. 57: 201–209.

Bossart-Whitaker, P., Carson, M., Babu, Y.S., Smith, C.D., Laver, W.G. and Air, G.M. 1993. Three-dimensional structure of influenza A N9 neuraminidase and its complex with the inhibitor 2-deoxy 2,3-dehydro-N-acetyl neuraminic acid. J. Mol. Biol 232: 1069–1083.

Brooks, B.R., Bruccoleri, R.E., Olafson, B.D., States, D.J., Swaminathan, S. and Karplus, M. 1983. CHARMM: a program for macromolecular energy minimization and dynamics calculations. J. Comput. Chem. 4: 187–217.

Burmeister, W.P., Henrissat, B., Bosso, C., Cusack, S. and Ruigrok, R.W.H. 1993. Influenza B virus neuraminidase can synthesize its own inhibitor. Structure 1: 19–26.

Burmeister, W.P., Ruigrok, R.W. and Cusack, S. 1992. The 2.2 Å resolution crystal structure of influenza B neuraminidase and its complex with sialic acid. EMBO J. 11: 49–56.

Carson, M. 1991. Ribbons 2.0. J. Appl. Crystallogr. 24: 958–961.

Chothia, C. and Finkelstein, A.V. 1990. The classification and origins of protein folding patterns. Annu. Rev. Biochem. 59: 1007–1039.

Cohen, S.S. 1979. Comparative biochemistry and drug design for infectious diseases. Science 205: 964–971.

Connolly, M. 1983. Solvent-accessible surfaces of proteins and nucleic acids. Science 221: 709–713.

Cushman, D.W., Cheung, H.S., Sabo, E.F. and Ondetti, M.A. 1977. Design of potent competitive inhibitors of angiotensin-converting enzyme: carboxyalkanoyl and mercaptoalkanoyl amino acids. Biochemistry 16: 5484–5491.

DesJarlais, R.L., Seibel, G.L., Kuntz, I.D., Furth, P.S., Alvarez, J.C., Ortiz de Montellano, P.R., DeCamp, D.L., Babe, L.M. and Craik, C.S. 1990. Structure-based design of nonpeptide inhibitors specific for the human immunodeficiency virus 1 protease. Proc. Natl. Acad. Sci. U.S.A. 87: 6644–6648.

Ealick, S.E., Babu, Y.S., Bugg, C.E., Erion, M.D., Guida, W.G., Montgomery, J.A. and Secrist, J.D. 1993. Application of X-ray crystallographic methods in the design of purine nucleoside phosphorylase inhibitors. Ann. N.Y. Acad. Sci. 685: 237–247.

Ehrlich, P. 1897. Die Wertbemessung des Diphtherieheilserums und deren theoretische Grundlagen. Klin. Jahrb. 60: 299. [English translation in Ehrlich, P. 1956. The Collected Papers of Paul Ehrlich, Vol. 292, pp. 107–125. Pergamon, London.]

Eisen, M.B., Wiley, D.C., Karplus, M. and Hubbard, R.E. 1994. HOOK: a program for finding novel molecular architectures that satisfy the chemical and steric requirements of a macromolecule binding site. Proteins Struct. Funct. Genet. 19: 199–221.

Fischer, E. 1894. Einfluss der configuration auf die wirkung der enzyme. Ber. Deutsch. Chem. Ges. 27: 2985–2993.

Gilson, M.K. and Honig, B. 1988. Calculation of the total electrostatic energy of a macromolecular system: solvation energies, binding energies, and conformational analysis. Proteins Struct. Funct. Genet. 4: 7–18.

Gilson, M.K., Sharp, K.A. and Honig, B.H. 1988. DelPhi. J. Comput. Chem. 9: 327–335.

Goodford, P.J. 1985. A computational procedure for determining energetically favorable binding sites on biologically important macromolecules. J. Med. Chem. 28: 849–857.

Greer, J. 1990. Comparative modeling methods: application to the family of the mammalian serine proteases. Proteins Struct. Funct. Genet. 7: 317–334.

Hayden, F.G., Belshe, R.B., Clover, R.D., Hay, A.J., Oakes, M.G. and Soo, W. 1989. Emergence and apparent transmission of rimantadine-resistant influenza A virus in families. New Engl. J. Med. 321: 1696–1702.

Haywood, A.M. 1994. Virus receptors: binding, adhesion strengthening, and changes in viral structure. J. Virol. 68: 1–5.

Janakiraman, M.N., White, C.L., Laver, W.G., Air, G.M. and Luo, M. 1994. Structure of influenza virus neuraminidase B/Lee/40 complexed with sialic acid and a de-hydro-analog at 1.8 Å resolution: implications for the catalytic mechanism. Biochemistry 33: 8172–8179.

Jedrzejas, M.J., Singh, S., Brouillette, W.J., Laver, W.G., Air, G.M. and Luo, M. 1995. Structures of aromatic inhibitors of influenza virus neuraminidase. Biochemistry 34: 3144–3151.

Jones, T.A. 1978. A graphics model building and refinement system for macromolecules. J. Appl. Crystallogr. 11: 268–272.

Jones, T.A. and Thirup, S. 1986. Using known substructures in protein model building and crystallography. EMBO J. 5: 819–822.

Kendrew, J., Dickerson, R., Strandberg, B., Hart, R., Davies, D., Phillips, D. and Shore, V. 1960. Structure of myoglobin: a three-dimensional Fourier synthesis at 2 Å resolution. Nature 185: 422–427.

Kohlstaedt, L.A., Wang, J., Friedman, J.M., Rice, P.A. and Steitz, T.A. 1992. Crystal structure at 3.5 A resolution of HIV-1 reverse transcriptase complexed with an inhibitor. Science 256: 1783–1790.

Kuntz, I.D., Blaney, J.M., Oatley, S.J., Langridge, R. and Ferrin, T.E. 1982. A geometric approach to macromolecule-ligand interactions. J. Mol. Biol. 161: 269–288.

Lapatto, R., Blundell, T., Hemmings, A., Overington, J., Wilderspin, A., Wood, S., Merson, J.R., Whittle, P.J., Danley, D.E., Geoghegan, K.F., Hawrylik, S.J., Lee, S.E., Scheld, K.G. and Hobart, P.M. 1989. X-ray analysis of HIV-1 proteinase at 2.7 Å resolution confirms structural homology among retroviral enzymes. Nature 342: 299–302.

Lentz, M.R., Webster, R.G. and Air, G.M. 1987. Site-directed mutation of the active site of influenza neuraminidase and implications for the catalytic mechanism. Biochemistry 26: 5351–5358.

Liu, C. and Air, G.M. 1993. Selection and characterization of a neuraminidase-minus mutant of influenza virus and its rescue by cloned neuraminidase genes. Virology 194: 403–407.

Meng, E.C., Gschwend, D.A., Blaney, J.M. and Kuntz, I.D. 1993. Orientational sampling and rigid-body minimization in molecular docking. Proteins Struct. Funct. Genet. 17: 266–278.

Miranker, A. and Karplus, M. 1991. Functionality maps of binding sites: a multiple copy simultaneous search method. Proteins Struct. Funct. Genet. 11: 29–34.

Montgomery, J.A., Niwas, S., Rose, J.D., Secrist, J.A., Babu, Y.S., Bugg, C.E., Erion, M.D., Guida, W.C. and Ealick, S.E. 1993. Structure-based design of inhibitors of purine nucleoside phosphorylase. 1. 9-(arylmethyl) derivatives of 9-deazaguanine. J. Med. Chem. 36: 55–69.

Murphy, B.R. and Webster, R.G. 1990. Orthomyxoviruses. In Virology, Vol. 1 (Fields, B.N. and Knipe, D.M., Eds.), pp. 1091–1152, Raven Press. New York.

Murphy, M.E., Moult, J., Bleackley, R.C., Gershenfeld, H., Weissman, I.L. and James, M.N. 1988. Comparative molecular model building of two serine proteinases from cytotoxic T lymphocytes. Proteins Struct. Funct. Genet. 4: 190–204.

Navia, M.A., Fitzgerald, P.M., McKeever, B.M., Leu, C.T., Heimbach, J.C., Herber, W.K., Sigal, I.S., Darke, P.L. and Springer, J.P. 1989. Three-dimensional structure of aspartyl protease from human immunodeficiency virus HIV-1. Nature 337: 615–620.

Ondetti, M.A., Rubin, B. and Cushman, D.W. 1977. Design of specific inhibitors of angiotensin-converting enzyme: new class of orally active antihypertensive agents. Science 196: 441–444.

Palese, P. and Schulman, J.L. 1977. Chemoprophylaxis. In Virus Infections of the Upper Respiratory Tract (Oxford, J.S., Ed.), pp. 189–205. CRC Press, Cleveland, OH.

Palese, P., Tobita, K., Ueda, M. and Compans, R.W. 1974. Characterization of temperature-sensitive influenza virus mutants defective in neuraminidase. Virology 61: 397–410.

Perutz, M., Rossmann, M., Cullis, A., Muirhead, H., Will, G. and North, A. 1960. Structure of haemoglobin: a three-dimensional Fourier synthesis at 5.5 Å resolution, obtained by X-ray analysis. Nature 185: 416–427.

Ponder, J.W. and Richards, J. 1987. Tertiary templates for proteins: use of packing criteria in the enumeration of allowed sequences for different structural classes. J. Mol. Biol. 193: 775–791.

Ramachandran, G.N., Ramakrishnan, C. and Sasisekharan, V. 1963. Stereochemistry of polypeptide chain configurations. J. Mol. Biol. 7: 95–99.

Reich, S.H., Fuhry, M.A., Nguyen, D., Pino, M.J., Welsh, K.M., Webber, S., Janson, C.A., Jordan, S.R., Matthews, D.A., Smith, W.W., Bartlett, C., Booth, C.L., Herrmann, S.M., Howland, E.F., Morse, C.A., Ward, R.W. and White, J. 1992. Design and synthesis of novel 6,7-imidazotetrahydroquinoline inhibitors of thymidylate synthase using iterative protein crystal structure analysis. J. Med. Chem. 35: 847–858.

Richardson, C.D., Scheid, A. and Choppin, P.W. 1980. Specific inhibition of paramyxovirus and myxovirus replication by oligopeptides with amino acid sequences similar to those at the N-termini of the F1 or HA2 viral polypeptides. Virology 105: 205–222.

Rossmann, M.G. and Rueckert, R.R. 1987. What does the molecular structure of viruses tell us about viral functions? Microbiol. Sci. 4: 206–214.

Rusinko, A.I., Skell, J.M., Baldwin, R. and Pearlman, R.S. 1988. CONCORD (University of Texas at Austin). Tripos Associates, St. Louis.

Silverstein, A.M. 1989. A History of Immunology. Academic Press, San Diego.

Spinelli, S., Liu, Q.Z., Alzari, P.M., Hirel, P.H. and Poljak, R.J. 1991. The three-dimensional structure of the aspartyl protease from the HIV-1 isolate BRU. Biochimie 73: 1391–1396.

Spiropoulou, C.F., Morzunov, S., Feldmann, H., Sanchez, A., Peters, C.J. and Nichol, S.T. 1994. Genome structure and variability of a virus causing hantavirus pulmonary syndrome. Virology 200: 715–723.

Sun, E. and Cohen, F.E. 1993. Computer-assisted drug discovery—a review. Gene 137: 127–132.

Varghese, J.N., McKimm-Breschkin, J.L., Caldwell, J.B., Kortt, A.A. and Colman, P.M. 1992. The structure of the complex between influenza virus neuraminidase and sialic acid, the viral receptor. Proteins Struct. Funct. Genet. 14: 327–332.

Verlinde, L.M.J. and Hol, W.G.J. 1994. Structure-based drug design: progress, results and challenges. Structure 2: 577–587.

von Itzstein, M., Wu, W.Y., Kok, G.B., Pegg, M.S., Dyason, J.C., Jin, B., Van Phan, T., Smythe, M.L., White, H.F., Oliver, S.W., Colman, P.M., Varghese, J.N., Ryan, D.M., Woods, J.M., Bethell, R.C., Hotham, V.J., Cameron, J.M. and Penn, C.R. 1993. Rational design of potent sialidase-based inhibitors of influenza virus replication [see comments]. Nature 363: 418–423.

Weiner, P.K. and Kollman, P.A. 1981. AMBER. J. Comput. Chem. 2: 287–303.

Weis, W., Brown, J.H., Cusack, S., Paulson, J.C., Skehel, J.J. and Wiley, D.C. 1988. Structure of the influenza virus hemagglutinin complexed with its receptor, sialic acid. Nature 333: 426–431.

Wlodawer, A., Miller, M., Jaskolski, M., Sathyanarayana, B.K., Baldwin, E., Weber, I.T., Selk, L.M., Clawson, L., Schneider, J. and Kent, S.B. 1989. Conserved folding in retroviral proteases: crystal structure of a synthetic HIV-1 protease. Science 245: 616–621.

16.

The Use of Structural Information in the Design of Picornavirus Capsid Binding Agents

GUY D. DIANA, MARK A. MCKINLAY,
& ADI TREASURYWALA

The cure for the common cold has continued to evade the best efforts of a number of research groups. Although the infection is generally mild and self-limiting, it remains a significant cause of lost productivity and physician visits. In the United States alone, it was estimated some years ago that the common cold is responsible for 27 million physician visits and 161 million days of restricted activity (Sperber and Hayden, 1988). Approximately 50% of colds result from rhinovirus infections (Gwaltney, 1985). The 100+ distinct rhinovirus serotypes form the largest subfamily of picornaviruses (Hamparian et al., 1987). The serotypes are divided into a major group, consisting of approximately 90% of the serotypes, and a minor group. The major group is characterized by its preference for binding to domains 1 and 2 of intercellular adhesion molecule-1 (ICAM-1) on the human host cell (Abraham and Colono, 1984; Greve et al., 1989; Staunton et al., 1989; Tomassini et al., 1989; Uncapher et al., 1991), while the minor group binds to the human low density lipoprotein receptor (Hofer, 1994). An effective antirhinovirus drug should be active against the majority of both groups of serotypes.

Although compounds capable of inhibiting rhinovirus replication in vitro have been found, demonstrating clinical efficacy has been more difficult. Several approaches have been examined, including the use of interferon (Monto et al., 1989; Turner et al., 1989), bradykinin antagonists (Higgins et al., 1990), receptor blockade (Hayden et al., 1988), and a host of synthetic antiviral agents. With the exception of one partially successful prophylactic clinical trial with pirodavir (Hayden et al., 1992), none of these attempts has resulted in an approved drug.

The majority of the active compounds that have been found through screening inhibit replication via binding to the outer protein shell (icosahedral capsid) of the virus which surrounds single stranded positive-sense mRNA. All of these compounds that have been studied bind to a hydrophobic pocket and inhibit attachment to the host cell receptor, and/or inhibit disassembly of the virion. To be a useful agent, a compound must possess potency against the vast majority of the 100+ rhinovirus serotypes. The results of X-ray crystallography as well as sequencing studies have shown that each rhinovirus serotype binding pocket appears to be unique. Consequently, a potent broad-spectrum agent must be capable of binding to a variety of sites. Designing compounds that achieve this goal has presented a challenge to synthetic and computational chemists alike.

Capsid Binding Agents

In the course of our work in the pursuit of antirhinovirus agents, we have investigated compounds that bind to the viral capsid and that have a dual mechanism of action. This class of compounds resulted from a project directed toward the synthesis of juvenile hormone mimetics, and was discovered by routine screening using an equine rhinovirus plaque reduction assay (Diana et al., 1977). This work subsequently led to the synthesis of WIN 51711 (Fig. 16.1), which was found to have activity both in vitro and in vivo against human rhino and enteroviruses. However, because of crystalluria observed in some subjects, clinical studies were terminated. Further exploration of this series of compounds resulted in the preparation of WIN 54954, which was evaluated clinically against three rhinovirus serotypes and an enterovirus, coxsackie A21. No statistically significant clinical effect was seen against the rhinoviruses, but WIN 54954 had efficacy against coxsackievirus.

In the case of the major rhinovirus class, WIN compounds are presumed to inhibit the ability of the virus to bind to ICAM-1 by inducing conformational changes in the cell receptor binding site. Several compounds from this series complexed to human rhinovirus (HRV) 14 displace certain polypeptide chains in the floor of the cell receptor binding site by approximately 4.5 Å, leading to a decreased affinity of the virus for its receptor (Pevear et al., 1989; McKinlay et al., 1992).

In the minor group of serotypes, WIN compounds bind to a very similar location on the virus (Kim et al., 1993). However, inhibition of replication of these serotypes occurs by inhibiting uncoating rather than by preventing attachment (Fox et al., 1986; Zeichardt et al., 1987). Crystallographic analysis of HRV14 (major group) and HRV1A (minor group) have shown distinct differences in the drug binding site, with the HRV14 binding site being longer and narrower. The dimensions of the respective binding sites determine the types of drugs that will optimally bind to each pocket. Furthermore, because the sensitivity of each serotype to a compound is dependent upon its binding

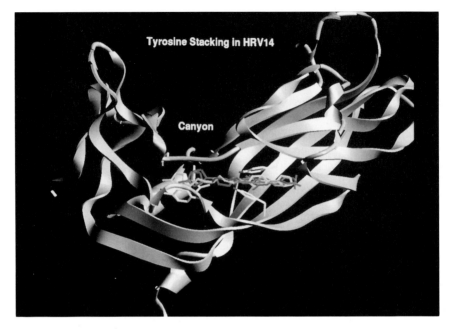

FIGURE 16.1 Ribbon representation of the protomeric unit of the capsid protein of HRV14. The protomer consists of four viral proteins. Several WIN compounds are shown bound to VP1 with the oxazoline ring in the hydrophobic pocket formed from an eight-stranded, antiparallel β-barrel. The phenyl rings are shown in a stacking conformation with Tyr 128 and Tyr 152.

affinity, the size and shape of the molecule are critical factors in designing agents that will fit into the binding sites.

X-ray Crystallography Studies

WIN Compound Binding Site in Human Rhinovirus 14

The three-dimensional structure of HRV14 was reported by Rossmann et al. in 1985. Subsequently, studies were performed with crystals of HRV14 bound with WIN 51711 and an analog, WIN 52084 (Smith et al., 1986). These studies revealed that the drugs were bound to the capsid protein in a hydrophobic pocket formed by an eight-stranded antiparallel β-barrel, below a depression or "canyon." The canyon is the cell receptor binding site, which recognizes domains 1 and 2 of ICAM-1 (Fig. 16.1).

Binding Mode

WIN 51711 binds with the oxazoline ring deep in the "toe" of the pocket and with the isoxazole ring in the proximity of a pore beneath the canyon (Fig.

16.2). The oxazolinylphenyl group lies between Tyr 128 and Tyr 152 in close proximity such that hydrophobic interactions occur. WIN 52084, however, was in the reverse orientation, with the methyl group attached to the oxazoline ring pointing toward a hydrophobic area formed by Leu 106 and Ser 107. Subsequent crystallographic studies with HRV14 have shown that all of the WIN compounds examined are bound with the isoxazole ring in the pore area and

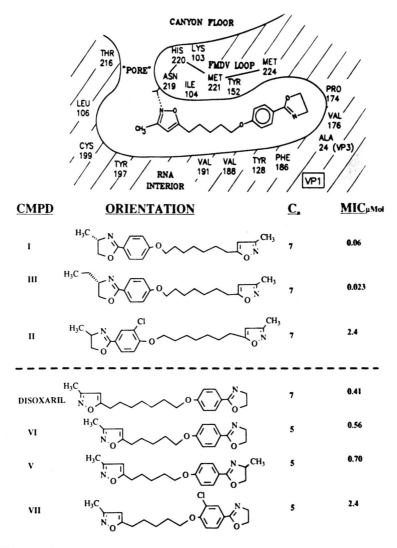

FIGURE 16.2 A schematic of the binding site of HRV14 with WIN 52035 bound in the pocket. With the exceptions noted, all of the compounds examined by X-ray crystallography are oriented in the pocket with the isoxazole ring in the ''heel'' portion of the binding site at the base of the pore. The nitrogen of the isoxazole ring is within hydrogen bonding distance of Asn 219.

the oxazoline ring in the toe of the binding site. Exceptions to this are those compounds, including WIN 52084, with a seven-membered chain between the isoxazole and phenoxy rings, and with alkyl substituents attached to the oxazoline ring, which are bound in the opposite orientation (Badger et al., 1988) (Figure 16.2). Because the structural difference between WIN 51711 and WIN 52084 resides in the group on the oxazoline ring, the hydrophobic interaction of this aliphatic group with a hydrophobic pocket in the pore of the binding site was examined. This interaction appears to determine the orientation of the molecule. The nonrandom binding of these compounds was encouraging because it permitted the modeling of synthetic targets.

Use of the Structure of Human Rhinovirus 14 in Drug Design

Although the structure of HRV14 was known at atomic resolution (Rossmann et al., 1985), designing molecules with a wide spectrum of activity against several different serotypes was unsuccessful. It was later found that the compound binding sites are unique to each serotype. Thus molecules could be modeled that fit the HRV14 WIN pocket, but when the compounds were synthesized and tested, they generally exhibited good activity only against HRV14. This limited spectrum of activity has subsequently been explained by genetic divergence of HRV14 from other serotypes, which results in sequence variations within the pocket. These differences determine the size of the binding site. This variability in binding site size presented a problem for designing broad-spectrum agents based on the HRV14 structure alone. Currently, the X-ray structures of several other human rhinovirus serotypes have been elucidated (Oliviera et al., 1993; V.L. Giranda, unpublished results) and the determination of the crystal structures of compounds bound to the virus is in progress. This information will assist in the design of future inhibitors.

Development of a Screening Strategy

In order to successfully discover a broad-spectrum antirhinovirus agent, it was important to select representative serotypes that could be utilized in a primary screen. This situation is the exact inverse of the traditional structure-based drug design approach. In the more common paradigm, one is interested in constructing a molecule that will bind with high specificity to the target molecule, whose structure is known, but will not bind to any other related targets. In the antirhinovirus case, we were interested in designing a molecule that would bind with high affinity to the known target structure and at the same time bind to all the other closely related targets (HRV serotypes). In addition, the binding was required to be specific for rhinoviruses but not for enzymes or other macromolecules in the human body. Because the eight-stranded β-barrel present in the rhinovirus compound binding site is not unique to rhinovirus capsid proteins, nonspecificity could present a problem. However, toxicologic and pathologic studies in animals as well as clinical studies with WIN 51711 and WIN

54954 indicated no toxic effects, which suggested that these compounds were highly selective for rhinoviruses.

Structure-Based Approach

A database was available of several hundred "similar" molecules that had previously been tested against a variety of HRV serotypes and that potentially contained information that could aid in the future design of an "ideal compound." The problem that emerged was how to effectively utilize this accumulated antirhinovirus data for the design of a broad-spectrum compound. An additional problem was how to most efficiently determine the extent of activity. Screening a compound against all 100+ serotypes was impractical and time consuming. The problem was then to select a reasonable number of representative serotypes for the initial evaluation, which would enhance the screening process and predict the spectrum of activity with some degree of certainty. Consequently, it was necessary to select serotypes that would be representative of the total number. Information from the current database was generated from 15 serotypes, (1A, 1B, 2, 6, 14, 15, 21, 22, 25, 30, 41, 50, 67, 86, and 89) using the available minimum inhibitory concentration (MIC) data for the drug (the concentration required to reduce the number of viral plaques in the assay by 50%). The initial approach was therefore centered on determining from the available data the structural properties of an agent that would inhibit the replication of a maximum number of these serotypes. From this, the mean MIC against all 15 serotypes for each of the agents tested was calculated, as well as a MIC_{80}, or that concentration of agent that would inhibit 80% of the serotypes tested. The question was, how good were these 15 serotypes at predicting generally useful structural properties?

Statistical Approach

To assess the predictive power of the 15 serotypes in the primary screen, the MIC data for a number of compounds was analyzed by a Pearson correlation. This measures the closeness of a linear relationship between two variables. If one of the variables can be expressed exactly as a linear function of the other, then the correlation is 1 or -1 depending upon whether the two variables are directly or inversely related, respectively. A correlation of 0 between two variables means that each variable has no linear predictive ability for the other. The correlation is normally squared and is a number between 0 and 1. The activity for several compounds against each of the 15 serotypes was correlated with their activity against the remaining 14, to determine if any similarities existed. The aim was to use these statistics to place serotypes into groups or families based on their comparable sensitivity to the WIN compounds.

From this and subsequent clustering analyses using several different techniques, two groups or families were identified (Andries et al., 1991). One family contained serotypes 2, 21, 22, 30, 50, and 89. The other contained serotypes

6, 15, 22, and 86. One interesting fact that emerged was that there were several serotypes, such as 1a, 1b, 41, and 67, that correlated with each other but did not demonstrate sensitivities comparable to the serotypes in the other two groups. Consequently, these serotypes represented a third unique family. Another finding was that HRV14 was not highly correlated with most other serotypes in the study and consequently was not a member of any of the three families (i.e., was not a ''representative'' serotype, at least of these 15). Thus it was understandable why the initial structure-based approach had failed.

Another method has been applied to other subsets of the data in order to achieve the goal of selecting the most representative serotypes for screening. This approach is, to our knowledge, unique in the industry and involves the use of genetic algorithms (Jaeger et al., 1995) for the uncovering of subtle correlations between the observations for the activity of a class of compounds against a battery of tests. This program attempts to optimize a property based on Darwinian rules of evolution, which are selection, crossover, and mutation. In this case, the parameter used was binding to the set of serotypes (A.M. Treasurywala and E.P. Jaeger, unpublished results). These studies have allowed us to select a set of seven serotypes that represent one half of all the known serotypes extremely well, and have furnished us with a very accurate and efficient screen.

Structure–Activity Studies

Crystal structures of other serotypes were needed, particularly those that were representative of specific clusters. This task was undertaken by Rossmann and coworkers and has resulted in the elucidation of the structures of two additional rhinovirus serotypes (Kim et al., 1989; Oliviera et al., 1993). This work has been extremely valuable in explaining the structure–activity relationships found for these serotypes, particularly with regard to the importance of the length of the molecule. Longer molecules have consistently demonstrated greater activity against HRV14 than against HRV1A because of the greater length of the HRV14 binding site. Conversely, shorter molecules were more potent against HRV1A (Mallamo et al., 1992; Kim et al., 1993).

The crystallographic results with HRV1A and HRV14 led to a study to investigate properties of compounds required for broad-spectrum activity. The results suggested that compounds with shorter chains linking the two rings at the ends of the molecules had a broader spectrum of activity. This was in agreement with observations resulting from traditional structure–activity studies (Diana et al., 1987, 1989). The synthesis of compounds with only three carbons separating the ends of the molecule provided analogs with the best spectrum of activity (Mallamo et al., 1992). This study has subsequently been repeated with a new and expanded set of compounds. Although there were some minor differences, the results were generally in agreement with the findings described earlier.

The Nature of Interactions within the Binding Site

It is clear that antirhinovirus drug design would not have been possible without the availability of crystal structures of the virus complexed with various small molecules. The wealth of data that these studies have provided is enormous. The use of X-ray crystal structures is almost a prerequisite for those involved in drug design. Examination of the binding pocket in rhinovirus 14, which was revealed by the early crystallographic work, showed that the predominant interactions between the drug and the pocket were hydrophobic. The only potential hydrogen bond was between the side chain of Asn 219 and the nitrogen in the isoxazole ring of the drug molecule. This potentially important interaction was studied by site-directed mutagenesis of Asn 219 (see Hydrogen Bonding) and shown to be inconsequential. These results were very significant because they suggested that utilizing this putative hydrogen bond in drug design would be nonproductive. As a result, hydrogen bonding potential with the Asn 219 residue was not designed into new molecules.

The X-ray crystallographic structure determination of various WIN compounds bound to HRV14 allowed the examination of their interactions with various hydrophobic and hydrophilic residues in the binding site. This information was particularly useful for the design of more potent compounds, because studies with HRV14 had suggested that there was a correlation of binding with antiviral activity (Fox et al., 1991).

Hydrogen Bonding

The position of Asn 219 (Fig. 16.1) within 3.5 Å of the isoxazole ring of the WIN compounds led to speculation that hydrogen bonding was occurring. However, the nitrogen of the isoxazole ring (and the oxazoline ring in the case of WIN 52084) is a weak base, and hydrogen bonding to these nitrogens would be expected to make only a minor contribution to the binding energy (Dean, 1989). Lau and Pettit (1989) used a computational method to measure the hydrogen bonding of the oxazoline nitrogen of WIN 52084 to Asn 219 and concluded that the contribution to the total binding energy was negligible. Confirmatory evidence to support this conclusion was provided by site-directed mutagenesis of Asn 219 to an alanine (Hadfield et al., 1996). The mutation removed the hydrogen bonding capability at this position. The resulting serotype was as sensitive to the WIN compounds as the wild-type virus, confirming that this interaction was insignificant.

Hydrophobic Interactions

One of the first compounds bound to HRV14 to be examined by X-ray crystallography was WIN 52084, which possesses an asymmetric center at the 4 position of the oxazoline ring. Although a racemic mixture of WIN 52084 was used in the crystallography studies, only one isomer appeared to bind to HRV14, suggesting that there was isomeric specificity with respect to binding.

The evaluation of each isomer against HRV14 revealed that the *S* isomer demonstrated a 10-fold greater potency than the *R* form. Furthermore, increasing the length of the alkyl chain attached to the oxazoline ring resulted in a gradual increase in potency, which suggested that a hydrophobic effect was important (Diana et al., 1988).

The oxazoline ring is capable of free rotation about the phenyl ring. X-ray studies indicated that the torsion angle between the oxazoline and phenyl rings was approximately 15°, which placed the methyl group on the oxazoline ring

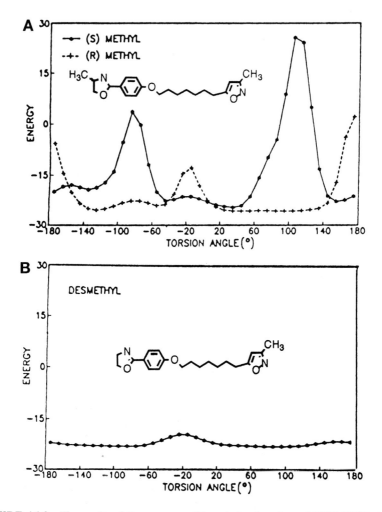

FIGURE 16.3   The results of the energy profiling study of analogs of WIN 52084 showing the relationship of energy to torsion angle. Steep valleys with low energy occur with the *S* isomers of WIN 52084 shown in *A*, the ethyl analog shown in *D*, and the gem dimethyl analog shown in *C*. The plot for the desmethyl analog in *B* is flat, suggesting free, unhindered rotation of the oxazoline ring about the phenyl ring. Figure continues.

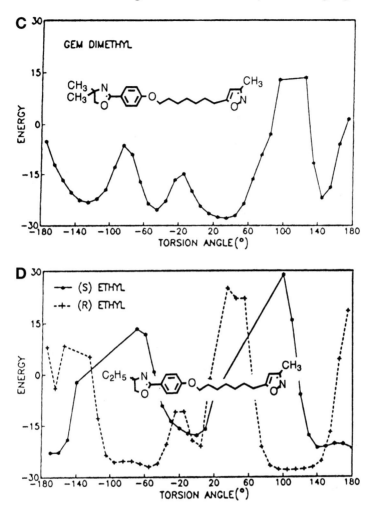

FIGURE 16.3 *Continued*

of WIN 52084 (*S*) in close proximity to a hydrophobic pocket formed by Ser 107, Leu 106, and Cys 198. The results of an energy profiling study using the program CHEM-X showed energy minima at −45° to +50° for the *S* isomers of WIN 52084 (Fig. 16.3*A*), with a low-lying area from 10° to 30°, and at approximately 10° for the *S*-ethyl homolog (Fig. 16.3*D*). In each case, the low-lying areas were flanked by high-energy peaks. However, although the *R* isomers displayed valleys in this area, they are not relatively low lying. The low-lying areas in the case of the *S* isomers suggest restricted rotation. The relative positions of the oxazoline and phenyl rings in these low-energy conformations would place the methyl or ethyl groups in a suitable position to interact favorably with the hydrophobic pocket and consequently increase the binding energy of the compound. Although the correlation of the torsion angle with biologic activity may be tenuous, it does suggest that the twist angle between the two ring systems could be important in determining biologic activity for this series of compounds.

Phenyl Stacking

With the exceptions cited, all of the compounds that have been examined by X-ray crystallographic studies exhibit a stacking of the phenyl ring with Tyr 128 and Tyr 152 in HRV14 such that the phenyl ring of the compounds resides between the aromatic rings of the tyrosines (Fig. 16.1). This type of aromatic–aromatic interaction is quite common in protein interactions (Burley and Petsko, 1985, 1988; Singh and Thornton, 1985; Levitt and Perutz, 1988; Anderson et al., 1993), and in many cases displays an electrostatic component (Karlstrom et al., 1983; Fowler and Moore, 1988). Furthermore, such interactions would be expected to contribute significantly to the binding energy (Serrano et al., 1990). The nature of this stacking was of interest because it would be possible to alter the electrostatic properties by modifying the substituents attached to the phenyl ring of the WIN compounds to enhance the interactions and consequently improve the binding. Several parameters were evaluated by computational methods to explore the nature of the stacking. These studies showed no evidence of electrostatic interactions, but rather led to the conclusion that the phenyl–phenyl stacking was hydrophobic.

The lack of an electrostatic effect associated with the phenyl–phenyl interactions may be due to the inability of the phenyl ring of these compounds to adapt an end-surface, or perpendicular, orientation as a result of space constraints within the pocket. A true perpendicular orientation of the three phenyl rings is not possible because of the relative positions of the tyrosines. The closer the rings approach this state, the greater the possibility of an electrostatic effect. An energy profiling study was performed with several monosubstituted analogs to determine low-energy orientations. As the phenyl ring was rotated from a planar to a perpendicular conformation with respect to Tyr 128 and Tyr 152, a corresponding increase in energy was observed, suggesting that a planar orientation was preferred.

Electrostatic Effects

Comparative Molecular Field Analysis Studies

CoMFA (comparative molecular field analysis) is a program designed to relate steric and electrostatic properties of molecules to biologic activity (Cramer et al., 1988). An essential requirement is that the compounds included in the study have a conformationally rigid structure. Alternatively, X-ray structures can be used in the study. This program was applied to a series of seven compounds whose structures have been determined while bound to HRV14 (Diana et al.,

1992). The steric and electrostatic environment of each compound was explored at 1-Å intervals using a proton probe and the values tabulated. Using a partial least-squares method, a regression analysis was performed to correlate these data with the biologic activity of each compound. The results of the study revealed that there was a strong steric component that was highly correlated with biologic activity. However, the electrostatic component correlated weakly. The results were displayed as three-dimensional contour maps (see Color Plate 7). These maps show qualitative agreement with the quantitative structure–activity relationship analysis portion of the program in that no significant positive correlation (red) exists between the electrostatic parameters and the biologic activity (see Color Plate 19, middle); however, there is a strong positive correlation for the steric parameters (see Color Plate 19, top).

The Use of X-Ray Crystallography and Molecular Dynamics in Drug Design

Volume Map Studies

A substantial number of related WIN compounds have been examined by X-ray crystallography while bound to HRV14, and the results indicated that they have a preferred orientation in the binding site. It therefore appeared possible to develop a model for compounds active against this serotype that would be based on the common molecular size and shape (Diana et al., 1990). The same series of compounds that were used in the CoMFA experiment, and a series of inactive compounds with related structures, were employed in this study. The structures, which were stored using the program SYBYL (Tripos Associates, Inc.), were overlaid according to their orientations in the binding site. The orientations of the inactive compounds were assumed from the similarity in structure to the active series, and the van der Waals volume occupied by each bound structure was color coded (see Color Plate 19, bottom). The resulting "volume maps" revealed that, although some bulky groups could be accommodated on the phenyl ring, the introduction of excessively bulky groups in this position led to inactivity. In the case of the active compounds, the ability to fill space in the pore area of the binding site (Fig. 16.1) was a contributing factor to good biologic activity. These results emphasized that it is important that compounds fully occupy the binding site in order to maximize the hydrophobic interactions contributing to the binding energy. However, excessive bulk requiring major conformational changes within the pocket is undesirable because this would require the expenditure of energy.

Molecular Biology Approach

The analysis of drug-resistant mutants arising spontaneously in vitro and the construction of site-specific mutants of rhinovirus have defined the parameters contributing to the binding of small hydrophobic compounds within VP1. This was of particular importance in demonstrating the inconsequential nature of the putative hydrogen bond involving Asn 219, as previously discussed.

Viruses resistant to the antiviral activity of the capsid binding agents exist in vitro at a frequency of between 10^{-4} and 10^{-5}. All of the mutations conferring high-level resistance mapped to two amino acids in the drug binding pocket of HRV14 (Heinz et al., 1989). In all cases, the mutations resulted in bulkier side chains, suggesting that resistance resulted from new steric interactions that interfered with compound binding. Mutations resulting in resistance to low concentrations of compound were also found with HRV14. These mapped to the canyon floor, and in some cases provided a virus that bound with greater affinity to the cellular receptor, ICAM-1 (Heinz et al., 1989). Some of these mutants were able to overcome the inhibition of attachment to ICAM-1 through a mutation that compensated for the conformational change in the canyon normally induced by compound binding (Shepard et al., 1993). The mutants capable of overcoming the attachment block were still inhibited to some extent in the uncoating process. These results suggest that, although conformational requirements in the canyon are critical for viral attachment, there is a certain degree of conformational flexibility that in this case reestablishes the ability of the virus to bind to the receptor.

When HRV1A mutants resistant to the capsid binding agents were isolated, unlike HRV14, all were found to be dependent on antiviral agents for growth (D.C. Pevear, personal communication). The viruses appeared to be less stable in the absence of drug and spontaneously uncoated, suggesting that the drugs restored the mutants' "wild-type" stability. This result added support to the hypothesis that successful drugs fill the drug binding pocket and block uncoating by "overstabilizing" the virion.

Studies utilizing an infectious cDNA of HRV1A have been conducted to evaluate the effects of single amino acid substitutions in the drug binding pocket on antiviral sensitivity. Because HRV1A is more homologous to other rhinovirus serotypes than is HRV14, it was possible to develop drug binding pockets that were characteristic of other serotypes in an otherwise HRV1A virus. These were used to answer the question of the role of the pocket in determining antiviral sensitivity. Although there are numerous differences in the amino acids comprising the capsids of HRV1B and HRV1A, a virus was created that demonstrated a HRV1B phenotype in terms of sensitivity to capsid binding antiviral agents. Two amino acid substitutions in the drug binding pocket of HRV1A were required to create a HRV49 pocket, and once again the phenotype of HRV49 was observed despite the fact that, except for the two amino acid changes, the virus was HRV1A.

The conclusion from these studies is that the differences in the amino acids lining the drug binding pocket play a predominant role in determining the antiviral sensitivity of a given serotype. By clustering the serotypes into groups using the statistical approach, a "consensus pocket" representing these groups could be constructed that would greatly assist in the design of compounds possessing the greatest potency and spectrum of activity.

The Future

It is clear that several factors, many of which we are only now beginning to understand, play an important role in the binding of a drug molecule to the WIN pocket. Not the least of these factors is the overall density of conformational states of the unbound hydrophobic molecule in the aqueous phase. Much of the overall binding energy is the result of the entropic gain in "squeezing" a hydrophobic molecule, such as the compounds that have been described, out of the aqueous environment. This effect is almost totally ignored in any "classical" structure-based design effort, where the focus is traditionally on the bound state of the molecule. Insight into this effect is expected to aid not only in the design of better antipicornaviral drugs, but also in other areas of structure-based drug design. Other factors affecting the overall binding free energy of a molecule may include the degree and type of aggregation of the drug molecule in the aqueous phase, a phenomenon about which very little is currently known. It is to be expected that varying amounts of aggregation may greatly affect the gain in free energy as the molecule binds to the virion.

The overall binding of a drug molecule to a picornavirion is not as rapid a process as the turnover of a substrate by an enzyme. It has been shown that the kinetics of binding would be at least several orders of magnitude slower than a typical enzyme turnover number (Fox et al., 1991), suggesting that there may be a preequilibrium step, or perhaps more than one, that occurs before the molecule descends into the WIN pocket. These observations suggest that there are other factors that affect the overall apparent binding energy that have been ignored so far simply because they are nonquantifiable. Identifying these factors may result in a deeper understanding of the general process by which small drug molecules bind to macromolecules and, it is hoped, will catalyze the attainment of a higher level of success in the design of better binding drug molecules.

The examination of a static crystal structure of a drug bound to a macromolecule is a tremendous feat in itself and is an absolutely indispensable starting point for any structure-based design. However, this is a starting point and not the end of the discovery process. The effort to design antipicornaviral agents that fit into a particular WIN pocket by simply observing the X-ray structure of a bound inhibitor cannot be totally successful. Several methods to examine the binding processes of a drug molecule to such a pocket are currently available, including Brownian dynamics and Monte Carlo methods. It is expected that tracing plausible paths by which molecules enter and bind to the WIN pockets will yield new insights into the problem of drug design. This will lead to the synthesis of the next generation of molecules that may have only a remote structural resemblance to the present ones, and that may interfere with some hitherto unknown step in the recognition and binding process.

The usefulness of such a structure-based approach is exemplified by the following experience. Mutants of HRV14 that were resistant to WIN compounds were isolated in the laboratory of Professor R. Rueckert. In one of these mutants, residue 199 in VP1 had mutated from a valine to a tryptophan. This mutant was insensitive to the drug even at high concentrations and was modeled by making a similar mutational change in the crystal structure of the original virus. It became apparent after some cursory conformational analysis of the new side chain that the tryptophan was at the heel of the binding site below the pore leading into the hydrophobic pocket. If the mode of entry of the drug was through the pore, then it appeared that the tryptophan blocked the entry.

Another explanation is possible. Modeling of the drug into the mutant with the new side chain in a low-energy conformation showed that, even if it could get past the block, there is insufficient space in the pocket to accommodate the compound. Alternatively, the mechanism of entry may not involve the drug worming its way down the pore into the pocket, but rather a breathing, munching motion in which the entire top part of the WIN pocket is opened, allowing the drug molecule to settle into it. Such large conformational changes are not altogether unknown for certain enzymes. This theory, named the "Munch Hypothesis," would explain the inactivity of the drug simply on the basis that it could not fit into the pocket of the mutant because of space constraints within the binding site rather than steric interactions with the bulky side chain that may occur during entry.

Corroborating evidence for the latter was subsequently provided when a compound with only five carbons, rather than seven, linking the two ends of the molecule was synthesized (Diana et al., 1985). When this molecule was modeled into the space of the mutant WIN pocket, it fit quite well, suggesting that there was no barrier to binding within the pocket if the drug could get there. This compound was indeed found to be as active against the mutant virus as against the wild-type, suggesting that the Munch Hypothesis was a plausible explanation for the mode of entry of these compounds and that the tryptophan is not blocking entry of the drug.

Finally, there is some evidence for the existence of naturally occurring material in the compound binding site in native HRV1A and HRV16 (Oliviera et al., 1993), as well as in poliovirus (Hogle et al., 1985). This has been termed *natural pocket factor* and appears to be displaced by the WIN compounds. The identification of this material is presently unknown. However, it has been speculated that this material increases thermal stability of the virus and that the pocket factor is extruded prior to the cell binding process. It appears that the WIN compounds also lend a degree of thermal stability to the virus. As a result of the greater degree of binding as compared to pocket factor, they cannot be expelled and so prevent uncoating or binding.

In conclusion, four main problems remain to be solved before the next major advance can be made in this fascinating odyssey to design more effective agents against human rhinoviral infections:

1. How does the drug get into its "final resting place" in the WIN pocket?
2. How can we predict the shape of the WIN pocket of one serotype from the knowledge of another?
3. Although the binding mode of this series of compounds can be predicted in HRV14, how can we predict the orientation of a molecule in this series in the WIN pocket of other serotypes?

4. Can the compound binding energy be measured with a high degree of reliability?

The detailed solution of these problems is the subject of active research in the pharmaceutical industry.

Acknowledgments

The authors would like to express their thanks to the following individuals for their contribution to this work: Edward Jaeger, Paul Kowalczyk, Daniel Pevear, Frank Dutko, John Mallamo and David Cutcliffe.

References

Abraham, G. and Colonno, R.J. 1984. Many rhinovirus serotypes share the same cellular receptor. J. Virol. 180: 814–817.

Anderson, D.E., Hurley, J.H., Nicholson, H., Baase, W.A. and Mathews, B.W. 1993. Hydrophobic core repacking and aromatic-aromatic interaction in the thermostable mutant of T4 lysozyme Ser 117 to Phe. Protein Sci. 2: 1285–1290.

Andries, K., Dewindt, J., Snoeks, R., Willebrords, R., Stokbroekx, R. and Lewi, P. J. 1991. A comparative test of fifteen compounds against all known rhinovirus serotypes as a basis for a more rational screening program. Antiviral Res. 16: 213–225.

Badger, J., Minor, I., Kremer, M.J., Oliviera, M.A., Smith, T.J., Griffith, J.P., Gueren, D.M.A., Krishnasawamy, S., Luo, M., Rossmann, M.G., McKinlay, M. A., Diana, G.D., Dutko, F.J., Fancher, M., Rueckert, R.R. and Heinz, B.A. 1988. Structural analysis of a series of antiviral agents complexed with human rhinovirus 14. Proc. Natl. Acad. Sci. U.S.A. 85: 3304–3308.

Burley, S.K. and Petsko, G.A. 1985. Aromatic-aromatic interaction: a mechanism of protein structure stabilization. Science 229: 23–28.

Burley, S.K. and Petsko, G.A. 1988. Weakly polar interaction in proteins. Adv. Protein Chem. 39: 125–192.

Cramer, R.D. III, Paterson, D. E. and Jeffrey, D. B. 1988. Comparative field analysis (CoMFA). 1. Effect of shape on binding of steroids to carrier proteins. J. Am. Chem. Soc. 110:5959–5967.

Dean, P.M. 1989. Molecular Foundation of Drug Receptor Interactions, pp. 91–93. Cambridge University Press, Cambridge, England.

Diana, G.D., Cutcliffe, D., Oglesby, R.C., Otto, M.J., Mallamo, J.P., Akullian, V. and McKinlay, M.A. 1989. Synthesis and structure-activity studies of some disubstituted phenylisoxazoles against human picornavirus. J. Med. Chem. 32: 450–455.

Diana, G.D., Kowalczk, P., Treasurywala, A.M., Oglesby, R.C., Pevear, D.C. and Dutko, F.J. 1992. CoMFA analysis of the interaction of antipicornavirus compounds in the binding pocket of human rhinovirus 14. J. Med. Chem. 35: 1002–1008.

Diana, G.D., McKinlay, M.A., Brisson, C.J., Zalay, E.S., Miralles, J.V. and Salvador, U.J. 1985. Isoxazoles with antipicornavirus activity. J. Med. Chem. 28: 748–752.

Diana, G.D., Oglesby, R.C., Akullian, V., Carabateas, P.M., Cutcliffe, D., Mallamo, J.P., Otto, M.J., McKinlay, M.A., Maliski, E.G. and Michalec, S.J. 1987. Structure-activity studies of 5-[4-(4,4-dihydro-2-oxazolyl)phenoxy]-3-methylisoxazoles: inhibitors of picornavirus uncoating. J. Med. Chem. 30: 383–388.

Diana, G.D., Otto, M.J., Treasurywala, A.M., McKinlay, M.A., Oglesby, R.C., Maliski, E.G., Rossmann, M.G. and Smith, T.J. 1988. Enantiomeric effects of homologues of disoxaril on the inhibitory activity against human rhinovirus 14. J. Med. Chem. 31: 540–544.

Diana, G.D., Salvador, U.J., Zalay, E.S., Johnson, R.E., Collins, J.C., Johnson, D., Hinshaw, W.B., Lorenz, R.R., Theilking, W.H. and Pancic, F. 1977. Antiviral activity of some β-diketones. 1. Aryl alkyl diketones: in vitro activity against both RNA and DNA viruses. J. Med. Chem. 20: 750–761.

Diana, G.D., Treasurywala, A.M., Bailey, T.R., Oglesby, R.C., Pevear, D.C. and Dutko, F.J. 1990. A model for compounds active against human rhinovirus 14 based on X-ray crystallography. J. Med. Chem. 33: 1306–1311.

Fowler, P.W. and Moore, G.J. 1988. Calculation of the magnitude and orientation of electrostatic interactions between small aromatic rings in peptides and proteins: implications for angiotensin II. Biochem. Biophys. Res. Commun. 153: 1296–1300.

Fox, M.P,., McKinlay, M.A., Diana, G.D. and Dutko, F.J. 1991. Binding affinities of structurally related human rhinovirus capsid binding compounds are related to their activities against human rhinovirus type 14. Antimicrob. Agents Chemother. 35: 1040–1047.

Fox, M.P., Otto, M.J., Shave, W.J. and McKinlay, M.A. 1986. Prevention of rhinovirus and poliovirus uncoating by Win 51711, a new antiviral drug. Antimicrob. Agents Chemother. 30: 110–115.

Greve, J.M., Davis, G., Meyer, A.M., Forte, C.P., Connolly, Y.S., Marlor, C.W., Kamarck, M.E. and McClelland, A. 1989. The major human rhinovirus receptor is ICAM-1. Cell 56: 839–847.

Gwaltney, J.M. Jr. 1985. Principles and practices of Infectious Diseases, 2nd ed. John Wiley & Sons, New York.

Hadfield, A.T., Oliviera, M.A., Kim, K.H., Minor, I., Kremer, M.J., Heinz, B.A., Shepard, D., Pevear, D.C., Rueckert, R.R. and Rossmann, M.G. 1996. Structural studies on human rhinovirus 14 drug resistant compensation mutants. Submitted to J. Mol. Biol.

Hamparian, V.V., Colonno, R.J., Cooney, M.K., Dick, E.C., Gwaltney, J.M., Hughes, J.H., Jordon, W.S. Jr., Kapikian, A.Z., Mogabgab, W.J., Monto, A., Phillips, C.A., Rueckert, R.R., Scheible, J.H., Stott, E.J. and Tyrell, D.A.J. 1987. A collaboration report: rhinoviruses—extension of the numbering system from 89 to 100. Virology. 159: 191–192.

Hayden, F. G., Andries, K. and Jannsen, P.A. 1992. Safety and efficacy of intranasal pirodavir (R77975) in experimental rhinovirus infection. Antimicrob. Agents Chemother. 36: 727–732.

Hayden, F.G., Gwaltney, J.M. Jr. and Colonno, R.J. 1988. Modification of experimental rhinovirus colds by receptor blockage. Antiviral Res. 9: 233–247.

Heinz, B.A., Reuckert, R.R., Shepard, D.A., Dutko, J.J., McKinlay, M.A., Fancher, M., Rossmann, M.G., Badger, J. and Smith, T.J. 1989. Genetic and molecular analysis of spontaneous mutants of human rhinovirus-14 that are resistant to an antiviral compound. J. Virol. 63: 2476–2485.

Higgins, P.G., Barrow, G.I. and Tyrell, D.A.J. 1990. A study of the efficacy of the bradykinin antagonist, NPC, in rhinovirus infections in human volunteers. Antiviral Res. 14: 339–344.

Hofer, F. 1994. Members of the low density lipoprotein family mediate cell entry of a minor-group common cold virus. Proc. Natl. Acad. Sci. U.S.A. 91: 1839–1842.

Hogle, J.M., Chow, M. and Filman, D.J. 1985. Three-dimensional structure of poliovirus at 2.9 Å resolution. Science. 229: 1358–1465.

Jaeger, E.F., Pevear, D.C., Felock, P.J., Russo, G.M. and Treasurywala, A.M. 1995. Genetic algorithm based method to design a primary screen for antirhinovirus agents in computer aided molecular design: Applications in agrochemicals, materials, and pharmaceuticals. ACS Symposium Series 589. pp. 139–155. American Chemical Society, Washington, D.C.

Karlstrom, G., Linse, P., Wallqvist, A. and Johnson, B. 1983. Intermolecular potentials for the H_2O-C_6H_6 and the C_6H_6-C_6H_6 systems calculated in an ab initio SCF CI approximation. J. Am. Chem. Soc. 105: 3777–3782.

Kim, S., Smith, T.J., Chapman, M.S., Rossmann, M.G., Pevear, D.C., Dutko, F.J., Felock, P.J., Diana, D.G. and McKinlay, M.A. 1989. Crystal structure of human rhinovirus serotype 1A (HRV-1A). J. Mol. Biol. 210: 91–111.

Kim, S., Willingmann, P., Gong, Z.X.C., Kremer, M.J., Chapman, M.S., Minor, I., Oliviera, M.A., Rossmann, M.G., Andries, K., Diana, G.D., Dutko, F.J., McKinlay, M.A. and Pevear, D.C. 1993. A comparison of the antirhinoviral drug binding pocket in HRV 14 and HRV 1A. J. Mol. Biol. 230: 206–227.

Lau, W.F. and Pettitt, M.B. 1989. Selective elimination of interactions: a method for assessing thermodynamic contributions to ligand binding with application to rhinovirus antivirals. J. Med. Chem. 32: 2542–2547.

Levitt, M. and Perutz, M.F. 1988. Aromatic rings act as hydrogen bond acceptors. J. Mol. Biol. 201: 751–754.

Mallamo, J.P., Diana, G.D., Pevear, D.C., Dutko, F.J., Chapman, M.S., Kyung, H.K., Minor, I., Oliviera, M. and Rossmann, M.G. 1992. Conformationally restricted analogues of disoxaril: a comparison of the activity against human rhinovirus types 14 and 1A. J. Med. Chem. 35: 4690–4695.

McKinlay, M.A., Pevear, D.C. and Rossmann, M.G. 1992. Treatment of picornavirus cold by inhibitors of viral uncoating and attachment. Annu. Rev. Microbiol. 46: 635–654.

Monto, A.S., Schwartz, S.A. and Albrecht, J.K. 1989. Ineffectiveness of postexposure prophylaxis of rhinovirus infection with low-dose intranasal alpha 2b interferon in families. Antimicrob. Agents Chemother. 33: 387–390.

Oliviera, M.A., Zhao, R., Lee, W.M., Kremer, M.J., Minor, I., Rueckert, R.R., Diana, G.D., Pevear, D.C., Dutko, F. J., McKinlay, M.A. and Rossmann, M.G. 1993. The structure of human rhinovirus 16. Structure 1: 51–68.

Pevear, D.C., Fancher, M.J., Felock, P.J., Rossmann, M.G., Miller, M.S., Diana, G.D., Treasurywala, A.M., McKinlay, M.A. and Dutko, F.J. 1989. Conformational changes in the floor of the canyon blocks adsorption to HeLa cell receptors. J. Virol. 63: 2002–2007.

Rossmann, M.G., Arnold, E., Erickson, J. W., Frankenberger, E. A., Griffith, J.P., Hecht, H.J., Johnson, J. E., Kamer, G., Luo, M., Mosser, A. G., Rueckert, R. R., Sherry, B. and Vriend, G. 1985. Structure of a human common cold virus and functional relationship to other picornaviruses. Nature. 317: 145–153.

Serrano, L., Bycroft, M. and Fersht, A.R. 1990. Aromatic-aromatic interactions and protein stability investigation by double-mutant cycles. J. Mol. Biol. 218: 465–475.

Shepard, D. A., Heinz, B. A. and Rueckert, R.R. 1993. Win compounds inhibit both attachment and eclipse of human rhinovirus 14. J. Virol. 67: 2245–2254.

Singh, J. and Thornton, J.M. 1985. The interaction between phenylalanine rings in proteins. FEBS Lett. 191: 1–6.

Smith, T.J., Kremer, M.J., Luo, M., Vriend, G., Arnold, E., Kamer, G., Rossmann, M.G., McKinlay, M.A., Diana, G.D. and Otto, M.J. 1986. The site of attachment in human rhinovirus 14 of antiviral agents that inhibit uncoating. Science 233: 409–419.

Sperber, S.J. and Hayden, F.G. 1988. Chemotherapy of rhinovirus colds. Antimicrob. Agents Chemother. 32: 409–419.

Staunton, D.E., Merluzzi, V.J., Rothlein, R., Barton, R., Marlin, S.D. and Springer, T.A. 1989. A cell adhesion molecule ICAM-1 is the major surface receptor for rhinoviruses. Cell. 56: 849–853.

Tomassini, T.E., Graham, D., DeWitt, C.M., Lineberger, D.W., Rodkey, J.A. and Colonno, R.J. 1989. cDNA cloning reveals that the major group of rhinovirus receptors on HeLa cells is intracellular adhesion molecule-1. Proc. Natl. Acad. Sci. U.S.A. 86: 4907–4911.

Turner, R.B., Durcan, J.F., Albrecht, J.K. and Crandall, A.S. 1989. Safety and tolerance of ocular administration of recombinant alpha interferon. Antimicrob. Agents Chemother. 33: 396–397.

Uncapher, G.R., DeWitt, C.M. and Colonno, R.J. 1991. The major human rhinovirus serotype families contain all but one human rhinovirus serotype. Virology. 180: 814–817.

Zeichardt, H., Otto, M.J., McKinlay, M.A., Willingmann, P. and Habermehl, K.O. 1987. Inhibition of poliovirus uncoating by disoxaril (WIN 51711). Virology. 60: 281–285.

17.

Rational Design of HIV Protease Inhibitors

C. NICHOLAS HODGE, T.P. STRAATSMA,

J. ANDREW McCAMMON, &

ALEXANDER WLODAWER

Replication of the human immunodeficiency virus (HIV) requires the action of a virally encoded protease to produce components of progeny virions within infected cells (Kohl et al., 1988). Therefore much effort has been devoted to developing specific inhibitors of protease in the hope that these may prove valuable in the management of HIV infections. Fortunately, the structures of protease and many of its inhibitor complexes have been determined by X-ray crystallography. This chapter describes how such structural data are being used in computer modeling and computer simulation studies to speed the discovery of clinically useful protease inhibitors. The first section outlines the crystallographic results that have been obtained for protease and protease–inhibitor complexes. A more detailed account has been published (Wlodawer and Erickson, 1993). Subsequent sections describe how computer modeling has helped in the discovery of a promising new class of protease inhibitors, and how free energy simulations can help in deciding which inhibitors within a given class will have the highest affinity for protease.

HIV-1 Protease Structure

The first indication that retroviral genomes might encode an aspartic protease came from analyses of comparative sequences. The deduced protein signature sequence Asp-Thr-Gly, as well as some other weak but critical homology with the eukaryotic aspartic proteases, led Toh et al. (1985) to suggest that HIV and

451

other retroviruses had proteases that resembled pepsin. Subsequently, a model of HIV-1 protease based on the structure of endothiapepsin was proposed by Pearl and Taylor (1987). Experimentally, the structure of the Rous sarcoma virus (RSV) protease confirmed this hypothesis (Miller et al., 1989a). This structure was followed almost immediately by the crystal structure determination of recombinant HIV-1 protease (Navia et al., 1989), independently verified using the synthetic enzyme (Wlodawer et al., 1989) and later using recombinant HIV-1 protease (Lapatto et al., 1989; Spinelli et al., 1991).

The general topology of the HIV-1 protease monomer is similar to that of a single domain in pepsin-like aspartic proteases (Fig. 17.1). The N-terminal

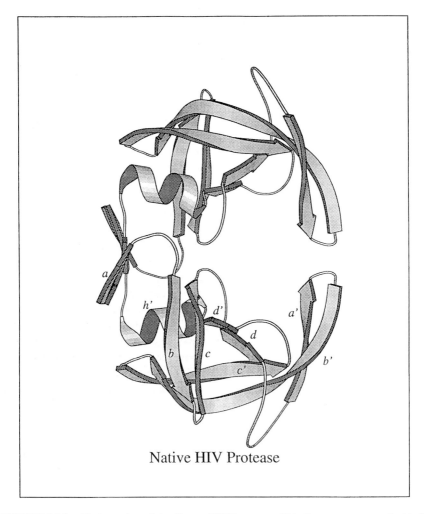

Native HIV Protease

FIGURE 17.1   Chain tracing of the dimer of HIV protease. This figure was prepared with the program MOLSCRIPT (Kraulis, 1991).

half of the chain consists of β-strands *a, b, c*, and *d* and a broad loop that corresponds to the *h* helix found in RSV protease. The second half of the molecule is topologically related to the first half by an approximate intramolecular twofold axis. The helix *h'* is very well defined and is followed by a straight C-terminal β-strand *q*, which forms the inner part of the dimer interface. Four of the β-strands in the molecular core are organized into a ψ-shaped sheet characteristic of all aspartic proteases.

The active site triad (Asp 25, Thr 26, and Gly 27) is located in a loop whose structure is stabilized by a network of hydrogen bonds (Fig. 17.2). The carboxylate groups of Asp 25 from both chains are nearly coplanar and show close contacts involving the OD1 atoms. The network is quite rigid because of an interaction called the "fireman's grip" (Blundell et al., 1985), in which each Thr 26 OG1 accepts a hydrogen bond from the Thr 26 main chain amide group of the opposing loop. Thr 26 also donates a hydrogen bond to the carbonyl oxygen atom of residue 24 on the opposite loop. Although the central features of the catalytic site are very similar between retroviral and cellular aspartic proteases, the residue following the triad differs, with alanine invariably present in retroviral proteases, whereas serine or threonine are most common in the pepsins. Another difference is the presence of only one flap in the pepsins, whereas two twofold-related flaps are present in HIV-1 protease. The flap is a β-hairpin that covers the active site and participates in the binding of inhibitors and substrates. A water molecule is bound to the two aspartates in both RSV and HIV proteases (Miller et al., 1989a; Wlodawer et al., 1989).

Although the emerging picture of HIV-1 protease apoenzyme was of considerable interest, it was clear that structures of the complexes with inhibitors

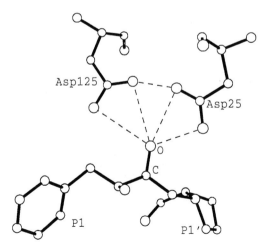

FIGURE 17.2 Active site of the HIV-1 protease, with the central part of the inhibitor JG-365 (Swain et al., 1990) shown interacting with the catalytic aspartates. Potential hydrogen-bonded interactions of the oxygen (○) in the nonscissile insert are marked with dashed lines. The carbon atom that is usually in the *S* conformation is marked as C.

were necessary for a fuller explanation of the activity, and to provide leads for drug design. The first reported X-ray crystal structure of an HIV protease–inhibitor complex was with the reduced amide hexapeptide, MVT-101 (Miller et al., 1989b). This inhibitor is not very potent but, because of wide dissemination of the coordinates, the structure has been used in many theoretical and modeling calculations (Wlodawer and Erickson, 1993).

Structures of HIV-1 protease complexed with hydroxyethylene-containing peptide mimics have been reported for inhibitors of various sizes, from tri- to octapeptide analogs, and side chain compositions (Jaskólski et al., 1991; Dreyer et al., 1992; Graves et al., 1992; Thompson et al., 1992). The hydroxyl group on the inhibitor is approximately centered about the two active site aspartic acid carboxylates, close to the position occupied by the water molecule in the active site of all uncomplexed aspartic proteases. Hydrogen bonding analysis of the hydroxyethylene inhibitor complexes leads to the conclusion that one of the two aspartates must be protonated in order to bind to the inhibitor hydroxyl group. The two short "tripeptide" mimics, Ro-31-8588 (Graves et al., 1992) and L-689,502 (Thompson et al., 1992), are notable for their subnanomolar K_i values for HIV protease. Both molecules are blocked at their N termini by bulky *t*-butoxy groups coupled to the P_1 amino acid via a carbamate linkage, and their backbone structures are very similar. (P_1 denotes the first inhibitor residue on the N-terminal side of the bond that would be hydrolyzed in a substrate, P_2 the second, and so forth. Similarly, P_1', P_2', and so on denote inhibitor residues on the C-terminal side, and S_1, S_1', and so on denote binding sites for these residues in protease.) The ether oxygen in this linkage was found to be critical for potency (Graves et al., 1992).

A series of four structures of hydroxyethyene-containing hexapeptide inhibitors that differed only in the P_1' side chain gave an explanation for structure–activity relationship (SAR) data that indicated a decreasing dependence of K_i on inhibitor length as the size of the P_1' side chain increases (Dreyer et al., 1992). Structures of complexes were solved with inhibitors containing Phe-[CH(OH)]CH$_2$]-X-, in which X was Gly, Ala, NorVal, or Phe. The conformations of all four inhibitors were similar except that in the case of the X = Gly compound the backbone curled back slightly to place the P_3' Val side chain partially into the unoccupied S_1' pocket. SAR studies with this and a similar series indicated that, as the size of the P_1' substituent increases, the relative potency gain that is realized by the addition of a P_3' substituent decreases. Two other structures of hydroxyethylene-based inhibitors were solved in a hexagonal crystal form, and showed clear disorder of their mode of binding (Murthy et al., 1992). The structure of a decapeptide hydroxyethylene inhibitor that mimicked the cleavage site of an opportunistic target of HIV-1 protease, namely a peptide of lactate dehydrogenase, has also been reported (Tomaszek et al., 1992).

An *S* hydroxyl group (Fig. 17.2) is the preferred stereochemical configuration in all the hydroxyethylene inhibitors described so far in the literature. The reason for this preference is that the *S* hydroxyl group is positioned to interact with both active site carboxylate side chains, whereas for the *R* hydroxyl group to make similar interactions would necessitate major conformational rearrangements in the inhibitor backbone. These changes would most likely lead to a higher energy inhibitor conformation, weaker nonbonded interactions with the enzyme, or both. These possibilities were investigated by computer simulation studies, as described later. It should be noted that the S hydroxyl group is in *anti* configuration with respect to the preceding R group of the P_1 substituent.

A dihydroxyethylene insert has been used to prepare inhibitors of high affinity for HIV protease. The crystal structure of a complex of HIV protease with the diol U-75875 indicates that the N-terminal hydroxyl group interacts approximately equally well with both active site aspartates, and in a manner similar to the isosteric S hydroxyl group of the monohydroxy inhibitors (Thanki et al., 1992). For U-75875, the second hydroxyl group was able to form a hydrogen bond with only one of the two aspartates, Asp 25. Thus, the two hydroxyl groups are not equivalent in the bound conformation, and the contribution of the second hydroxyl group to inhibitor potency remains unclear.

The hydroxyethylamine insert represents a significant departure from the reduced amide and hydroxyethylene isosteres because an extra bond is now being inserted between the P_1 and P_1' residues. The structure of a complex with the heptapeptide [CH(OH)-CH$_2$-N] analog JG-365 indicated that, in spite of the added bond and the constrained prolyl group, the inhibitor backbone approximated the conformations observed for both the reduced amide and hydroxyethylene-based inhibitor structures (Swain et al., 1990). The JG-365 preparation was a racemic mixture of the R and S configurations at the hydroxyl carbon atom. However, only the S isomer was observed in the crystal structure. These results indicated that the $S(anti)$ isomer was preferred, and the crystal structure revealed that the S hydroxyl group superimposed onto the S hydroxyl group for the hydroxyethylene-containing inhibitors.

In marked contrast to the preference for the *anti* hydroxyl group configuration in all previously described HIV protease inhibitors (and also for aspartic protease peptidomimetic inhibitors) that contained hydroxyethylene inserts, the R, or *syn*, hydroxyl group configuration is preferred in hydroxyethylamine inserts for inhibitors that lack a P_3' residue. One such compound, Ro-31-8959, is the first HIV protease inhibitor to be administered to humans, and it is currently approved as an anti-AIDS drug under the name Sequinavir. The R hydroxyl group compounds must be binding in a substantially different mode than that for the S hydroxyl groups seen in the previous structures. The predicted binding mode for Ro-31-8959 with HIV protease was verified experimentally with the crystal structure determination of the complex (Krohn et al., 1991). As predicted, the Diq and t-butyl groups bound in the S_1' and S_2' subsites, respectively. The Diq carbonyl group is hydrogen bonded via the buried water to the flaps. However, the nitrogen atom of the Diq moiety had the opposite configuration to that of the proline ring in JG-365, and the nitrogen of the t-butylamide is displaced considerably ($\sim$1.8 Å) from the normal (P_2') N position and as a result cannot form a hydrogen bond with the Gly 27 carbonyl oxygen. More importantly, the position of the backbone in this region prohibits further chain extension into the S_3' subsite. Owing to the large bulk of the Diq group, a hydroxyl group in the S configuration cannot be as readily accommodated with this moiety in the P_1' position as it can for the case where proline occupies the P_1' site. Thus, the crystallographic analysis provides a clear structural basis for interpreting the SAR of the hydroxyethylamine inhibitors.

The insert based on the unnatural amino acid statine can be considered to be a truncated hydroxyethylene transition isostere of Leu-Gly in which the P_1 amide nitrogen is absent. In addition, the symmetrically hydrogen-bonded P_1 and P_2' amide nitrogens are separated by only five bonds as opposed to six as in a peptide substrate or in hydroxyethylene or reduced amide isosteres. Acetyl-pepstatin, a general aspartic protease inhibitor, is reported to inhibit HIV protease with a K_i of 20 nmol. The crystal structure of acetyl-pepstatin complexed with HIV protease has been solved at 2.0- to 2.5-Å resolution (Fitzgerald et al., 1990; Swain et al., 1992). The pattern of hydrogen bonding and subsite interactions for acetyl-pepstatin was similar to what had been observed for the hydroxyethylene inhibitor–HIV protease complexes, with the main difference being the lack of a P_1' residue occupying the S_1' subsite.

All of the inhibitors discussed thus far are based on the peptidic substrates, and their main chain runs in the same direction. However, because HIV protease is a homodimer of two identical subunits, it was postulated that strictly symmetric or quasi-symmetric inhibitors might show improved potency and/or specificity. A number of structures of such inhibitors have been solved so far. Some of them, such as A-74704, were constructed by running the polypeptide in the N-to-C direction (Erickson et al., 1990), whereas others, such as L-700,417, run C to N (Bone et al., 1991). The nonscissile insert may contain one or two hydroxyl groups, and thus the inhibitors may have fully symmetric or quasi-symmetric character. Another class of symmetric inhibitors contain phosphinate inserts (Abdel-Meguid et al., 1993). Nevertheless, the mode of binding of the inhibitor does not necessarily follow the symmetry exactly, and some symmetric inhibitors were reported to bind to the protease in an asymmetric fashion (Dreyer et al., 1993).

Although most of the structural information available in the literature is for HIV-1 protease, crystals of HIV-2 protease complexed with the inhibitors are also available. This enzyme is roughly 50% identical in sequence to HIV-1 protease, and some of the complexes studied were the same in both cases (e.g., for the inhibitor U-75875; Mulichak et al., 1993). Another structure utilized a reduced peptide bond inhibitor BI-LA-398, reminiscent of MVT-101 (Tong et al., 1993). None of these structures has an unexpected binding mode, and thus the structural data for both proteases may be interpreted together.

Much less has been published about the binding of nonpeptidic inhibitors of HIV protease. One such compound is UCSF-8 (Rutenber et al., 1993), a modification of the drug haloperidol. This compound binds in a very different mode than the peptidic compounds, and indeed in a way different than postulated by the predictions that led to its discovery. Nonpeptide compounds based on cyclic urea, however, bind in a manner similar to the peptidic inhibitors, while showing much improved pharmacokinetic properties (Lam et al., 1994). These compounds are discussed later in more detail.

A number of general statements can be made on the basis of all available structures of the complexes of HIV protease with inhibitors, especially peptidic ones. The protease side chains comprising the pockets S_1 and S_1', with the exception of the active site aspartates, are mostly hydrophobic. The side chains of the active site aspartates and the main chain hydroxyl group of those inhibitors that contain such a central group are involved in polar contacts. Almost all of the inhibitors discussed here have hydrophobic moieties at P_1 and P_1'.

Although the S_2 and S_2' pockets are hydrophobic, both hydrophilic and hydrophobic residues can occupy these sites. The P_2 and P_2' hydrophobic side chains are observed in different orientations for the different inhibitors, forming contacts with different groups in the enzyme binding pocket. For inhibitors containing Asn or Gln, the amide side chains are also stabilized by polar contact with the carbonyl oxygen of the previous residues in the inhibitor. There are also polar contacts between some P_2/P_2' amide groups and polar side chains of the protease.

Distal to S_2/S_2', the subsites are less well defined, and diverse side chains are accommodated in the S_3 and S_3' subsites. In addition, some inhibitors do not have groups that occupy an S_3 subsite. Where P_3 groups do extend into the S_3 subsite, they usually can be superimposed on one another, although the side chain orientations vary considerably. The subsites S_4, S_4', S_5, and S_5' do not represent real pockets, but rather loose binding regions on the surface of the enzyme, and are quite poorly defined.

HIV protease undergoes a considerable structural change upon inhibitor binding. The change is particularly apparent in the flap region (residues 45 to 55), where the displacement in α-carbon coordinates may be as large as 7 Å, with a quarter of the α-carbon atoms differing by over 1 Å. However, the enzyme structure is well conserved in the different complexes, with rms changes usually being in the range of 0.6 Å. These differences are well within the range of agreement for protein structures refined independently, or crystallized in different space groups.

All of the peptidic inhibitors are bound in the protease active site in an extended conformation. When they are superimposed upon one another, their functional elements are in overall alignment. The contacts between the main chain of the inhibitor and the protease are very similar for all the complexes, with the nonhydrolyzable scissile bond analog of each inhibitor aligned with the active site aspartate carboxyl groups (Asp 25/125). In all cases except for MVT-101 (which has a reduced peptide bond as a nonscissile moiety), the hydroxyl group at the nonscissile junction is positioned between the protease aspartate carboxyl groups within hydrogen bonding distance to at least one carboxylate oxygen of each aspartate. The hydrogen bonds are made mostly between the main chain atoms of both the enzyme and the inhibitor, and follow a similar pattern in most complexes.

A feature common to almost all complexes of HIV-1 protease is a buried water molecule that bridges the P_2 and P_1' carboxyl groups of the inhibitor and

Ile 50 and Ile 150 amide groups of the flaps. This water is approximately tetrahedrally coordinated and is completely inaccessible to solvent (Wlodawer et al., 1989). It has been suggested that functional substitution of this water might lead to novel synthetic targets for HIV protease inhibitors (Thompson et al., 1992; Erickson et al., 1990). One approach to such inhibitors is described later.

Computer Modeling and the Discovery of New Inhibitors

The insight into the unique molecular structure of protease described in the preceding section was instrumental in guiding the research covered in this section: the use of modeling and visualization methods at DuPont Merck in the design of nonpeptidyl inhibitors as clinically useful acquired immunodeficiency syndrome (AIDS) therapies. Once it became clear that HIV required protease for replication, many industrial and academic laboratories began the search for agents that would both block the activity of the enzyme and serve as leads for a potential treatment for AIDS. Discovery of useful drugs that exert their therapeutic effect through the inhibition of aspartyl protease enzymes presents an especially difficult challenge. In order to understand the solutions described in this section, the major problems that must be overcome in bringing any drug, but particularly large peptidyl inhibitors of aspartyl proteases, from a scientifically interesting tool to a clinically useful drug are summarized here.

1. *Specificity*—Proteases are ubiquitous and frequently cleave at many sites. It is important to inhibit only the target protease in the presence of many essential, but mechanistically similar, host proteases.
2. *Oral availability and tissue distribution*—Peptidyl inhibitors exhibit notoriously poor pharmacokinetic properties (Plattner and Norbeck, 1990). Except for acute diseases, a useful drug must be delivered orally, or possibly nasally or transdermally. Even in cases in which an injectable agent is acceptable, peptides tend to exhibit poor half-lives and tissue distribution. Designing structures that retain the affinity and specificity of peptidyl inhibitors while imparting favorable pharmacokinetic properties is the fundamental challenge of peptidomimetic drug research (Olson et al., 1993).
3. *Stability*—Any drug candidate must be reasonably stable, both in vivo in terms of half-life, and chemical stability to formulation and storage. Peptides are generally chemically stable but are subject to cleavage by peptidases in vivo.
4. *Novelty*—The average cost of discovery, testing, and obtaining regulatory approval for a drug now exceeds $250 million. To recoup this investment, a clear proprietary right to the structures of interest is essential.
5. *Accessibility*—Even for life-threatening illnesses, treatment must be affordable to be useful. Successful drugs require reasonably inexpensive process chemistry that can be scaled up to meet patient needs.

6. *Safety*—The drug cannot interfere with the myriad other biologic processes in a way that compromises the health or comfort of the patient. This again requires a highly specific, stable chemical species and is a challenge in any drug discovery project.

A great deal of research on the inhibition of aspartyl proteases in general (Rich, 1986), and many years of attempts to design clinically useful renin inhibitors as antihypertensives in particular (Greenlee and Siegel, 1991, and references therein), provided medicinal chemists with a significant head start in the discovery of protease inhibitors. For example, a number of scissile bond replacements had been reported that bind with high affinity to the active site of renin and other aspartyl proteases, provided that appropriate flanking amino acids were appended. Groups at several laboratories examined these isosteres, discovered variations, and developed highly potent and specific inhibitors of protease; in addition to the work described in the previous section, important early examples include those of Dreyer et al. (1989), McQuade et al. (1990), Roberts et al. (1990), and Lyle et al. (1991).

When the National Cancer Institute group first described the C2-symmetric dimeric structure of RSV protease (Miller et al., 1989a), and it was suggested that HIV protease contained similar features (Pearl and Taylor, 1987), researchers at several locations independently drew the same conclusion: The symmetry could be used to advantage in the design of specific inhibitors of HIV protease. Although peptidyl substrates are by nature asymmetric, it was reasoned that a symmetric inhibitor might uniquely occupy a symmetric active site, and exhibit weaker affinity for monomeric human aspartyl proteases such as renin. Several reports highlight the successful application of the symmetry concept to the design of potent, specific inhibitors of HIV protease with outstanding in vitro antiviral activity (Erickson et al., 1990; Büdt et al., 1991; Jadhav et al., 1991; Babine et al., 1992; Jadhav and Woerner, 1992). Unfortunately, but not unexpectedly, neither the symmetric nor the asymmetric classes of compounds provided favorable pharmacokinetic profiles in animals. Continued optimization of acyclic peptidyl molecules has resulted in recent reports of compounds with reasonable bioavailability, although clinical data are not yet known. This success appears to be due to systematic decreases in molecular weight and modulation of solubility of the acyclic peptidyl molecules (Norbeck et al., 1992; Alteri et al., 1993; Dorsey et al., 1993).

Although knowledge of the three-dimensional features of the enzyme, of its mechanism, and of substrate specificity led to the extremely potent antiviral agents described earlier, effective compounds for the scientific community's primary goal, an effective treatment for AIDS, were not identified. Lam et al. (1994a) therefore set the objective of designing a completely novel series of HIV protease inhibitors with the following properties: subnanomolar potency against the enzyme, greater than 10^4 specificity ratio for mammalian aspartyl proteases, submicromolar antiviral efficacy, novelty, safety, accessibility, and greater than 20% oral bioavailability. They believed that the understanding of the SARs from the diol peptidyl inhibitors, the increasing availability of detailed information on the enzyme and complexed peptidyl inhibitors at the molecular level, and their close integration of synthetic chemistry with computational and modeling expertise provided a better chance of success in this venture than in competing with larger laboratories in attempting to optimize peptides. If successful, a breakthrough would occur in understanding HIV protease inhibition and a new class of molecules to treat AIDS would be obtained.

The minimum essential features for binding to HIV protease were defined and a model of a symmetric inhibitor in the active site of protease was generated from published crystal structures (Fig. 17.3). Several three-dimensional models ("pharmacophores") were constructed to represent the relative spatial orientation of critical groups (Lam et al., 1994). These models included the structural water present in all solved structures that serves to anchor the mobile flaps of protease to the P_1/P_1' amide carbonyl oxygens of linear peptidyl inhibitors (e.g., Miller et al., 1989b; Erickson, 1990), as well as the lipophilic phenyl rings and the hydroxyl group (Fig. 17.4).

Again the availability of structural information at the molecular level was important to the progress of the research. The distance constraints that were employed in these models were derived from crystal structures of inhibitors complexed to the active site. Although these distances can sometimes be inferred from activity relationships in the absence of receptor structural information (Waller and Marshall, 1993), such approaches do not always provide unambiguous spatial resolution and are prohibitive computationally when applied to ligands with many degrees of freedom, such as these hexapeptide analogs. The position of, or even necessity for, the structural water that was successfully incorporated into nonpeptidyl inhibitors (box D in Fig. 17.4) would

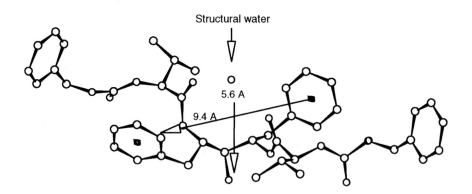

Structural water

5.6 A

9.4 A

Peptidyl inhibitor in HIVPR active site

FIGURE 17.3 Abbott inhibitor A74704 in HIV protease active site. Note the relative spatial orientations of the structural water molecule, the two interior phenyl rings, and the hydroxyl group at the bottom center of the molecule (distances in the crystal structure are shown). A model of a diol inhibitor (P9941; Lam et al., 1994) with similar binding features was used to generate the distance model shown in Figure 17.4.

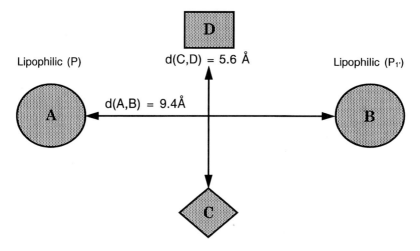

Bidentate H-bond acceptor

Lipophilic (P)   d(C,D) = 5.6 Å   Lipophilic (P₁′)

D

d(A,B) = 9.4Å

A   B

C

H-bond donor/acceptor

FIGURE 17.4   One of several "pharmacophores" or distance models that retain only the essential binding features of the inhibitor from Figure 17.3. The groups A, B, C, and D represent the two phenyl rings, the structural water, and the hydroxyl group, respectively. A new technique, three-dimensional database searching, was used to locate the three-dimensional structures of known organic molecules from the Cambridge Crystallographic Database (Allen, 1991) that matched these critical features and distances within certain constraints (Lam et al., 1994).

likely never have been inferred from a model based only on the binding affinity of inhibitors to enzyme.

Distances from a model that contained the important features of the peptidyl inhibitors were used to search a library of three-dimensional structures of organic molecules for fits that matched within certain constraints (Bures et al., 1992; Martin, 1992; Lam et al., 1994). The Cambridge Crystallographic Database yielded a number of compounds that satisfied these constraints. Visualizing these compounds superimposed onto inhibitor in the active site of protease (Fig. 17.5) yielded conceptual structures that quickly evolved into real targets (Lam et al., 1994), as summarized in Figure 17.6. Cyclic ureas (IV in Fig. 17.6), the most attractive and accessible structures that the team identified, were synthesized from available diol precursors (Lam et al., 1993, 1996) and found to bind to the enzyme with affinity constants in the low micromolar range. To understand and improve this binding, low-energy conformations of this molecule were calculated using distance geometry (Blaney et al., 1984) and quenched dynamics (O'Connor et al., 1992) techniques. These studies confirmed the conformational features that were envisioned during the design process and subsequently supported by the observation of identical conformations in single-crystal X-ray structures of free inhibitors. The synthesis and remarkable activity of XK216, with a molecular weight of only 407, a K_i of 3 nmol/

Early inhibitor based on distance model

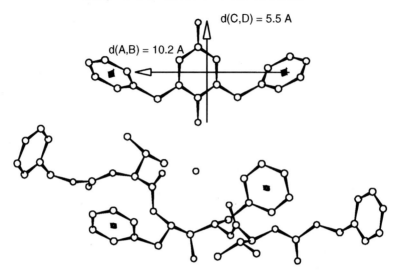

FIGURE 17.5   Elaboration of an initial "hit" from the Cambridge Crystallographic Database (biphenylphenol I; Fig. 17.6) led to an early conceptual HIV protease ligand. The computed conformation and distances are shown and matched with A74704.

l, and an antiviral IC_{90} of 4 μmol/l, confirmed these speculations. This structure met many of our initial objectives, including outstanding oral availability, and the hunt was on for a compound with optimum properties as a drug candidate and to elaborate the scope of compound space that this concept embraced.

The contribution of computational and visualization techniques to this project did not end with the identification of the lead series. As described elsewhere, a simple, rapid, and informative binding model was developed to assist the efforts of medicinal chemists in cyclic urea optimization, and in the design of new scaffolds based on the pharmacophore concept shown earlier (Hodge et al., 1994). An empirical force field (Dauber-Osguthorpe et al., 1988; Maple et al., 1988) was found to be effective at reproducing the experimentally de-

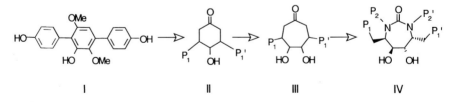

FIGURE 17.6   Conceptual evolution of cyclic ureas I to IV from initial hit 1. The cyclic ureas evolved by adding features to the initial hit that were expected to improve binding, increase synthetic accessibility, and provide desirable in vivo properties such as oral availability.

termined low-energy conformations of cyclic ureas when a quenched dynamics protocol similar to that reported by O'Connor et al. (1992) was employed. Simulations of the molecular motions of these inhibitors at high temperatures were calculated for 200 ps, with intermediate structures collected and energy minimized every picosecond. Generally, the experimentally determined conformation was located within 20 to 30 ps. Rank ordering of the strain energies of the calculated structures consistently placed the experimentally observed conformation at or among the lowest energy structures. Solution nuclear magnetic resonance and single-crystal X-ray both confirmed that N,N'-substituted cyclic ureas exist preferentially in the axial-equatorial-equatorial-axial conformation regardless of stereochemistry at the hydroxyl oxygens. These molecules adopt the same conformation when bound to protease.

This simple technique was extremely useful in screening structural modifications to cyclic ureas as candidates for synthesis. If a proposed structure fell into low-energy conformations that projected the key functionality into the correct regions of space (see Figs. 17.3, 17.4, and 17.5), then it was considered an attractive candidate if it met other medicinal chemistry requirements discussed earlier. In contrast, if significant strain enthalpy (>2 kcal/mol) were required to force the proposed new molecule into a shape that was favorable for interaction with the enzyme, then it was considered to be a lower priority target, particularly if it presented a difficult synthetic protocol. The technique was intended to ask a single question: Can the proposed structure adopt the shape of a cyclic urea, in an energetically favorable way? The formulation was based on the observation that cyclic ureas effectively define a high-affinity binding region of the protease active site. When asked with this in mind, the question was an effective tool for prioritizing synthesis.

Because this method correctly predicted free inhibitor geometries, we also evaluated the inhibitors within the protease active site in order to understand the effect of modified structures on the enzyme, and the effect of the active site environment on inhibitor conformation. Manual best-fit docking of the proposed compound into the empty active site of protease was followed by high-temperature dynamics as described previously with enzyme atoms fixed and inhibitor atoms unconstrained. Every picosecond, the entire structure was partially relaxed by energy minimization with enzyme heavy atoms constrained, then with enzyme backbone heavy atoms constrained, and finally with all atoms free to move. The resulting low-energy structures were ranked according to the energetic interactions between enzyme and inhibitor. This protocol yielded two low-energy bound conformations of cyclic ureas, one of which was almost identical to that observed in subsequently solved X-ray structures (Lam et al., 1994).

These studies suggested modified structures that could be tolerated in the active site, and allowed a more detailed evaluation of the interaction of proposed novel inhibitors that passed the first ''shape'' test described earlier. We reported a modification of this method to rationalize the differential sensitivity of protease mutants to inhibitors (Otto et al., 1993). It should be noted that the qualitatively accurate geometries that are located depend on biasing the starting position of cyclic urea into the binding orientation suggested by the design process. Other methods could be more successful in searching the active site for correct binding geometries from grossly perturbed starting orientations (Shoichet et al., 1992).

These methods and creative medicinal chemistry succeeded in optimizing the original lead to yield compounds that met all of our initial objectives, including oral availability, selectivity, and extremely potent antiviral efficacy, and that show early promise in AIDS therapy (Shum et al., 1993; Lam et al., 1994).

A variety of computational tools play critical roles in many phases of a successful drug discovery project: visualization of enzyme, visualization of enzyme complexed with inhibitor, database searching with distance constraints, evaluation of low-energy conformations of lead structures, ligand–receptor docking with and without energetic terms, and estimation of binding affinities of altered structures. The success of the HIV protease studies was highly dependent on at least four factors that are not present in the majority of targets for which structural information exists:

1. The inhibitors are fairly rigid.
2. The initial approximation to the binding orientation of the new inhibitors turned out to be correct and invariant with structural changes in the inhibitor. This is not always the case (Rutenber et al., 1993).
3. The enzyme structure is qualitatively invariant even when very different classes of inhibitors are bound (see preceding section).
4. With the exception of well-defined structural water molecules, solvent is excluded from the active site and does not appear to play an important role in inhibitor binding. The small influence of water is particularly true with smaller inhibitors such as cyclic ureas, which do not extend to the channel openings and are virtually free of contact with solvent.

Our ability to successfully apply the methods to a broader selection of ligand–receptor interactions depends critically on improving our handling of solvent and ion effects and predicting interactions of flexible ligands with flexible receptor structures. These improvements depend on being able to rapidly search large regions of many-atom conformational space with accurate (0.5 kcal/mol) evaluation of free energies of binding. The next section describes advances that are beginning to increase our ability to quantitatively predict the effect of a small change in structure, in the presence of solvent, on the energetics of ligand–receptor interaction.

Free Energy Simulations in the Refinement of Antiviral Compounds

Molecular simulations, in combination with techniques that allow for the evaluation of thermodynamic quantities such as free energies, have become an increasingly popular tool for the a priori calculation and analysis of relative binding affinities of different ligands for a given receptor. Although such calculations are still limited in their range of applicability and reliability, useful predictions are possible, especially for systems in which small modifications to a ligand have been made. Application of these methods in the design of novel inhibitory compounds, such as for the HIV protease, still requires some lead compound and experimental data on the structure of its complex with the receptor.

The relation of experimental and calculated thermodynamic results is usually presented in the form of a thermodynamic cycle, such as in Figure 17.7 (Tembe and McCammon, 1984). The relative free energy of binding of the two ligands L_1 and L_2 to receptor R is experimentally available from binding affinity experiments represented by processes 1 and 2 in Figure 17.7. The process of diffusion of a ligand in and out of the active site of an enzyme is typically too complicated, and certainly prohibitively expensive, to be studied by computer simulation. However, making the nonphysical change of mutating ligand L_1 into L_2 may not require the movement or displacement of a large fraction of the molecular system. This process is indicated by paths 3 and 4 in Figure 17.7, for the free ligands and the enzyme-complexed ligands, respectively. Because free energy (ΔG) is a state function, which means that the free energy difference between any two systems does not depend on the path taken to transform one into the other, the following equality holds:

$$\Delta G_2 - \Delta G_1 = \Delta G_4 - \Delta G_3 \qquad (17.1)$$

Relative equilibrium binding constants can thus be evaluated from

$$\frac{K_1}{K_2} = \exp\left(\frac{\Delta G_2 - \Delta G_1}{RT}\right) = \exp\left(\frac{\Delta G_4 - \Delta G_3}{RT}\right) \qquad (17.2)$$

where R is the universal gas constant and T is the absolute temperature. In the study of biomolecular systems, two techniques to evaluate free energy differences from trajectories obtained from molecular dynamics simulations, or ensembles of configurations obtained from Monte Carlo simulations, are widely used. The thermodynamic perturbation technique is based on an equation originally derived by Zwanzig (1954):

$$\Delta G = -RT \ln \left\langle \exp\left(-\frac{\Delta H}{RT}\right)\right\rangle_o \qquad (17.3)$$

Here the logarithmic term signifies an ensemble average, which, for sufficiently long molecular dynamics simulations, may be taken as the time average. Equation 17.3 can be described in simple terms. One carries out a molecular dynamics simulation of the reference system (e.g., $R:L_1$ in water), replaces L_1 by L_2 in frequent "snapshots" from the simulation, calculates the corresponding change in energy (ΔH), and then averages over the snapshots to obtain ΔG. More precisely, H represents the Hamiltonian operator of the system that describes the interactions of the system. ΔH describes the difference between the two systems for which the free energy difference is being evaluated. The subscript o is an indication that the free energy difference is evaluated from a

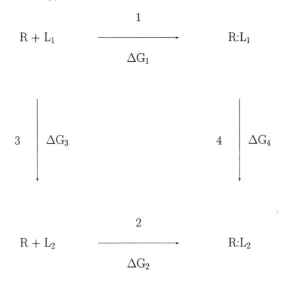

FIGURE 17.7 Thermodynamic cycle used in calculation of relative affinities of binding. Processes 1 and 2 are the physical binding of ligands L_1 and L_2 to the receptor R. The relative affinities of these ligands for R can be calculated by considering the nonphysical model processes, 3 and 4 (see text).

simulation of one of the two systems only. Another frequently used, and arguably more reliable, technique is thermodynamic integration, based on the simple integration

$$\Delta G = \int_0^1 \left\langle \frac{\partial H}{\partial \lambda} \right\rangle_\lambda d\lambda \tag{17.4}$$

where the Hamiltonian operator is made dependent on some control variable λ in such a way that the two systems of interest are properly described when $\lambda = 0$ and $\lambda = 1$, respectively. The relative accuracy and reliability of these and other techniques have been discussed in more detail in a number of reviews (Reddy et al., 1991; Rao et al., 1992; Straatsma and McCammon, 1991, Straatsma, 1994).

Only a few studies have been reported in which these free energy difference simulation techniques have been used for HIV protease inhibitors. The main reason is probably the difficulty of an adequate description of the behavior of the active site with the force fields that are available. In particular, the observation that the aspartyl residues of the active site form a planar arrangement of their carboxyl groups in all published crystal structures of inhibitors with HIV-1 protease presents a problem. In molecular simulations with standard force fields, such a planar conformation can only be obtained by imposing additional, rather ad hoc restraints. Such additional restraints have a marked influence on free energy differences calculated from these simulations. The reported free energy difference studies predominantly deal with the difference in binding affinity of identical peptide inhibitors that differ in stereochemistry at the scissile bond position. The following examples of such studies serve as an illustration of the basic difficulties of applying statistical thermodynamic techniques to the HIV protease–inhibitor system.

One of the first applications of free energy simulation techniques to HIV-1 protease inhibitors was reported by Reddy et al. (1991), who evaluated the relative free energy of binding of the hydroxyethylamine inhibitors Ac-Ser-Leu-Asn-Phe[CHOH-CH$_2$]Pro-Ile-Val-OMe and Ac-Ser-Leu-Asn-Phe[CHOH-CH$_2$]Pro-Ile-OMe. The calculated difference in free energy of binding to HIV-1 protease was obtained from the average of the results for calculations for both the R and S stereomers of the hydroxyethylamine moiety, and found to be about 14 kJ/mol. The calculated result is in remarkable agreement with the experimentally determined value of 16 kJ/mol.

Rao et al. (1992) employed thermodynamic perturbation calculations to evaluate the free energy difference between the two stereomeric forms of the hydroxyethylene isostere inhibitor Ala-Ala-Phe[CHOH-CH$_2$]Gly-Val-Val-OMe. This study found the S stereomer to bind more strongly by 14 kJ/mol, which is in reasonable agreement with the experimental value of 11 kJ/mol. The free energy difference for substitution of the P_1' Gly residue by Nle was evaluated to be 17 kJ/mol. In the same study, the difference in free energy of binding was calculated for Ac-Thr-Ile-Nle[CH$_2$-NH]Nle-Glu-Arg-NH$_2$ and the inhibitor obtained by substitution of Nle by Met at the P_1' position. This mutation was calculated to lead to increased binding by about 3 kJ/mol. The system was studied in a rectangular box with more than 2000 water molecules, and subject to periodic boundary conditions, which constitutes a system of considerable size for molecular simulation calculations. Consequently, the length of the simulations had to be limited to times of up to 100 ps. In these simulations, one of the active site aspartyl residues was protonated and the other was kept negatively charged. Because the two active site aspartyl diad carboxyl groups are observed to be in a coplanar arrangement in crystal structures of the enzyme–inhibitor complexes, distance constraints were employed to keep the active site conformation rigidly fixed during the simulations. Imposing such a constraint to the system made a difference of as much as 24 kJ/mol on the calculated free energy differences. This difference illustrates the difficulty of obtaining a proper description of the active site with the functions defined by the force field used, and should caution against overinterpretation of the results, even if they agree with experimentally determined values.

Tropsha and Hermans (1992) studied the difference in binding affinity of the diastereomeric forms of Ac-Ser-Leu-Asn-Phe[CHOH-CH$_2$]Pro-Ile-Val-OMe. The S stereomer binds more strongly by about 12 kJ/mol. Using energy minimization methods, these authors also determined that, when the S stereomer is bound to the enzyme, there is a strong preference for the Asp 125 residue to be protonated, in comparison with Asp 25. The simulations were performed for the enzyme–inhibitor complex in a 1-nm sphere around the Phe α-carbon. Simulations of the free inhibitor were done in a cubic box with periodic boundary conditions for a much simplified peptide Ac-Ala[CHOH-CH$_2$]Pro-Mam, which was additionally restrained to a conformation similar to the extended inhibitor in the enzyme complex. Because of these simplifications and limited simulation volumes, simulations of up to 300 ps could be performed, thereby increasing the statistical accuracy of the results. In view of the approximations made in this study, the agreement with the experimentally determined 11 kJ/mol for this mutation is remarkable.

A more detailed study on the effect of active site conformational restraints on calculated free energies of inhibitor binding to HIV-1 protease was presented by Straatsma (1994). In this study, the free energy difference was evaluated for the two stereomeric forms of the inhibitor Val-Ser-Gln-Asn-Leu[CHOH-CH$_2$]Val-Ile-Val, in which the peptide unit has been replaced by the hydroxyethylene moiety. With weak restraints to enforce planarity of the active site aspartyl side chains compared to earlier studies, the S stereomer showed increased binding by 32 kJ/mol. The effect of the restraints, however, was found to be quite different for S and R stereomers. Without these conformational restraints, the S stereomer is bound stronger by only 22 kJ/mol. This study illustrates that the use of conformational restraints may have a profound influence on calculated free energy differences. As a result, the predictive power of such calculations will remain rather limited until the force field can be modified to adequately describe the interactions in the active site without ad hoc additional terms. It may be that more elaborate terms, such as the inclusion of electronic polarization explicitly into the force field, will be necessary to achieve such a description. More realistic potential energy functions that include such features are being developed. Combined with the continuing increases in the power of computers, these developments will enable simulations to become key tools in drug discovery.

Acknowledgments

Charles J. Eyermann, Patrick Y. Lam, and Prabhakar K. Jadhav made critical contributions to the work at DuPont Merck.

Research at the University of California-San Diego and at the University of Houston was supported in part by the National Science Foundation (NSF), the National Institute of Health, and the Metacenter Program of the NSF Supercomputer Centers. Research at the National Cancer Institute (NCI) was sponsored in part by the NCI, under contract with ABL. The contents of this publication do not necessarily reflect the view or policies of the Department of Health and Human Services, nor does mention of trade names, commercial products, or organizations imply endorsement by the U.S. Government.

References

Abdel-Meguid, S.S., Zhao, B., Murthy, K.H.M., Winborne, E., Choi, J.-K., DesJarlais, R.L., Minnich, M.D., Culp, J.S., Debouck, C., Tomaszek, T.A., Meek, T.D. and

Dreyer, G.B. 1993. Inhibition of human immunodeficiency virus-1 protease by a C2-symmetric phosphinate: synthesis and crystallographic analysis. Biochemistry 32: 7972–7980.

Allen, F.H. 1991. The development of versions 3 and 4 of the Cambridge Structural Database System. J. Chem. Tuf. Comput. Sci. 31: 187.

Alteri, E., Bold, G., Cozens, R., Faessler, A., Klimkait, T., Lang, M., Lazdins, J., Poncioni, B., Roesel, J.L., Davies, J.E., Galloy, J.J., Johnson, O., Kennard, O., Macrae, C.F., Mitchel, E.M., Mitchel, G.F., Smith, J.M. and Watson, D.G. 1993. CGP-53437, an orally bioavailable inhibitor of HIV type 1 protease with potent antiviral activity. Antimicrob. Agents Chemother. 37: 2087–2092.

Babine, R.E., Zhang, N., Jurgens, A.R., Schow, S.R., Desai, P.R., James, J.C. and Semmelhack, M.F. 1992. The use of HIV-1 protease structure in inhibitor design. Biorg. Med. Chem. Lett. 2: 541–546.

Blaney, J.M., Crippen, G.M., Dearing, A. and Dixon, J.S. 1984. DGEOM, Quantum Chemistry Program Exchange 590 Indiana University.

Blundell, T., Jenkins, J., Pearl, L., Sewell, T., and Pedersen, V. 1985. The high resolution structure of endothiapepsin. In Aspartic Proteinases and Their Inhibitors (Kostka, V., Ed.), pp. 151–161. Walter de Gruyter, Berlin.

Bone, R., Vacca, J.P., Anderson, P.S. and Holloway, M.K. 1991. X-ray crystal structure of the HIV protease complex with L-700,417, an inhibitor with pseudo C2 symmetry. J. Am. Chem. Soc. 113: 9382–9384.

Büdt, K.-H., Stowasser, B., Knolle, J., Ruppert, D., Meichsner, C., Paessens, A. and Hansen, J. 1991. Retroviral protease inhibitors (EPA Publication No. 428849). Environmental Protection Agency, Washington, DC.

Bures, M.G., Hutchins, C.W., Maus, M., Kohlbrenner, W., Kadam, S. and Erickson, J.W. 1992. Using three-dimensional substructure searching to identify novel, non-peptidic inhibitors of HIV-1 protease. Tetrahedron Comp. Method 3: 673–680.

Dauber-Osguthorpe, P., Roberts, V.A., Osguthorpe, D.J., Wolff, J., Genest, M. and Hagler, A.T. 1988. Structure and energetics of ligand binding to proteins. Proteins Struct. Funct. Genet. 4: 31–37.

Dorsey, D.D., Levin, R.B., McDaniel, S.L., Vacca, J.P., Darke, P.L., Zugay, J.A., Emini, A., Schleif, W.A., Quintero, J.C. 1993. L-735,524: the rational design of a potent and orally available HIV-1 protease inhibitor. Paper presented at the 206th American Chemical Society National Meeting, Chicago.

Dreyer, G.B., Boehm, J.C., Chenera, B., DesJarlais, R.L., Hassell, A.M., Meek, T.D., Tomaszek, T.A. and Lewis, M. 1993. A symmetric inhibitor binds HIV-1 protease asymmetrically. Biochemistry 32: 937–947.

Dreyer, G.B., Lambert, D.M., Meek, T.D., Carr, T.J., Tomaszek, T.A. Jr., Fernandez, A.V., Bartus, H., Cacciavillani, E., Hassell, A.M., Minnich, M., Petteway, S.R. Jr., Metcalf, B.W. and Lewis M. 1992. Hydroxyethylene isostere inhibitors of human immunodeficiency virus-1 protease: structure-activity analysis using enzyme kinetics, X-ray crystallography, and infected T-cell assays. Biochemistry 31: 6646–6659.

Dreyer, G.B., Metcalf, B.W., Tomaszek, T.A., Carr, T.J., Chandler, A.C. III, Hyland, L., Fakhoury, S.A., Magaard, V.W., Moore, M.L., Strickler, J.E. and Meek, T.D. 1989. Inhibition of human immunodeficiency virus 1 protease in vitro: rational design of substrate analog inhibitors. Proc. Natl. Acad. Sci. U.S.A. 86: 9752–9756.

Erickson, J., Neidhart, D.J., VanDrie, J., Kempf, D.J., Wang, X.C., Norbeck, D.W., Plattner, J.J., Rittenhouse, J.W., Turon, M., Wideburg, N., Kohlbrenner, W.E., Simmer, R., Helfrich, R., Paul, D.A. and Knigge, M. 1990. Design, activity and 2.8 Å crystal structure of a C2-symmetric inhibitor complexed to HIV-1 protease. Science 249: 527–533.

Fitzgerald, P.M.D., McKeever, B.M., VanMiddleworth, J.F., Springer, J.P., Heimbach, J.C., Leu, C.-T., Herber, W.K., Dixon, R.A.F. and Darke, P.L. 1990. Crystallographic analysis of a complex between human immunodeficiency virus type 1 protease and acetyl-pepstatin at 2.0 Å. resolution. J. Biol. Chem. 265: 14209–14219.

Graves, B.J., Hatada, M.H., Miller, J.K., Graves, M.C., Roy, S., Cook, C.M., Kröhn, A., Martin, J.A. and Roberts, N.A. 1992. Three-dimensional x-ray crystal structure of HIV-1 protease complex with a hydroxyethylene-inhibitor. In Structure and Function of the Aspartic Proteinases (Bunn, B., Ed.), pp. 455–460. Plenum Press, New York.

Greenlee, W.K. and Siegel, P.K.S. 1991. Angiotensin/renin modulators: renin inhibitors. Annu. Rep. Med. Chem. 26: 64–66.

Hodge, C.N., Aldrich, P.E., Bacheler, L.T., Chang, C.H., Eyermann, C.J., Grubb, M., Jackson, D.A., Jadhav, P.K., Korant, B., Lam, P.Y.S., Maurin, M.B., Meek, J.L., Otto, M.J., Rayner, M.M., Sharpe, T.R., Shum, L., Winslow, D.L. and Erickson, S.V. 1996. Improved cyclic urea inhibitors of the HIV protease: synthesis, potency, resistance profile, human pharmacokinetics and X-ray crystal structure of DMP450. Chem. Biol. 3: 301–314.

Jadhav, P., McGee, L., Shenvi, A. and Hodge, C.N. 1991. New 1,4-Di:amino-2,3-di: hydroxybutane Derivatives. World Patent WO 9118-866-A.

Jadhav, P.K. and Woerner, F. 1992. Synthesis of C-2 symmetric protease inhibitors from D-mannitol. Biorg. Med. Chem. Lett. 2: 353–356.

Jaskólski, M., Tomasselli, A.F., Sawyer, T.K., Staples, D.G., Heinrikson, R.L., Schneider, J., Kent, S.B.H. and Wlodawer, A. 1991. Structure at 2.5 Å. resolution of chemically synthesized human immunodeficiency virus type 1 protease complexed with a hydroxyethylene-based inhibitor. Biochemistry 30: 1600–1609.

Kohl, N.E., Emini, E.A., Schleif, W.A., Davis, L.J., Heimbach, J.C., Dixon, R.A., Scolnick, E.M. and Sigal, I.S. 1988. Active human immunodeficiency virus protease is required for viral infectivity. Proc. Natl. Acad. Sci. U.S.A. 85: 4686–4690.

Kraulis, P.J. 1991. MOLSCRIPT: a program to produce both detailed and schematic plots of protein structures. J. Appl. Crystallogr. 24: 946–950.

Krohn, A., Redshaw, S., Ritchie, J.C., Graves, B.J. and Hatada, M.H. 1991. Novel binding mode of highly potent HIV-proteinase inhibitors incorporating the (R)-hydroxyethylamine isostere. J. Med. Chem. 34: 3340–3342.

Lam, P.Y., Eyermann, C.J., Hodge, C.N., Jadhav, P.K., and Delucca, G.V. 1993. World Patent Application WO 9307128.

Lam, P.Y.S., Jadhav, P.K., Eyermann, C.J., Hodge, C.N., Ru, Y., Bacheler, L.T., Meek, J.L., Otto, M.J., Rayner, M.M., Wong, Y.N., Chang, C.-H., Weber, P.C., Jackson, D.A., Sharpe, T.R. and Erickson-Viitanen, S. 1994. Rational design of potent, bioavailable, nonpeptide cyclic ureas as HIV protease inhibitors. Science 263: 380–384.

Lam, P.Y., Ru, Y., Hodge, C.N., Jadhav, P.K. and Eyermann, C.J. 1996. Cyclic HIV protease inhibitors. Synthesis, conformational analysis, P2/P2' structure-activity relationship and molecular recognition of cyclic ureas. J. Med. Chem. 39: 3514–3525.

Lapatto, R., Blundell, T., Hemmings, A., Overington, J., Wilderspin, A., Wood, S., Merson, J.R., Whittle, P.J., Danley, D.E., Geoghegan, K.F., Havrylik, S.J., Lee,

S.E., Scheld, K.G. and Hobart, P.M. 1989. X-ray analysis of HIV-1 proteinase at 2.7 Å. resolution confirms structural homology among retroviral enzymes. Nature 342: 299–302.

Lyle, T.A., Wiscount, C.M., Guare, J.P., Thompson, W.J., Anderson, P.S., Darke, P.L., Zugay, J.A., Emini, E.A., Schleif, W.A., Qunitero, J.C., Dixon, R.A.F., Sigal, I.S. and Huff, J.S. 1991. Benzocycloalkyl amines as novel C-termini for HIV protease inhibitors. J. Med. Chem. 34: 1228–1230.

Maple, J.R., Dinur, U. and Hagler, A.T. 1988. Derivations of force fields for molecular mechanics and dynamics from ab initio energy surfaces. Proc. Natl. Acad. Sci. U.S.A. 85: 5350–5354.

Martin, Y.C. 1992. 3D database searching in drug design. J. Med. Chem. 35: 2145–2154.

McQuade, T.J., Tomasselli, A.G., Liu, L., Karacostas, V., Moss, B., Sawyer, T.K., Heinrikson, R.L. and Tarpley, W.G. 1990. A synthetic HIV-1 protease inhibitor with antiviral activity arrests HIV-like particle maturation. Science 247: 454–456.

Miller, M., Jaskólski, M., Rao, J.K.M., Leis, J. and Wlodawer, A. 1989a. Crystal structure of a retroviral protease proves relationship to aspartic protease family. Nature 337: 576–579.

Miller, M., Schneider, J., Sathyanarayana, B.K., Toth, M.V., Marshall, G.R., Clawson, L., Selk, L., Kent, S.B.H. and Wlodawer, A. 1989b. Structure of complex of synthetic HIV-1 protease with a substrate-based inhibitor at 2.3 Å resolution. Science 246: 1149–1152.

Mulichak, A.M., Hui, J.O., Tomasselli, A.G., Heinrikson, R.L., Curry, K.A., Tomich, C.-S., Thaisrivongs, S., Sawyer, T.K. and Watenpaugh, K.D. 1993. The crystallographic structure of the protease from human immunodeficiency virus type 2 with two synthetic peptidic transition state analog inhibitors. J. Biol. Chem. 268: 13103–13109.

Murthy, K.H.M., Winborne, E.L., Minnich, M.D., Culp, J.S. and Debouck, C. 1992. The crystal structures at 2.2-Å. resolution of hydroxyethylene-based inhibitors bound to human immunodeficiency virus type 1 protease show that the inhibitors are present in two distinct orientations. J. Biol. Chem. 267: 22770–22778.

Navia, M.A., Fitzgerald, P.M.D., McKeever, B.M., Leu, C.-T., Heimbach, J.C., Herber, W.K., Sigal, I.S., Darke, P.L. and Springer, J.P. 1989. Three-dimensional structure of aspartyl protease from human immunodeficiency virus HIV-1. Nature 337: 615–620.

Norbeck, D.W., Kempf, D., Knigge, M., Vasovananda, S., Clement, J., Kohlbrenner, W., Sham, H. 1992. Orally bioavailable HIV proteinase inhibitors for experimental AIDS therapy. Paper presented at the 32nd Interscience Conference on Antimicrobial Agents and Chemotherapy.

O'Connor, S.D., Smith, P.E., Al-Obeidi, F. and Pettitt, B.M. 1992. Quenched molecular dynamics simulations of tuftsin and proposed cyclic analog. J. Med. Chem. 35: 2870–2881.

Olson, G.L., Bolin, D.R., Bonner, M.P., Bos, M., Cook, C.M., Fry, D.C., Graves, B.J., Hatada, M., Hill, D.E., Kahn, M., Madison, V.S., Rusiecki, V.K., Sarabu, R., Sepinwall, J., Vincent, G.P. and Voss, M.E. 1993. Concepts and progress in the development of peptide mimetics. J. Med. Chem. 36: 3039–3049.

Otto, M.J., Garber, S., Winslow, D.L., Reid, C.D., Aldrich, P.E., Jadhav, P.K., Patterson, C.E., Hodge, C.N. and Cheng, Y.-S.E. 1993. In vitro isolation and identification of human immunodeficiency virus (HIV) variants with reduced sensitivity to C-2 symmetrical inhibitors of HIV type 1 protease. Proc. Natl. Acad. Sci. U.S.A. 90: 7543–7546.

Pearl, L.H. and Taylor, W.R. 1987. A structural model for the retroviral proteases. Nature 329: 351–354.

Plattner, J.J. and Norbeck, D.W. 1990. Obstacles to drug discovery from peptide leads. In Drug Discovery Technologies (Clark, C.R. and Moos, W.H., Eds.), pp. 92–126. Ellis Horwood, Chichester, England.

Rao, B.G., Tilton, R.F. and Singh U.C. 1992. Free energy perturbation studies on inhibitor binding to HIV-1 proteinase. J. Am. Chem. Soc. 114: 4447–4452.

Reddy, M.R., Viswanadhan, V.N. and Weinstein, J.N. 1991. Relative differences in the binding free energies of human immunodeficiency virus-1 protease inhibitors—a thermodynamic cycle perturbation approach. Proc. Natl. Acad. Sci. U.S.A. 88: 287–291.

Rich, D.H. 1986. Inhibitors of aspartic proteinases. Res. Monogr. Cell Tissue Physiol. 12: 179–217.

Roberts, N.A., Martin, J.A., Kinchington, D., Broadhurst, A.V., Craig, J.C., Duncan, I.B., Galpin, S.A., Handa, B.K., Kay, J., Krohn, A., Lambert, R.W., Merret, J.H., Mills, J.S., Parkes, K.E.B., Redshaw, S., Ritchie, A.J., Taylor, D.L., Thomas, G.J. and Machin, P.J. 1990. Rational design of peptide-based HIV proteinase inhibitors. Science 248: 358–361.

Rutenber, E., Fauman, E.B., Keenan, R.J., Fong, S., Furth, P.S., Ortiz de Montellano, P.R., Meng, E., Kuntz, I.D., DeCamp, D.L., Salto, R., Rose, J.R., Craik, C.-B. and Stround, R.M. 1993. Structure of a non-peptide inhibitor complexed with HIV-1 protease: developing a cycle of structure-based drug design. J. Biol. Chem. 268: 15343–15346.

Shoichet, B.K., Bodian, D.L. and Kuntz, I.D. 1992. Molecular docking using shape descriptors. J. Comput. Chem. 13: 380–397.

Shum, L., Robinson, C.A. and Pieniaszek, H.J. Jr. 1993. Protein binding of DMP-323 in human, monkey, dog, rat, and mouse plasma. AAPS (American Association of Pharmaceutical Scientists) Eighth Annual Meeting and Exposition. Pharm. Res. 10(Suppl.), Orlando, Florida.

Spinelli, S., Liu, Q.Z., Alzari, P.M., Hirel, P.H. and Poljak, R.J. 1991. The three-dimensional structure of the aspartyl protease from the HIV-1 isolate BRU. Biochimie 73: 1391–1396.

Straatsma, T.P. 1994. HIV-1 proteinase inhibitor binding: the effect of active site conformational restraints on calculated free energies of ligand binding. NATO ASI Series, Ser C, 426 (Computational Approaches in Supramolecular Chemistry), 495–513.

Straatsma, T.P. and McCammon, J.A. 1991. Theoretical calculations of relative affinities of binding. Methods Enzymol. 202: 497–511.

Straatsma, T.P. and McCammon, J.A. 1992. Computational alchemy. Annu. Rev. Phys. Chem. 43: 407–435.

Straatsma, T.P., Zacharias, M. and McCammon, J.A. 1993. HIV-1 proteinase inhibitor binding. The effect of active site conformational restraints on calculated free energies of ligand binding. In Computer Simulation of Biomolecular Systems, Vol. II (van Gunsteren, W.F. and Wiikinson, A.J., Eds.). ESCOM, Leiden, Netherlands.

Swain, A.L., Gustchina, A. and Wlodawer, A. 1992. Comparison of three inhibitor complexes of human immunodeficiency virus protease. In Structure and Function of the Aspartic Proteinase: Genetics, Structure and Mechanisms (Dunn, B., Ed.), pp. 433–441. Plenum Press, New York.

Swain, A.L., Miller, M.M., Green, J., Rich, D.H., Schneider, J., Kent, S.B. and Wlodawer, A. 1990. X-ray crystallographic structure of a complex between a synthetic protease of human immunodeficiency virus 1 and a substrate-based hydroxyethylamine inhibitor. Proc. Natl. Acad. Sci. U.S.A. 87: 8805–8809.

Tembe, B.L. and McCammon, J.A. 1984. Ligand-receptor interactions. Comput. Chem. 8: 281–283.

Thanki, N., Rao, J.K.M., Foundling, S.I., Howe, W.J., Moon, J.B., Hui, J.O., Tomasselli, A.G., Heinrikson, R.L., Thaisrivongs, S. and Wlodawer, A. 1992. Crystal structure of a complex of HIV-1 protease with a dihydroxyethylene-containing inhibitor: comparisons with molecular modeling. Protein Sci. 1: 1061–1072.

Thompson, W.J., Fitzgerald, P.M.D., Holloway, M.K., Emini, E.A., Darke, P.L., McKeever, B.M., Schleif, W.A., Quintero, J.C., Zugay, J.A., Tucker, T.J., Schwering, J.E., Homnick, C.F., Nunberg, J., Springer, J.P. and Huff, J.R. 1992. Structure-activity studies and antiviral properties of hydroxyethylene transition state inhibitors of HIV-1 protease. J. Med. Chem. 35: 1685–1701.

Toh, H., Ono, M., Saigo, K. and Miyata, T. 1985. Retroviral protease-like sequence in the yeast transposon Ty1. Nature 315: 691–692.

Tomaszek, T.A., Moore, M.L., Stricker, J.E., Sanchex, R.L., Dixon, J.S., Metcalf, B.W., Hassell, A., Dreyer, G.B., Brooks, I., Debouck, C., Meek, T.D. and Lewis, M. 1992. Biochemistry 31: 10153–10168.

Tong, L., Pav, S., Pargellis, C., Do, F., Lamarre, D. and Anderson, P.C. 1993. Crystal structure of human immunodeficiency virus (HIV) type 2 protease in complex with a reduced amide inhibitor and comparison with HIV-1 protease structures. Proc. Natl. Acad. Sci. U.S.A. 90: 8387–8391.

Tropsha, A. and Hermans, J. 1992. Application of free energy simulations to the binding of a transition state analog inhibitor to HIV protease. Protein Eng. 5: 29–33.

van Gunsteren, W.F., Beutler, T.C., Fraternali, F., King, P.M., Mark, A.E. and Smith, P.E. 1993. Molecular dynamics and stochastic dynamic simulations: a primer. In Computer Simulation of Biomolecular Systems: Theoretical and Experimental Applications, Vol. II. (van Gunsteren, W.F., Weiner, P.K. and Wilkinson, A., Eds.). ESCOM, Leiden, Netherlands.

Waller, C.L. and Marshall, G.R. 1993. Three-dimensional quantitative structure-activity relationship of angiotensin converting enzyme and thermolysin inhibitors. II. A comparison of CoMFA models incorporating molecular orbital fields and desolvation free energies based on active-analog and complementary-receptor-field alignment rules. J. Med. Chem. 36: 2390–2403.

Wlodawer, A. and Erickson, J.W. 1993. Structure-based inhibitors of HIV-1 protease. Annu. Rev. Biochem. 62: 543–585.

Wlodawer, A., Miller, M., Jaskólski, M., Sathyanarayana, B.K., Baldwin, E., Weber, I.T., Selk, L.M., Clawson, L., Schneider, J. and Kent, S.B.H. 1989. Conserved folding in retroviral proteases: crystal structure of synthetic HIV-1 protease. Science 245: 616–621.

Zwanzig, R.W. 1954. High-temperature equation of state by a perturbation method. I. Nonpolar gases. J. Chem. Phys. 22: 1420–1426.

Index